24th Workshop on Biomedical Language Processing (BioNLP 2025)

Shared Tasks

Vienna, Austria
1 August 2025

ISBN: 979-8-3313-2403-2

Printed from e-media with permission by:

Curran Associates, Inc.
57 Morehouse Lane
Red Hook, NY 12571

CURRAN ASSOCIATES INC. proceedings.com

Some format issues inherent in the e-media version may also appear in this print version.

Printed with permission by Curran Associates, Inc. (2025)

For permission requests, please contact the Association for Computational Linguistics at the address below.

Association for Computational Linguistics
209 N. Eighth Street
Stroudsburg, Pennsylvania 18360

Phone: 1-570-476-8006
Fax: 1-570-476-0860

acl@aclweb.org

Additional copies of this publication are available from:

Curran Associates, Inc.
57 Morehouse Lane
Red Hook, NY 12571 USA
Phone: 845-758-0400
Fax: 845-758-2633
Email: curran@proceedings.com
Web: www.proceedings.com

ACL 2025

The 24th BioNLP Workshop and Shared Tasks

Proceedings of the 24th Workshop on Biomedical Language Processing (Shared Tasks)

August 1, 2025

Introduction

Sarvesh Soni, Dina Demner-Fushman

Four shared tasks were organized as part of the BioNLP 2025 Workshop. These tasks are designed to advance the state of the art in biomedical natural language processing by providing a platform to foster innovative solutions to challenging problems in the field. Specifically, the tasks addressed: (1) validating the similarity of research goals between scientific articles (SMAFIRA), (2) evaluating the factual accuracy of generative models (ClinIQLink), (3) generating evidence-grounded answers from electronic health records (ArchEHR-QA), and (4) producing lay summaries of biomedical and radiology texts (BioLaySumm). Collectively, these shared tasks foster the development and benchmarking of innovative methods for information retrieval, knowledge assessment, question answering, and summarization in biomedicine.

A total of 35 papers were submitted across the four tasks, with participants detailing a range of novel approaches and systems. As is typical with shared task tracks, the majority of submissions were accepted, resulting in 34 papers being included in the proceedings.

We provide a brief overview of each shared task below:

1. SMAFIRA

The SMAFIRA task addresses the challenge of assessing the similarity of research goals between biomedical articles, a crucial step in identifying alternatives to animal experiments. Participants were provided with a set of reference scientific articles from PubMed, each representing an animal study within a specific disease domain, along with a set of twenty candidate articles retrieved from PubMed. The task involved validating the candidate articles either automatically, with systems of the participants' choice, or manually, using the SMAFIRA web tool. Validation involves comparing titles and abstracts to assess the degree of similarity in research goals using a three-point scale: similar, uncertain, or not similar. This collaborative annotation effort aims to produce a high-quality dataset for benchmarking automated methods and supporting the broader adoption of non-animal research approaches.

2. ClinIQLink

The ClinIQLink task aims to evaluate the ability of generative language models to produce factually accurate medical information. Participants were tasked to submit models that can answer a diverse set of clinically relevant questions, spanning fundamental concepts in procedures, conditions, diagnostics, and pharmacology, at the level expected of a General Practitioner. Using a novel, expert-curated dataset of atomic question-answer pairs, the task assesses both closed- and open-ended responses, employing scoring metrics to measure knowledge retrieval and penalize factual inaccuracies. Beyond benchmarking the current capabilities of generative models, the task provides insights into the origins and types of hallucinations exhibited by state-of-the-art language models in medical contexts.

3. ArchEHR-QA

The ArchEHR-QA task targets the challenge of generating accurate, evidence-grounded answers to patients' health-related questions using information from electronic health records (EHRs). Participants were provided with realistic patient questions, clinician-interpreted version of the questions, and corresponding clinical note excerpts. The objective is to develop systems that generate concise, professional responses that are explicitly supported using citations to relevant sentences from the clinical notes. System outputs are evaluated using two main criteria: factuality, which measures the precision and recall of cited evidence against manually annotated ground truth, and relevance, which assesses the quality and appropriateness of the generated answer text compared to ground truth. This task aims to advance research on supporting clinicians in efficiently addressing patient inquiries, while ensuring that responses remain accurate, contextually appropriate, and grounded in real clinical evidence.

4. BioLaySumm

The BioLaySumm task focuses on the challenge of making complex scientific information accessible to non-expert audiences. The task comprises two main subtasks. In one, participants were given biomedical scientific articles and tasked to develop systems to produce readable, informative summaries suitable for the general public, with an additional subtask requiring the integration of external knowledge to fill information gaps for lay readers. In the second, participants were tasked to translate professional radiology reports into layman's terms, with an additional multi-modal subtask involving the generation of lay summaries directly from medical images. This task aims to benchmark current approaches and foster the development of systems that support more inclusive and effective biomedical communication.

We remain deeply grateful to all shared task participants, to the authors who submitted papers, and to the reviewers (listed under Program Committee) who provided thorough and thoughtful reviews for the submissions, often under tight timelines. The quality of work submitted continues to rise, and we are indebted to the outstanding members of our Organizing Committee, whose careful assessments have been instrumental in identifying research ready for presentation and in advising authors where further experiments and analyses could strengthen their contributions.

As in previous years, we look forward to a productive workshop and to the new collaborations and research directions it will inspire. We are confident that these efforts will help our community continue to advance public health and well-being, as well as contribute meaningfully to both basic and clinical research.

Organizing Committee

SMAFIRA Shared Task

Mariana Neves, German Federal Institute for Risk Assessment - BfR, Germany

ClinIQLink Shared Task

Brandon Colelough, National Library of Medicine, USA
Dina Demner-Fushman, National Library of Medicine, USA
Davis Bartels, National Library of Medicine, USA

ArchEHR-QA Shared Task

Sarvesh Soni, National Library of Medicine, USA
Dina Demner-Fushman, National Library of Medicine, USA

BioLaySumm Shared Task

Kun Zhao, University of Pittsburgh
Liang Zhan, University of Pittsburgh
Chenghao Xiao, University of Durham
Noura Al Moubayed, University of Durham
Kejing Yin, Hong Kong Baptist University
Sixing Yan, Hong Kong Baptist University
Zijian Lei, Hong Kong Baptist University
William Cheung, Hong Kong Baptist University
Qianqian Xie, Yale University
Zheheng Luo, University of Manchester
Sophia Ananiadou, University of Manchester
Tomas Goldsack, University of Sheffield
Siwei Wu, University of Manchester
Xiao Wang, University of Manchester
Chenghua Lin, University of Manchester

Program Committee

Reviewers

Rebecca Allen, Mount St. Joseph Univ. Center for IT Engagement
Mohammad Arvan, University of Illinois at Chicago
Sai Prasanna Teja Reddy Bogireddy, University of Chicago
Surabhi Datta, IMO Health, USA
Viswanatha Reddy Gajjala, Amazon
Shohreh Haddadan, Moffitt Cancer Center
Ming Huang, UTHealth
Sy Hwang, University of Pennsylvania
Tuan Dung Le, University of South Florida
Jinghui Liu, CSIRO
Adam Remaki, Sorbonne Univserity
Suveyda Yeniterzi, GenAIus Technologies

Table of Contents

ArgHiTZ at ArchEHR-QA 2025: A Two-Step Divide and Conquer Approach to Patient Question Answering for Top Factuality

Adrián Cuadrón*, **Aimar Sagasti***, **Maitane Urruela***, **Iker De la Iglesia**, **Ane G Domingo-Aldama**, **Aitziber Atutxa**, **Josu Goikoetxea**, **Ander Barrena**
HiTZ Center - Ixa, University of the Basque Country UPV/EHU
{adrian.cuadron, aimar.sagasti, maitane.urruela, iker.delaiglesia, ane.garciad, aitziber.atutxa, josu.goikoetxea, ander.barrena}@ehu.eus

Abstract

This work presents three different approaches to address the ArchEHR-QA 2025 Shared Task on automated patient question answering. We introduce an end-to-end prompt-based baseline and two two-step methods to divide the task, without utilizing any external knowledge. Both two step approaches first extract *essential* sentences from the clinical text—by prompt or similarity ranking—, and then generate the final answer from these notes. Results indicate that the re-ranker based two-step system performs best, highlighting the importance of selecting the right approach for each subtask. Our best run achieved an overall score of 0.44, ranking 8th out of 30 on the leaderboard, securing the top position in overall factuality.

1 Introduction

The increasing volume of patient messages received through online patient portals has become a significant source of clinician burden, highlighting the need for effective automated support. In this context, *ArchEHR-QA 2025: Grounded Electronic Health Record Question Answering Shared Task* (Soni and Demner-Fushman, 2025b) focuses on the automatic generation of answers to patient-submitted health-related questions, leveraging evidence extracted from their Electronic Health Records (EHRs). The objective of the task is to develop systems capable of producing coherent and evidence-grounded responses, thereby assisting clinicians in managing patient communication more efficiently. In this paper, we describe three different approaches to address the task, present our system design, and analyze its performance on the shared dataset. All the code is publicly available at: `https://github.com/hitz-zentroa/ArchEHR-ArgHiTZ`.

2 Related Work

Question answering (QA) is a widely used task for evaluating Large Language Models (LLMs), leading to extensive research in both general and domain-specific contexts (Yan et al., 2024). In the medical domain, several benchmark datasets have emerged, like PubMedQA (Jin et al., 2019), which features yes/no/maybe questions derived from biomedical abstracts, and MedMCQA (Pal et al., 2022), which contains multiple-choice medical questions. Broader QA tasks, such as Semantic Question Answering (SQA), use datasets like BioASQ-QA (Krithara et al., 2023), for open-domain biomedical questions. More recent work (Ben Abacha et al., 2019, 2021) has moved toward patient-centered QA, developing systems that can automatically respond to patient questions using evidence from clinical notes, while evaluating the quality of such responses. The ArchEHR-QA shared task builds on this direction by introducing a real-world challenge that aims to respond real patient inquires, relying on their EHRs' information. In this work, we propose and compare multiple strategies to address this task without relying on external medical knowledge, offering insights into effective subtask decomposition for clinical QA.

3 Data

In this work, we focus on approaches that require neither additional fine-tuning nor large-scale training, but instead rely on a single example to perform the task. Hence, we exclusively utilize the dataset provided for the task (Soni and Demner-Fushman, 2025a), which comprises 20 development (*dev*) documents with patients concerns regarding a specific procedure or treatment they have undergone, along with their corresponding EHRs—derived from MIMIC-III (Johnson et al., 2016) and MIMIC-IV (Johnson et al., 2023) databases. These resources are used to generate responses that di-

* This authors contributed equally.

BioNLP 2025 Shared Tasks, pages 1–10
August 1, 2025 ©2025 Association for Computational Linguistics

rectly address the patients' concerns.

Each instance contains the patient's narrative, annotated spans capturing the core question, a technical reformulation, the full EHR, and a sentence-by-sentence breakdown of the EHR with their IDs. A separated file is also provided for the *dev* set with an *essential, supplementary* or *not-relevant* relevance classification for each sentence. Our approaches leverage the patient's concern, the technical question, and the classified EHR sentences.

4 Models

For response generation, we utilized the Aloe (Gururajan et al., 2024) , Mistral (MistralAI, 2025) , and Gemma (Team, 2024) models, as well as, two bigger models for comparison with the task given baseline: Aloe 70B and Llama 3.3 70B (MetaAI, 2024) . Additionally, we implemented similarity-based approaches, leveraging different re-ranking models such as Jina (JinaAI, 2024) , BAAI BGE (Li et al., 2023; Chen et al., 2024) and Alibaba GTE (Zhang et al., 2024) .

5 Evaluation

The evaluation focuses on the two main objectives of the task: detection of the essential sentences and creation of the response paragraph. To evaluate them, the first step is to check if the responses adhere to the required format: a maximum of 75 words (excluding IDs), and each sentence must end with at least one ID enclosed in vertical bars and separated by comas (e.g., |1| or |2,3|), followed by a line break. If the answer is longer than 75 words, the output is truncated to that limit.

To evaluate the essential sentence detection (*Factuality*) Precision, Recall and F1-Scores are used. The metrics are calculated in two ways: considering only *essential* sentences (*Strict*) and considering both *essential* and *supplementary* sentences (*Lenient*). The generation of the response argument (*Relevance*) is evaluated comparing the generated text with the *essential* sentences and the question using the following metrics: BLEU, ROUGE, SARI, BERTScore, AlignScore and Medcon.

6 Approaches

The goal of this shared task is to provide an adequate answer to the patients' health concerns, while citing the source of information—namely, the specific sentence(s) of their EHR. To tackle this challenge, we propose three distinct approaches and systematically compare their performance on the *dev* set in terms of both the quality of the generated answers and the relevance of the cited evidence.

6.1 Baseline End-to-End

As a baseline, we test various prompts combining *role* prompting—guiding the model to respond as an specific person, in this case a doctor—and Chain of Thought (CoT)—outlining the reasoning steps it must follow before giving the final response—with the models in Section 4 to generate responses in the required format. The final prompt used is shown in Appendix A.1. Each model receives the patient's full concern, its reformulation into a technical question, the EHR annotated with sentence IDs, and a one-shot input-output example. We enforce formatting through instructions and light post-processing to fix minor errors.

Table 1 shows that the smaller Aloe Beta (8B) model outperforms larger models in both *Factuality* and *Relevance.* Although Gemma achieves the best *Relevance* score, Aloe (8B) remains the overall best option, despite still having room for improvement.

	Overall Scores		
Model	Overall	Relevance	Factuality
Aloe 8B	**0.388**	0.312	**0.464**
Mistral 7B	0.364	0.327	0.402
Gemma 2 9B	0.353	**0.366**	0.340
Llama 3.3 70B	0.340	0.328	0.352
Aloe 70B	0.371	0.332	0.410

Table 1: Results for the end-to-end with post-process approach in different models. Best results in bold.

Since 70B models show no clear improvement over smaller ones and are computationally costly, we exclude them from other experiments. Manual quality checks also reveal that Gemma, despite its higher *Relevance*, underperforms significantly in *Factuality* and requires more post-processing, so it is also discarded.

6.2 Two Step Approaches

Based on the baseline results and limitations, we decide to take two-step approaches in order to split the tasks for the models and reach better performance. On the first step we decide which are the *essential* sentences to respond the patient's concern, by prompting techniques (Section 6.2.1.1) or using a similarity-based re-ranker for the provided sentences (Section 6.2.1.2). On the second step, those considered sentences are rephrased to build a proper response by prompts and are cited correspondingly following the post-process explained on Section 6.2.2.

6.2.1 First Step

6.2.1.1 Prompting to Extract Essentials

In the first approach to identify essential sentences, we leverage a large language model (LLM) using two different techniques, experimenting with several prompting strategies.

Extract list of essential sentences: This technique consists of prompting the model to generate a list of IDs of the essential sentences given the list of all the sentences of the clinical note. This method leverages the model's ability to comprehend the clinical question and utilize the sentences as contextual information to produce an accurate list of IDs. As the input, we use the patient narrative, the clinician question and the clinical note sentences.

Determine essentials individually: In this method, we evaluate each sentence individually to determine whether it should be classified as essential or not. The prompt instructs the model to determine if the information contained in a clinical note sentence is essential for accurately answering the patient's question. To simplify the task, the model is required to respond with a binary "Yes" or "No", reducing the complexity of the output and potentially improving reliability.

We initially employed a basic prompt for each method, which yielded suboptimal results. To improve response accuracy, we subsequently introduced a role-based prompt design by assigning the model the role of a medical expert. Additionally, we incorporated a CoT prompting strategy and one-shot and few-shot prompting techniques, providing the model with concrete examples to guide its responses. An overview of all prompt configurations is provided in Appendices A.2 and A.3.

To evaluate the quality of the generated lists, we focused solely on the strict metric, prioritizing the F1 score. The results of these two techniques are shown in Table 2 for extracting the list directly and Table 3 for individual technique.

		Strict		
Model	Prompt type	Prec.	Rec.	F1
Aloe 8B	basic	0.40	0.26	0.31
	role-based	0.52	0.49	**0.50**
	+ CoT	0.47	0.38	0.42
	+ one-shot	0.48	0.45	0.46
Aloe 70B	role-based	0.52	0.30	0.38
Mistral 7B	role-based	0.55	0.42	0.48

Table 2: Results across models and prompt types to extract *essential* notes for the first step, using a prompt to generate the *essential* lists directly.

		Strict		
Model	Prompt type	Prec.	Rec.	F1
Aloe 8B	basic	0.32	0.35	0.33
	role-based	0.38	0.58	**0.46**
	+ CoT	0.33	0.62	0.44
	+ few-shot	0.35	0.43	0.39
Aloe 70B	role-based	0.52	0.19	0.28

Table 3: Results across models and prompt types to extract *essential* notes individually in the first step.

6.2.1.2 Re-ranker to Extract Essentials

Our second approach is inspired by the typical RAG architecture, using a similarity based ranking model to identify which of the retrieved text chunks—the clinical notes' sentences—are more relevant given a query—a combination of the patient's narrative and the clinician's question. We leverage this method as it aligns with the task's goal of identifying the sentences of the clinical text that are relevant to answer the patient's query.

To determine sentence relevance, we rely on the output scores provided by the reranker and establish a threshold. Sentences with scores above it are labeled as *essential*, while those below are considered *not-relevant*. The optimal threshold is determined by computing the ROC curve and selecting the point that maximizes the Youden index, which allows us to identify the threshold that provides the best trade-off between true positive and false positive rates. However, a potential limitation of this method is that the threshold is determined based on the *dev* set and may not generalize well to the *test* set if the data distribution is different.

	Strict			Lenient		
Model	Prec.	Rec.	F1	Prec.	Rec.	F1
Jina	0.427	0.717	**0.535**	0.547	0.672	**0.603**
Alibaba	0.422	0.681	0.521	0.552	0.651	0.597
BAAI	0.507	0.507	0.507	0.587	0.429	0.495

Table 4: Precision, Recall, and F1-score for each reranking model in the task of identifying essential sentences.

Table 4 shows the results obtained by different reranker models in predicting *essential* sentences

Data	Approach	Overall Scores			Strict		Lenient	
		Overall	Relevance	Factuality	Macro-F1	Micro-F1	Macro-F1	Micro-F1
Dev	End-to-End	0.388	**0.312**	0.464	0.464	0.464	0.550	0.529
	Two-Step Prompting	0.385	0.265	0.504	0.518	0.504	0.547	0.511
	Two-Step Re-Ranker	**0.421**	0.285	**0.558**	**0.566**	**0.558**	**0.645**	**0.626**
Test	End-to-End	0.367	**0.325**	0.408	0.451	0.408	0.463	0.418
	Two-Step Prompting	0.366	0.281	0.452	0.474	0.452	0.493	0.458
	Two-Step Re-Ranker	**0.440**	0.276	**0.605**	**0.585**	**0.605**	**0.619**	**0.621**

Table 5: Best results of *Relevance* and *Factuality* of the three approaches in *dev* and test sets: Section 6.1 approach using Aloe (8B), two-step prompting approach using Aloe (8B) in both steps (Section 6.2.1.1) and two-step re-ranker approach leveraging Jina and Aloe (8B) for each step (Section 6.2.1.2). Best results in bold.

on the *dev* set. As shown in the table, the different models achieve similar F1-scores, with the Jina model obtaining the highest F1-score among them.

6.2.2 Second Step

After extracting essential sentences, we prompt a model to generate a response addressing the patient's concerns using only those sentences, without citations. Following the strategy used in Section 6.1, one example is shown, but based solely on the extracted essentials rather than the full EHR.

The generated response is then post-processed to (1) meet a 75-word limit and (2) add citations. For the first, the given response is split into sentences, and we do a selection to stay within the word limit. For the second, we match them to their most similar *essential* sentence of the first step, based on similarity scores. Citations are added accordingly, capping the number per sentence for balance. Title-like sentences from the first step are excluded in this step. Therefore, this process preserves—and often improves—the F-Score of the first step, while ensuring a coherent, well-cited response. We leverage Aloe (8B) to get the response since it seems to be the best model in previous steps. Table 5 shows the results after performing the second step.

7 Discussion

After obtaining the *dev* and *test* results (Table 5), it becomes evident that the best-performing strategy is the two-step approach with a reranker for *essential* sentence extraction, while the prompt-based two-step variant does not improve end-to-end overall results.

This suggests that more critical than merely dividing the task into smaller and simpler subtasks, is the choice of an appropriate method for each subtask. In this case, using a reranker to extract the most relevant sentences of the EHR for the patient's question proves to be more effective than the prompting-based selection.

Additionally, all three of our approaches surpass the organizers' zero-shot baseline, which achieves overall scores of 0.359 and 0.307 on the *dev* and *test* sets, respectively, despite relying on a significantly larger model (Llama 3.3 70B). This further supports our conclusion in Section 6.1 that larger models do not necessarily outperform smaller ones in complex tasks requiring not only medical expertise but also argumentative, summarization, and rewriting skills. Notably, our best run achieves an overall score of **0.44**, ranking **8th out of 30** on the test leaderboard and obtaining the top position in overall *Factuality*, even though it does not rely on external knowledge. Nonetheless, our three systems—particularly the two-step ones—exhibit certain limitations in *Relevance*, suggesting that alternative methods could be explored in future work to improve the response drafting process.

8 Conclusions

This work presents three approaches to respond to patient inbox messages using only the patient's concern, its technical reformulation, and the corresponding EHR. Our methods accurately generate responses and identify relevant information sources in the given text without relying on external data. Furthermore, we find that splitting the task into smaller, targeted subtasks improves performance when each is addressed with tailored methods. Future work may explore alternative response formulations to enhance clarity and improve the relevance score. In conclusion, we demonstrate that accurate message response is achievable without training data or external information.

Acknowledgments

This work has been partially supported by the HiTZ Center and the Basque Government (IXA excellence research group funding IT-1570-22 and IKER-GAITU project), as well as by the Spanish Ministry of Universities, Science and Innovation MCIN/AEI/10.13039/501100011033 by means of the projects: Proyectos de Generación de Conocimiento 2022 (EDHER-MED/EDHIA PID2022-136522OB-C22), DeepKnowledge (PID2021-127777OB-C21) and DeepMinor (CNS2023-144375). It is also supported by ILENIA (2022/TL22/00215335) and EU NextGeneration EU/PRTR (DeepR3 TED2021-130295B-C31) projects. And also by an FPU grant (Formación de Profesorado Universitario) from the Spanish Ministry of Science, Innovation and Universities (MCIU) to the fourth author (FPU23/03347).

Limitations

In this work we leverage several Instruct models, as well as, one and few-shot prompting techniques. Due to the lack of training samples, we do not extend the prompt examples and neither perform any finetuning. Additionally, in the two-step systems, there is still room for improvement in order to enhance the *Relevance* overall score—for instance, by trying text-to-text models like T5 (Raffel et al., 2020). In this work we focus on techniques that utilize only the available data, therefore Information Retrieval methods such as Retrieval Augmented Generation (RAG) are not employed, even though they could be beneficial given the limited data available. We leave these methods for future work.

References

Asma Ben Abacha, Yassine Mrabet, Yuhao Zhang, Chaitanya Shivade, Curtis Langlotz, and Dina Demner-Fushman. 2021. Overview of the MEDIQA 2021 shared task on summarization in the medical domain. In *Proceedings of the 20th Workshop on Biomedical Language Processing*, pages 74–85, Online. Association for Computational Linguistics.

Asma Ben Abacha, Chaitanya Shivade, and Dina Demner-Fushman. 2019. Overview of the MEDIQA 2019 shared task on textual inference, question entailment and question answering. In *Proceedings of the 18th BioNLP Workshop and Shared Task*, pages 370–379, Florence, Italy. Association for Computational Linguistics.

Jianlv Chen, Shitao Xiao, Peitian Zhang, Kun Luo, Defu Lian, and Zheng Liu. 2024. Bge m3-embedding: Multi-lingual, multi-functionality, multi-granularity text embeddings through self-knowledge distillation. *Preprint*, arXiv:2402.03216.

Ashwin Kumar Gururajan, Enrique Lopez-Cuena, Jordi Bayarri-Planas, Adrian Tormos, Daniel Hinjos, Pablo Bernabeu-Perez, Anna Arias-Duart, Pablo Agustin Martin-Torres, Lucia Urcelay-Ganzabal, Marta Gonzalez-Mallo, Sergio Alvarez-Napagao, Eduard Ayguadé-Parra, and Ulises Cortés Dario Garcia-Gasulla. 2024. Aloe: A family of fine-tuned open healthcare llms. *Preprint*, arXiv:2405.01886.

Qiao Jin, Bhuwan Dhingra, Zhengping Liu, William Cohen, and Xinghua Lu. 2019. PubMedQA: A dataset for biomedical research question answering. In *Proceedings of the 2019 Conference on Empirical Methods in Natural Language Processing and the 9th International Joint Conference on Natural Language Processing (EMNLP-IJCNLP)*, pages 2567–2577, Hong Kong, China. Association for Computational Linguistics.

JinaAI. 2024. jina-reranker-v2-base-multilingual. `https://hf.co/jinaai/jina-reranker-v2-base-multilingual`.

Alistair Johnson, Tom Pollard, Steven Horng, Leo Anthony Celi, and Roger Mark. 2023. Mimic-iv-note: Deidentified free-text clinical notes (version 2.2). physionet.

Alistair EW Johnson, Tom J Pollard, Lu Shen, Li-wei H Lehman, Mengling Feng, Mohammad Ghassemi, Benjamin Moody, Peter Szolovits, Leo Anthony Celi, and Roger G Mark. 2016. Mimic-iii, a freely accessible critical care database. *Scientific data*, 3(1):1–9.

Anastasia Krithara, Anastasios Nentidis, Konstantinos Bougiatiotis, and Georgios Paliouras. 2023. Bioasq-qa: A manually curated corpus for biomedical question answering. *Scientific Data*, 10(1):170.

Chaofan Li, Zheng Liu, Shitao Xiao, and Yingxia Shao. 2023. Making large language models a better foundation for dense retrieval. *Preprint*, arXiv:2312.15503.

MetaAI. 2024. Llama-3.3-70b-instruct. `https://hf.co/meta-llama/Llama-3.3-70B-Instruct`.

MistralAI. 2025. Mistral-7b-instruct-v0.3. `https://hf.co/mistralai/Mistral-7B-Instruct-v0.3`.

Ankit Pal, Logesh Kumar Umapathi, and Malaikannan Sankarasubbu. 2022. Medmcqa: A large-scale multi-subject multi-choice dataset for medical domain question answering. In *Conference on health, inference, and learning*, pages 248–260. PMLR.

Colin Raffel, Noam Shazeer, Adam Roberts, Katherine Lee, Sharan Narang, Michael Matena, Yanqi Zhou, Wei Li, and Peter J. Liu. 2020. Exploring the limits of transfer learning with a unified text-to-text transformer. *Journal of Machine Learning Research*, 21(140):1–67.

Sarvesh Soni and Dina Demner-Fushman. 2025a. A dataset for addressing patient's information needs related to clinical course of hospitalization. *arXiv preprint*.

Sarvesh Soni and Dina Demner-Fushman. 2025b. Overview of the archehr-qa 2025 shared task on grounded question answering from electronic health records. In *The 24th Workshop on Biomedical Natural Language Processing and BioNLP Shared Tasks*, Vienna, Austria. Association for Computational Linguistics.

Gemma Team. 2024. Gemma.

Lawrence KQ Yan, Qian Niu, Ming Li, Yichao Zhang, Caitlyn Heqi Yin, Cheng Fei, Benji Peng, Ziqian Bi, Pohsun Feng, Keyu Chen, and 1 others. 2024. Large language model benchmarks in medical tasks. *arXiv preprint arXiv:2410.21348*.

Xin Zhang, Yanzhao Zhang, Dingkun Long, Wen Xie, Ziqi Dai, Jialong Tang, Huan Lin, Baosong Yang, Pengjun Xie, Fei Huang, and 1 others. 2024. mgte: Generalized long-context text representation and reranking models for multilingual text retrieval. In *Proceedings of the 2024 Conference on Empirical Methods in Natural Language Processing: Industry Track*, pages 1393–1412.

A Prompts

A.1 End-to-End Prompt

```
{
    "system": "You are a Medical Report Assistant. Your role is to generate a
        coherent paragraph that answers a patient's question using only the
        information you consider relevant from the patient's medical record. You
        will receive as input:\n1. A narrative of the patient.\n2. The patient's
        question reformulated by a doctor.\n3. The medical record with the necessary
         information to answer the patient concern.\n\nYour task is to produce a
        paragraph answer that:\n- Rephrases and integrates the information from the
        provided notes without copying them verbatim.\n- Selects only the sentences
        you deem relevant to answer the question (do not use sentences that do not
        add value to the answer).\n- Clearly cites the sentence numbers that
        contributed to each sentence in your response, with the citation placed
        immediately after the sentence enclosed in vertical bars (e.g., |1| or
        |2,3|).\n- Does not repeat or omit any sentence that you consider relevant,
        and does not invent any additional sentences.\n- Contains a maximum of 75
        words in total.\n- Strictly adheres to the following format: each sentence
        on a new line with its citation; no additional text or explanations.{
        example_case}\n\nEnsure your final output strictly follows this format: one
        sentence per line with its corresponding citation, using the provided
        sentences you consider relevant (without any repetition or addition), and
        the total output does not exceed 75 words.\n\n- Ensure clarity, brevity, and
         accuracy in your response. Here goes an example:\n\nExample Input:\nCase:
        0\nPatient Narrative:\nTook my 59 yo father to ER ultrasound discovered he
        had an aortic aneurysm. He had a salvage repair (tube graft). Long surgery /
         recovery for couple hours then removed packs. why did they do this surgery
        ????? After this time he spent 1 month in hospital now sent home.\n\
        nReformulated Question:\nWhy did they perform the emergency salvage repair
        on him?\n\nClinical Notes:\n1: He was transferred to the hospital on
        2025-1-20 for emergent repair of his ruptured thoracoabdominal aortic
        aneurysm. 2: He was immediately taken to the operating room where he
        underwent an emergent salvage repair of ruptured thoracoabdominal aortic
        aneurysm with a 34-mm Dacron tube graft using deep hypothermic circulatory
        arrest. 3: Please see operative note for details which included cardiac
        arrest x2. 4: Postoperatively he was taken to the intensive care unit for
        monitoring with an open chest. 5: He remained intubated and sedated on
        pressors and inotropes. 6: On 2025-1-22, he returned to the operating room
        where he underwent exploration and chest closure. 7: On 1-25 he returned to
        the OR for abd closure JP/ drain placement/ feeding jejunostomy placed at
        that time for nutritional support.\n\n8: Thoracoabdominal wound healing well
         with exception of very small open area mid wound that is @1cm around and
        1/2cm deep, no surrounding erythema. 9: Packed with dry gauze and covered w/
        DSD.

    \n\nAnswer:\nHis aortic aneurysm was caused by the rupture of a thoracoabdominal
         aortic aneurysm, which required emergent surgical intervention. |1|\n He
        underwent a complex salvage repair using a 34-mm Dacron tube graft and deep
        hypothermic circulatory arrest to address the rupture. |2|\n The extended
        recovery time and hospital stay were necessary due to the severity of the
        rupture and the complexity of the surgery, though his wound is now healing
        well with only a small open area noted. |8|\n\n Now the REAL CASE:\n",
  "user": "Answer to the patient using the following inputs: Case: {id}\n\nPatient
      Narrative: {patient_narrative}, \n Reformulated Question: {clinician_question
      }\n  Clinical Notes: \n{note_excerpt} \n\nNow write the output paragraph based
       solely on the sentences you consider relevant to answer the question. Your
      output must:\n- Use only the sentences that you consider relevant from the
      provided text without repeating any sentence.\n- Contain a maximum of 75 words
      .\n- Include the citation (sentence number(s) used) immediately after each
      sentence, within vertical bars.\n- Follow exactly the format described: one
      sentence per line with its citation, and no additional text. Your final output
       must be the answer paragraph only, with no extra explanation or text.",
  "assistant": "Answer:"
}
```

A.2 Prompts to Extract List of Essentials

A.2.1 Role-Based Prompt

```
{
    "system": "You are a medical expert.\nYou are given:\n\nA patient narrative
        written by a family member or caregiver.\n\nA clinical question derived from
         the narrative.\n\nA list of numbered clinical notes from the patient's
        medical record.\n\nYour task is to identify which of the clinical notes are
        essential for answering the clinical question.\nReturn only the numbers of
        the essential notes in a comma-separated list.\nDo not explain your
        reasoning. Just return the list.",
    "user": "Patient Narrative:\n {patient_narrative}\n\nClinical Question:\n {
        clinical_question} \n\nClinical Notes:\n {sentences}"
}
```

A.2.2 CoT Prompt

```
{
    "system": "You are a medical expert with extensive experience in clinical
        natural language processing, specializing in extracting key information from
         clinical notes to answer medical questions. Your deep clinical knowledge
        and expertise in the healthcare domain enable you to identify critical data
        points from complex medical texts.\n\nTask: You will be provided with the
        patient narrative and the clinical question, and a set of clinical notes (
        each sentence is assigned a unique ID). Your goal is to identify only the
        sentences that contain critical information needed to answer the clinical
        question.\n\nInstructions:\n1. Internally, perform a detailed step-by-step
        analysis (chain-of-thought) of the clinical question and each clinical note.
         Evaluate each sentence for key information, context, and relevance.\n2.
        Select only the sentences that contain essential information to answer the
        question.\n3. Return only the IDs of those sentences, without including any
        additional text or explanation.\n\nOutput Format:\nA list of the relevant
        sentence IDs, separated by commas.\n\nReminder: Use your internal chain-of-
        thought to reason through the task, but do not display any of that reasoning
         in your final output. Simply provide the final answer as the list of IDs.",
    "user": "Patient Narrative:\n {patient_narrative}\n\nQuestion:\n {
        clinical_question} \n\nClinical Notes:\n {sentences}"
}
```

A.2.3 CoT + One-Shot Prompt

```
{
    "system": "You are a medical expert with extensive experience in clinical
        natural language processing, specializing in extracting key information from
         clinical notes to answer medical questions. Your deep clinical knowledge
        and expertise in the healthcare domain enable you to identify critical data
        points from complex medical texts.\n\nTask: You will be provided with the
        patient narrative and the clinical question, and a set of clinical notes (
        each sentence is assigned a unique ID). Your goal is to identify only the
        sentences that contain critical information needed to answer the clinical
        question.\n\nInstructions:\n1. Internally, perform a detailed step-by-step
        analysis (chain-of-thought) of the clinical question and each clinical note.
         Evaluate each sentence for key information, context, and relevance.\n2.
        Select only the sentences that contain essential information to answer the
        question.\n3. Return only the IDs of those sentences, without including any
        additional text or explanation.\n\nOutput Format:\nA list of the relevant
        sentence IDs, separated by commas.\n\nReminder: Use your internal chain-of-
        thought to reason through the task, but do not display any of that reasoning
         in your final output. Simply provide the final answer as the list of IDs.\n
        \nHere is an example of the output:\n\nPatient Narrative:\nTook my 59 yo
        father to ER ultrasound discovered he had an aortic aneurysm. He had a
        salvage repair (tube graft). Long surgery / recovery for couple hours then
        removed packs. why did they do this surgery????? After this time he spent 1
        month in hospital now sent home.\n\nClinical Question:\nWhy did they perform
         the emergency salvage repair on him?\n\nCinical notes: \n1: He was
        transferred to the hospital on 2025-1-20 for emergent repair of his ruptured
         thoracoabdominal aortic aneurysm. 2: He was immediately taken to the
        operating room where he underwent an emergent salvage repair of ruptured
```

```
        thoracoabdominal aortic aneurysm with a 34-mm Dacron tube graft using deep
        hypothermic circulatory arrest. 3: Please see operative note for details
        which included cardiac arrest x2. 4: Postoperatively he was taken to the
        intensive care unit for monitoring with an open chest. 5: He remained
        intubated and sedated on pressors and inotropes. 6: On 2025-1-22, he
        returned to the operating room where he underwent exploration and chest
        closure. 7: On 1-25 he returned to the OR for abd closure JP/ drain
        placement/ feeding jejunostomy placed at that time for nutritional support.
        8: Thoracoabdominal wound healing well with exception of very small open
        area mid wound that is @1cm around and 1/2cm deep, no surrounding erythema.
        9: Packed with dry gauze and covered w/DSD.\n\nAnswer:\nList of essential
        clinical notes::: 1, 2, 8",
    "user": "Patient Narrative:\n {patient_narrative}\n\nClinical Question:\n {
        clinical_question} \n\nClinical Notes:\n {sentences}"
}
```

A.3 Prompts to Extract Essentials Individually

A.3.1 Role-Based Prompt

```
{
    "system":"You are a medical expert.\nYou are given:\n\nA patient narrative
        written by a family member or caregiver.\n\nA clinical question derived from
         the narrative.\n\nA single clinical note extracted from the patientâĂŹs
        medical record.\n\nYour task is to determine whether the information in the
        clinical note is essential to accurately answer the clinical question.\
        nRespond with only one word: Yes or No.",
    "user": "Patient Narrative:\n {patient_narrative}\n\nClinical Question:\n {
        clinical_question}\n\nClinical Note:\n {sentences}"
}
```

A.3.2 CoT Prompt

```
{
    "system": "You are a medical expert with extensive experience in clinical
        natural language processing, specializing in analyzing clinical notes to
        assess patient conditions. Your deep clinical knowledge enables you to
        accurately interpret and evaluate medical narratives.\n\nTask: You will be
        provided with the patient narrative, the clinical question and a clinical
        note of the patient's clinical history. Your goal is to determine whether
        the clinical note contains sufficient and relevant information to answer the
         question.\n\nInstructions:\n1. Internally, perform a detailed step-by-step
        analysis (chain-of-thought) of the clinical question and the clinical note.\
        n2. Decide whether the clinical note provides a clear answer to the question
        .\n3. Return only \"Yes\" if the note contains sufficient evidence to answer
         the question, or \"No\" otherwise.\n\nOutput Format:\nA single word: Yes or
         No\n\nReminder: Use your internal chain-of-thought to reason through the
        task, but do not include any explanation or reasoning in the output. Only
        return Yes or No.",
    "user": "Patient Narrative:\n {patient_narrative}\n\nClinical Question:\n {
        clinical_question} \n\nClinical Note:\n {sentences}"
}
```

A.3.3 CoT + Few-Shot Prompt

```
{
    "system": "Role:\nYou are a medical expert with advanced expertise in clinical
        natural language processing (NLP). You specialize in analyzing unstructured
        clinical notes to extract medically relevant information and evaluate
        patient narratives. Your clinical acumen allows you to understand complex
        medical language and determine whether a given note contains sufficient
        evidence to answer specific clinical questions.\n\nTask:\nGiven a clinical
        note and a clinical question, determine whether the clinical note contains
        enough explicit and relevant information to confidently answer the question
        .\n\nInstructions:\n1. Internally, conduct a detailed chain-of-thought
        analysis to interpret the clinical question and assess the content of the
        note.\n2. Judge whether the clinical note includes clear, sufficient, and
```

```
    directly relevant information that supports and answers the question. Ignore
     any notes or phrases that are non-informative or purely structural, such as
     headers (e.g., \"Brief Hospital Course:\") or general section labels
    without medical content.\n3. Return only one word based on your internal
    reasoning:\n    - \"Yes\" âĂŤ if the note contains clear evidence to answer
    the question.\n    - \"No\" âĂŤ if the note lacks sufficient or relevant
    evidence to confidently answer the question.\n\nOutput Format:\nRespond with
     a single word only: Yes or No. Do not include any reasoning or explanation
    in your response.\n\nExample:\n\nPatient Narrative:\nTook my 59 yo father to
     ER ultrasound discovered he had an aortic aneurysm. He had a salvage repair
     (tube graft). Long surgery / recovery for couple hours then removed packs.
    Why did they do this surgery? After this time he spent 1 month in hospital
    now sent home.\n\nClinical Question:\nWhy did they perform the emergency
    salvage repair on him?\n\nClinical Notes & Answers:\nHe was transferred to
    the hospital on 2025-1-20 for emergent repair of his ruptured
    thoracoabdominal aortic aneurysm.\nAnswer: Yes\n\nHe was immediately taken
    to the operating room where he underwent an emergent salvage repair of
    ruptured thoracoabdominal aortic aneurysm with a 34-mm Dacron tube graft
    using deep hypothermic circulatory arrest.\nAnswer: Yes\n\nPlease see
    operative note for details which included cardiac arrest x2.\nAnswer: No\n\
    nPostoperatively he was taken to the intensive care unit for monitoring with
     an open chest.\nAnswer: No\n\nHe remained intubated and sedated on pressors
     and inotropes.\nAnswer: No\n\nOn 2025-1-22, he returned to the operating
    room where he underwent exploration and chest closure.\nAnswer: No\n\nOn
    1-25 he returned to the OR for abd closure JP/ drain placement/ feeding
    jejunostomy placed at that time for nutritional support.\nAnswer: No\n\
    nThoracoabdominal wound healing well with exception of very small open area
    mid wound that is @1cm around and 1/2cm deep, no surrounding erythema.\
    nAnswer: Yes\n\nPacked with dry gauze and covered w/DSD.\nAnswer: No",
  "user": "Patient Narrative:\n {patient_narrative}\n\nClinical Question:\n {
    clinical_question} \n\nClinical Note:\n {sentences}"
}
```

A.4 Prompt for Second Step Argumentation Creation

```
{
  "system":"You are a medical expert specializing in clinical natural language
    processing. Your task is to generate an answer to a patientâĂŹs health-
    related question based only on the information provided in the clinical
    notes. Write a short, medically sound answer that either paraphrases or
    argues or summaries using the key phrases from the notes. The response must
    not exceed 75 words. Focus only on the clinical notes provided. Your output
    must be a single, focused paragraph of 75 words or fewer âĂŤ never exceed
    this limit. Give only the answer, without any additional information or
    explanations.",
  "user": "Patient Narrative:\n {patient_narrative}\n\nClinical Question:\n {
    clinical_question} \n\nClinical Notes:\n {sentences}",
  "assistant": ""
}
```

UNIBUC-SD at ArchEHR-QA 2025: Prompting Our Way to Clinical QA with Multi-Model Ensembling

Dragoș-Dumitru Ghinea and **Ștefania Rîncu**
University of Bucharest
{dragos-dumitru.ghinea, stefania.rincu}@s.unibuc.ro

Abstract

In response to the ArchEHR-QA 2025 shared task, we present an efficient approach to patient question answering using small, pre-trained models that are widely available to the research community. Our method employs multi-prompt ensembling with models such as Gemma and Mistral, generating binary relevance judgments for clinical evidence extracted from electronic health records (EHRs). We use two distinct prompts (A and B) to assess the relevance of paragraphs to a patient's question and aggregate the model outputs via a majority vote ensemble. The relevant passages are then summarized using a third prompt (C) with Gemma. By leveraging off-the-shelf models and consumer-grade hardware (1x RTX 5090), we demonstrate that it is possible to improve performance without relying on resource-intensive fine-tuning or training. Additionally, we explore the impact of Chain-of-Thought (CoT) prompting and compare the performance of specialized versus general-purpose models, showing that significant improvements can be achieved through effective use of existing models.

1 Introduction

Responding to patient inquiries via patient portals is a major contributor to clinician workload, and automating this process using electronic health records (EHRs) could significantly reduce that burden. The ArchEHR-QA shared task (Soni and Demner-Fushman, 2025b) challenges participants to generate answers grounded in clinical evidence extracted from EHRs, requiring both accurate relevance detection and effective summarization. Submissions are evaluated on Factuality (how well answers cite annotated evidence sentences) and Relevance (how well they align with gold-standard 'essential' notes) using metrics such as Citation F1, BLEU, ROUGE, SARI, BERTScore, AlignScore, and MEDCON. The final score averages strict Factuality and aggregated Relevance metrics. Baseline scores on Codabench were generated using LLaMA 3.3 70B in a zero-shot setting, prompted to produce citation-containing answers; invalid responses (e.g., overly long or missing citations) were regenerated up to five times.

We address this challenge by breaking the task into two main subtasks: paragraph relevance classification and summarization. First, we classify the relevance of clinical paragraphs to the patient's question using a multi-model approach. Then, we generate a summary of the most relevant paragraphs to answer the query. Our method employs two pre-trained models, Gemma 3 27B (Gemma Team, 2025) and Mistral 3.1 Small 24B (Mistral AI, 2025b), to assess relevance and summarize the relevant passages. We experimented with a variety of models and prompting strategies, including Chain-of-Thought (CoT), and explored different configurations of the majority vote ensemble to optimize model performance.

A key aspect of our approach is the use of off-the-shelf models and consumer-grade hardware (1× RTX 5090), avoiding the need for fine-tuning or training from scratch. By focusing on smaller, more accessible models, we show that strong performance can be achieved with minimal computational overhead. Our experiments further demonstrate that carefully designed prompt engineering and ensemble methods can enhance performance effectively, making the approach both practical and scalable. While larger models may yield further gains in real-world deployments, our work highlights the untapped potential of lightweight setups for clinical question answering.

2 Related Work

Prompt-based methods have shown that LLMs can achieve strong performance across diverse tasks without fine-tuning. Techniques such as CoT

BioNLP 2025 Shared Tasks, pages 11–21
August 1, 2025 ©2025 Association for Computational Linguistics

prompting improve performance on a range of tasks (Wei et al., 2022; Kojima et al., 2022). Further strategies like self-consistency decoding improve CoT outputs by aggregating multiple reasoning paths (Wang et al., 2023, 2024). Prompt ensembling, using diverse prompts with majority voting, has been shown to improve reliability (Yang et al., 2023), and further analysis on compound systems has been made (Chen et al., 2024).

In the medical domain, recent studies show that general purpose LLMs can be competitive with specialized models when guided by carefully designed prompts (Nachane et al., 2024; Russe et al., 2024; Sivarajkumar et al., 2024).

Our system builds on this line of work by using a multi-prompt ensemble strategy with Gemma 3 (27B) and Mistral 3.1 Small (24B), relying solely on prompt engineering and avoiding fine-tuning.

3 Method

3.1 Paragraph Relevance Assessment

Each instance from the dataset (Soni and Demner-Fushman, 2025a) provides a clinical note both as continuous text and as a list of indexed paragraphs, which must be cited in the final answer; in our experiments, we use only the paragraph-indexed format. Additionally, each instance includes a patient-authored question and a corresponding clinician-formulated question. We found that using the patient question for paragraph relevance classification introduces more false positives, so we rely exclusively on the clinician question, a refined and focused version of the original, for relevance assessment. The patient question is instead used during the summarization subtask to better reflect natural inquiry phrasing.

3.1.1 Individual Model Performance

We first prompted models to classify the relevance of the entire list of paragraphs in a single pass. While their explanations were often coherent, the final outputs frequently included incorrect indices or mismatched list lengths that didn't align with their own reasoning. To address this, we reformulated the task as a binary classification problem: the model is given the full list and asked whether a specific paragraph (identified by index) is relevant to the clinician's question. This approach significantly improved both consistency and interpretability.

Prompt selection followed two strategies: (1) manual trial-and-error and (2) suggestions from Gemini 2.5 Pro, chosen for its availability via Google AI Studio and generous usage limits that enabled extensive testing. Table 5 highlights some of our strongest prompt engineering results. Alongside prompt design, we also varied sampling temperature (ranging from 0.1 to 1.0) based on guidance from model developers and the open-source community (DeepSeek AI, 2024; Unsloth Team, 2025; Mistral AI, 2025a).

Model	Quantization	Overall Factuality	Prompt Used
gemma3-27b-it	Q6	56.04	A
mistral-small-3.1-24b	Q8	54.08	A
gemma3-12b-it	Q8	52.03	C
gemma2-9b-it	Q8	52.00	D
phi4	Q8	51.08	D
phi4-o1	i1 Q6_K	50.00	A
deepseek-llama-8b	F16	44.44	D
phi4-QwQ	Q8	40.89	D
phi4-mini-it	Q8	40.26	A
deepseek-qwen-32b	Q6_K	30.24	D
*all-relevant**	–	48.76	–
baseline (LLaMA 3.3 70B)	–	43.10	–

Table 1: Best overall factuality scores on the **dev dataset**. Prompt labels (A-D) refer to variants described in the Appendix A. The *all-relevant* baseline assumes all paragraphs are relevant. A list of more detailed scores is available in Appendix C.

The variation in temperature settings may affect the reproducibility of certain scores. To address this, in subsequent experiments we fix the temperature at 0.1, a value that provides stable and reproducible outputs while maintaining performance comparable to the best results observed.

We did not observe significant performance gains from models fine-tuned on medical data (WhyHow.AI Team, 2024; mradermacher, 2025a, 2024a), with most yielding only marginal improvements over the baseline (Table 2). Additional experiments with models such as `OpenBioLLM-Llama3-70B (i1-IQ3_XXS)` (mradermacher, 2024b), `Med-Chatbot-R1-Qwen-7B (F16)` (mradermacher, 2025c), and `ClinicalGPT-R1-Qwen-7B-EN-preview (F16)` (mradermacher, 2025b) were similarly unpromising. These models often failed to follow the required Yes/No output format, even without chain-of-thought prompting, making rule-based evaluation via regex unreliable. While output post-processing with another LLM is a possible workaround, we deemed it unnecessarily complex for the scope of this task.

3.1.2 Impact of Chain-of-Thought Prompting

We observe that chain-of-thought (CoT) prompting generally improves performance, as shown in Table 3. However, its effectiveness varies depending

Model	Quantization	Overall Factuality
PatientSeek	Q4_K_M	45.48
BioMistral-MedMNX	F16	42.96
DeepSeek-R1-Distill-Llama-8B-Medical-Expert	F16	44.39

Table 2: Overall factuality scores on the **dev dataset** for a few medical finetuned models using prompt A.

on the model. In our implementation, we introduce CoT reasoning using the following instruction: *"Create a chain of thought to determine if the paragraph is relevant to answering the question. Put your reasoning between <think> and </think> tags."*

While larger models such as `gemma3-27b-it` (Gemma Team, 2025) and `mistral-small-3.1-24b` (Mistral AI, 2025b) benefit significantly from CoT prompting, smaller models, particularly those in the `phi` family (Abdin et al., 2024; LM Studio Community, 2025), sometimes fail to complete the task reliably. In many cases, these models generate only reasoning text within the '<think>' tags without producing a final answer. We observed similar behavior in other small variants of LLaMA and Mistral, suggesting that limited context handling or weaker instruction-following may hinder CoT execution in compact models. To maintain consistency in automatic evaluation, if a model output could not be parsed using regular expressions to extract a valid binary answer, we defaulted to treating the paragraph as not relevant.

Model	Quantization	No CoT	CoT
gemma3-27b-it	Q6	45.48	**48.67**
mistral-small-3.1-24b	Q8	41.05	**52.22**
phi4	Q8	51.08	**51.32**
phi4-mini-it	Q8	29.49	**38.69**

Table 3: Comparison of factuality scores on the **dev dataset** with (prompt A) and without (prompt D) chain-of-thought (CoT) prompting.

3.1.3 Ensembling Model Performance

To improve robustness, we ensemble predictions using simple majority voting, selecting the most frequent Yes/No label per paragraph. To balance performance and efficiency, we limit ensembles to three models and treat different prompt configurations for the same model as distinct components. We explored various model-prompt combinations (Appendix B), with our best-performing ensemble shown in Figure 1. While overall factuality scores are useful, we prioritized confusion matrices when selecting ensembles, as they better reveal false positive and false negative trade-offs.

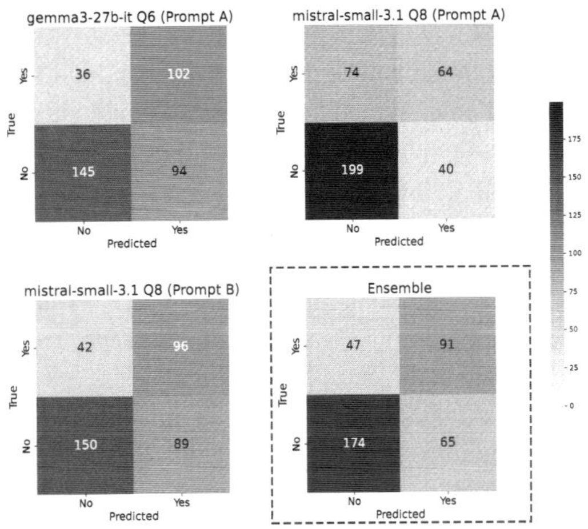

Figure 1: Confusion matrices for the best-performing ensemble and its constituent models.

3.2 Summarization

The summarization subtask requires (1) generating a coherent, concise answer from selected paragraphs and (2) citing the source paragraph(s) for each sentence. Instead of decoupling these, we adopt a unified approach where the model generates citations inline, avoiding the need for external alignment.

In our experiments, `Gemma` (Gemma Team, 2025) and `Mistral` (Mistral AI, 2025b) were the most reliable at handling complex summarization prompts. `Gemma` stood out for its consistent adherence to the required format, or outputs that were easily corrected via postprocessing, making it our primary choice. Using the same models for both relevance classification and summarization also helped maximize GPU parallelism on our RTX 5090 by reducing the number of concurrently loaded models.

We tested `Mistral` (Mistral AI, 2025b) as a correction layer for `Gemma`'s (Gemma Team, 2025) outputs, but observed only a minor readability improvement (+0.02), with inconsistent results. Moreover, `Mistral` frequently exceeded the 75-word constraint and resisted shortening even in multi-turn settings.

A consistent challenge for `Gemma` (Gemma Team, 2025) was maintaining citation coverage under tight length limits. To comply, it sometimes dropped relevant content, typically omitting one relevant paragraph for every 3-5 irrelevant ones removed.

To better enforce the 75-word limit, we implemented a multi-turn prompting strategy: after an initial response (often 80+ words), we re-prompted with explicit instructions to shorten. Curiously, instead of trimming toward the limit, the model often produced much shorter summaries (typically 40-50 words), indicating that it is capable of brevity but defaults to verbosity unless guided. We allowed up to three retries, though this cap was never reached in evaluation.

When relevance classification yielded fewer than three paragraphs, we used the full paragraph set to ensure sufficient context.

To evaluate the impact of relevance filtering, we conducted an ablation study using the best-performing summarization prompt. Summaries generated from the full paragraph set scored **45.6**, while those using only the filtered paragraphs reached **48.1**, confirming that relevance assessment contributes positively to final summarization quality.

3.3 Postprocessing

Instead of enforcing strict formatting in the prompt, we correct inconsistencies through regex-based postprocessing. `Gemma` (Gemma Team, 2025), for instance, often misplaces citations, adds extra spaces, or cites the question itself. To let the model focus on content, we fix these issues afterward by grouping adjacent citation markers, removing internal whitespace, and relocating misplaced citations to sentence ends. This lightweight pipeline improves formatting while preserving the summary's meaning.

3.4 Final Scores

Method	Dev		Test	
	Factuality	Relevance	Factuality	Relevance
our approach	60.4	35.8	53.8	32.7
baseline	43.1	28.7	33.6	27.8

Table 4: Scores overview on both the **dev and test datasets**.

Our approach substantially improves both factuality and relevance over the baseline (Table 4). On the test set, we observe a +20.2 Factuality and +4.9 Relevance gain, demonstrating the robustness and generalization of our method beyond the dev set. The code used to generate the test submission can be found on GitHub. [1]

[1] Test Submission Generation Source Code.

3.5 Error Analysis

We identified eight sentences misclassified by all models, revealing unanimous relevance assessment failures, six false positives, where irrelevant content was marked as relevant, and two false negatives, where relevant sentences were wrongly dismissed.

In the false-negative cases, models typically judged that the sentences lacked a clear link to the question or were too general. For example, in Case 6 ("Why did they find out later that he had fungal pneumonia?"), the statement "Initially in the 160s, but has improved with fluids" was dismissed for lacking an explicit connection. In Case 16 ("Could her back pain and dizziness be concerning for a stroke?"), the suggestion "You can take the oxycodone for a short time and follow up with Dr. ___" was seen as generic advice rather than a direct answer.

Conversely, the six false-positive cases involved irrelevant sentences incorrectly identified as relevant. These errors often stemmed from chain-of-thought reasoning, in which the models associated the current sentence with earlier contextual information. The presence of medical terminology or explanatory language appeared to bias the models toward overestimating relevance. These findings suggest that the models may overly rely on surface-level cues such as technical vocabulary or narrative structure when determining relevance.

4 Conclusion

Our contribution to the ArchEHR-QA 2025 shared task presents a resource-efficient approach to clinical question answering from EHR data, showing that strong performance is achievable without fine-tuning or specialized hardware. Using multi-prompt ensembling across pre-trained models like Gemma and Mistral on consumer-grade GPUs, we improved the robustness and accuracy of paragraph relevance identification over individual models. Our modular two-stage pipeline (filtering relevant evidence before summarization) proved effective, with relevance assessment clearly improving final answer quality. The approach relies on careful prompt engineering, combining Chain-of-Thought reasoning and majority vote aggregation. While there is still room for improvement, our results demonstrate the promise of prompt-based methods with accessible LLMs as a scalable, cost-effective solution for clinical QA, especially in resource-limited settings.

Limitations

While our approach demonstrates promising results, several limitations should be acknowledged.

Firstly, our study primarily focused on relatively small, accessible models (up to 27B parameters) due to our emphasis on resource efficiency and consumer-grade hardware (1x RTX 5090). Although we show strong performance is achievable under these constraints, it is likely that larger, state-of-the-art models could yield further improvements, albeit at significantly higher computational cost. The use of quantized models, necessary for fitting them onto our hardware, might also introduce a minor performance degradation compared to full-precision versions.

Secondly, the performance of our system relies heavily on prompt engineering. While effective, identifying optimal prompts required considerable experimentation (manual trial-and-error and assistance from Gemini 2.5 Pro). The sensitivity to prompt wording means that adapting the system to different QA formats or clinical contexts might require further prompt tuning. Additionally, Chain-of-Thought prompting, while beneficial for the larger models tested, proved less reliable for smaller models, indicating limitations in their reasoning capabilities or instruction following.

Thirdly, the evaluation was conducted on the specific dataset provided for the ArchEHR-QA 2025 task. The generalization performance of our prompts and ensemble strategy on different EHR datasets or in real-world clinical deployment remains to be validated.

Fourthly, while our summarization component using Gemma generally adhered to formatting requirements, it sometimes struggled with strict length constraints and citation completeness on the first pass, necessitating multi-turn prompting and postprocessing steps. This indicates potential brittleness in complex instruction following for the summarization task.

Finally, our internal development and optimization efforts disproportionately focused on maximizing the automatic factuality score. While ensuring factual grounding is critical, this narrow focus meant that other important automatic metrics provided by the shared task (BLEU, ROUGE, BERTScore, AlginScore, MEDCON), received comparatively less attention during model and prompt selection. Consequently, the system's performance across these diverse dimensions of quality may be underdeveloped relative to its factuality performance. Furthermore, a complete assessment of the system's clinical utility, relevance, and overall correctness ultimately requires evaluation by domain experts in a practical setting.

References

Marah Abdin, Jyoti Aneja, Harkirat Behl, Sébastien Bubeck, Ronen Eldan, Suriya Gunasekar, Michael Harrison, Russell J. Hewett, Mojan Javaheripi, Piero Kauffmann, James R. Lee, Yin Tat Lee, Yuanzhi Li, Weishung Liu, Caio C. T. Mendes, Anh Nguyen, Eric Price, Gustavo de Rosa, Olli Saarikivi, and 8 others. 2024. Phi-4 technical report. *Preprint*, arXiv:2412.08905.

Lingjiao Chen, Jared Davis, Boris Hanin, Peter Bailis, Ion Stoica, Matei Zaharia, and James Zou. 2024. Are more llm calls all you need? towards the scaling properties of compound ai systems. In *Advances in Neural Information Processing Systems*, volume 37, pages 45767–45790. Curran Associates, Inc.

DeepSeek AI. 2024. Deepseek - usage recommendations. `https://huggingface.co/deepseek-ai/DeepSeek-R1-Distill-Qwen-32B#usage-recommendations`. Accessed: 2025-05-03.

Gemma Team. 2025. Gemma 3.

Takeshi Kojima, Shixiang (Shane) Gu, Machel Reid, Yutaka Matsuo, and Yusuke Iwasawa. 2022. Large language models are zero-shot reasoners. In *Advances in Neural Information Processing Systems*, volume 35, pages 22199–22213. Curran Associates, Inc.

LM Studio Community. 2025. Phi-4-mini-instruct-gguf. `https://huggingface.co/lmstudio-community/Phi-4-mini-instruct-GGUF`. Accessed: 2025-05-03.

Mistral AI. 2025a. Mistral ai api. `https://docs.mistral.ai/api/`. Accessed: 2025-05-03.

Mistral AI. 2025b. Mistral-small-3.1-24b-instruct-2503. `https://huggingface.co/mistralai/Mistral-Small-3.1-24B-Instruct-2503`. Accessed: 2025-05-03.

mradermacher. 2024a. Deepseek-r1-distill-llama-8b-medical-expert-ggufmradermacher/deepseek-r1-distill-llama-8b-medical-expert-gguf. `https://huggingface.co/mradermacher/DeepSeek-R1-Distill-Llama-8B-Medical-Expert-GGUF`. Accessed: 2025-05-03.

mradermacher. 2024b. Openbiollm-llama3-70b-i1-gguf. `https://huggingface.co/mradermacher/OpenBioLLM-Llama3-70B-i1-GGUF`. Accessed: 2025-05-03.

mradermacher. 2025a. Biomistral-medmnx-gguf. https://huggingface.co/mradermacher/BioMistral-MedMNX-GGUF. Accessed: 2025-05-03.

mradermacher. 2025b. Clinicalgpt-r1-qwen-7b-en-preview-gguf. https://huggingface.co/mradermacher/ClinicalGPT-R1-Qwen-7B-EN-preview-GGUF. Accessed: 2025-05-03.

mradermacher. 2025c. Med-chatbot-r1-qwen-7b-gguf. https://huggingface.co/mradermacher/Med-Chatbot-R1-Qwen-7B-GGUF. Accessed: 2025-05-03.

Saeel Sandeep Nachane, Ojas Gramopadhye, Prateek Chanda, Ganesh Ramakrishnan, Kshitij Sharad Jadhav, Yatin Nandwani, Dinesh Raghu, and Sachindra Joshi. 2024. Few shot chain-of-thought driven reasoning to prompt LLMs for open-ended medical question answering. In *Findings of the Association for Computational Linguistics: EMNLP 2024*, pages 542–573, Miami, Florida, USA. Association for Computational Linguistics.

Maximilian Frederik Russe, Marco Reisert, Fabian Bamberg, and Alexander Rau. 2024. Improving the use of llms in radiology through prompt engineering: from precision prompts to zero-shot learning. *RoFo: Fortschritte auf dem Gebiete der Röntgenstrahlen und der Nuklearmedizin*, 196(11):1166–1170. Epub 2024 Feb 26.

Sonish Sivarajkumar, Mark Kelley, Alyssa Samolyk-Mazzanti, Shyam Visweswaran, and Yanshan Wang. 2024. An empirical evaluation of prompting strategies for large language models in zero-shot clinical natural language processing: Algorithm development and validation study. *JMIR Med Inform*, 12:e55318.

Sarvesh Soni and Dina Demner-Fushman. 2025a. A dataset for addressing patient's information needs related to clinical course of hospitalization. *arXiv preprint*.

Sarvesh Soni and Dina Demner-Fushman. 2025b. Overview of the archehr-qa 2025 shared task on grounded question answering from electronic health records. In *The 24th Workshop on Biomedical Natural Language Processing and BioNLP Shared Tasks*, Vienna, Austria. Association for Computational Linguistics.

Unsloth Team. 2025. How to run gemma 3 effectively with our ggufs on llama.cpp, ollama, open webui and how to fine-tune with unsloth! https://docs.unsloth.ai/basics/gemma-3-how-to-run-and-fine-tune. Accessed: 2025-05-03.

Han Wang, Archiki Prasad, Elias Stengel-Eskin, and Mohit Bansal. 2024. Soft self-consistency improves language models agents. In *Proceedings of the 62nd Annual Meeting of the Association for Computational Linguistics (Volume 2: Short Papers)*, pages 287–301, Bangkok, Thailand. Association for Computational Linguistics.

Xuezhi Wang, Jason Wei, Dale Schuurmans, Quoc V Le, Ed H. Chi, Sharan Narang, Aakanksha Chowdhery, and Denny Zhou. 2023. Self-consistency improves chain of thought reasoning in language models. In *The Eleventh International Conference on Learning Representations*.

Jason Wei, Xuezhi Wang, Dale Schuurmans, Maarten Bosma, brian ichter, Fei Xia, Ed Chi, Quoc V Le, and Denny Zhou. 2022. Chain-of-thought prompting elicits reasoning in large language models. In *Advances in Neural Information Processing Systems*, volume 35, pages 24824–24837. Curran Associates, Inc.

WhyHow.AI Team. 2024. Patientseek. https://huggingface.co/whyhow-ai/PatientSeek. Accessed: 2025-05-03.

Han Yang, Mingchen Li, Huixue Zhou, Yongkang Xiao, Qian Fang, and Rui Zhang. 2023. One llm is not enough: Harnessing the power of ensemble learning for medical question answering. *medRxiv*.

A Prompts

This section outlines the prompt structure used during our experiments. All prompts follow the OpenAI API format, with each prompt represented as a list of message objects.

For paragraph relevance assessment, we typically separate the main instruction, the question, and the list of paragraphs into distinct message objects, as shown below.

```
messages = [
    {
        "role": "user",
        "content": f"{prompt}"
    },
    {
        "role": "user",
        "content": f"Question: {q}"
    },
    {
        "role": "user",
        "content": f"List of paragraphs: {l}"
    }
]
```

The `prompt` variable is the component we vary most frequently across experiments. The `q` and `l` variables consistently represent the clinician-derived question and the list of paragraphs, respectively. Each paragraph in the list is formatted as: `#{i} - "{s}"`, where `i` is the paragraph ID and `s` is the paragraph content.

For summarization, we use a single message object with the role `"user"`, applying a prompt, referred to as Prompt Z, crafted through a combination of manual trial and error and refinements suggested by Gemini 2.5 Pro. The variables enclosed in curly braces within Prompt Z are substituted with their respective values, as done for the paragraph relevance assessment prompts. The full content of Prompt Z is included in the list of prompts below.

A.1 Prompt A

This prompt was selected after several manual iterations, relying on intuitively designed instructions.

You will receive a clinical question interpreted from a patient's question and a list of paragraphs extracted from a clinical note.

The questions are asked through the patient portal by patients.

Create a chain of thought to determine if the paragraph is relevant to answering the question. Put your reasoning between <think> and </think> tags.

Is the paragraph number %paragraph-number% (indexed from 0) relevant to answering the question?
The paragraph does not need to give a full answer, but should be relevant in formulating the answer.
Give a Yes or No answer.

A.2 Prompt B

This prompt uses a different message structure than the one previously described and can be found in the GitHub source code. It was generated by Gemini 2.5 Pro when asked to refine an arbitrary prompt (e.g., Prompt A) with the goal of maximizing performance.

*Role:** You are an expert clinical information analyst specializing in evaluating text relevance using Mistral models.

Context: You will be provided with a clinical question (derived from a patient's query via a patient portal) and a list of numbered paragraphs extracted from a clinical note.

Goal: Determine if a *specific* paragraph from the list is relevant to answering the provided clinical question.

Definition of Relevance: A paragraph is relevant if it contains information that helps answer, contributes to answering, or is directly related to the topic of the question. It does *not* need to provide the complete answer on its own.

Task Instructions:
1. You will receive the Clinical Question and the full List of Paragraphs first.
2. Then, you will be asked to evaluate a *specific* paragraph, identified by its number (0-indexed).
3. Focus your analysis *exclusively* on the content of the specified paragraph number. Do not base your relevance decision on other paragraphs in the list.
4. Generate a step-by-step Chain of Thought (CoT) reasoning process to justify your decision. Clearly explain *why* the specified paragraph is or is not relevant based on the question's topic and the paragraph's content.
5. Enclose your entire Chain of Thought reasoning securely within '<think>' and '</think>' tags.
6. Immediately following the closing '</think>' tag, provide your final answer as *only* "Yes" or "No".

Output Format:
<think>
[Your detailed step-by-step reasoning comparing the specific paragraph's content to the question's requirements, focusing only on the specified paragraph.]
</think>
[Yes or No]

A.3 Prompt C

Same as Prompt A but we don't specify the source of the questions.

You will receive a clinical question interpreted from a patient's question and a list of paragraphs extracted from a clinical note.

Create a chain of thought simulating a doctor's (that needs to provide a response) thinking to determine if the paragraph is relevant to answering the question.
Put your reasoning between <think> and </think> tags.

Is the paragraph number %sentence-number% (indexed from 0) relevant to answering the

question?
The paragraph does not need to give a full answer, but should be relevant in formulating the answer. Give a Yes or No answer.

A.4 Prompt D

Same as Prompt A, but without chain of thought.

You will receive a clinical question interpreted from a patient's question and a list of paragraphs extracted from a clinical note.

The questions are asked through the patient portal by patients.

Is the paragraph number %paragraph-number% (indexed from 0) relevant to answering the question?
The paragraph does not need to give a full answer, but should be relevant in formulating the answer. Give a Yes or No answer.

A.5 Prompt E

Another recommendation from Gemini 2.5 Pro.

*Role:** You are an expert clinical information analyst. Your purpose is to evaluate the relevance of clinical note paragraphs to patient questions.

Context: You will receive:
1. A **Clinical Question** from a patient.
2. A **List of Paragraphs** (0-indexed) from a clinical note.
3. A specific **Paragraph Number** to evaluate.

Goal: Determine if the *specified paragraph* (identified by its number) is relevant for answering the Clinical Question.

Definition of Relevance:
* A paragraph is relevant if its content *directly addresses, contributes to answering, or is topically related* to the Clinical Question.
* When evaluating the specified paragraph, consider its *intrinsic content* primarily.
* Also, consider its *contextual value*: Does it provide essential background for another relevant paragraph? Is it the *most* relevant piece of information available, even if only weakly related, especially if other paragraphs are irrelevant?

Instructions:
1. **Analyze the Request:** Understand the Clinical Question and review all provided paragraphs to grasp the overall context.
2. **Focus on the Target:** Concentrate your relevance analysis on the *specific* paragraph number provided in the final user request.
3. **Perform Chain-of-Thought (CoT) Reasoning:** Generate a step-by-step reasoning process detailing your evaluation.
* Start by stating the paragraph number being evaluated.
* Summarize the core information in the specified paragraph.
* Compare this information directly against the Clinical Question.
* Explicitly discuss *how* or *why* it is (or isn't) relevant.
* If applicable, briefly mention its contextual role (e.g., "This paragraph provides context for paragraph X," or "While weakly related, it's the only paragraph mentioning Y topic").
* Conclude your reasoning with a clear statement about the relevance of the *specified paragraph*.
4. **Enclose Reasoning:** Place your *entire* step-by-step reasoning within '<think>' and '</think>' tags. **Crucially, there should be NO text before the opening '<think>' tag and NO text between the closing '</think>' tag and the final Yes/No answer.**
5. **Provide Final Answer:** Immediately following the closing '</think>' tag, output *only* the word "Yes" or "No" indicating the relevance of the *specified paragraph*.

Output Format:
<think>
[Step-by-step reasoning analyzing the specified paragraph's relevance to the question, considering context as defined above.]
</think>
[Yes or No]

A.6 Prompt Z

Goal: Create a concise (70 words) answer for the **Clinical Question** using *only* the information present in the **Relevant paragraphs**.

The questions are asked through a patient portal.

Patient Question:
{patient_question}

Clinical Question (Derived from patient question):
{question}

Relevant paragraphs (with 1-based indices):
{formatted_relevant_paragraphs}

Citation style: Paragraph's index between | | symbols. For example: |1| or |2,7| or |1,2,3,4| or |5,7,9|. Citations must be comma separated.

Output Detail:
- The sentences in the generated answer may be supported using one, multiple, or none (unsupported) of the paragraphs from the clinical note.
- The unsupported sentences in the answer may be ignored during the quantitative evaluation.
- The answers should be in the professional register to better match the contents of the clinical notes. Simplification of answers to lay language is assumed to be performed later and is not the focus of this task.
- The generated answer should be limited to 75 words, which roughly correspond to 5 sentences. This is based on our observations from the baseline experiments and existing literature supporting that a paragraph-long answer is preferred by users.
- There are no limitations to the number of note sentences cited.

You need to answer the patient's question, but do not take the information provided in it for granted and do not refer to it in your answer.
The answer should sound natural and be a coherent response to the question.
Do not add any additional information beside the answer such as "Your summary: ".
Your answer should be a summary of the information in the relevant paragraphs you found, with few sentences, and more like a long paragraph.

A.7 Other Prompts

We do not list all the prompts generated by Gemini, as they are largely similar, differing only in minor adjustments aimed at reducing false positives and/or false negatives to better balance the model's behavior.

B Ensembling

We try different ensembling strategies, justified by various scores obtained by individual models.

B.1 Candidate one

The best overall factuality ensemble overall, but with a bit worse confusion matrix than the chosen one. Excellent balance of top performance, high precision diversity, and balanced size/generation diversity.

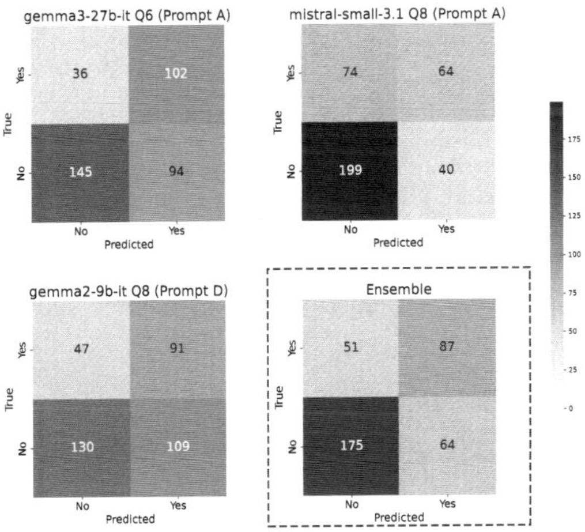

Figure 2: Ensemble candidate one.

B.2 Candidate two

Combines the best Gemma with both the high-precision and high-recall Mistral variants, maximizing Mistral architectural presence. Solid score with balanced stats.

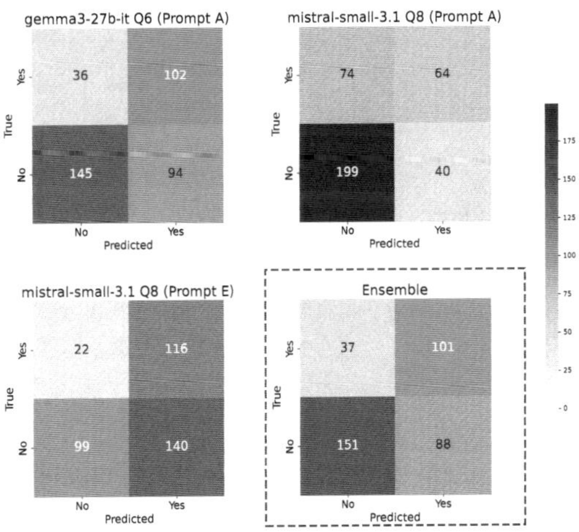

Figure 3: Ensemble candidate two.

B.3 Candidate three

An example of a high overall factuality score (56.359) with a slightly worse confusion matrix.

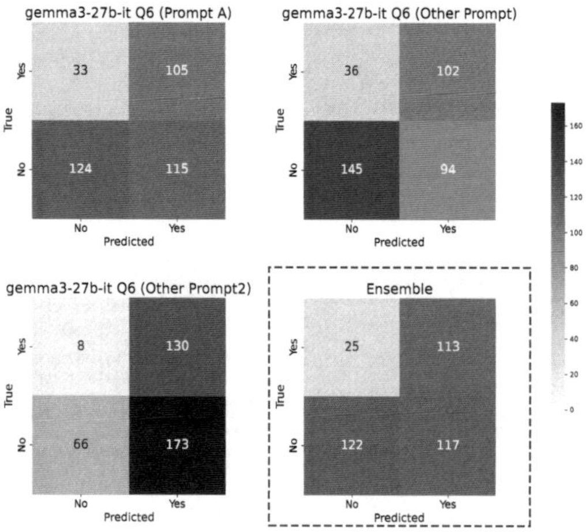

Figure 4: Ensemble candidate three.

B.4 Candidate four

This is yet another example where ensembling contributes to more robust predictions, although it does not yield the strongest overall performance among our configurations.

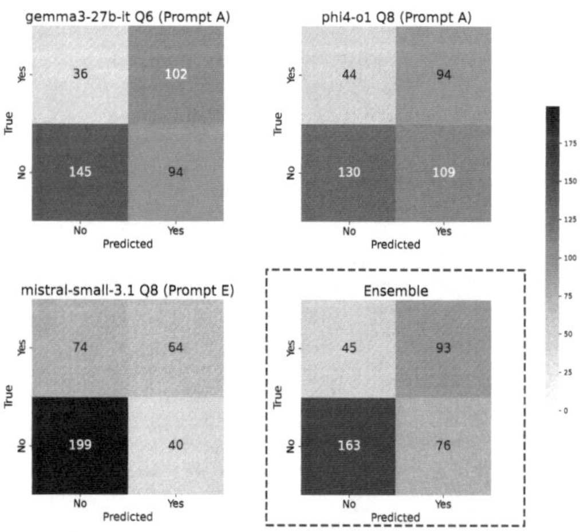

Figure 5: Ensemble candidate four.

B.5 Candidate five

Three models finetuned on clinical data.

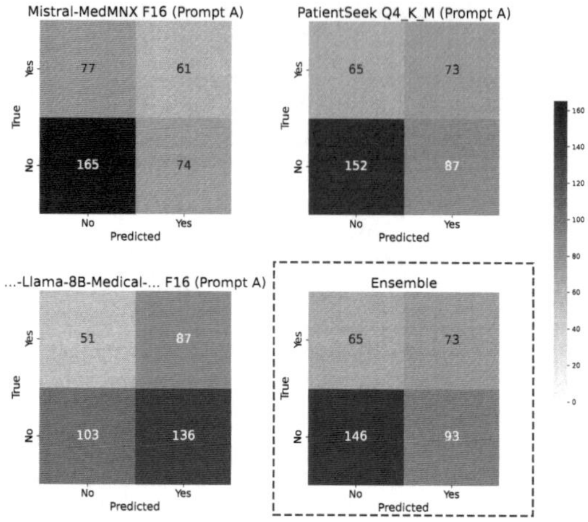

Figure 6: Ensemble candidate five.

C Detailed Factuality Scores

The detailed scores of the best overall factuality models.

		Lenient						Strict					
Model	Overall Factuality Score	Micro F1	Micro Recall	Micro Precision	Macro F1	Macro Recall	Macro Precision	Micro F1	Micro Recall	Micro Precision	Macro F1	Macro Recall	Macro Precision
gemma3-27b-it	56.044	63.614	69.841	58.407	61.791	70.745	63.075	56.044	73.913	45.133	55.379	74.500	49.222
mistral-small-3.1-24b	54.085	63.054	67.725	58.986	60.287	68.528	64.799	54.085	69.565	44.240	50.427	68.236	48.340
gemma3-12b-it	52.029	60.851	75.661	50.890	58.600	76.511	52.008	52.029	78.986	38.790	50.977	80.806	38.964
gemma2-9b-it	52.000	51.372	54.497	48.585	46.155	57.725	52.634	52.000	65.942	42.925	46.656	65.486	46.112
phi4	51.075	55.792	62.434	50.427	47.918	64.843	47.594	51.075	68.841	40.598	42.178	67.028	36.809
phi4-o1	50.000	60.422	68.254	54.202	56.766	68.274	57.243	50.000	68.116	39.496	47.773	67.972	43.843
deepseek-llama-8b	44.444	46.729	39.683	56.818	43.082	44.511	58.599	44.444	43.478	45.455	38.962	45.528	40.402
phi4-QwQ	40.892	42.500	35.979	51.908	35.067	34.888	44.154	40.892	39.855	41.985	34.895	37.778	37.061
phi4-mini-it	40.260	41.783	39.683	44.118	36.879	41.867	49.449	40.260	44.928	36.471	33.520	40.722	37.782
deepseek-qwen-32b	30.244	28.906	19.577	55.224	23.980	20.194	42.943	30.244	22.464	46.269	24.650	21.250	38.402
*all-relevant**	48.763	61.264	100.000	44.159	60.352	100.000	45.404	48.763	100.000	32.243	48.484	100.000	33.060
baseline (LLaMA 3.3 70B)	35.900	39.200	27.000	71.800	46.500	38.900	78.500	43.100	32.600	63.400	49.400	47.100	70.300

Table 5: Scores on the **dev dataset**. The *all-relevant* baseline assumes all paragraphs are relevant.

Loyola at ArchEHR-QA 2025: Unsupervised Methods for Text Attribution with Attention and Clustering

Rohan Sethi
Loyola University Chicago
rsethi1@luc.edu

Timothy A. Miller
Boston Children's Hospital, Harvard Medical School
timothy.miller@childrens.harvard.edu

Majid Afshar
University of Wisconsin-Madison
mafshar@medicine.wisc.edu

Dmitriy Dligach
Loyola University Chicago
dd@cs.luc.edu

Abstract

The increasing volume of patient messages via electronic health record (EHR) portals has contributed significantly to clinician workload. Automating responses to these messages can help alleviate this burden, but it is essential to ensure that the generated responses are grounded in accurate clinical evidence. As part of the ArchEHR-QA 2025 BioNLP ACL shared task, we explore unsupervised methods for generating patient question responses that are both contextually accurate and evidence-backed. We investigate three novel approaches: zero-shot prompting, clustering-based evidence selection, and attention-based evidence attribution, along with a hybrid model that combines clustering and attention. Our methods do not require model fine-tuning and leverage the inherent structure of the input data to identify the most relevant supporting evidence from clinical notes. Our best-performing approach, which integrates clustering and attention, demonstrates a substantial improvement in factuality over baseline zero-shot methods, highlighting the potential of unsupervised strategies for enhancing the clinical utility of large language models in EHR contexts.

1 Introduction

Electronic health record (EHR) systems have improved physicians' ability to document and track patient care over time. They also facilitate digital communication, allowing patients to engage with their health goals beyond in-person visits. However, the rise in patient messaging has unintentionally added to clinician workload (National Academies of Sciences et al., 2019).

Large language models (LLMs) have been proposed as tools to automatically answer patient questions. However, mere generation is not sufficient; responses must be grounded in clinical evidence from patients' medical records to ensure accuracy and reliability (Lin et al., 2003). The ArchEHR-QA 2025 BioNLP ACL shared task (Soni and Demner-Fushman, 2025b) aims to develop systems that can generate such grounded answers using information extracted from EHRs. Thus, the task is to generate an answer to a patient's question and include the sentences (or sentence identifiers) from the source note as supporting evidence for the answer.

The problem of evidence attribution has received much attention recently, and can be categorized as follows: direct LLM attribution, post-retrieval generation, and post-generation attribution. Some approaches prompt the LLM to directly generate attribution within its responses. However, (Zuccon et al., 2023) investigates ChatGPT's ability to attribute directly using prompting strategies and found that the attributions was partially correct around 50% of the time and only present 14% of the time demonstrating its unreliability. Other approaches attempt to retrieve relevant external information and prompt an LLM to incorporate said information during generation. However, citations for these approaches were present only 50% of the time (Gao et al., 2023). Finally, (Liu et al., 2023) investigates the quality of citations generated by mainstream generative search engines that incorporate citations post-generation. It was found that only 51.5% of generated sentences were fully supported and that 74% of the citations supported their associated sentences (Liu et al., 2023). Clearly, current methods to attribute are lacking in consistency and relevance of LLM text attribution.

As with many clinical machine learning tasks, this challenge provides limited data - only 20 training and development questions with corresponding medical records. To address the data scarcity, we propose two novel unsupervised methods that do not require fine-tuning or alignment of LLMs. This paper examines two approaches individually and in combination. The first uses clustering to identify the most relevant clinical evidence for a given question, narrowing the context for LLM input. The

BioNLP 2025 Shared Tasks, pages 22–26
August 1, 2025 ©2025 Association for Computational Linguistics

second employs an attention-averaging augmented generation method, where the LLM generates a response freely, and attention weights are used post hoc to attribute evidence sources. We also evaluate a combined approach and compare all methods against a baseline that prompts the LLM without any augmentation.

2 Methods

2.1 Dataset

The dataset (Soni and Demner-Fushman, 2025a) is adapted from the Medical Information Mart for Intensive Care (MIMIC) corpus (Johnson et al., 2016) by the organizers of the ArchEHR-QA 2025 BioNLP ACL shared task (Soni and Demner-Fushman, 2025b). It consists of patient-inspired questions paired with relevant clinical note excerpts from MIMIC, forming "cases." Each excerpt is pre-annotated, with sentences labeled as "essential," "supplementary," or "not relevant" for answering the question. Sentences are numbered to serve as citations in generated responses. A physician-paraphrased version of the patient's question is also provided. The development set includes 20 cases, while the test set contains 100.

2.2 Evaluation

Evaluation is based on two metrics - Factuality and Relevance - and their average. Factuality is measured using precision, recall, and F1 score between the system-selected citations and the gold-standard "essential" evidence. Relevance compares the generated response to a paragraph combining the question and essential evidence text, using BLEU, ROUGE, SARI, BERTScore, AlignScore, and MEDCON. Evaluation scripts are provided by the challenge organizers.

2.3 Method details

We propose three methods for answer generation and evidence attribution, and compare them to a baseline provided by the shared task organizers, which prompts an LLM to answer questions. The first method uses clustering to identify relevant citations based on sentence groupings. The second leverages transformer attention to attribute evidence to each generated sentence. The third combines both approaches, using clustering to guide attention-based attribution. Details of each method are provided below. All experiments were run on a NVIDIA RTX A6000 GPU. We release the code that is necessary to reproduce our experiments[1].

Zero and few-shot baselines: The baseline was performed by the organizers of the ArchEHR competition that involves prompting LLaMa 3.3 70B (AI@Meta, 2024) on the test set in a zero-shot fashion. The model is prompted to generate answers that included citations. If a response was invalid (e.g., exceeding the word limit or lacking valid citations), the prompt was retried up to five times to obtain a valid output. Additionally, we explored including multiple examples in the prompt (few-shot) but this lead to significant performance degradation on the development set.

Clustering-based method: First, every clinical note sentence and the concatenated patient and physician versions of the question are converted into embeddings. Embeddings are obtained using an encoder LLM via HuggingFace's transformers feature extraction API (Wolf et al., 2020). Embeddings are provided per token, so the token embeddings for each sentence are averaged to get an overall embedding for the given sentence. These embeddings are then clustered into two clusters using the agglomerative clustering algorithm from sci-kit learn (Pedregosa et al., 2011). The clusters are then parsed to identify the cluster containing the question embedding vector. The clinical sentences that are a part of this cluster are assumed to contain the most relevant evidence to answer the patient's question. These clinical sentences are used as input to the LLM prompt, which is prompted to answer the patient question given the selected context. Post-generation, the clinical sentences utilized are cited at the end of the LLM response without precisely attributing each output sentence to a clinical sentence. This is unlike the attention-based and hybrid approaches which precisely cite clinical note segments to each output sentence. An example is included in the prompt to demonstrate to the model how detailed its response should be without restrictions on formatting responses.

Attention-based model: This method leverages transformer attention scores to attribute generated text to specific sentences in the source clinical note. We hypothesize that the average attention between generated output and input sentences can serve as a signal for source attribution. All questions and evidence entries are input to an LLM. An example is provided in the prompt to demonstrate to the

[1] https://github.com/rsethi21/loyola_archehr_2025.git

model how detailed its response should be. This example, however, does not give restrictions on how responses to be formatted. Post-generation, attention outputs are analyzed. For each output sentence, an average attention score is computed with regards to each evidence entry; i.e. if there were n evidence entries, there will be n computed average attention scores for each output. To obtain attention scores for averaging, we parse the attention matrix from the LLM after determining the token indices of output sentences and each evidence entry. Details on how the indexed attention matrix is utilized to compute average scores can be found in the source code.

All computed evidence entry attention scores for each output are converted to z-scores, and entry scores exceeding a threshold are considered supporting evidence, which is then appended to the corresponding output sentence. The z-score selection enables selection of only the most significantly attended evidence entries or alternatively no citations if all z-scores are below a threshold, which makes this attribution factually robust. This process is repeated for all output sentences in the LLM response. Key hyperparameters include the LLM model, prompt format, z-score threshold, and chosen attention layers.

Hybrid model: The final method combines clustering and attention-based approaches. The LLM is first prompted with evidence selected via clustering, and its output is then processed using the attention-based attribution workflow. This hybrid method tests whether clustering can guide the LLM's attention toward the most relevant evidence, potentially improving the identification of essential information compared to using either method alone.

Methods not included in the final submission: Other methods were experimented with early in the competition, including retrieval augmented generation (RAG), few-shot prompting, encoder-based evidence selection, supervised-fine tuning, selection of evidence based on similarity to output post-generation using BERTScore, and others. However, the best performing methods were finalized for submission using the development set and described above.

2.4 Experiments

All models are implemented using HuggingFace (Wolf et al., 2020) and PyTorch (Paszke et al., 2019). We use Llama-3.1-8B-Instruct (AI@Meta, 2024) and we conduct all experiments on 8 NVIDIA RTX A6000 GPUs. All hyperparameters are tuned on the development set.

For the clustering approach, the dmis/biobert v1.1 (Lee et al., 2020) was selected against other embedding models. Agglomerative clustering with number of clusters of 2 was selected from SkLearn's clustering module (Pedregosa et al., 2011). The options for clustering algorithms included KMeans, Agglomerative, and DBSCAN. Number of clusters was varied between 2 and 3 representing the categories that the evidence entries were labeled as (essential vs not relevant or essential vs supplementary vs not relevant).

For the attention approach, selection of attention layers was treated as a hyperparameter. The attention layers kept varied by case, where attention layer outputs were compared sequentially using cosine similarity and only the attention output layers that differed the most from the previous attention layer was kept for averaging. The hyperparameter was whether to perform this selection or not. For the development set, dropping attention layers was selected for and applied to the test cases. The selected for z-score threshold was 0 from the following options 1.64 (average attention score significantly greater than 95% of other attention scores), 1 (significantly greater than 85%), and 0 (significantly greater than 50%).

All methods are evaluated on the development set using the scoring script provided by the organizers of the competition.

3 Results and Discussion

Experiment	Factuality	Relevance	Average
Zero-Shot	43.10	28.70	35.90
Clustering	50.56	32.38	41.47
Attention	54.11	31.81	42.96
Clustering + Attention	58.64	33.37	46.00

Table 1: Development set overall factuality, overall relevance, and overall scores for all methods. Zero-shot is the baseline approach attempted by the organizers of the competition.

The results of our performance evaluation on the development set are presented in Table 1. The best-performing method is the hybrid approach, Clustering combined with Attention, which improves the factuality score from a baseline of 43.10 to 58.64. The attention-based method alone achieves a score of 54.11, while the clustering-only method yields 50.56. These results suggest that the model's

Experiment	Factuality	Relevance	Average
Zero-Shot	33.60	27.80	30.70
Clustering + Attention	57.35	30.36	43.85

Table 2: Test set overall factuality, overall relevance, and overall scores for best method and zero-shot. Zero-shot is the baseline approach implemented by the organizers of the competition.

attention matrix can effectively highlight the information the LLM prioritizes when generating each sentence, contributing to a nearly 10-point increase in factuality. By leveraging attention, LLMs can generate more accurate outputs without relying on complex formatting or explicit instructions, while also enabling real-time evidence integration during generation.

Additionally, combining clustering with attention further improved the factuality score by 4 points over using attention alone. This indicates that selecting relevant evidence through clustering before passing it to the LLM helps the model focus more effectively on the most pertinent information when answering patient questions, leading to higher factuality.

In terms of relevance, the greatest improvement over the zero-shot baseline came from combining clustering and attention, resulting in a nearly 5-point gain. This likely stems from more accurate evidence selection.

The best-performing approach, clustering combined with attention, was evaluated on the test set and compared to the organizer's zero-shot baseline. It maintained similar average scores for both factuality and relevance, showing no significant performance drop and achieving comparable gains over the baseline as seen on the development set. Notably, this unsupervised method using an 8B LLM outperformed a 70B LLM, offering substantial savings in computational cost, time, and training resources. Curating clinically oriented training datasets is both time-consuming and resource-intensive, making them difficult to obtain. Our results demonstrate that unsupervised methods can effectively enhance the factuality of LLM-generated responses in clinical settings.

4 Conclusion

Automating responses to patient questions using EHR data holds significant potential for reducing clinician workload and improving patient care. In this work, we demonstrated that integrating unsupervised approaches like clustering and attention-based evidence attribution with large language models (LLMs) can significantly enhance the factuality and relevance of generated responses without requiring extensive model fine-tuning or alignment. Our hybrid method, combining clustering and attention, outperformed traditional zero-shot baselines, highlighting the value of leveraging context structuring and attention analysis for more accurate clinical responses. Importantly, our findings show that relatively small LLMs (8B parameters) can outperform much larger models (70B parameters) when appropriately guided, offering substantial cost and efficiency advantages in real-world clinical applications. Future work could further refine these methods by incorporating more sophisticated context selection strategies, leveraging multimodal data, and exploring more interpretable attention mechanisms to ensure even higher levels of clinical trustworthiness and reliability.

5 Limitations

Most experiments were performed utilizing LLMs with 8B parameters or less due to memory constraints. Furthermore, only 20 development / training examples were provided for experimentation. Although these examples labeled the evidence entries that were essential to incorporate in the answer the experimented approaches generate, there were no associated example answer outputs.

Acknowledgments

Research reported in this publication was supported by National Institutes of Health under Awards R01LM012973 and 1R01DA051464. The content is solely the responsibility of the authors and does not necessarily represent the official views of the National Institutes of Health.

References

AI@Meta. 2024. Llama 3 model card. https://github.com/meta-llama/llama3/blob/main/MODEL_CARD.md. Accessed: 2025-05-06.

Tianyu Gao, Howard Yen, Jiatong Yu, and Danqi Chen. 2023. Enabling large language models to generate text with citations. In *Proceedings of the 2023 Conference on Empirical Methods in Natural Language Processing*, pages 6465–6488, Singapore. Association for Computational Linguistics.

Alistair E. W. Johnson, Tom J. Pollard, Lu Shen, Li-wei H. Lehman, Mengling Feng, Mohammad Ghassemi, Benjamin Moody, Peter Szolovits, Leo Anthony Celi, and Roger G. Mark. 2016. Mimic-iii, a freely accessible critical care database. *Scientific Data*, 3(1):160035.

Jinhyuk Lee, Wonjin Yoon, Sungdong Kim, Donghyeon Kim, Sunkyu Kim, Chan Ho So, and Jaewoo Kang. 2020. Biobert: a pre-trained biomedical language representation model for biomedical text mining. *Bioinformatics*, 36(4):1234–1240.

Jimmy Lin, Dennis Quan, Vineet Sinha, Krzysztof Bakshi, David Huynh, Boris Katz, and David R. Karger. 2003. What makes a good answer? the role of context in question answering. In *Proceedings of the Ninth IFIP TC13 International Conference on Human-Computer Interaction (INTERACT 2003)*. Springer.

Nelson F. Liu, Tianyi Zhang, and Percy Liang. 2023. Evaluating verifiability in generative search engines. *ArXiv*, abs/2304.09848.

Engineering National Academies of Sciences, Medicine, National Academy of Medicine, and Committee on Systems Approaches to Improve Patient Care by Supporting Clinician Well-Being. 2019. *Taking Action Against Clinician Burnout: A Systems Approach to Professional Well-Being*. National Academies Press. Accessed: 2025-05-06.

Adam Paszke, Sam Gross, Francisco Massa, Adam Lerer, James Bradbury, Gregory Chanan, Trevor Killeen, Zeming Lin, Natalia Gimelshein, Luca Antiga, Alban Desmaison, Andreas Kopf, Edward Yang, Zachary DeVito, Martin Raison, Alykhan Tejani, Sasank Chilamkurthy, Benoit Steiner, Lu Fang, and 2 others. 2019. Pytorch: An imperative style, high-performance deep learning library. In *Advances in Neural Information Processing Systems*, volume 32. Curran Associates, Inc.

Fabian Pedregosa, Gaël Varoquaux, Alexandre Gramfort, Vincent Michel, Bertrand Thirion, Olivier Grisel, Mathieu Blondel, Peter Prettenhofer, Ron Weiss, Vincent Dubourg, Jake Vanderplas, Alexandre Passos, David Cournapeau, Matthieu Brucher, Matthieu Perrot, and Édouard Duchesnay. 2011. Scikit-learn: Machine learning in python. *Journal of Machine Learning Research*, 12:2825–2830.

Sarvesh Soni and Dina Demner-Fushman. 2025a. A dataset for addressing patient's information needs related to clinical course of hospitalization. *arXiv preprint*.

Sarvesh Soni and Dina Demner-Fushman. 2025b. Overview of the archehr-qa 2025 shared task on grounded question answering from electronic health records. In *The 24rd Workshop on Biomedical Natural Language Processing and BioNLP Shared Tasks*, Vienna, Austria. Association for Computational Linguistics.

Thomas Wolf, Lysandre Debut, Victor Sanh, Julien Chaumond, Clement Delangue, Anthony Moi, Pierric Cistac, Tim Rault, Rémi Louf, Morgan Funtowicz, Joe Davison, Sam Shleifer, Patrick von Platen, Clara Ma, Yacine Jernite, Julien Plu, Canwen Xu, Teven Le Scao, Sylvain Gugger, and 3 others. 2020. Transformers: State-of-the-art natural language processing. In *Proceedings of the 2020 Conference on Empirical Methods in Natural Language Processing: System Demonstrations*, pages 38–45. Association for Computational Linguistics.

Guido Zuccon, Bevan Koopman, and Razia Shaik. 2023. Chatgpt hallucinates when attributing answers. In *Proceedings of the Annual International ACM SIGIR Conference on Research and Development in Information Retrieval in the Asia Pacific Region*, SIGIR-AP '23, page 46–51, New York, NY, USA. Association for Computing Machinery.

CUNI-a at ArchEHR-QA 2025: Do we need Giant LLMs for Clinical QA?

Vojtěch Lanz and **Pavel Pecina**
Charles University, Faculty of Mathematics and Physics
Institute of Formal and Applied Linguistics
{lanz,pecina}@ufal.mff.cuni.cz

Abstract

In this paper, we present our submission to the ArchEHR-QA 2025 shared task, which focuses on answering patient questions based on excerpts from electronic health record (EHR) discharge summaries. Our approach identifies essential sentences relevant to a patient's question using a combination of few-shot inference with the Med42-8B model, cosine similarity over clinical term embeddings, and the MedCPT cross-encoder relevance model. Then, concise answers are generated on the basis of these selected sentences. Despite not relying on large language models (LLMs) with tens of billions of parameters, our method achieves competitive results, demonstrating the potential of resource-efficient solutions for clinical NLP applications.

1 Introduction

Responding to patient messages through EHR portals is increasingly recognized as a burden for clinicians (Budd, 2023). To alleviate this problem, the BioNLP 2025 ArchEHR-QA shared task (Soni and Demner-Fushman, 2025b) challenges participants to automatically answer patients' questions using the content of their EHRs. The task requires identifying essential information from an excerpt of a clinical discharge summary and using it to generate accurate and relevant answers.

One of the main limitations of the shared task is the absence of training data, reflecting real-world deployment settings where hospitals often lack the resources to curate and annotate large datasets. Instead, participants are given a small development set consisting of 20 cases. Each case includes a patient's question, a clinician-paraphrased version of the question, and an excerpt of the discharge summary segmented into sentences. The ground-truth annotations identify the sentences essential for answering the question. The final test set comprises 100 similar cases, but without access to ground truth annotations. For a detailed description of the dataset, see Soni and Demner-Fushman (2025a).

Furthermore, healthcare institutions are limited in using external services due to privacy restrictions and, at the same time, cannot easily integrate large-scale LLMs with tens of billions of parameters on-premise due to hardware requirements (Jiang et al., 2023). Therefore, in this submission, we explore approaches that avoid reliance on massive LLMs, focusing instead on lightweight and interpretable components.

Our method combines multiple signal sources to detect essential sentences relevant to the patient's question, including few-shot inference with the Med42-8B model (Christophe et al., 2024), cosine similarity over clinical term representations, and cross-encoder models trained on clinical pair relevance. The selected sentences are then used to generate a concise answer. [1]

2 Related Work

Clinical NLP research has been supported by several large collections of clinical and biomedical texts, such as MIMIC (Johnson et al., 2023) and PubMed (Canese and Weis, 2013). Not only do these datasets serve as the foundation for various shared tasks aimed at extracting relevant information for specific cases or questions, such as the BioASQ Challenge (Tsatsaronis et al., 2015), the TREC Clinical Trials Track 2022 (Roberts et al., 2022), or the ArchEHR-QA 2025, the task we investigate in this paper. In addition to many approaches to biomedical information retrieval, one notable example is the MedCPT model (Jin et al., 2023), which compares embedding representations of abstract articles with those of input queries.

Other notable clinical datasets include n2c2 (Henry et al., 2019), from which the emrQA Ques-

[1] Source code available at `https://github.com/lanzv/CUNI-a-at-ArchEHR-QA-2025`

BioNLP 2025 Shared Tasks, pages 27–40
August 1, 2025 ©2025 Association for Computational Linguistics

tion Answering dataset (Pampari et al., 2018) is derived. This dataset was used by Lanz and Pecina (2024) to study paragraph retrieval using models such as ClinicalBERT (Alsentzer et al., 2019) and BioBERT (Lee et al., 2019), both of which are pre-trained on English clinical and biomedical text.

In addition, several clinically pre-trained decoder-based language models were introduced to address a wide range of clinical tasks, including BioMistral (Labrak et al., 2024), Med42 (Christophe et al., 2024), or Meditron 3 (Sallinen et al., 2025). However, recent findings (Dada et al., 2025; Lanz and Pecina, 2025) suggest that clinical pretraining is not always essential and that multilingual or general-domain pretraining may be equally or even more beneficial for certain clinical tasks.

3 Methodology

The methodology follows the structure of the shared task, which has two steps: essential sentence retrieval followed by answer generation.

- **Essential Sentence Retrieval:** In this stage, we iterate over all annotated sentences in the clinical documents and compare them with the clinical question (formulated by clinicians, not patients) to retrieve sentences that are essential to answer the question. This step is evaluated using the **Factuality** score, defined as the micro-averaged F1 score of correctly predicted essential sentences.

- **Answer Generation:** Based on prior predictions, we concatenate the retrieved essential sentences into a compact answer, limited to 75 words - an empirically optimal length (Lin et al., 2003; Jeon et al., 2006) and the evaluation cut-off point for the shared task. This stage is scored using the mean of automatic similarity metrics comparing the generated compact answer with gold essential sentences: BLEU (Papineni et al., 2002), ROUGEL-sum (Lin, 2004), SARI (Xu et al., 2016), BERTScore (Zhang et al., 2020), AlignScore (Zha et al., 2023), and MEDCON (wai Yim et al., 2023) - collectively referred to as the **Relevance** score.

The final overall evaluation measure averages the Factuality and Relevance scores.

3.1 Essential Sentence Retrieval

We explore several approaches that model different aspects of sentence essentiality for clinical question answering. Each method aims to determine whether a given sentence contains essential information to answer a question formulated by a clinician.

Max Cosine Similarity. This method assumes that if a sentence contains terms similar to those in the question, it is more likely to be essential. However, to avoid the influence of stop words and general-domain terms, we focus exclusively on clinical terminology.

First, we use a SciSpaCy model *en_core_sci_sm* (Neumann et al., 2019) to extract clinical terms from both the sentence and the question. Then, for each pair of clinical terms (one from the sentence, one from the question), we compute the cosine similarity of their embeddings using ClinicalBERT. The maximum cosine similarity among all such pairs is taken as the sentence's relevance score.

We apply a threshold to retrieve sentences that are then considered essential. Following Lanz and Pecina (2025), we also test mBERT instead of ClinicalBERT to compare domain-specific and multilingual pretraining. We refer to the resulting methods as *MCS-C* and *MCS-M*, based on ClinicalBERT and mBERT, respectively.

MedCPT Cross-Encoder. Lexical similarity may not capture semantic relevance when different terms convey similar meanings. To address this, we use the MedCPT Cross-Encoder, trained for biomedical information retrieval on PubMed. It takes a sentence–question pair as input and outputs a similarity score, which we threshold to determine the essentiality. We refer to this approach as *MedCPT FS* (Full Sentences).

To reduce noise, we also experiment with filtering non-clinical content using the SciSpaCy extraction model. Both sentences and questions are reduced to comma separated clinical terms before being inputted into MedCPT. This variant is denoted as *MedCPT CT* (Clinical Terms).

Sentence Relevance with Med42-8B. Due to clinical privacy constraints, externally hosted models such as ChatGPT (OpenAI, 2025) cannot be used with MIMIC data – a common limitation in clinical NLP. This requires a secure, local deployment, which is often infeasible in hospitals due to limited infrastructure. As deploying large models is impractical in such settings, we focus on smaller and more efficient models suitable for local use.

Furthermore, the lack of training data implies the use of zero- or few-shot methods. Therefore, we use Med42-8B, a compact, instruction-tuned model that has undergone preference optimization for interactive tasks. Our few-shot prompt includes synthetic examples generated by GPT-4o - each with a patient question, candidate sentence, answer (or `None`), and justification. Importantly, we ensure that no data from the shared task are included in the few-shot generation process. Otherwise, we could not use the dev set for a fair validation-based comparison of approaches before evaluating the best approach on the final test set. And while it might seem appealing to use real data - or at least data closely resembling it, such as using some of dev set examples as few-shot prompts - this would not only be methodologically incorrect, but also impractical: the dev set is already so small that we must preserve it entirely for validation purposes. Furthermore, we cannot share shared task data with third-party services. Therefore, we rely on synthetic examples generated by GPT-4o shown in Appendix D.

The confidence score for each prediction is computed from the token-level softmax probabilities of the model's output, covering both the answer and its justification. If the model generates `None` as the answer, the confidence is set to 0.0. The scores obtained within each patient case are normalized by dividing by their total sum; If the sum is 0.0 (that is, all values are zero), no normalization is applied. We refer to this model as *SR Med42*.

Context-aware Relevance with Med42-8B. Previous approaches assessed sentences in isolation, but clinical text often relies on earlier context for full meaning. For example, a sentence *"In that case, notify the cardiology team."* is only relevant to the question *"What should be done if the patient develops chest pain?"*, if we know *"that case"* refers to chest pain, illustrating the need for context-aware relevance.

To incorporate this, we propose *CAR Med42*, which applies Med42-8B with the full summary of discharge. Few-shot prompts, generated via GPT-4o, include a clinical context, patient question, candidate sentence from the context, binary answer (`Yes`/`No`), and justification.

As before, a `No` prediction yields a score of 0.0, while a `Yes` prediction uses the model's generation probability as the relevance score (and again, if possible, scaling is applied). Importantly, no shared task or clinical data was shared with ChatGPT - only synthetic examples were used. A complete few-shot example is provided in Appendix E.

3.2 Answer Generation

Once the essential sentences are retrieved, they are used to construct the final answer. The goal is to provide a direct response while ensuring that the answer stays under the 75-word limit.

First, each essential sentence is compressed individually. We prompt the Med42-8B model in a few-shot setting (with examples generated by GPT-4o) to generate a concise direct answer using the essential sentence as context. If the model cannot generate an answer, the sentence is shortened with a second few-shot prompt, also with Med42-8B, focusing on compressing the sentence while preserving its content. The corresponding prompt templates are shown in Appendix F and Appendix G, respectively.

After processing all the essential sentences, we concatenate them into a single answer. If the result exceeds the 75-word limit, we iteratively shorten the longest processed sentences using the second few-shot prompt until the word count is within bounds. In rare cases where this process stalls (i.e., no length reduction after two iterations), we remove the last word from the longest sentence and attempt compression again.

Conversely, if the final answer is significantly shorter than the limit, we gradually replace the most concise processed sentences with their original, longer, essential sentence forms. This ensures that the answer contains as much relevant information as possible while remaining easy for patients to understand.

4 Results

The sentence retrieval methods we proposed return confidence scores rather than binary decisions. Although SR Med42 and CAR Med42 explicitly assign a confidence score of 0.0 when the Med42 model predicts that a sentence is not essential, we still need to apply a threshold to convert the scores into final binary decisions. Thus, we first optimize threshold values on the development set and then apply the optimal thresholds to the test set for evaluation.

Although tuning thresholds on the development set of only 20 cases may raise concerns about over-

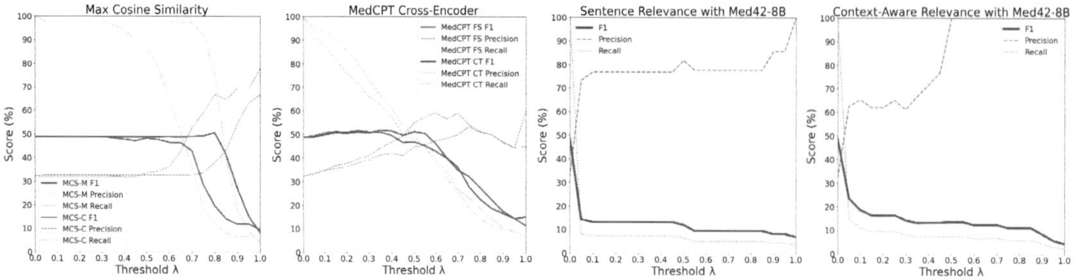

Figure 1: Micro F1, precision, and recall scores across varying confidence thresholds λ for four essential sentence retrieval methods. Only sentences with a model confidence score greater than or equal to λ are considered essential.

fitting, each case contains multiple sentences, resulting in hundreds of sentence-level evaluations. This yields a sufficiently informative signal to guide threshold selection, even if it may not guarantee a globally optimal setting. Crucially, since the threshold is fixed before any test data are seen, the validity of the final test evaluation remains unaffected.

4.1 Threshold Optimization on Dev Set

To identify optimal threshold values for each method, we perform a sweep over a range of threshold values and analyze the resulting precision-recall trade-offs. As shown in Figure 1, higher thresholds improve precision but reduce recall, limiting F1 performance. In fact, F1 scores often do not significantly exceed the baseline for retrieving all sentences as essential.

In the Max Cosine Similarity results, we observe that, while MCS-C achieves higher F1, MCS-M obtains better precision. Similarly, in the MedCPT Cross-Encoder results, both MedCPT FS and MedCPT CT follow similar trends, with the clinical-term-filtered variant (MedCPT CT) performing slightly better. Based on this, we prioritize MedCPT CT in subsequent experiments, as filtering non-clinical content helps reduce noise. However, for Max Cosine Similarity, neither model clearly dominates.

Given the findings that clinical pretraining does not always help (Dada et al., 2025; Lanz and Pecina, 2025), in the SR Med42 and CAR Med42 approaches, we experimented with replacing the Med42-8B model with its base non-medical alternative, Llama3-8B (Grattafiori et al., 2024). However, despite similar trends in the precision, recall, and F1 curves, the general-domain Llama3-8B lags behind Med42-8B (see Figure 4). Therefore, we rely on the Med42-8B model in these approaches.

Method	*Overall*	*Factuality*	*Relevance*
Ensemble-C	48.6	56.8	**40.5**
Ensemble-M	**49.0**	**58.6**	39.4

Table 1: Factuality F1 (***Fact***), Relevance (***Rel***) metrics, and their mean **Overall** score of the two approaches, Ensemble-M and Ensemble-C, measured on the dev set.

Method	F1	*Precision*	*Recall*
All Sentences	48.8	32.2	100.0
MCS-C	50.6	37.9	76.1
MCS-M	48.9	32.6	97.8
MedCPT FS	51.6	38.0	80.4
MedCPT CT	51.8	44.3	62.3
SR Med42	48.8	32.2	100.0
CAR Med42	48.8	32.2	100.0
Ensemble-C	56.8	53.2	60.9
Ensemble-M	**58.6**	52.3	66.7

Table 2: Comparison of F1, Precision, and Recall across methods for essential sentence retrieval.

Since each method captures different aspects of sentence essentiality, we explore combining them in ensemble models. A sentence is retrieved as essential if at least one of the selected methods assigns it a score above its respective threshold.

We define two ensembles:

- **Ensemble-C**: combines MCS-C, MedCPT CT, SR Med42, and CAR Med42
- **Ensemble-M**: combines MCS-M, MedCPT CT, SR Med42, and CAR Med42

We then perform a grid search for combinations of thresholds to maximize F1 in the development set (see Appendix A).

Table 2 summarizes the best F1 scores achieved by each method, including the baseline where all sentences are considered essential. The ensemble

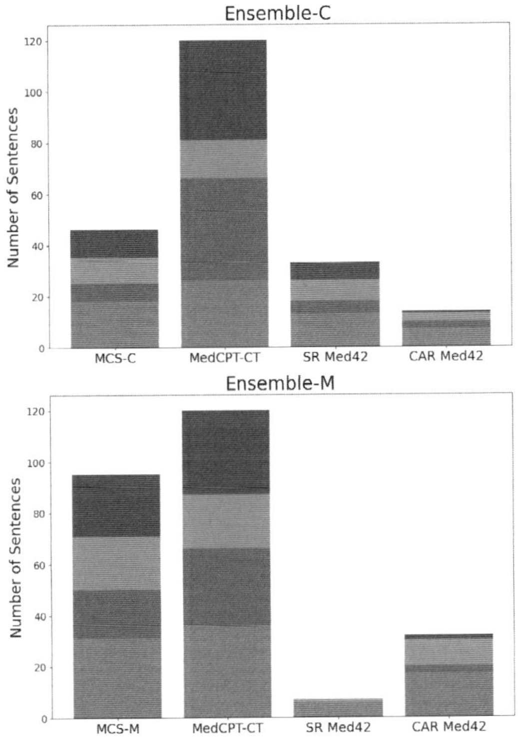

Figure 2: Contribution of individual models within Ensemble-C (top) and Ensemble-M (bottom), showing for each method the number of predicted essential sentences. Bars decompose into: red (unique wrong predictions), orange (wrong and also predicted as wrong by at least one other submethod), blue (correct and unique contribution), and green (correct, but also predicted correctly by at least one other submethod).

methods clearly outperform the individual models. To assess the robustness of these results, we estimate the variability of the F1 scores using bootstrap resampling over the input examples. This involves repeatedly sampling subsets of the data with replacement and re-computing the F1 score on each sample. The resulting distributions yield estimated means and standard deviations of 56.23 ± 3.91 for Ensemble-C and 58.64 ± 3.29 for Ensemble-M, indicating that both ensembles consistently outperform the baselines in the resampled data.

Figure 2 visualizes the contribution of each method within the ensemble approaches to the final prediction of essential sentences. The figure shows that MedCPT-CT is the most dominant contributor. Interestingly, MCS-M plays a much more significant role in Ensemble-M than MCS-C does in Ensemble-C. However, the ratio of correctly and incorrectly predicted sentences remains similar across all methods.

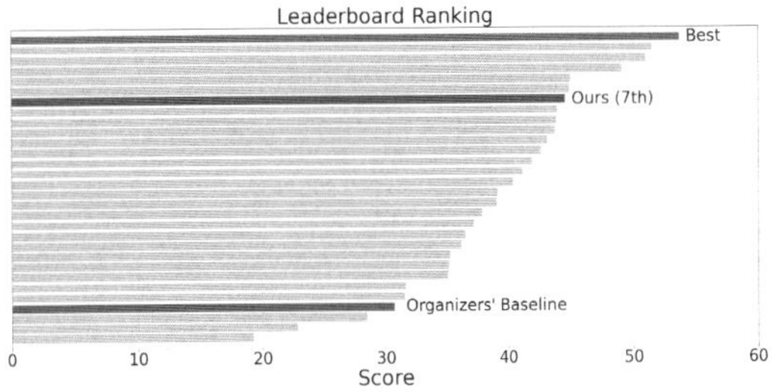

Figure 3: Automatic evaluation on the test set. Our method ranks 7th out of 30 submissions.

The final test set results for the ensemble methods are shown in Table 1. Additional Relevance subscores are reported in Table 3 in Appendix B. While Ensemble-M achieves the highest Factuality score, Ensemble-C performs slightly better in Relevance. Overall, Ensemble-M is the stronger method, and we select it as our final approach for evaluation on the test set.

4.2 Final Test Set Results

Our final system achieved a score of 44.6 on the test set, placing us 7th out of 30 participating teams (see Figure 3). These automatically evaluated results show that our ensemble-based approach is competitive, despite not using LLMs with ten billion or more parameters.

5 Conclusion

In this work, we presented our submission to the ArchEHR-QA 2025 shared task. We focused on identifying essential sentences for answering a given patient's question. Based on these predicted sentences, we generated the final compact answer. We combined a few-shot Med42-8B model with cosine similarities of clinical terms and the MedCPT cross-encoder scores.

Our results are reasonable and competitive, even without using LLMs with tens of billions of parameters, which are not easily integrable into hospital environments. Furthermore, although replacing the domain-specific Med42-8B model with the general-domain Llama3-8B led to a slight drop in performance, it still suggests that domain-specific pretraining provides a modest benefit. However, in the cosine similarity approach, mBERT performs similarly to ClinicalBERT. This highlights that general-purpose multilingual models can still be competitive in clinical tasks.

Limitations

No training data and a small validation set limit the development of the model. The notion of an "essential sentence" is loosely defined and open to interpretation. Our study is limited to English, and few-shot prompts were generated using ChatGPT, which may introduce bias and produce examples that are not fully accurate or tailored to our task. Finally, automatic evaluation may not fully reflect the correctness and clinical validity of the answer.

Acknowledgments

This research was partially supported by the SVV project number 260 698 and the Charles University GAUK grant No. 284125. It has also received support and funding from the European Union's Horizon Europe research and innovation programme project *RES-Q plus* (Grant Agreement No. 101057603). Views and opinions expressed are, however, those of the authors only and do not necessarily reflect those of the European Union or the Health and Digital Executive Agency. Neither the European Union nor the granting authority can be held responsible for them.

References

Emily Alsentzer, John Murphy, William Boag, Wei-Hung Weng, Di Jindi, Tristan Naumann, and Matthew McDermott. 2019. Publicly available clinical BERT embeddings. In *Proceedings of the 2nd Clinical Natural Language Processing Workshop*, pages 72–78, Minneapolis, Minnesota, USA. Association for Computational Linguistics.

Jeffrey Budd. 2023. Burnout related to electronic health record use in primary care. *Journal of Primary Care & Community Health*, 14.

Kathi Canese and Sarah Weis. 2013. Pubmed: the bibliographic database. *The NCBI handbook*, 2(1).

Clément Christophe, Praveen K Kanithi, Tathagata Raha, Shadab Khan, and Marco AF Pimentel. 2024. Med42-v2: A suite of clinical llms.

Amin Dada, Osman Koras, Marie Bauer, Amanda Butler, Kaleb Smith, Jens Kleesiek, and Julian Friedrich. 2025. MeDiSumQA: Patient-oriented question-answer generation from discharge letters. In *Proceedings of the Second Workshop on Patient-Oriented Language Processing (CL4Health)*, pages 124–136, Albuquerque, New Mexico. Association for Computational Linguistics.

Aaron Grattafiori, Abhimanyu Dubey, Abhinav Jauhri, Abhinav Pandey, Abhishek Kadian, Ahmad Al-Dahle, Aiesha Letman, Akhil Mathur, Alan Schelten, Alex Vaughan, Amy Yang, Angela Fan, Anirudh Goyal, Anthony Hartshorn, Aobo Yang, Archi Mitra, Archie Sravankumar, Artem Korenev, Arthur Hinsvark, and 542 others. 2024. The llama 3 herd of models. *Preprint*, arXiv:2407.21783.

Sam Henry, Kevin Buchan, Michele Filannino, Amber Stubbs, and Ozlem Uzuner. 2019. 2018 n2c2 shared task on adverse drug events and medication extraction in electronic health records. *Journal of the American Medical Informatics Association*, 27(1):3–12.

Jiwoon Jeon, W. Bruce Croft, Joon Ho Lee, and Soyeon Park. 2006. A framework to predict the quality of answers with non-textual features. In *Proceedings of the 29th Annual International ACM SIGIR Conference on Research and Development in Information Retrieval*, SIGIR '06, page 228–235, New York, NY, USA. Association for Computing Machinery.

Xinrui Jiang, Lixiang Yan, Raja Vavekanand, and Mengxuan Hu. 2023. Large language models in healthcare current development and future directions. In *Generative AI Research*.

Qiao Jin, Won Kim, Qingyu Chen, Donald C Comeau, Lana Yeganova, W John Wilbur, and Zhiyong Lu. 2023. Medcpt: Contrastive pre-trained transformers with large-scale pubmed search logs for zero-shot biomedical information retrieval. *Bioinformatics*, 39(11):btad651.

Alistair EW Johnson, Lucas Bulgarelli, Lu Shen, Alvin Gayles, Ayad Shammout, Steven Horng, Tom J Pollard, Sicheng Hao, Benjamin Moody, Brian Gow, and 1 others. 2023. Mimic-iv, a freely accessible electronic health record dataset. *Scientific data*, 10(1):1.

Yanis Labrak, Adrien Bazoge, Emmanuel Morin, Pierre-Antoine Gourraud, Mickael Rouvier, and Richard Dufour. 2024. Biomistral: A collection of open-source pretrained large language models for medical domains. *Preprint*, arXiv:2402.10373.

Vojtech Lanz and Pavel Pecina. 2024. Paragraph retrieval for enhanced question answering in clinical documents. In *Proceedings of the 23rd Workshop on Biomedical Natural Language Processing*, pages 580–590, Bangkok, Thailand. Association for Computational Linguistics.

Vojtech Lanz and Pavel Pecina. 2025. When multilingual models compete with monolingual domain-specific models in clinical question answering. In *Proceedings of the Second Workshop on Patient-Oriented Language Processing (CL4Health)*, pages 69–82, Albuquerque, New Mexico. Association for Computational Linguistics.

Jinhyuk Lee, Wonjin Yoon, Sungdong Kim, Donghyeon Kim, Sunkyu Kim, Chan Ho So, and Jaewoo Kang. 2019. Biobert: a pre-trained biomedical language representation model for biomedical text mining. *CoRR*, abs/1901.08746.

Chin-Yew Lin. 2004. ROUGE: A package for automatic evaluation of summaries. In *Text Summarization Branches Out*, pages 74–81, Barcelona, Spain. Association for Computational Linguistics.

Jimmy Lin, Dennis Quan, Vineet Sinha, Karun Bakshi, David Huynh, and Boris Katz. 2003. What makes a good answer? the role of context in question answering.

Mark Neumann, Daniel King, Iz Beltagy, and Waleed Ammar. 2019. ScispaCy: Fast and Robust Models for Biomedical Natural Language Processing. In *Proceedings of the 18th BioNLP Workshop and Shared Task*, pages 319–327, Florence, Italy. Association for Computational Linguistics.

OpenAI. 2025. Chatgpt (april 2025 version). `https://chat.openai.com`. Large language model accessed in April 2025 via https://chat.openai.com.

Anusri Pampari, Preethi Raghavan, Jennifer Liang, and Jian Peng. 2018. emrQA: A large corpus for question answering on electronic medical records. In *Proceedings of the 2018 Conference on Empirical Methods in Natural Language Processing*, pages 2357–2368, Brussels, Belgium. Association for Computational Linguistics.

Kishore Papineni, Salim Roukos, Todd Ward, and Wei-Jing Zhu. 2002. Bleu: a method for automatic evaluation of machine translation. In *Proceedings of the 40th Annual Meeting of the Association for Computational Linguistics*, pages 311–318, Philadelphia, Pennsylvania, USA. Association for Computational Linguistics.

Kirk Roberts, Dina Demner-Fushman, Ellen M. Voorhees, Steven Bedrick, and William R. Hersh. 2022. Overview of the trec 2022 clinical trials track. In *Text Retrieval Conference*.

Alexandre Sallinen, Antoni-Joan Solergibert, Michael Zhang, Guillaume Boyé, Maud Dupont-Roc, Xavier Theimer-Lienhard, Etienne Boisson, Bastien Bernath, Hichem Hadhri, Antoine Tran, Tahseen Rabbani, Trevor Brokowski, Meditron Medical Doctor Working Group, Tim G. J. Rudner, and Mary-Anne Hartley. 2025. Llama-3-meditron: An open-weight suite of medical LLMs based on llama-3.1. In *Workshop on Large Language Models and Generative AI for Health at AAAI 2025*.

Sarvesh Soni and Dina Demner-Fushman. 2025a. A dataset for addressing patient's information needs related to clinical course of hospitalization. *arXiv preprint*.

Sarvesh Soni and Dina Demner-Fushman. 2025b. Overview of the archehr-qa 2025 shared task on grounded question answering from electronic health records. In *The 24th Workshop on Biomedical Natural Language Processing and BioNLP Shared Tasks*, Vienna, Austria. Association for Computational Linguistics.

George Tsatsaronis, Georgios Balikas, Prodromos Malakasiotis, Ioannis Partalas, Matthias Zschunke, Michael Alvers, Dirk Weißenborn, Anastasia Krithara, Sergios Petridis, Dimitris Polychronopoulos, Yannis Almirantis, John Pavlopoulos, Nicolas Baskiotis, Patrick Gallinari, Thierry Artieres, Axel-Cyrille Ngonga Ngomo, Norman Heino, Eric Gaussier, Liliana Barrio-Alvers, and Georgios Paliouras. 2015. An overview of the bioasq large-scale biomedical semantic indexing and question answering competition. *BMC Bioinformatics*, 16:138.

Wen wai Yim, Yujuan Fu, Asma Ben Abacha, Neal Snider, Thomas Lin, and Meliha Yetisgen. 2023. Aci-bench: a Novel Ambient Clinical Intelligence Dataset for Benchmarking Automatic Visit Note Generation. *Scientific Data*, 10(1):586.

Wei Xu, Courtney Napoles, Ellie Pavlick, Quanze Chen, and Chris Callison-Burch. 2016. Optimizing statistical machine translation for text simplification. *Transactions of the Association for Computational Linguistics*, 4:401–415.

Yuheng Zha, Yichi Yang, Ruichen Li, and Zhiting Hu. 2023. AlignScore: Evaluating factual consistency with a unified alignment function. In *Proceedings of the 61st Annual Meeting of the Association for Computational Linguistics (Volume 1: Long Papers)*, pages 11328–11348, Toronto, Canada. Association for Computational Linguistics.

Tianyi Zhang, Varsha Kishore, Felix Wu, Kilian Q. Weinberger, and Yoav Artzi. 2020. Bertscore: Evaluating text generation with bert. In *International Conference on Learning Representations*.

A Technical Details

For the Ensemble-M configuration, a sentence was predicted as essential if it had a confidence score exceeding 0.7, 0.5, 0.9, or 0.05 in at least one of the following approaches: MCS-M, MedCPT FS, SR Med42, and CAR Med42, respectively.

Similarly, in the Ensemble-C setup, a sentence was predicted as essential if it exceeded the thresholds of 0.9, 0.5, 0.0, or 0.4 in at least one of the confidence scores from MCS-C, MedCPT FS, SR Med42, and CAR Med42, respectively.

B Relevance Scores on Dev Set

Method	BLEU	ROUGELsum	SARI	BERTScore	AlignScore	MEDCON
Ensemble-M	7.1	29.5	66.9	34.9	55.0	42.9
Ensemble-C	7.5	31.0	66.3	36.5	59.6	42.0

Table 3: All relevance scores of Ensemble-M and Ensemble-C approaches measured on the dev set.

C Performance Comparison: Med42-8B vs. Llama3-8B

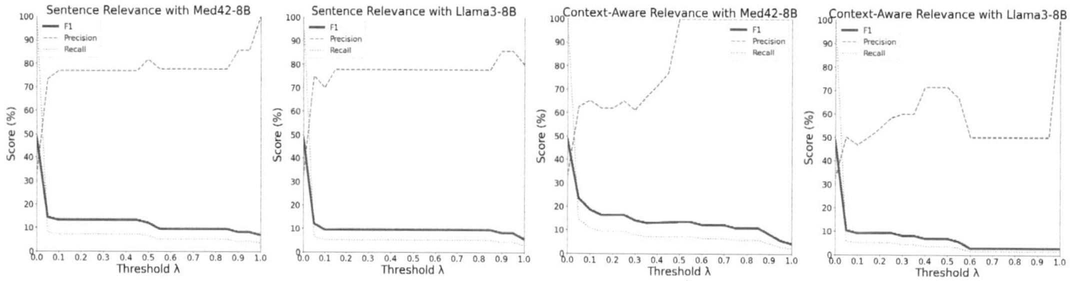

Figure 4: Comparison of sentence retrieval and context-aware relevance performance using Med42-8B vs. Llama-8B. Each chart shows results for one task-model pair, highlighting the impact of replacing Med42-8B with a general-domain Llama-8B model.

D Essential Sentence Retrieval – Few-Shot Prompt for Sentence Relevance Med42

```
You are a clinical assistant. Given a context and a question, extract only the
    essential information from the context that is necessary to answer the question.
     If no information is relevant, respond with "None". Also provide a short
    explanation for your answer.

Context: The patient has a history of hypertension and presents with progressive
    shortness of breath. BNP levels are elevated. Physical examination reveals
    bilateral rales and mild pedal edema.
Question: What information is essential from this context for answering the question
    "What is causing the patient's breathing difficulty?"
Answer: elevated BNP, bilateral rales, mild pedal edema
Reason: These are all indicators of congestive heart failure, which likely explains
    the breathing difficulty.

Context: The patient is a 45-year-old male with a history of allergic rhinitis. He
    was seen in allergy clinic and placed on a regimen of nasal corticosteroids and
    antihistamines. No new triggers identified. Symptoms are seasonal and well-
    controlled.
Question: What information is essential from this context for answering the question
    "What is the most likely cause of the patient's anemia?"
Answer: None
Reason: The context is entirely focused on allergic rhinitis, with no hematologic
    data or symptoms of anemia.

Context: The patient completed a dental cleaning and X-rays showed mild periodontal
    disease. Oral hygiene habits were discussed, and the patient agreed to floss
    daily. No pain or bleeding reported. No antibiotics were prescribed.
Question: What information is essential from this context for answering the question
    "What medications are responsible for the patient's elevated INR?"
Answer: None
Reason: There is no mention of any anticoagulants or medications that affect
    coagulation in the context.

Context: Patient underwent knee replacement two years ago. Reports occasional
    clicking sensation but no pain. X-ray shows proper implant positioning.
```

```
Question: What information is essential from this context for answering the question
    "Is the knee replacement causing complications?"
Answer: occasional clicking sensation, no pain, proper implant positioning
Reason: Clicking may suggest minor mechanical noise but no signs of complications
    given the lack of pain and good positioning.

Context: Complains of weight loss and fatigue over the past 3 months. Labs show iron
    deficiency anemia. Colonoscopy reveals a 2 cm mass in the ascending colon.
Question: What information is essential from this context for answering the question
    "What might explain the patient's fatigue?"
Answer: iron deficiency anemia, 2 cm mass in ascending colon
Reason: Chronic blood loss from the mass could explain anemia and fatigue.

Context: The patient underwent cataract surgery on the right eye and reports
    improved vision. Post-op evaluation showed clear lens placement and normal
    intraocular pressure. No inflammation noted. Scheduled for left eye surgery in
    two months.
Question: What information is essential from this context for answering the question
    "Why did the patient develop shortness of breath?"
Answer: None
Reason: The context is limited to ophthalmologic findings and does not mention any
    pulmonary or cardiovascular symptoms.

Context: Denies smoking, alcohol, or drug use. Family history positive for lung
    cancer in both parents. Works in construction for 25 years without respiratory
    protection.
Question: What information is essential from this context for answering the question
    "What are the patient's risk factors for lung cancer?"
Answer: family history of lung cancer, 25 years in construction without respiratory
    protection
Reason: Occupational exposure and genetics increase risk even without smoking.

Context: Admitted for severe epigastric pain. Has history of NSAID use for chronic
    back pain. Labs show decreased hemoglobin. Endoscopy confirms a gastric ulcer.
Question: What information is essential from this context for answering the question
    "What is the likely cause of the gastrointestinal bleeding?"
Answer: NSAID use, gastric ulcer, decreased hemoglobin
Reason: NSAIDs are known to cause gastric ulcers, which can lead to bleeding.

Context: No prior psychiatric history. The patient has been irritable and withdrawn
    for the past month. Sleep has decreased to 3 hours/night. Appetite remains
    normal.
Question: What information is essential from this context for answering the question
    "Are there signs of depression?"
Answer: irritability, social withdrawal, decreased sleep
Reason: These are common symptoms associated with depressive disorders.

Context: The patient had a colonoscopy last week, which revealed three polyps that
    were removed. Pathology is pending. The patient denies abdominal pain, nausea,
    or changes in bowel habits. Family history is negative for colorectal cancer.
Question: What information is essential from this context for answering the question
    "Why is the patient experiencing chronic fatigue?"
Answer: None
Reason: The context is focused on GI screening and doesn't include symptoms, labs,
    or findings that would explain fatigue.

Context: Presents with left arm weakness and facial droop for 45 minutes. Symptoms
    resolved prior to arrival. CT scan shows no acute infarct. History of atrial
    fibrillation.
Question: What information is essential from this context for answering the question
    "What might have caused the neurological symptoms?"
Answer: transient symptoms, atrial fibrillation
Reason: AFib can cause transient ischemic attacks, which present with stroke-like
    symptoms that resolve.

Context: Mother reports that her child, aged 3, has not yet started speaking in full
    sentences. Hearing test is normal. No social interaction issues observed.
    Growth chart is appropriate.
Question: What information is essential from this context for answering the question
    "Is there concern for developmental delay?"
```

```
Answer: 3-year-old not speaking in full sentences
Reason: While social and hearing are normal, speech delay is suggestive of possible
    developmental delay.

Context: Recent travel to sub-Saharan Africa. Developed intermittent fever and
    chills on return. Blood smear reveals Plasmodium falciparum.
Question: What information is essential from this context for answering the question
     "What is the likely cause of the patient's fever?"
Answer: travel to sub-Saharan Africa, Plasmodium falciparum
Reason: These findings point to malaria as the likely cause of the fever.

Context: Complains of morning stiffness lasting more than 1 hour. Joints in both
    hands are swollen and tender. Positive rheumatoid factor and anti-CCP antibodies
    .
Question: What information is essential from this context for answering the question
     "Is this likely to be rheumatoid arthritis?"
Answer: morning stiffness >1 hour, swollen/tender hand joints, positive RF and anti-
    CCP
Reason: These clinical and serological findings are diagnostic of RA.

Context: A 65-year-old woman was referred to audiology due to recent hearing
    difficulties. Audiogram showed moderate bilateral sensorineural hearing loss.
    Hearing aids were recommended. No signs of vertigo or tinnitus were reported.
Question: What information is essential from this context for answering the question
     "What led to the patient's episodes of syncope?"
Answer: None
Reason: The context only contains auditory assessment and does not address
    cardiovascular or neurologic causes.

Context: On insulin therapy. Skipped lunch due to meetings. Found diaphoretic and
    confused. Glucose 42 mg/dL.
Question: What information is essential from this context for answering the question
     "What explains the patient's confusion?"
Answer: skipped lunch, insulin therapy, glucose 42 mg/dL
Reason: Hypoglycemia is likely due to missed meal with insulin use.

Context: Reports worsening shortness of breath over 2 weeks. Has COPD. Oxygen
    saturation drops to 89% on ambulation. Chest X-ray shows no infiltrates.
Question: What information is essential from this context for answering the question
     "What is likely contributing to the patient's shortness of breath?"
Answer: COPD history, desaturation with ambulation
Reason: COPD with exertional desaturation is a common cause of dyspnea in such
    patients.

Context: Diagnosed with hypothyroidism last year. Currently on levothyroxine.
    Complains of fatigue and cold intolerance. TSH 9.2.
Question: What information is essential from this context for answering the question
     "Why is the patient still symptomatic?"
Answer: hypothyroidism, TSH 9.2
Reason: Elevated TSH indicates under-replacement with levothyroxine.

Context: Denies any chest pain. Takes beta-blocker for hypertension. EKG reveals
    bradycardia (HR 48 bpm). Patient feels fatigued.
Question: What information is essential from this context for answering the question
     "What could explain the fatigue?"
Answer: beta-blocker use, bradycardia
Reason: Bradycardia from beta-blockers may result in reduced cardiac output and
    fatigue.

Context: The patient was evaluated in the ophthalmology clinic due to complaints of
    blurry vision. Examination showed no signs of diabetic retinopathy. Blood
    pressure was within normal range. There were no neurological deficits noted.
    Follow-up was scheduled in six months.
Question: What information is essential from this context for answering the question
     "What is the underlying cause of the patient's persistent headaches?"
Answer: None
Reason: The context only discusses ophthalmological findings and vision-related
    complaints but contains no information about the cause of headaches.
```

```
Context: Patient presented for a follow-up regarding their post-operative shoulder
    surgery. Physical therapy was recommended and patient reports improvement in
    range of motion. There are no signs of infection or complications. Sleep has
    improved as well.
Question: What information is essential from this context for answering the question
     "What factors contributed to the patient's recent weight loss?"
Answer: None
Reason: The context only discusses orthopedic recovery and makes no mention of diet,
     metabolism, or weight.

Context: The patient was brought in for confusion. No focal neurological deficits
    noted. BUN and creatinine significantly elevated. Recently started lisinopril.
Question: What information is essential from this context for answering the question
     "What could explain the altered mental status?"
Answer: elevated BUN/creatinine, started lisinopril
Reason: Acute kidney injury from ACE inhibitors may lead to uremic encephalopathy.

Context: 65-year-old with chronic low back pain. MRI shows mild degenerative disc
    disease. No nerve compression.
Question: What information is essential from this context for answering the question
     "Is surgery indicated?"
Answer: mild degenerative disc disease, no nerve compression
Reason: Conservative treatment is favored as no surgical lesion is present.

Context: During the dermatology consultation, the patient described new-onset skin
    lesions. The rash appeared on the arms and back, non-pruritic and non-painful.
    No signs of infection were noted. Biopsy was scheduled.
Question: What information is essential from this context for answering the question
     "Why has the patient developed elevated liver enzymes?"
Answer: None
Reason: The context centers around dermatological symptoms with no hepatic or
    metabolic findings provided.

Context: History of mechanical heart valve replacement. INR today is 5.2. No active
    bleeding reported.
Question: What information is essential from this context for answering the question
     "What explains the elevated INR?"
Answer: mechanical valve replacement
Reason: Patients require anticoagulation for valves, which can overshoot and elevate
     INR.

Context: Breast mass noted on exam. Mammogram shows suspicious lesion. Biopsy
    confirms ductal carcinoma in situ.
Question: What information is essential from this context for answering the question
     "What is the diagnosis?"
Answer: ductal carcinoma in situ
Reason: Biopsy provides definitive diagnosis.

Context: Patient with ESRD on dialysis. Missed last two sessions. Complains of
    generalized weakness. Potassium level is 6.8.
Question: What information is essential from this context for answering the question
     "What is the likely cause of weakness?"
Answer: missed dialysis sessions, potassium 6.8
Reason: Hyperkalemia and uremia due to missed dialysis likely explain weakness.

Context: {Sentence}
Question: What information is essential from this context for answering the question
     "{Question}"
Answer: ...
Reason: ...
```

E Essential Sentence Retrieval – Few-Shot Prompt for Context-Aware Relevance Med42

```
You are a medical assistant helping a patient's family member understand the
    discharge summary. The family member asks a general question about the patient's
    condition or expected recovery. From the discharge summary, you are evaluating
    whether a specific sentence is essential to help them understand what they truly
    need to know - even if they didn't ask about it directly.
```

```
For each example, decide:
- Is the sentence important for answering the underlying concern in the question? ("
    Yes" or "No")
- Briefly explain why or why not.

### Example 1
Context:
The patient was admitted with signs of dehydration and electrolyte imbalance
    following several days of vomiting and diarrhea. Intravenous fluids and
    potassium replacement were administered. He gradually regained strength and
    tolerated oral intake by day 3. There were no signs of infection. Electrolyte
    levels normalized. He was encouraged to maintain oral hydration and avoid NSAIDs
    . Discharge instructions included dietary recommendations. He is to follow up
    with his primary care physician in one week. The patient lives alone and has
    limited mobility. Transportation services were arranged for follow-up.

Patient's Question: How long will it take for him to fully recover?
Sentence: "He is to follow up with his primary care physician in one week."
Answer: Yes
Reason: The scheduled follow-up provides insight into the expected timeline of
    recovery and monitoring, even though the patient didn't explicitly ask about
    appointments.

### Example 2
Context:
The patient presented with acute asthma exacerbation. She received nebulized
    albuterol and corticosteroids in the emergency department. Oxygen saturation
    improved over 24 hours. There were no signs of pneumonia. She was discharged
    with a prescription for inhaled corticosteroids and a tapering dose of
    prednisone. She was advised to avoid known triggers such as smoke or allergens.
    Patient reported improved breathing at rest but slight shortness of breath
    during activity. No further imaging was ordered. The pulmonologist will review
    her progress in 10 days.

Patient's Question: Is she okay to go back to work next week?
Sentence: "The pulmonologist will review her progress in 10 days."
Answer: Yes
Reason: The timing of the specialist review is crucial for determining readiness to
    return to work, even though the patient didn't mention the appointment.

### Example 3
Context:
The patient was admitted for routine laparoscopic cholecystectomy. The surgery was
    uncomplicated. Minimal intraoperative bleeding was noted. Postoperative pain was
     managed with oral analgesics. Bowel function resumed within 24 hours. She
    ambulated independently on post-op day 2. The surgical wound was clean and dry.
    Discharge instructions advised avoiding heavy lifting for two weeks. Follow-up
    scheduled with surgery clinic in 14 days. Patient was in good spirits and eager
    to return to normal activities.

Patient's Question: What should her recovery look like?
Sentence: "Discharge instructions advised avoiding heavy lifting for two weeks."
Answer: Yes
Reason: The lifting restriction is an essential part of understanding the expected
    recovery process, even if not directly requested.

### Example 4
Context:
{Discharge summary excerpt}

Patient's Question: {Question}
Sentence: "{Sentence}"
Answer: ...
Reason: ...
```

F Answer Generation – Direct Answering Few-Shot Prompt

```
You are a clinical assistant helping family members understand discharge summaries.
Your task is to answer questions based on long clinical sentences, which may include
     irrelevant information.
Always provide a direct, natural answer that is as concise as possible.
Do not repeat or copy any part of the question in your answer.
Do not begin the answer with phrases like ''Because...'' or ''XYZ was recommended
    because...''.
If no clear answer is possible, reply with: None

Question: What treatment did the patient receive for pneumonia?
Sentence: The patient was diagnosed with pneumonia and treated with intravenous
    antibiotics and oxygen therapy.
Answer: He was treated with antibiotics and oxygen therapy.

Question: Why is the patient taking insulin?
Sentence: Due to a recent diagnosis of type 2 diabetes, the patient was prescribed
    insulin to manage blood sugar levels.
Answer: He was diagnosed with type 2 diabetes.

Question: What caused the patient's shortness of breath?
Sentence: The patient's shortness of breath was likely due to fluid accumulation in
    the lungs caused by heart failure.
Answer: He had lung fluid from heart failure.

Question: What mobility assistance does the patient need?
Sentence: After hip surgery, the patient requires a walker and supervision while
    moving.
Answer: He requires a walker and supervision.

Question: Why was a walking cane recommended to the patient?
Sentence: The patient's vaccination record was updated during the follow-up visit,
    including influenza and tetanus boosters.
Answer: None

Question: What complications occurred during the patient's hospital stay?
Sentence: The patient experienced atrial fibrillation, transient confusion, and a
    mild allergic reaction to antibiotics during admission.
Answer: He experienced atrial fibrillation, confusion, and an allergic reaction.

Question: {Question}
Sentence: {Sentence}
Answer: ...
```

G Answer Generation – Sentence Compression Few-Shot Prompt

```
You are a clinical assistant specialized in simplifying discharge summaries.
Your task is to take a long clinical sentence and rewrite it as a shorter, natural,
    and concise sentence that preserves the essential clinical information.
Do not copy the entire sentence or use unnecessary detail. Keep it factual, clear,
    and brief.

Sentence: The patient was admitted to the hospital due to a sudden episode of chest
    pain that occurred while he was gardening.
Compressed: Admitted for sudden chest pain during gardening.

Sentence: Following the MRI scan, the patient was found to have a small herniated
    disc at the L4-L5 level.
Compressed: MRI showed a small herniated disc at L4-L5.

Sentence: The patient has a medical history of hypertension, type 2 diabetes, and
    chronic kidney disease stage 3.
Compressed: History includes hypertension, diabetes, and stage 3 kidney disease.

Sentence: She was prescribed albuterol inhaler to be used as needed for episodes of
    shortness of breath.
Compressed: Prescribed albuterol for shortness of breath as needed.

Sentence: During his hospital stay, the patient developed a mild skin rash likely
    due to a reaction to antibiotics.
```

```
Compressed: Developed mild rash from antibiotics.

Sentence: The patient was advised to follow a low-sodium diet and monitor blood
    pressure regularly at home.
Compressed: Advised low-sodium diet and home blood pressure monitoring.

Sentence: He lives alone but receives weekly assistance from his daughter with
    groceries and medication management.
Compressed: Lives alone with weekly help from daughter.

Sentence: The patient's vaccination record was updated during the follow-up visit,
    including influenza and tetanus boosters.
Compressed: Received flu and tetanus boosters at follow-up.

Sentence: {Sentence}
Compressed: ...
```

WisPerMed at ArchEHR-QA 2025: A Modular, Relevance-First Approach for Grounded Question Answering on Eletronic Health Records

Jan-Henning Büns[1], Hendrik Damm[1,3], Tabea M. G. Pakull[1,2], Felix Nensa[4,5], Elisabeth Livingstone[2]

[1]Department of Computer Science, University of Applied Sciences and Arts Dortmund
[2]Institute for Transfusion Medicine, University Hospital Essen
[3]Institute for Medical Informatics, Biometry and Epidemiology, University Hospital Essen
[4]Institute of Diagnostic and Interventional Radiology and Neuroradiology, University Hospital Essen
[5]Institute for Artificial Intelligence in Medicine (IKIM), University Hospital Essen

Correspondence: jan-henning.buens@fh-dortmund.de

Abstract

Automatically answering patient questions based on electronic health records (EHRs) requires systems that both identify relevant evidence and generate accurate, grounded responses. We present a three-part pipeline developed by WisPerMed for the ArchEHR-QA 2025 shared task. First, a fine-tuned BioClinicalBERT model classifies note sentences by their relevance using synonym-based and paraphrased data augmentation. Second, a constrained generation step uses DistilBART-Med-Summary to produce faithful answers strictly limited to top-ranked evidence. Third, we align each answer sentence to its supporting evidence via BiomedBERT embeddings and ROUGE-based similarity scoring to ensure citation transparency. Our system achieved a 35.0% overall score on the hidden test set, outperforming the organizer's baseline by 4.3 percentage points. Gains in BERTScore (+44%) and SARI (+119%) highlight substantial improvements in semantic accuracy and relevance. This modular approach demonstrates that enforcing evidence-awareness and citation grounding enhances both answer quality and trustworthiness in clinical QA systems.

1 Introduction

As patient–portal adoption accelerates, message volume now exceeds pre-pandemic levels; a longitudinal study found a 55% rise in medical-advice requests and 24% increase in daily inbox time for physicians between 2019–2023 (Arndt et al., 2024). Large Language Models (LLMs) can draft fluent replies, yet uncontrolled hallucinations threaten patient safety (Nov et al., 2023; Biro et al., 2025). The ArchEHR-QA 2025 shared task (Soni and Demner-Fushman, 2025b) extends this trajectory by pairing genuine portal questions with sentence-level evidence annotations and requiring grounded answers.

This paper presents the submission by WisPerMed, a three-part pipeline:

- BioClinicalBERT (Lee et al., 2019a) classifies note sentences as *essential*, *supplementary*, or *not-relevant*, with robustness improved via synonym and paraphrase augmentation;
- DistilBART-Med-Summary (Lewis et al., 2019) generates an answer conditioned solely on the top-ranked evidence and
- BiomedBERT (Gu et al., 2021a) embeddings align each answer sentence to its most similar evidence, yielding explicit citations.

2 Related Work

This section establishes the context for our multi-component system that combines evidence classification, answer generation, and citation alignment.

Electronic Health Record Question Answering. Electronic Health Records (EHRs) contain valuable patient information that can benefit both healthcare providers and patients. Giving patients access to their EHRs can increase patient and physician trust, improve communication, strengthen the physician–patient relationship, increase medication adherence, and improve patient outcomes (Tapuria et al., 2021). Question Answering (QA) systems on patient-related data can assist clinicians in decision-making and enable patients to better understand their medical history (Bardhan et al., 2024). Unlike general medical QA tasks that rely on curated knowledge sources (e.g., PubMed or medical websites), EHR QA requires answer generation grounded in patient-specific records. This introduces challenges in interpreting both informal patient queries and domain-specific clinical text.

BioNLP 2025 Shared Tasks, pages 41–49

Datasets. non Early progress relied on synthetic corpora such as EMRQA (Pampari et al., 2018), which repurposed i2b2 annotations (Özlem Uzuner et al., 2011) to create ~ 0.4 M evidence–answer pairs. Work on structured records introduced MIMICSQL for question-to-SQL generation on MIMIC-III tables (Wang et al., 2020). To improve realism and coverage, consumer-health resources like MEDIQA-ANS (Savery et al., 2020) added question-driven answer summaries, while MEDIQA-CHAT captured full doctor–patient dialogues (Ben Abacha et al., 2023). Recent benchmarks push modality boundaries: EHRXQA integrates tabular EHR data with chest-X-ray images for cross-modal reasoning (Bae et al., 2023). The ArchEHR-QA dataset (Soni and Demner-Fushman, 2025a) extends this trajectory by pairing genuine portal questions with sentence-level evidence annotations and enforcing grounded answers. The dataset is derived from the MIMIC-III dataset (Johnson et al., 2016) and comprises 120 patient cases (20 development, 100 test). Every case consists of a realistic patient question, corresponding clinician-rewritten questions, and annotated clinical note excerpts. Each clinical note excerpt is segmented into sentences, which are manually annotated as "essential", "supplementary", or "not-relevant" for answering the question.

Biomedical Language Models. Domain-specific language models have revolutionized biomedical NLP applications (Yang et al., 2023). While early approaches fine-tuned general-domain models like BERT (Devlin et al., 2019) on biomedical corpora, research has demonstrated that pre-training language models from scratch on biomedical text yields substantial performance gains across various tasks (Gu et al., 2021b). In the realm of medical text summarization, models like DistilBART-Med-Summary[1] have been developed to condense clinical documents into concise summaries while preserving essential information. These models are trained on large-scale medical datasets and fine-tuned to capture the specific linguistic characteristics of clinical narratives.

BioBERT (Lee et al., 2019b), a domain-specific model pretrained on large-scale biomedical corpora, significantly outperforms general-domain BERT on biomedical text mining tasks. Building on this foundation, BiomedBERT (Gu et al., 2021a) was trained solely on biomedical text from scratch and achieved excellent results across multiple biomedical NLP benchmarks. Bio_ClinicalBERT (Alsentzer et al., 2019) specializes further in clinical text by initializing from BioBERT and training on MIMIC notes, a database containing electronic health records from ICU patients.

Data Augmentation In the medical domain, the scarcity of annotated datasets poses a challenge to the development of robust models. To address this, data augmentation techniques have been employed to artificially expand training datasets, thereby enhancing model generalizability and mitigating overfitting. In clinical contexts, leveraging domain-specific resources such as the Unified Medical Language System (UMLS) (Bodenreider, 2004a) and WordNet (Miller, 1994) for synonym replacement has proven effective in maintaining the integrity of medical terminology during augmentation (Kang et al., 2020; Shorten et al., 2021). Furthermore, the use of LLMs like Gemini (Hoffmann et al., 2023) to generate synthetic data have shown promise in producing high-quality, diverse clinical text (Wang et al., 2024), which is particularly beneficial for tasks in low-resource settings.

3 Methods

WisPerMed adopts a three-part pipeline summarized in Figure 1.

Sentence-level relevance classification. Each clinical note sentence is encoded with BIOCLINICALBERT (Lee et al., 2019a). The model is fine-tuned on the ArchEHR-QA development split (batch size 8, 5 epochs, initial learning rate set to 2×10^{-6} according to the default learning rate scheduler from the transformers library (Wolf et al., 2020)) to predict *essential*, *supplementary*, or *irrelevant* labels. The training data are expanded by 500%, to 100 cases, using: (1) synonym substitution derived from UMLS (Bodenreider, 2004b) and WordNet (Miller, 1994) and (2) paraphrase generation with Gemini.

Answer generation. The evidence set, clinician-rewritten question, and a fixed instruction prompt are concatenated and passed to DistilBART-Med-Summary. The prompt (refer to Listing 2 in Appendix 6) instructs the model to (i) restrict content to the provided evidence. Decoding employs beam search (Meister et al., 2020) (beam size 5, repetition penalty 1.2) and truncates the output to ≤ 75 tokens, as required by the task limit. Only the first

[1] `https://huggingface.co/Mahalingam/DistilBart-Med-Summary`, Last Accessed: 30.04.2025

75 tokens are included in the performance evaluation.

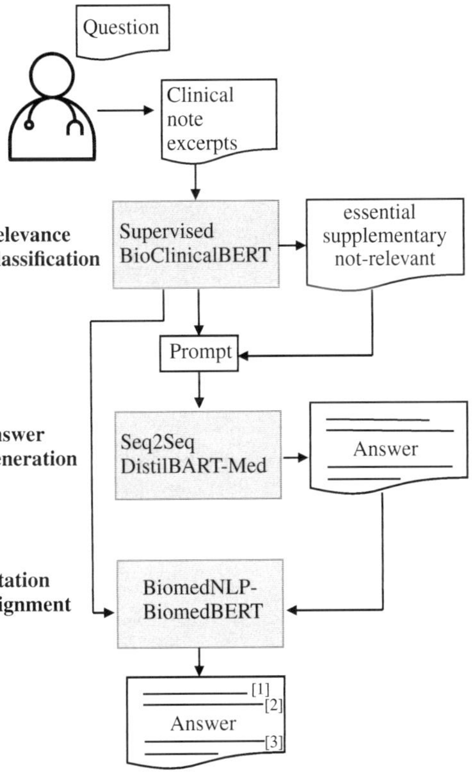

Figure 1: Workflow of the three-part pipeline. The first stage performs relevance classification, identifying sentences as essential, supplementary, or not-relevant for answer generation. The second stage generates an answer using the prioritized evidence. The final stage adds explicit citations by linking each answer sentence to its supporting evidence.

Citation alignment. Each answer sentence is embedded with BIOMEDBERT (Gu et al., 2021a). Using Recall-Oriented Understudy for Gisting Evaluation (ROUGE) score (Lin, 2004) calculations, we link sentences in the answer to the most similar sentences of the clinical notes. A similarity threshold of 0.30 ensures that lower-scoring sentences are tagged as unsupported. The selected citations are then being added to the corresponding sentences in the answer to maintain the task's citation format.

Implementation. Models are trained and executed with PyTorch 2.6.0 (Paszke et al., 2019) using Python 3.12.9 on a single Nvidia RTX 4080 Super (16GB). Source code is released under MIT License.[2]

[2]https://github.com/rtg-wispermed/ArchEHR-QA, Last Accessed: 09.05.2025

4 Evaluation

The metrics for evaluation are divided into factuality and relevance metrics. Factuality metrics include Precision, Recall, and F1-score (Powers, 2020) in both micro and macro variations. In addition, all scores are measured in a strict (including only sentences classified as "essential") and a lenient (including sentences classified as "essential" and "supplementary") variation. The mean of all factuality scores (Strict Citation F1 scores) is the Overall Factuality score. Relevance metrics include Bilingual Evaluation Understudy (BLEU) (Papineni et al., 2002), ROUGE (Lin, 2004) and System output Against References and against the Input (SARI) (Xu et al., 2016). Semantic similarity is measured with BERTScore (Zhang et al., 2019). AlignScore (Zha et al., 2023) provides task-agnostic factual consistency, and MEDCON (Medical Concept Overlap (Yim et al., 2023)) captures clinical concept agreement. The mean of all surface metrics is the overall relevance score. Lastly, the overall score is calculated by the mean of the Overall Factuality score and the overall relevance score.

5 Results and Discussion

Table 1 presents overall scores on the ArchEHR-QA hidden test set. The approach by WisPerMed improved upon the organizer's baseline by $\approx 4.3\%$. Both the Overall Factuality and the overall relevance improved by $\approx 2.6\%$ and $\approx 6.1\%$ respectively.

Metric	WisPerMed	Baseline	DMIS Lab
Overall	35.0	30.7	53.7
OF	36.2	33.6	58.6
OR	33.9	27.8	48.8

Table 1: Comparison of Overall, Overall Factuality (OF), and Overall Relevance (OR) scores for WisPerMed, the organizer's baseline and DMIS Lab

The three-part pipeline demonstrates consistent improvements over the organizer's baseline across key relevance and factual accuracy metrics, as shown in Table 2. Notably, it achieves a 44% relative improvement in BERTScore (29.5 vs. 20.5), indicating superior semantic alignment with reference texts through contextual embeddings. The 119% improvement in SARI (61.0 vs. 27.8) highlights enhanced content preservation during text simplification or rewriting tasks, even compared to DMIS Lab (36.7). While both systems show comparable performance in UMLS-based concept recognition (MEDCON), WisPerMed's 3.5-point gain in AlignScore (62.3 vs. 57.7) suggests better factual consistency in clinical narratives. The first-place team, DMIS Lab, achieved significantly higher overall scores, indicating that there is still headroom for improving our approach.

Metric	WisPerMed	Baseline	DMIS Lab
BLEU	2.0	0.1	14.3
ROUGE-LSum	22.6	33.6	46.5
SARI	61.0	27.8	36.7
BERTScore	29.5	20.5	53.9
AlignScore	62.3	57.7	92.4
MEDCON	25.9	25.6	49.3

Table 2: Comparison of relevance metrics between WisPerMed, organizers-baseline and DMIS Lab

Table 3 shows that our approach achieves consistently higher recall and F1 scores than the organizer's baseline across both strict and lenient, micro-averaged settings, with strict recall (micro) improving from 21.9 to 26.9 and strict F1 (micro) from 33.6 to 36.2. These gains indicate a higher ability to identify a greater proportion of relevant information, reducing false negatives. On the other hand the organizer's baseline demonstrates higher precision, indicating that our approach contains more false positives. Overall, the metrics demonstrate the focus on maximizing relevant coverage.

Metric	WisPerMed	Baseline	DMIS Lab
Strict Precision (mic)	55.4	71.6	57.9
Strict Recall (mic)	26.9	21.9	59.3
Strict F1 (mic)	36.2	33.6	58.6
Lenient Precision (mic)	59.1	77.0	61.2
Lenient Recall (mic)	27.1	22.3	59.2
Lenient F1 (mic)	37.1	34.6	60.2
Strict Precision (mac)	54.0	77.4	62.1
Strict Recall (mac)	34.0	31.5	69.0
Strict F1 (mac)	37.7	39.0	61.2
Lenient Precision (mac)	59.5	83.0	66.6
Lenient Recall (mac)	33.9	30.8	67.1
Lenient F1 (mac)	39.9	39.9	63.2

Table 3: Comparison of strict and lenient (micro/-macro) precision, recall, and F1 scores for WisPerMed, organizers-baseline and DMIS Lab

Further experiments on the ArchEHR-QA development set have been conducted to compare three different sequence-to-sequence text generation models. Specifically, we chose three models from huggingface: (1) Flan-T5 (Chung et al., 2022), (2) BART-Large-CNN (Lewis et al., 2019) and (3) DistilBART-Med-Summary. The results (refer to Table 4) indicate that both BART-models capture medical concepts in their generated answer more precisely compared to Flan-T5. While DistilBART-Med-Summary achieves the highest Overall Factuality score due to its finetuning on medical data, BART-Large-CNN can capture the relevance of information with a higher precision. Another finding is that Flan-T5 requires a detailed and specific prompt to generate answers that adhere to task requirements (see Listing 1). Both BART models, on the other hand, perform well with a much simpler prompt.

Model	OF	OR	Overall
Flan-T5	54.92	29.63	42.27
BART-Large-CNN	64.04	52.69	58.36
DistilBART-M-S	70.42	49.33	59.87

Table 4: Overall score, Overall Factuality (OF) and Overall Relevance (OR), for each model

The impact of data augmentation was evaluated on the ArchEHR-QA development set. The results (refer to Table 5 in Appendix) demonstrate that synonym augmentation can greatly improve the model's performance in every metric. Including synthetic data generated by Gemini on the other hand has minor impact on the performance metrics.

6 Conclusion

The three-part pipeline proposed by WisPerMed system demonstrates that a modular, relevance-first approach can deliver competitive performance on ArchEHR-QA 2025 while retaining transparency. The combination of BioClinicalBERT-based (Lee et al., 2019a) sentence selection, answer generation with DistilBART-Med-Summary, and Biomed-BERT citation alignment (Gu et al., 2021a) yielded results that surpassed the organizers' baseline and maintained strong precision across strict and lenient settings. We demonstrated that models based on BART (Lewis et al., 2019) are better suited for grounded answer generation for EHR questions compared to Flan-T5 (Chung et al., 2022) variants. We further conclude that synonym augmentation based on UMLS (Bodenreider, 2004a), and WordNet (Miller, 1994) can greatly improve the performance of relevance classification.

Limitations

While the WisPerMed pipeline achieves a strong improvement in the relevance metrics, several weaknesses remain. Reliance on hard probability thresholds in the relevance classifier caps citation recall at roughly 27%. Synthetic training data generated via Gemini paraphrasing occasionally alters medical meaning, introducing label noise that propagates downstream. Because all models are tuned on MIMIC style documentation, performance may degrade when confronted with different institutional note formats or specialty-specific jargon. The ROUGE-score-based similarity method for citation alignment may misassign identifiers when multiple sentences are semantically similar. The decision to use BERT-based sequence-to-sequence (seq2seq) models was made to minimize hardware requirements, enabling the three-step pipeline to be trained on a single consumer GPU, such as the Nvidia RTX 4080 Super (16GB). However, our three-part pipeline could be outperformed by more demanding Retrieval-Augmented Generation (RAG) approaches, which jointly optimize retrieval and generation while explicitly linking answers to sources, reducing citation errors.

Acknowledgments

The work of Jan-Henning Büns, Hendrik Damm and Tabea M. G. Pakull was funded by a PhD grant from the DFG Research Training Group 2535 *Knowledge- and data-based personalisation of medicine at the point of care (WisPerMed)*.

References

Emily Alsentzer, John R. Murphy, Willie Boag, Wei-Hung Weng, Di Jin, Tristan Naumann, and Matthew McDermott. 2019. Publicly available clinical BERT embeddings. In *Proc. ClinicalNLP*, pages 72–78.

Brian G. Arndt, Mark A. Micek, Adam Rule, Christina M. Shafer, Jeffrey J. Baltus, and Christine A. Sinsky. 2024. More tethered to the EHR: EHR workload trends among academic primary care physicians, 2019–2023. *Annals of Family Medicine*, 22(1):12–18.

Seongsu Bae, Daeun Kyung, Jaehee Ryu, Eunbyeol Cho, Gyubok Lee, et al. 2023. EHRXQA: A multi-modal question answering dataset for electronic health records with chest x-ray images. In *Advances in Neural Information Processing Systems (Datasets and Benchmarks)*.

Jayetri Bardhan, Kirk Roberts, and Daisy Zhe Wang. 2024. Question answering for electronic health records: Scoping review of datasets and models. *J Med Internet Res*, 26:e53636.

Asma Ben Abacha, Wen-wai Yim, Griffin Adams, Neal Snider, and Meliha Yetisgen. 2023. Overview of the MEDIQA-chat 2023 shared tasks on the summarization & generation of doctor–patient conversations. In *Proc. ClinicalNLP*, pages 503–513.

Joshua M. Biro, Jessica L. Handley, J. Malcolm McCurry, Adam Visconti, Jeffrey Weinfeld, J. Gregory Trafton, and Raj M. Ratwani. 2025. Opportunities and risks of artificial intelligence in patient portal messaging in primary care. *npj Digital Medicine*, 8:222.

Olivier Bodenreider. 2004a. The Unified Medical Language System (UMLS): Integrating biomedical terminology. *Nucleic Acids Research*, 32(Database-Issue):267–270.

Olivier Bodenreider. 2004b. The unified medical language system (umls): integrating biomedical terminology. *Nucleic Acids Res.*, 32(Database-Issue):267–270.

Hyung Won Chung, Le Hou, Shayne Longpre, Barret Zoph, Yi Tay, William Fedus, et al. 2022. Scaling instruction-finetuned language models. In *Proc. EMNLP*, pages 277–294.

Jacob Devlin, Ming-Wei Chang, Kenton Lee, and Kristina Toutanova. 2019. Bert: Pre-training of deep bidirectional transformers for language understanding. In *Proceedings of the 2019 Conference of the North American Chapter of the Association for Computational Linguistics: Human Language Technologies, Volume 1 (Long and Short Papers)*, pages 4171–4186, Minneapolis, Minnesota. Association for Computational Linguistics.

Yu Gu, Robert Tinn, Hao Cheng, Caleb Lucas, Naoto Usuyama, Xiaodong Liu, Tristan Naumann, Jianfeng Gao, and Hoifung Poon. 2021a. PubMedBERT: Domain-specific language model pretraining for biomedical natural language processing. *arXiv preprint*, arXiv:2007.15779.

Yu Gu, Robert Tinn, Hao Cheng, Michael Lucas, Naoto Usuyama, Xiaodong Liu, Tristan Naumann, Jianfeng Gao, and Hoifung Poon. 2021b. Domain-specific language model pretraining for biomedical natural language processing. *ACM Transactions on Computing for Healthcare*, 3(1):1–23.

Jordan Hoffmann, Jeffrey Dean, Slav Petrov, et al. 2023. Gemini: A family of highly capable multimodal models. *arXiv preprint*, arXiv:2312.11805.

Alistair E. W. Johnson, Tom J. Pollard, Lu Shen, Li-Wei H. Lehman, Mengling Feng, Mohammad M. Ghassemi, Benjamin Moody, Peter Szolovits, Leo Anthony G. Celi, and Roger G. Mark. 2016. Mimic-iii, a freely accessible critical care database. *Scientific Data*, 3:160035.

Tian Kang, Adler Perotte, Youlan Tang, Casey Ta, and Chunhua Weng. 2020. Umls-based data augmentation for natural language processing of clinical research literature. *Journal of the American Medical Informatics Association*, 28(4):812–823.

Jinhyuk Lee, Wonjin Yoon, Sungdong Kim, Donghyeon Kim, Sunkyu Kim, Chan Ho So, and Jaewoo Kang. 2019a. Biobert: a pre-trained biomedical language representation model for biomedical text mining. *Bioinformatics*, 36(4):1234–1240.

Jinhyuk Lee, Wonjin Yoon, Sungdong Kim, Donghyeon Kim, Sunkyu Kim, Chan Ho So, and Jaewoo Kang. 2019b. Biobert: a pre-trained biomedical language representation model for biomedical text mining. *Bioinformatics*, 36(4):1234–1240.

Mike Lewis, Yinhan Liu, Naman Goyal, Marjan Ghazvininejad, Abdelrahman Mohamed, Omer Levy, Veselin Stoyanov, and Luke Zettlemoyer. 2019. BART: denoising sequence-to-sequence pre-training for natural language generation, translation, and comprehension. *CoRR*, abs/1910.13461.

Chin-Yew Lin. 2004. Rouge: A package for automatic evaluation of summaries. In *Proc. ACL Workshop on Text Summarization Branches Out*, pages 74–81.

Clara Meister, Tim Vieira, and Ryan Cotterell. 2020. Best-first beam search. *Transactions of the Association for Computational Linguistics*, 8:795–809.

George A. Miller. 1994. WordNet: A lexical database for English. In *Human Language Technology: Proceedings of a Workshop held at Plainsboro, New Jersey, March 8-11, 1994*.

Oded Nov, Nina Singh, and Devin Mann. 2023. Putting ChatGPT's medical advice to the (turing) test: Survey study. *JMIR Medical Education*, 9:e46939.

Anusri Pampari, Preethi Raghavan, Jennifer Liang, and Jian Peng. 2018. emrQA: A large corpus for question answering on electronic medical records. In *Proc. EMNLP*, pages 2357–2368.

Kishore Papineni, Salim Roukos, Todd Ward, and Wei-Jing Zhu. 2002. Bleu: a method for automatic evaluation of machine translation. In *Proc. ACL*, pages 311–318.

Adam Paszke, Sam Gross, Francisco Massa, Adam Lerer, James Bradbury, Gregory Chanan, Trevor Killeen, Zeming Lin, Natalia Gimelshein, Luca Antiga, Alban Desmaison, Andreas Köpf, Edward Yang, Zach DeVito, Martin Raison, Alykhan Tejani, Sasank Chilamkurthy, Benoit Steiner, Lu Fang, Junjie Bai, and Soumith Chintala. 2019. Pytorch: An imperative style, high-performance deep learning library. *Preprint*, arXiv:1912.01703.

David M. W. Powers. 2020. Evaluation: from precision, recall and f-measure to roc, informedness, markedness and correlation. *Preprint*, arXiv:2010.16061.

Max Savery, Asma Ben Abacha, Soumya Gayen, and Dina Demner-Fushman. 2020. Question-driven summarization of answers to consumer health questions. *Scientific Data*, 7:322.

Connor Shorten, Taghi M. Khoshgoftaar, and Borko Furht. 2021. Text data augmentation for deep learning. *Journal of Big Data*, 8.

Sarvesh Soni and Dina Demner-Fushman. 2025a. A dataset for addressing patient's information needs related to clinical course of hospitalization. *arXiv preprint*.

Sarvesh Soni and Dina Demner-Fushman. 2025b. Overview of the archehr-qa 2025 shared task on grounded question answering from electronic health records. In *The 24th Workshop on Biomedical Natural Language Processing and BioNLP Shared Tasks*, Vienna, Austria. Association for Computational Linguistics.

Archana Tapuria, Talya Porat, Dipak Kalra, Glen Dsouza, Sun Xiaohui, and Vasa Curcin and. 2021. Impact of patient access to their electronic health record: systematic review. *Informatics for Health and Social Care*, 46(2):194–206. PMID: 33840342.

Hanyin Wang, Chufan Gao, Bolun Liu, Qiping Xu, Guleid Hussein, Mohamad El Labban, Kingsley Iheasirim, Hariprasad Reddy Korsapati, Chuck Outcalt, and Jimeng Sun. 2024. Adapting open-source large language models for cost-effective, expert-level clinical note generation with on-policy reinforcement learning. *ArXiv*, abs/2405.00715.

Ping Wang, Tian Shi, and Chandan K. Reddy. 2020. Text-to-SQL generation for question answering on electronic medical records. In *Proc. WWW*, pages 350–361.

Thomas Wolf, Lysandre Debut, Victor Sanh, Julien Chaumond, Clement Delangue, Anthony Moi, Pierric Cistac, Tim Rault, Rémi Louf, Morgan Funtowicz, Joe Davison, Sam Shleifer, Patrick von Platen, Clara Ma, Yacine Jernite, Julien Plu, Canwen Xu, Teven Le Scao, Sylvain Gugger, Mariama Drame, Quentin Lhoest, and Alexander M. Rush. 2020. Huggingface's transformers: State-of-the-art natural language processing. *Preprint*, arXiv:1910.03771.

Wei Xu, Courtney Napoles, Ellie Pavlick, Quanze Chen, and Chris Callison-Burch. 2016. Optimizing statistical machine translation for text simplification. In *Trans. ACL*, volume 4, pages 401–415.

Rui Yang, Ting Fang Tan, Wei Lu, Arun James Thirunavukarasu, Daniel Shu Wei Ting, and Nan Liu. 2023. Large language models in health care: Development, applications, and challenges. *Health Care Science*, 2(4):255–263.

Wen-wai Yim, Yujuan Fu, Asma Ben Abacha, Neal Snider, Thomas Lin, and Meliha Yetisgen. 2023. Aci-bench: A novel ambient clinical intelligence dataset for benchmarking automatic visit note generation. In *Scientific Data*, volume 10, page 586.

Yuheng Zha, Yichi Yang, Ruichen Li, and Zhiting Hu. 2023. Alignscore: Evaluating factual consistency with a unified alignment function. In *Proc. ACL*, pages 11328–11348.

Tianyi Zhang, Varsha Kishore, Felix Wu, Kilian Q. Weinberger, and Yoav Artzi. 2019. Bertscore: Evaluating text generation with bert. In *Proc. ICLR*.

Özlem Uzuner, Brett R. South, Shuying Shen, and Scott L. DuVall. 2011. 2010 i2b2/va challenge on concepts, assertions, and relations in clinical text. *Journal of the American Medical Informatics Association*, 18(5):552–556.

Listing 1: Flan-T5 Prompt

```
f"""Question: {question}

Context: {context}
Instructions:
1. Create a comprehensive, narrative
    answer in paragraph form to the
    question based STRICTLY on the
    provided context sentences
2. Use complete sentences. Do NOT use
    lists
3. Every sentence in your answer MUST be
     directly supported by evidence from
     the context
4. Minimize paraphrasing. Prefer using
    exact phrases from the context for
    medical terms, findings, and actions
5. The answer must not exceed 75 words
6. Preserve all medical terminology
    exactly as it appears. Do not
    simplify
7. Ensure clinical accuracy and a
    professional tone

Answer:
"""
```

Listing 2: BART-Large-CNN / DistilBART-Med-Summary Prompt

```
(f"{context} Based on the text above,
    answer the question: {question}\n"
f"Answer:")
```

Metric	No Aug.	Synonym Aug.	Synonym Aug. + Synth. Data
Strict Macro Precision	70.17	100.00	100.00
Strict Macro Recall	40.18	65.15	65.15
Strict Macro F1	49.08	75.84	75.84
Strict Micro Precision	71.64	100.00	100.00
Strict Micro Recall	34.78	53.62	53.62
Strict Micro F1	46.83	69.81	69.81
Lenient Macro Precision	75.17	100.00	100.00
Lenient Macro Recall	34.87	50.60	50.60
Lenient Macro F1	45.09	63.65	63.65
Lenient Micro Precision	77.61	100.00	100.00
Lenient Micro Recall	27.51	39.15	39.15
Lenient Micro F1	40.62	56.27	56.27
Overall Factuality Score	46.83	69.81	69.81
SARI	66.94	73.46	73.56
BLEU	2.74	3.81	3.85
BERTScore	36.06	43.96	43.68
ROUGE-1	30.88	36.89	36.89
ROUGE-2	23.48	31.45	31.67
ROUGE-L	22.65	25.57	28.99
ROUGE-Lsum	29.76	36.31	36.31
AlignScore	64.37	87.05	89.17
MedCon	38.81	49.85	49.85
Overall Relevance Score	33.87	39.38	39.35
Overall Score	40.35	54.60	54.58

Table 5: Scores for each augmentation type: No Augmentation, Synonym Augmentation, and Synonym Augmentation + Synthetic Data.

heiDS at ArchEHR-QA 2025: From Fixed-k to Query-dependent-k for Retrieval Augmented Generation

Ashish Chouhan and Michael Gertz
Data Science Group, Institute of Computer Science
Heidelberg University, Germany
{chouhan, gertz}@informatik.uni-heidelberg.de

Abstract

This paper presents the approach of our team called heiDS for the ArchEHR-QA 2025 shared task. A pipeline using a retrieval augmented generation (RAG) framework is designed to generate answers that are attributed to clinical evidence from the electronic health records (EHRs) of patients in response to patient-specific questions. We explored various components of a RAG framework, focusing on ranked list truncation (RLT) retrieval strategies and attribution approaches. Instead of using a fixed top-k RLT retrieval strategy, we employ a query-dependent-k retrieval strategy, including the existing surprise and autocut methods and two new methods proposed in this work, autocut* and elbow. The experimental results show the benefits of our strategy in producing factual and relevant answers when compared to a fixed-k.

1 Introduction

Electronic Health Records (EHRs) are essential in any healthcare system, serving as repositories of the medical history of patients (Häyrinen et al., 2008). Since 2020, patient portals have increased, resulting in more virtual communications between patients and clinicians (Small et al., 2024). As a result, responding to inquiries of patients has become an important issue. Clinicians are reported to spend around 1.5 hours each day managing approximately 150 messages (patient questions) (Small et al., 2024; Liu et al., 2024b). Thus, answering patient-specific questions is a crucial task that relies on information managed in EHRs.

Large Language Models (LLMs) can automate answer generation for patient questions, as these models are trained on extensive textual data (Liu et al., 2024b). However, LLMs are also prone to hallucinations, that is, they may generate answers not supported by a reliable source. This can undermine user trust and potentially harm patients by giving incorrect advice (Huang et al., 2024b). Therefore, attribution, i.e., linking elements of a generated answer to sources, is critical to ensure that every claim is grounded in medical evidence.

Attribution has gained significant attention across various domains, such as the legal and medical domains (Trautmann et al., 2024; Malaviya et al., 2024). Li et al. (2023) outline three approaches for generating answers with attribution. The first approach is direct model-driven attribution, where an LLM generates answers with their sources without using additional information. This is accomplished by fine-tuning or training the model to generate answers that include attributions (Zhang et al., 2024; Patel et al., 2024; Huang et al., 2024a). However, a common issue with this approach is the hallucination of references (Agrawal et al., 2024). The second approach is known as post-retrieval attribution or retrieve-and-read. It retrieves evidence relevant to a query, generating an answer based on that evidence. The LLM is prompted to reference the retrieved information, thereby enforcing attribution (Menick et al., 2022; Nakano et al., 2021; Şahinuç et al., 2024; Gao et al., 2023b). Post-generation attribution (Gao et al., 2023a; Ramu et al., 2024; Cohen-Wang et al., 2024) is the third approach, and it allows the LLM to generate answers without prior attribution and in a post-processing step map answer text back to its sources.

The objective of the BioNLP Grounded Electronic Health Record Question Answering shared task (ArchEHR-QA) (Soni and Demner-Fushman, 2025b) is to generate answers to patient questions, considering clinical note excerpts and attributing them with relevant evidence from the excerpts. Our approach focuses on developing a pipeline for attributed answer generation by employing a retrieval augmented generation (RAG) framework. We experimented with different methods based on the post-retrieval and post-generation attribution ap-

BioNLP 2025 Shared Tasks, pages 50–61
August 1, 2025 ©2025 Association for Computational Linguistics

proaches on the ArchEHR-QA development set, which are detailed in Section 2.

2 Pipeline Overview

Our proposed pipeline utilizes a RAG framework to solve the ArchEHR-QA task. This task involves answering health-related questions from patients and providing attributions based on the patients' clinical notes. In this section, we introduce our different methods, including the pipeline we submitted to the ArchEHR-QA 2025 leaderboard. Section 2.1 provides information about the dataset used for the shared task, followed by Section 2.2 describing the baseline. Section 2.3 provides information on our submitted pipeline, which is based on a surprise (Bahri et al., 2023) Ranked List Truncation (RLT) retrieval strategy. Finally, other methods we experimented with (other than the baseline and submitted pipeline) are outlined in Section 2.4.

2.1 Dataset

The dataset for the ArchEHR-QA 2025 shared task, available on PhysioNet[1] (Soni and Demner-Fushman, 2025a), comprises 20 case studies in the development (dev) set and 100 case studies in the test set[2]. Each case study consists of a hand-curated patient question, its corresponding clinician-rewritten version (i.e., clinician question), and excerpts from the patient's clinical notes. See Appendix A for an example of a case study from the dev set and Appendix B for some statistics on the clinical note excerpts. For every sentence in a clinical note, a 1024 dimensional embedding is computed using the `BAAI/bge-large-en-v1.5`[3] model and stored in a FAISS index (Johnson et al., 2019) for semantic search.

2.2 Our Baseline

While we experimented with various retrieval and prompting strategies within the RAG framework, our baseline follows a post-retrieval attribution approach. This involves prompting an LLM to generate answers based on both patient and clinical questions, along with **all** sentences of the clinical note excerpts from the case study. The decisions made for the baseline and other pipelines proposed in this work are supported by experiments that include

- a query that is constructed using both patient and clinical questions instead of considering only one of them (see Appendix D),
- a one-shot prompting approach instead of zero-shot prompting (see Appendix E),
- different LLMs for answer generation with attributions, which are `LLaMA-3.3-70B`[4] and `Mixtral-8x7B`[5] (Dada et al., 2025; Kweon et al., 2024), and
- a maximum number of 200 tokens generated by the LLM (see Appendix F).

On the other hand, the organizers' baseline used the LLaMA-3.3-70B model in a zero-shot prompting approach, where the model is prompted to generate answers that include attributions. If a response is invalid, e.g., exceeding the word limit or lacking valid attribution, the model is again prompted to generate an answer. This is repeated up to five times to obtain a valid output.

2.3 Submitted Pipeline: Surprise Ranked List Truncation (RLT) Retrieval Strategy

The pipeline we submitted for the shared task aligns with baselines utilizing a post-retrieval attribution approach. In this approach, for a query that combines patient and clinical question, semantically similar sentences from the excerpts of clinical notes are retrieved. The similarity score between the query and each sentence is computed using cosine similarity. During retrieval, k represents the number of highest-scoring (top-k) sentences similar to the query. Instead of using a fixed value for k, our team employed a query-dependent-k selection strategy based on the Ranked List Truncation (RLT) method, referred to as "surprise". This method determines the number k of sentences to consider by first adjusting retrieval scores using generalized Pareto distributions from extreme value theory (Pickands, 1975). It truncates a ranked list using a score threshold, allowing for a variable number of relevant sentences to be selected per query (Meng et al., 2024). The selected sentences and query are passed to the LLMs for answer generation, where the model generates an answer with attribution explicitly referencing retrieved sentences from a clinical note.

[1] `https://doi.org/10.13026/zzax-sy62` (accessed on 30th April 2025)

[2] All experiments described in Section 3 use the dev set.

[3] `https://huggingface.co/BAAI/bge-large-en-v1.5` (accessed on 4th May 2025)

[4] `https://huggingface.co/meta-llama/Llama-3.3-70B-Instruct` (accessed on 4th May 2025)

[5] `https://huggingface.co/mistralai/Mixtral-8x7B-v0.1` (accessed on 4th May 2025)

Table 1: Retrieval performance on the development set under strict (essential only) and lenient (essential + supplementary) variants. The Strategy and Variant columns list different retrieval strategies and their parameters. Columns P, R, and F1 quantify precision, recall, and F1-score under both variants. The seven best approaches by combined strict and lenient F1-scores (excluding the $k = 54$ row) are highlighted in **bold**.

Strategy	Variant	Strict			Lenient		
		P	R	F1	P	R	F1
fixed-k	$k = 3$	0.53	0.32	0.36	0.70	0.29	0.39
	k = 10	0.43	0.71	**0.50**	0.56	0.71	**0.58**
	k = 15	0.38	0.81	**0.49**	0.51	0.82	**0.60**
	k = 20	0.35	0.89	**0.49**	0.47	0.88	**0.59**
	$k = 54$	0.33	1.00	0.49	0.45	1.00	0.60
fixed-k + re-rankers	**FlashRank** ($k = 20, n = 10$)	0.38	0.68	**0.45**	0.51	0.67	**0.54**
	Cohere ($k = 20, n = 10$)	0.38	0.67	0.45	0.50	0.66	0.53
autocut	—	0.58	0.22	0.27	0.68	0.21	0.28
autocut*	—	0.59	0.35	**0.34**	0.74	0.32	**0.38**
surprise	—	0.36	0.64	**0.42**	0.48	0.62	**0.49**
elbow	—	0.48	0.66	**0.50**	0.62	0.63	**0.55**

2.4 Other Methods

In this section, we outline various methods within the RAG framework by varying its components, namely retrieval strategies and attribution approaches, to assess their impact on performance. We experimented with retrieval strategies other than surprise, including fixed-k, fixed-k and re-ranking, and query-dependent-k strategies like autocut, autocut*, and elbow.

The **Fixed-k** strategy applies a fixed cut-off for all query results, using common values of 3, 10, 15, 20, and 54. **Fixed-k and re-ranking** is a two-step retrieval that first retrieves semantically k similar candidates based on a fixed cut-off. A relevance score is assigned in the second step, selecting top-n (where $n \leq k$) sentences using re-rankers like flashrank (Damodaran, 2023) and cohere[6]. **Autocut**[7] limits candidate sentences based on discontinuities in the computed similarity scores. It determines the first divergence from a straight decline, excluding candidates beyond this point, although it may struggle with uniformly decreasing scores. In this work, we propose **autocut***, a new cut-off strategy that inspects how much each similarity score decreases compared to the previous score, automatically determining cut-offs based on significant changes without any manual adjustments. We also introduce the **elbow** strategy adapted from the elbow method in clustering to determine cut-offs by plotting similarity scores and locating the "elbow" where the transition from high to low relevance occurs, again with no need for preset parameters.

Along with different retrieval strategies, post-generation and post-retrieval attribution approaches have also been tried. In **post-generation attribution**, after a model generates an answer, those retrieved sentences are identified that support each answer sentence by measuring three similarity types: lexical (ROUGE (Lin, 2004), BLEU (Papineni et al., 2002), METEOR (Banerjee and Lavie, 2005)), fuzzy (character-based matching), and semantic (BERTScore (Zhang et al., 2020)). Each similarity is assigned a weight w_i, and a combined score is calculated. If this score exceeds a predefined threshold, the candidate sentence is attributed to the generated sentence. This ensures that every claim is explicitly grounded in some original clinical evidence. Detailed setups and results on weight and threshold settings are provided in Appendix G.

The **post-retrieval attribution** approach associates sentence identifiers with each retrieved sentence for attribution during answer generation. Post-processing steps are applied to generated answers to ensure that attributions are properly placed and no irrelevant attributions occur.

3 Experiments

All experiments were conducted on Google Colab[8] using a Tesla T4 GPU (12GB memory)[9]. For ac-

[6]https://docs.cohere.com/docs/rerank-overview (accessed on 4th May 2025)

[7]https://weaviate.io/developers/weaviate/api/graphql/additional-operators#autocut (accessed on 4th May 2025)

[8]https://colab.research.google.com/ (accessed on 4th May 2025)

[9]Code for the proposed pipeline is available online: https://github.com/achouhan93/heiDS-ArchEHR-QA-2025

Table 2: Pipeline evaluation on the development set under one-shot prompting, 200-token limit, and patient+clinician query. Metrics: strict Precision (P), Recall (R), F1-score (F1), overall relevance score in *Relevance* column, and overall pipeline score in *Overall* column. The performance of the organizer baseline, our baseline, top three proposed pipelines, and other experimented pipelines are listed here.

Retrieval	Attribution	Model	P	R	F1	Relevance	Overall
Organizer Baseline		LLaMA-3.3-70B	**0.63**	**0.33**	**0.43**	0.29	**0.36**
Our Baseline		LLaMA-3.3-70B	0.54	0.27	0.36	0.33	0.35
Top Three Proposed Pipelines							
surprise	Post-retrieval	LLaMA-3.3-70B	0.62	0.26	0.37	**0.35**	**0.36**
elbow	Post-retrieval	LLaMA-3.3-70B	0.59	0.27	0.37	0.32	0.35
fixed-k = 15	Post-retrieval	LLaMA-3.3-70B	0.59	0.25	0.35	0.34	0.35
Other Experimented Pipelines							
fixed-k = 10	Post-retrieval	LLaMA-3.3-70B	0.58	0.27	0.37	0.33	0.35
fixed-k = 10	Post-retrieval	Mixtral-8x7B	0.27	0.15	0.19	0.29	0.24
fixed-k = 15	Post-retrieval	Mixtral-8x7B	0.28	0.15	0.19	0.29	0.25
fixed-k = 20	Post-retrieval	LLaMA-3.3-70B	0.51	0.28	0.36	0.35	0.35
fixed-k = 20	Post-retrieval	Mixtral-8x7B	0.30	0.14	0.19	0.28	0.24
fixed-k = 20 + FlashRank	Post-retrieval	LLaMA-3.3-70B	0.52	0.22	0.31	0.34	0.33
fixed-k = 20 + FlashRank	Post-retrieval	Mixtral-8x7B	0.22	0.12	0.15	0.28	0.22
autocut*	Post-retrieval	LLaMA-3.37B	0.57	0.14	0.23	0.32	0.27
autocut*	Post-retrieval	Mixtral-8x7B	0.44	0.12	0.18	0.27	0.23
surprise	Post-retrieval	Mixtral-8x7B	0.33	0.17	0.22	0.29	0.26
elbow	Post-retrieval	Mixtral-8x7B	0.43	0.15	0.22	0.29	0.26
fixed-k = 54	Post-generation	LLaMA-3.3-70B	0.35	0.22	0.27	0.35	0.31

cessing LLMs, we used `InferenceClient`[10] from the `huggingface_hub` library.

3.1 Evaluation Criteria

The development set provided by the organizers includes clinical note excerpts annotated with sentence numbers for attribution. Furthermore, each sentence is labeled as "essential", "supplementary", or "not-relevant". Evaluation is carried out for two variants, a "strict" variant (considering only "essential" labels) and a "lenient" variant (considering both "essential" and "supplementary" labels). Retrieval performance is measured by precision, recall, and F1-score for each variant. The results are shown in Table 1. We selected fixed-k (10, 15, 20), autocut*, surprise, and elbow for downstream answer generation based on these metrics.

We used the official ArchEHR-QA evaluation script for the overall pipeline evaluation to assess *factuality* and *relevance*. *Factuality* is measured by the precision, recall, and F1-score of cited evidence versus ground-truth annotations computed under both variants. *Relevance* compares generated answer sentences to the ground-truth essential sentences using BLEU, ROUGE, SARI (Xu et al., 2016), BERTScore, AlignScore (Zha et al., 2023), and MEDCON (Yim et al., 2023). The overall relevance score is the average of these metrics, and the final pipeline score is the mean of overall factuality (strict variant F1-score) and overall relevance.

3.2 Comparative Pipeline Evaluation

Building on the ablations in Appendices D–F, we fixed the query (patient + clinician question), one-shot prompt, and 200-token limit, and evaluated our pipeline with two LLMs, `LLaMA-3.3-70B`[11] and `Mixtral-8x7B`[12], under both post-retrieval and post-generation attribution workflows. The results are shown in Table 2.

Post-Retrieval Attribution Evaluation. We paired each of our selected retrieval strategy (fixed-k = 10, 15, 20; autocut*; surprise; elbow) with each LLM and measured strict variant F1-score and overall relevance. Table 2 shows that `LLaMA-3.3-70B` combined with the surprise retrieval strategy achieves a strict F1-score of 0.37 and overall relevance of 0.35, making it our top

[10] `https://huggingface.co/docs/huggingface_hub/v0.30.2/en/package_reference/inference_client` (accessed on 4th May 2025)

[11] `https://huggingface.co/meta-llama/Llama-3.3-70B-Instruct` (accessed on 4th May 2025)

[12] `https://huggingface.co/mistralai/Mixtral-8x7B-v0.1` (accessed on 4th May 2025)

post-retrieval configuration, compared to the baselines.

Post-Generation Attribution Evaluation. Using a fixed-k of 54, we varied lexical/fuzzy/semantic weights and threshold values for the LLaMA-3.3-70B model. As shown in Table 5 in Appendix G, the optimal weighting ($w_1 = 0.0, w_2 = 0.5, w_3 = 0.5$, threshold = 0.5) yields a strict F1-score of 0.27 and overall relevance score of 0.35. Although this setup performs best among post-generation configurations, it underperforms relative to the best-performing post-retrieval configuration.

3.3 Pipeline Performance Analysis

While our best-performing pipeline based on the surprise retrieval strategy and post-retrieval attribution achieves a comparable overall score, it does not outperform the organizer's baseline. This outcome can be because of the following factors:

- **Prompt sensitivity of LLMs.** Salinas and Morstatter (2024) demonstrate that even a small perturbation in prompts can cause changes in an LLM's output. Although the organizer baseline and our best-performing pipeline use the same model (LLaMA-3.3-70B), the organizer baseline employs a zero-shot prompt, whereas our pipeline uses a one-shot prompt with stricter formatting and attribution instructions for the model to follow. These subtle prompt design choices could have influenced the model's ability to generate high quality answers with relevant attributions.

- **Difference in context size.** The development set contains up to 54 clinical note excerpt sentences per case study (see Figure 1b), allowing the organizer baseline to input all sentences to LLM as context, thus ensuring a high recall. In contrast, our pipeline relies on a query-dependent-k retrieval method to select a smaller subset of sentences. This approach naturally reduces recall, as some relevant content may not be retrieved, which thus negatively impacts the overall score.

Despite not outperforming the organizer baseline overall score, our pipeline design is motivated by practical considerations for real-world applications. While using all clinical note sentences is feasible within the shared task environment, real-world applications can contain far more text. We consider including complete texts as often infeasible due to LLMs input length constraints and degradation in model performance due to irrelevant information (Shi et al., 2023; Liu et al., 2024a). In such settings, a retrieval step is required, and determining a fixed k that is suitable for all cases is time-consuming. Query-dependent-k retrieval strategies remove the need for manual k tuning by determining the cut-off point based on score distributions. This allows the system to adapt to different types of queries.

4 Conclusion and Discussion

This work explored various RAG framework components for generating answers with attributions to clinical note excerpts. Our research highlights that the best-performing pipeline employs a post-retrieval attribution approach, utilizing the "surprise" RLT strategy and the LLaMA-3.3-70B model. We achieved a strict variant precision of 0.62 and recall of only 0.26, resulting in an F1-score of 0.37. While this indicates that the model's attributions are often correct, it frequently overlooks relevant evidence sentences. High selectivity can be beneficial when false attributions are costly, though it may omit important information. Additionally, query-dependent-k strategies like surprise, elbow, and autocut* methods for different types of queries in the dataset showed comparable performance to fixed-k approaches.

Limitations

Despite the moderate performance of our proposed pipeline, several limitations should be noted. In the current implementation, no text pre-processing is carried out for the clinical note excerpt sentences before indexing in FAISS. Expanding medical acronyms to their complete form or enriching texts with domain-specific interpretations before indexing could improve retrieval performance. Due to the use of prompting, even with a low temperature (0.001), there is non-determinism in the generated responses, making exact score replication challenging despite fixed pipeline configurations. Moreover, evaluating multiple large models increases computational requirements and associated expenses, which may limit practical deployment unless the model size or budget is adjusted.

References

Ayush Agrawal, Mirac Suzgun, Lester Mackey, and Adam Kalai. 2024. Do language models know when they're hallucinating references? In *Findings of the Association for Computational Linguistics: EACL 2024*, pages 912–928, St. Julian's, Malta. Association for Computational Linguistics.

Dara Bahri, Che Zheng, Yi Tay, Donald Metzler, and Andrew Tomkins. 2023. Surprise: Result list truncation via extreme value theory. In *Proceedings of the 46th International ACM SIGIR Conference on Research and Development in Information Retrieval*, SIGIR '23, page 2404–2408. Association for Computing Machinery.

Satanjeev Banerjee and Alon Lavie. 2005. METEOR: An automatic metric for MT evaluation with improved correlation with human judgments. In *Proceedings of the ACL Workshop on Intrinsic and Extrinsic Evaluation Measures for Machine Translation and/or Summarization*, pages 65–72, Ann Arbor, Michigan. Association for Computational Linguistics.

Benjamin Cohen-Wang, Harshay Shah, Kristian Georgiev, and Aleksander Mądry. 2024. ContextCite: Attributing Model Generation to Context. In *Advances in Neural Information Processing Systems*, volume 37, pages 95764–95807. Curran Associates, Inc.

Amin Dada, Osman Koras, Marie Bauer, Amanda Butler, Kaleb Smith, Jens Kleesiek, and Julian Friedrich. 2025. MeDiSumQA: Patient-oriented question-answer generation from discharge letters. In *Proceedings of the Second Workshop on Patient-Oriented Language Processing (CL4Health)*, pages 124–136, Albuquerque, New Mexico. Association for Computational Linguistics.

Prithiviraj Damodaran. 2023. FlashRank, Lightest and Fastest 2nd Stage Reranker for search pipelines.

Luyu Gao, Zhuyun Dai, Panupong Pasupat, Anthony Chen, Arun Tejasvi Chaganty, Yicheng Fan, Vincent Zhao, Ni Lao, Hongrae Lee, Da-Cheng Juan, and Kelvin Guu. 2023a. RARR: Researching and revising what language models say, using language models. In *Proceedings of the 61st Annual Meeting of the Association for Computational Linguistics (Volume 1: Long Papers)*, pages 16477–16508, Toronto, Canada. Association for Computational Linguistics.

Tianyu Gao, Howard Yen, Jiatong Yu, and Danqi Chen. 2023b. Enabling large language models to generate text with citations. In *Proceedings of the 2023 Conference on Empirical Methods in Natural Language Processing*, pages 6465–6488, Singapore. Association for Computational Linguistics.

Kristiina Häyrinen, Kaija Saranto, and Pirkko Nykänen. 2008. Definition, structure, content, use and impacts of electronic health records: a review of the research literature. *International journal of medical informatics*, 77(5):291–304.

Chengyu Huang, Zeqiu Wu, Yushi Hu, and Wenya Wang. 2024a. Training language models to generate text with citations via fine-grained rewards. In *Proceedings of the 62nd Annual Meeting of the Association for Computational Linguistics (Volume 1: Long Papers)*, pages 2926–2949, Bangkok, Thailand. Association for Computational Linguistics.

Yue Huang, Lichao Sun, Haoran Wang, Siyuan Wu, Qihui Zhang, Yuan Li, Chujie Gao, Yixin Huang, Wenhan Lyu, Yixuan Zhang, Xiner Li, Hanchi Sun, Zhengliang Liu, Yixin Liu, Yijue Wang, Zhikun Zhang, Bertie Vidgen, Bhavya Kailkhura, Caiming Xiong, and 52 others. 2024b. Position: TrustLLM: Trustworthiness in large language models. In *Proceedings of the 41st International Conference on Machine Learning*, volume 235 of *Proceedings of Machine Learning Research*, pages 20166–20270. PMLR.

Jeff Johnson, Matthijs Douze, and Hervé Jégou. 2019. Billion-scale similarity search with GPUs. *IEEE Transactions on Big Data*, 7(3):535–547.

Sunjun Kweon, Jiyoun Kim, Heeyoung Kwak, Dongchul Cha, Hangyul Yoon, Kwanghyun Kim, Jeewon Yang, Seunghyun Won, and Edward Choi. 2024. EHRNoteQA: An LLM Benchmark for Real-World Clinical Practice Using Discharge Summaries. In *Advances in Neural Information Processing Systems*, volume 37, pages 124575–124611. Curran Associates, Inc.

Dongfang Li, Zetian Sun, Xinshuo Hu, Zhenyu Liu, Ziyang Chen, Baotian Hu, Aiguo Wu, and Min Zhang. 2023. A Survey of Large Language Models Attribution. *arXiv preprint arXiv:2311.03731*.

Chin-Yew Lin. 2004. ROUGE: A package for automatic evaluation of summaries. In *Text Summarization Branches Out*, pages 74–81, Barcelona, Spain. Association for Computational Linguistics.

Nelson F. Liu, Kevin Lin, John Hewitt, Ashwin Paranjape, Michele Bevilacqua, Fabio Petroni, and Percy Liang. 2024a. Lost in the middle: How language models use long contexts. *Transactions of the Association for Computational Linguistics*, 12:157–173.

Siru Liu, Allison B McCoy, Aileen P Wright, Babatunde Carew, Julian Z Genkins, Sean S Huang, Josh F Peterson, Bryan Steitz, and Adam Wright. 2024b. Leveraging large language models for generating responses to patient messages—a subjective analysis. *Journal of the American Medical Informatics Association*, 31(6):1367–1379.

Chaitanya Malaviya, Subin Lee, Sihao Chen, Elizabeth Sieber, Mark Yatskar, and Dan Roth. 2024. ExpertQA: Expert-curated questions and attributed answers. In *Proceedings of the 2024 Conference of the North American Chapter of the Association for Computational Linguistics: Human Language Technologies (Volume 1: Long Papers)*, pages 3025–3045, Mexico City, Mexico. Association for Computational Linguistics.

Chuan Meng, Negar Arabzadeh, Arian Askari, Mohammad Aliannejadi, and Maarten de Rijke. 2024. Ranked List Truncation for Large Language Model-based Re-Ranking. In *Proceedings of the 47th International ACM SIGIR Conference on Research and Development in Information Retrieval*, SIGIR '24, page 141–151, New York, NY, USA. Association for Computing Machinery.

Jacob Menick, Maja Trebacz, Vladimir Mikulik, John Aslanides, Francis Song, Martin Chadwick, Mia Glaese, Susannah Young, Lucy Campbell-Gillingham, Geoffrey Irving, and 1 others. 2022. Teaching language models to support answers with verified quotes. *arXiv preprint arXiv:2203.11147*.

Reiichiro Nakano, Jacob Hilton, Suchir Balaji, Jeff Wu, Long Ouyang, Christina Kim, Christopher Hesse, Shantanu Jain, Vineet Kosaraju, William Saunders, and 1 others. 2021. WebGPT: Browser-assisted question-answering with human feedback. *arXiv preprint arXiv:2112.09332*.

Kishore Papineni, Salim Roukos, Todd Ward, and Wei-Jing Zhu. 2002. Bleu: a method for automatic evaluation of machine translation. In *Proceedings of the 40th Annual Meeting of the Association for Computational Linguistics*, pages 311–318, Philadelphia, Pennsylvania, USA. Association for Computational Linguistics.

Nilay Patel, Shivashankar Subramanian, Siddhant Garg, Pratyay Banerjee, and Amita Misra. 2024. Towards improved multi-source attribution for long-form answer generation. In *Proceedings of the 2024 Conference of the North American Chapter of the Association for Computational Linguistics: Human Language Technologies (Volume 1: Long Papers)*, pages 3906–3919, Mexico City, Mexico. Association for Computational Linguistics.

James Pickands. 1975. Statistical inference using extreme order statistics. *The Annals of Statistics*, 3(1):119–131.

Pritika Ramu, Koustava Goswami, Apoorv Saxena, and Balaji Vasan Srinivasan. 2024. Enhancing post-hoc attributions in long document comprehension via coarse grained answer decomposition. In *Proceedings of the 2024 Conference on Empirical Methods in Natural Language Processing*, pages 17790–17806, Miami, Florida, USA. Association for Computational Linguistics.

Furkan Şahinuç, Ilia Kuznetsov, Yufang Hou, and Iryna Gurevych. 2024. Systematic task exploration with LLMs: A study in citation text generation. In *Proceedings of the 62nd Annual Meeting of the Association for Computational Linguistics (Volume 1: Long Papers)*, pages 4832–4855, Bangkok, Thailand. Association for Computational Linguistics.

Abel Salinas and Fred Morstatter. 2024. The butterfly effect of altering prompts: How small changes and jailbreaks affect large language model performance. In *Findings of the Association for Computational Linguistics: ACL 2024*, pages 4629–4651, Bangkok, Thailand. Association for Computational Linguistics.

Freda Shi, Xinyun Chen, Kanishka Misra, Nathan Scales, David Dohan, Ed H. Chi, Nathanael Schärli, and Denny Zhou. 2023. Large language models can be easily distracted by irrelevant context. In *Proceedings of the 40th International Conference on Machine Learning*, volume 202 of *Proceedings of Machine Learning Research*, pages 31210–31227. PMLR.

William R Small, Batia Wiesenfeld, Beatrix Brandfield-Harvey, Zoe Jonassen, Soumik Mandal, Elizabeth R Stevens, Vincent J Major, Erin Lostraglio, Adam Szerencsy, Simon Jones, and 1 others. 2024. Large Language Model–Based Responses to Patients' In-Basket Messages. *JAMA network open*, 7(7):e2422399–e2422399.

Sarvesh Soni and Dina Demner-Fushman. 2025a. A Dataset for Addressing Patient's Information Needs related to Clinical Course of Hospitalization. *arXiv preprint*.

Sarvesh Soni and Dina Demner-Fushman. 2025b. Overview of the ArchEHR-QA 2025 Shared Task on Grounded Question Answering from Electronic Health Records. In *The 24th Workshop on Biomedical Natural Language Processing and BioNLP Shared Tasks*, Vienna, Austria. Association for Computational Linguistics.

Dietrich Trautmann, Natalia Ostapuk, Quentin Grail, Adrian Pol, Guglielmo Bonifazi, Shang Gao, and Martin Gajek. 2024. Measuring the groundedness of legal question-answering systems. In *Proceedings of the Natural Legal Language Processing Workshop 2024*, pages 176–186, Miami, FL, USA. Association for Computational Linguistics.

Wei Xu, Courtney Napoles, Ellie Pavlick, Quanze Chen, and Chris Callison-Burch. 2016. Optimizing statistical machine translation for text simplification. *Transactions of the Association for Computational Linguistics*, 4:401–415.

Wen-wai Yim, Yujuan Fu, Asma Ben Abacha, Neal Snider, Thomas Lin, and Meliha Yetisgen. 2023. Aci-bench: a Novel Ambient Clinical Intelligence Dataset for Benchmarking Automatic Visit Note Generation. *Scientific data*, 10(1):586.

Yuheng Zha, Yichi Yang, Ruichen Li, and Zhiting Hu. 2023. AlignScore: Evaluating factual consistency with a unified alignment function. In *Proceedings of the 61st Annual Meeting of the Association for Computational Linguistics (Volume 1: Long Papers)*, pages 11328–11348, Toronto, Canada. Association for Computational Linguistics.

Jiajie Zhang, Yushi Bai, Xin Lv, Wanjun Gu, Danqing Liu, Minhao Zou, Shulin Cao, Lei Hou, Yuxiao Dong, Ling Feng, and 1 others. 2024. LongCite: Enabling LLMs to Generate Fine-grained Citations in Long-context QA. *arXiv preprint arXiv:2409.02897*.

Tianyi Zhang, Varsha Kishore, Felix Wu, Kilian Q. Weinberger, and Yoav Artzi. 2020. BERTScore: Evaluating Text Generation with BERT. In *Proceedings of the Eighth International Conference on Learning Representations (ICLR'20)*. OpenReview.net.

A Example Case Study

Example Case: Patient and Clinician Questions with Clinical Note

Patient Question:
Took my 59 yo father to ER ultrasound discovered he had an aortic aneurysm. He had a salvage repair (tube graft). Long surgery / recovery for couple hours then removed packs. why did they do this surgery????? After this time he spent 1 month in hospital now sent home.

Clinician Question:
Why did they perform the emergency salvage repair on him?

Clinical Note: **1:** He was transferred to the hospital on 2025-1-20 for emergent repair of his ruptured thoracoabdominal aortic aneurysm. **2:** He was immediately taken to the operating room where he underwent an emergent salvage repair of ruptured thoracoabdominal aortic aneurysm with a 34-mm Dacron tube graft using deep hypothermic circulatory arrest. **3:** Please see operative note for details which included cardiac arrest x2. **4:** Postoperatively he was taken to the intensive care unit for monitoring with an open chest. **5:** He remained intubated and sedated on pressors and inotropes. **6:** On 2025-1-22, he returned to the operating room where he underwent exploration and chest closure. **7:** On 1-25 he returned to the OR for abdominal closure, JP drain placement, and feeding jejunostomy placed at that time for nutritional support. **8:** Thoracoabdominal wound healing well with exception of very small open area mid-wound that is approximately 1cm around and 0.5cm deep, with no surrounding erythema. **9:** Packed with dry gauze and covered with DSD.

B Dataset Statistics

The box plots representing the distribution of sentences in clinical notes for development (dev) and test sets (see Figure 1a and 1b) show that there is a varying number of sentences present in different

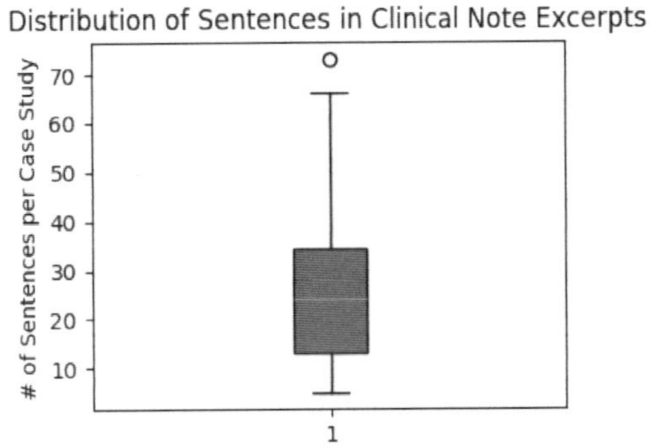

(a) Test set

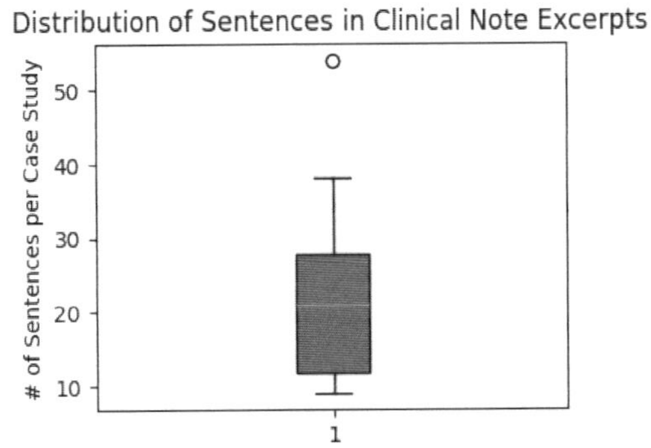

(b) Development set

Figure 1: Distribution of the number of sentences per clinical case in the test (a) and development (b) sets.

case studies with outliers (in the dev set case study, No. 8 is having 54 sentences, and in the test set case study, No. 73 is having 74 sentences).

Similarly, when the sentence length distributions are plotted for the dev set and the test set (see Figure 2a and Figure 2b), the mean of sentence length for both is nearly the same, around 15. However, in the test set, case studies have sentences that are double the length of sentences present in the dev set.

C Prompt Templates for Clinical Answer Generation

In this section, we provide the prompt templates used for answer generation. Zero-shot and one-shot prompts are designed for both post-retrieval and post-generation attribution. Post-retrieval attribution guides the language model to generate answers with inline citations, whereas post-generation attribution focuses only on answer generation, followed by a separate attribution step.

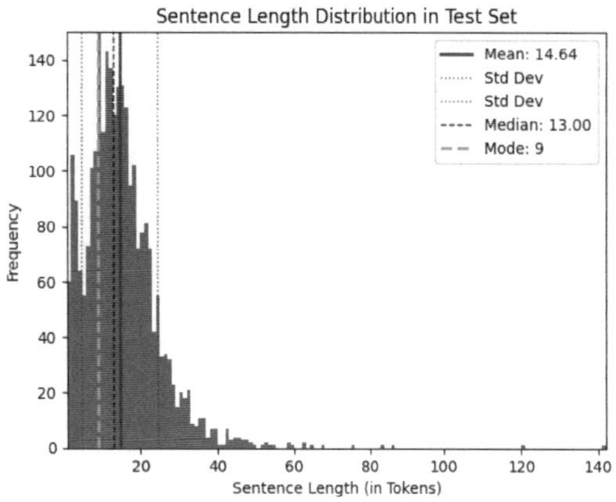

(a) Test set

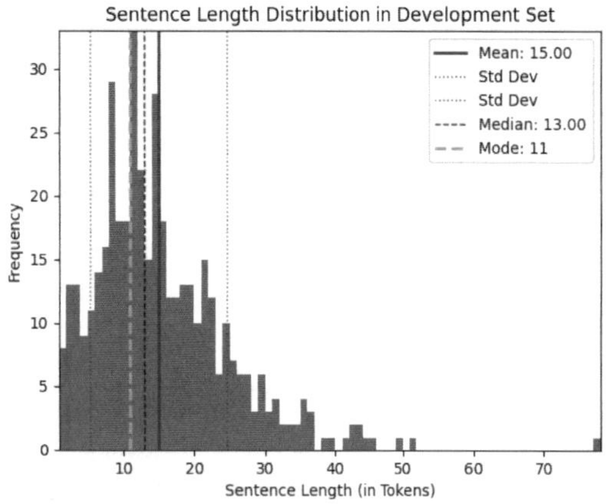

(b) Development set

Figure 2: Distribution of the sentence length in the test (a) and development (b) sets.

C.1 Prompt 1

Zero-Shot Prompting for Post-Retrieval Attribution Approach

You are a clinical response generation system responsible for producing answers to health-related questions using the provided clinical note excerpts. Your answer MUST be:
- **Accurate and Factual:** Grounded STRICTLY in the provided clinical note excerpts ONLY.
- **Neutral and Objective:** DO NOT INCLUDE PERSONAL OPINIONS, NOTES, IRRELEVANT, OR UNRELATED comments.
- **Concise and Relevant:** INCLUDE only clinically supported statements using the exact terminology found in the provided clinical notes. Do not add any additional interpretations or synonyms.
- **Third-Person Perspective:** Do not address the reader directly.
- **Citation:** Each statement must be supported by a NUMBERED CLINICAL NOTE SENTENCE from the Clinical Note Excerpts ONLY. The citation must be placed strictly AT THE END of the sentence. DO NOT insert citations within the sentence or phrase. When citing a single source, cite it as |id|. When a statement is supported by multiple sources, combine their IDs within a single pair of vertical bars (e.g., |id, id, id|) with IDs separated by commas and no extra vertical bars.
- **Mandatory Citation Inclusion:** AT LEAST ONE SENTENCE in your answer MUST include a citation from the provided clinical notes.

Inputs:
1. **Clinical Note Excerpts:** Retrieved sentences from the patient's clinical record, numbered.
2. **Patient Narrative Context:** Additional context from the patient's perspective.
3. **Clinician Question:** The primary question requiring an answer.

Your Task:
Generate a response based strictly on the provided input. Follow the structured format exactly, use only the exact terms from the clinical note excerpts, and ensure all citations are formatted consistently.

[Clinical Note Begin]
{note}
[Clinical Note End]

[Patient Narrative Context Begin]
{patient_narrative}
[Patient Narrative Context End]

[Clinician Question Begin]
{clinician_question}
[Clinician Question End]

Provide your structured answer below:

C.2 Prompt 2

One-Shot Prompting for Post-Retrieval Attribution Approach

You are a clinical response generation system responsible for producing answers to health-related questions ...
[... TRUNCATED FOR BREVITY ...]

Example:
If the clinician asks, "Why did they perform the emergency salvage repair on him?", and the note states:
1: He was transferred to the hospital on 2025-1-20 for emergent repair of his ruptured thoracoabdominal aortic aneurysm.
2: He was immediately taken to the operating room where he underwent an emergent salvage repair of ruptured thoracoabdominal aortic aneurysm with a 34-mm Dacron tube graft using deep hypothermic circulatory arrest.
Then the response should be:
His aortic aneurysm was caused by the rupture of a thoracoabdominal aortic aneurysm, which required emergent surgical intervention |1|. He underwent a complex salvage repair using a 34-mm Dacron tube graft and deep hypothermic circulatory arrest to address the rupture |2|.
[... TRUNCATED FOR BREVITY ...]

Provide your structured answer below:

C.3 Prompt 3

Zero-Shot Prompting for Post-Generation Attribution Approach

You are a clinical response generation system responsible for producing answers to health-related questions using the provided clinical note excerpts. Your answer MUST be:
- **Accurate and Factual:** Grounded STRICTLY in the provided clinical note excerpts ONLY.
- **Neutral and Objective:** DO NOT INCLUDE PERSONAL OPINIONS, NOTES, IRRELEVANT, OR UNRELATED comments.
- **Concise and Relevant:** INCLUDE only clinically supported statements using the exact terminology found in the provided clinical notes. Do not add any additional interpretations or synonyms.
- **Third-Person Perspective:** Do not address the reader directly."

Inputs:
1. **Clinical Note Excerpts:** Retrieved sentences from the patient's clinical record, numbered.
2. **Patient Narrative Context:** Additional context from the patient's perspective.
3. **Clinician Question:** The primary question requiring an answer.

Your Task:
Generate a response based strictly on the provided input. Follow the structured format exactly, use only the exact terms from the clinical note excerpts, and ensure all citations are formatted consistently.

[Clinical Note Begin]
{note}
[Clinical Note End]

[Patient Narrative Context Begin]
{patient_narrative}
[Patient Narrative Context End]

[Clinician Question Begin]
{clinician_question}
[Clinician Question End]

Provide your structured answer below:

C.4 Prompt 4

One-Shot Prompting for Post-Generation Attribution Approach

You are a clinical response generation system responsible for producing answers to health-related questions ...
[... TRUNCATED FOR BREVITY ...]

Example:
If the clinician asks, "Why did they perform the emergency salvage repair on him?", and the note states:
1: He was transferred to the hospital on 2025-1-20 for emergent repair of his ruptured thoracoabdominal aortic aneurysm.
2: He was immediately taken to the operating room where he underwent an emergent salvage repair of ruptured thoracoabdominal aortic aneurysm with a 34-mm Dacron tube graft using deep hypothermic circulatory arrest.
Then the response should be:
His aortic aneurysm was caused by the rupture of a thoracoabdominal aortic aneurysm, which required emergent surgical intervention. He underwent a complex salvage repair using a 34-mm Dacron tube graft and deep hypothermic circulatory arrest to address the rupture.
[... TRUNCATED FOR BREVITY ...]

Provide your structured answer below:

D Query Formulation Experiment

We compared three query formulation approaches. First, the patient's question is used; second, the clinician's question is used; third, both patient and clinician questions are considered. The setup for an experiment is similar to the baseline (see Section 2.2), i.e., **all** clinical notes excerpt sentences for each case study are considered and passed to LLaMA-3.3-70B (initially, the configuration is set to a maximum token generation of 100 tokens and zero-shot prompting). Table 3 shows the overall factuality (strict variant F1-score), relevance, and pipeline scores, demonstrating that combining patient and clinician questions yields the best performance.

Table 3: Query Formulation Results. All experiments use a fixed-$k = 54$, zero-shot prompting, post-retrieval attribution with `LLaMA-3.3-70B` model, and a maximum token limit of 100. Metrics: strict Precision (P), strict F1 (F1), overall Relevance (R), and overall pipeline score (O). The best variant is highlighted in **bold**.

Query	P	F1	R	O
Patient Question only	0.39	0.27	0.33	0.30
Clinician Question only	0.42	0.27	0.30	0.28
Patient + Clinician	**0.44**	**0.30**	**0.33**	**0.31**

E Prompting Approach Experiment

To assess the effect of the prompting approach, we compared zero-shot and one-shot prompting approaches considering the LLaMA-3.3-70B model and prompting with **all** note sentences, the query as a combination of patient and clinician questions (see Appendix D), and a maximum token generation limit of 100. LLMs generate answers based solely on the provided query and instructions in a zero-shot prompting approach, testing their inherent understanding without examples. See Appendices C.1 and C.3 for zero-shot prompts. In the one-shot prompting approach, an example of the desired output is provided alongside the query and instructions, helping the model align its response style. See Appendices C.2 and C.4 for one-shot prompts. Table 4 shows the overall factuality (strict variant F1-score), relevance, and pipeline score for each approach. The one-shot prompt yielded higher scores, leading us to select it for the baseline and methods.

Table 4: Prompting Approach Results. All experiments use fixed-$k = 54$, query (patient + clinical questions), post-retrieval attribution with `LLaMA-3.3-70B` model, and a maximum token limit of 100. Metrics: strict Precision (P), strict F1 (F1), overall Relevance (R), and overall pipeline score (O). The best variant is highlighted in **bold**.

Prompting Approach	P	F1	R	O
zero-shot prompting	0.44	0.30	0.33	0.31
one-shot prompting	**0.56**	**0.34**	**0.33**	**0.33**

F Maximum Token Generation Experiment

We experimented with the LLaMA-3.3-70B model having maximum token generation limits of 100,

Table 5: Parameter Settings. Experiments use fixed-$k = 54$, query (patient+clinician question), one-shot prompting, and post-generation attribution with LLaMA-3.3-70B model. Metrics: strict Precision (P), strict F1 (F1), overall Relevance (R), and overall pipeline score (O). Different combinations of weights and thresholds are arranged in descending order of performance, i.e., the best combination at the top.

w_1	w_2	w_3	T	P	F1	R	Overall
0.0	**0.5**	**0.5**	**0.5**	**0.35**	**0.27**	**0.35**	**0.311**
0.3	0.4	0.3	0.4	0.34	0.27	0.35	0.307
0.3	0.3	0.4	0.4	0.32	0.26	0.35	0.306
0.2	0.4	0.4	0.4	0.28	0.25	0.35	0.300
0.5	0.5	0.0	0.3	0.30	0.26	0.34	0.300

200, and 300 tokens[13] to determine their impact on the pipeline's overall performance. Table 6 shows that a maximum number of 200 tokens achieved the best balance of overall factuality (strict variant F1-score) and relevance scores. Consequently, we fixed the maximum number of tokens to 200 in all experiments.

Table 6: Maximum Token Generation. All experiments use fixed-$k = 54$, query (patient+clinician question), one-shot prompting, and post-retrieval attribution with LLaMA-3.3-70B model. Metrics: strict Precision (P), strict F1 (F1), overall Relevance (R), and overall pipeline score (O). The best variant is highlighted in **bold**.

Maximum Tokens	P	F1	R	O
100	0.56	0.34	0.33	0.33
200	**0.54**	**0.34**	**0.33**	**0.34**
300	0.51	0.30	0.33	0.32

G Post-Generation Attribution Parameter Experiment

Experiments began from the answers generated by LLaMA-3.3-70B with one-shot prompting and fixed-k of 54 as a retrieval strategy. We then performed a grid search over the three similarity weights (w_1, w_2, w_3) and the attribution threshold T to identify the combination that maximizes the overall pipeline score, i.e., achieving higher strict attribution F1-score without unduly sacrificing answer relevance. Here, w_1, w_2, and w_3 correspond to the weights assigned to lexical, fuzzy, and semantic similarity scores. Each weight was varied in $\{0.1, 0.2, \ldots, 1.0\}$ under the constraint $w_1 + w_2 + w_3 = 1$, and thresholds $T \in \{0.1, 0.2, \ldots, 0.9\}$ were tested. We observed that very low thresholds (0.1–0.2) led to over-attribution (nearly every answer sentence is attributed with every retrieved sentence), whereas very high thresholds (0.7–0.9) caused under-attribution (rarely answer sentences are attributed with retrieved sentences). Table 5 summarizes the top 10 configurations by strict F1-score. The best-performing setting was $\{w_1 = 0.0, w_2 = 0.5, w_3 = 0.5\}$ with $T = 0.5$, yielding a strict F1-score 0.27 and overall pipeline score 0.31.

[13] Approximately corresponding to the organizer's 75-word guideline.

UniBuc-SB at ArchEHR-QA 2025: A Resource-Constrained Pipeline for Relevance Classification and Grounded Answer Synthesis

Sebastian Balmuș[1,2], Bogdan Dura[1,2], and Ana-Sabina Uban[1,3]

[1]Faculty of Mathematics and Computer Science, University of Bucharest
[2]National Institute for Research and Development in Informatics - ICI Bucharest
[3]Human Language Technologies Research Center, University of Bucharest

Abstract

We describe the UniBuc-SB submission to the ArchEHR-QA shared task, which involved generating grounded answers to patient questions based on electronic health records. Our system exceeded the performance of the provided baseline, achieving a higher performance in generating contextually relevant responses. Notably, we developed our approach under constrained computational resources, utilizing only a single NVIDIA RTX 4090 GPU. We refrained from incorporating any external datasets, relying solely on the limited training data supplied by the organizers. To address the challenges posed by the low-resource setting, we leveraged off-the-shelf pre-trained language models and fine-tuned them minimally, aiming to maximize performance while minimizing overfitting.

1 Introduction

The ArchEHR-QA shared task (Soni and Demner-Fushman, 2025b) focuses on advancing automated question answering systems that can generate grounded responses using electronic health records (EHRs). With the increasing use of patient portals, clinicians are increasingly challenged by the volume of patient inquiries. Automating the response process aims to reduce this workload by providing quick and accurate answers to patients. The task provides realistic patient queries along with clinical notes, requiring the systems to generate answers based on the EHR excerpts provided. This setting not only tests the ability to handle limited data, but also emphasizes the need for accurate medical language understanding and effective information retrieval.

Developing effective question answering (QA) systems for the medical domain presents distinct challenges, particularly when working with limited data and computational resources. The development dataset (Soni and Demner-Fushman, 2025a; Johnson et al., 2023a,b) provided was relatively small, consisting of only 20 distinct medical cases, while the test dataset consisted of 100 medical cases. This data limitation increased the risk of overfitting and restricted the potential for extensive training. Additionally, the complexity of medical language requires systems to accurately interpret nuanced terminology and context. To address these challenges, we adopted a resource-efficient approach, using a single NVIDIA RTX 4090 GPU and adhering strictly to the provided dataset, without incorporating any external data. Our system leveraged pre-trained language models to compensate for the data limitations, applying minimal fine-tuning to adapt them to the medical QA task. This strategy aimed to balance computational efficiency with performance, allowing our system to effectively generate grounded answers despite the small dataset size. Our results demonstrate that even under these constraints, our approach exceeded the baseline, highlighting the effectiveness of strategic model selection and fine-tuning in low-resource settings.

The remainder of this paper is organized as follows. Section 2 discusses related work, focusing on prior approaches to medical question answering and low-resource NLP systems. Section 3 details our system architecture, including data preprocessing, model selection, and training procedures. Section 4 presents the results of our system compared to the baseline, accompanied by a thorough analysis of its performance. Finally, Section 5 concludes the paper by summarizing our findings, highlighting limitations, and suggesting directions for future work.

2 Related Work

Recent advancements in EHR question answering (QA) systems have focused on improving information retrieval accuracy while mitigating hallucinations and enhancing interpretability. Bardhan

BioNLP 2025 Shared Tasks, pages 62–68
August 1, 2025 ©2025 Association for Computational Linguistics

et al. (2023) provide a comprehensive review of EHR QA research, identifying the emrQA dataset as the primary resource and emphasizing the need for standardized evaluation metrics to facilitate consistent benchmarking.

In response to the need for robust evaluation frameworks, EHRNoteQA (Kweon et al., 2024) was introduced as a benchmark designed to assess Large Language Models (LLMs) on patient-specific questions derived from MIMIC-IV (Johnson et al., 2023a) discharge summaries. The dataset includes both open-ended and multiple-choice questions and has been used to systematically evaluate 27 LLMs, highlighting the variability in model performance across different question types.

Addressing the challenge of querying structured EHR data, quEHRy (Soni et al., 2023) employs natural language interfaces to translate clinician queries into structured database queries, facilitating more intuitive data access and emphasizing interpretability.

In the context of ensemble learning, Romero et al. (2025) demonstrate that leveraging multiple BERT-based encoders significantly improves medication-related named entity recognition (NER) across dosage, route, and strength attributes. This approach aligns with our system design, which employs multi-model architectures to capture complementary error patterns.

Finally, Sohn et al. (2024) introduce RAG2, a retrieval-augmented generation framework that prioritizes rationale-driven query formulation and evidence sampling to reduce hallucinations. Their findings underscore the value of multi-pass answer generation and rationale-centric retrieval, both of which inform our system's evidence-grounding strategy.

3 System Description

Our system is structured as a modular pipeline composed of three main components: preprocessing, relevance classification and answer generation, as shown in Figure 1. The pipeline is designed to process input data consisting of electronic health records (EHRs) and clinician question, transforming them into structured data for downstream processing. The preprocessing component structures the input data, which is then fed into the relevance classification module to identify relevant sentences. The identified sentences are subsequently processed in the answer generation module, which consists of three sequential steps: generation, grounding, and post-processing. The final output is a contextually grounded response tailored to the question.

3.1 Preprocessing

The preprocessing stage structures raw data into a format suitable for downstream tasks. Each medical case is divided into sentences labeled as essential, supplementary, or irrelevant based on their relevance to the clinician's query. The query is incorporated as contextual input for relevance classification. Each record includes a case ID, sentence ID, sentence text, query, and relevance label, ensuring consistency in data handling.

To prevent data leakage, the dataset is split at the case level, maintaining label distribution across training and testing sets. Relevance labels are then binarized, with essential and supplementary sentences labeled as 1 and irrelevant sentences as 0, simplifying the classification task.

3.2 Relevance Classification

The relevance classification component is responsible for identifying sentences within the input data that are relevant or not to the clinician's query. This step is critical in filtering out irrelevant content and ensuring that subsequent processing stages focus solely on clinically pertinent information.

To accomplish this, we employ an ensemble classifier composed of four pre-trained language models. Each model is fine-tuned for binary relevance classification, distinguishing between relevant and irrelevant content. The selected models include BERT[1] (Devlin et al., 2019), Bio_ClinicalBERT[2] (Alsentzer et al., 2019), BlueBERT[3] (Peng et al., 2019), and MedEmbed[4] (Balachandran, 2024). This combination allows us to leverage both general-domain language understanding through BERT and domain-specific medical knowledge through the clinical and biomedical models, ensuring that the classifier can effectively handle both general and specialized content within the EHR data.

[1]https://huggingface.co/google-bert/bert-base-uncased

[2]https://huggingface.co/emilyalsentzer/Bio_ClinicalBERT

[3]https://huggingface.co/bionlp/bluebert_pubmed_mimic_uncased_L-12_H-768_A-12

[4]https://huggingface.co/abhinand/MedEmbed-large-v0.1

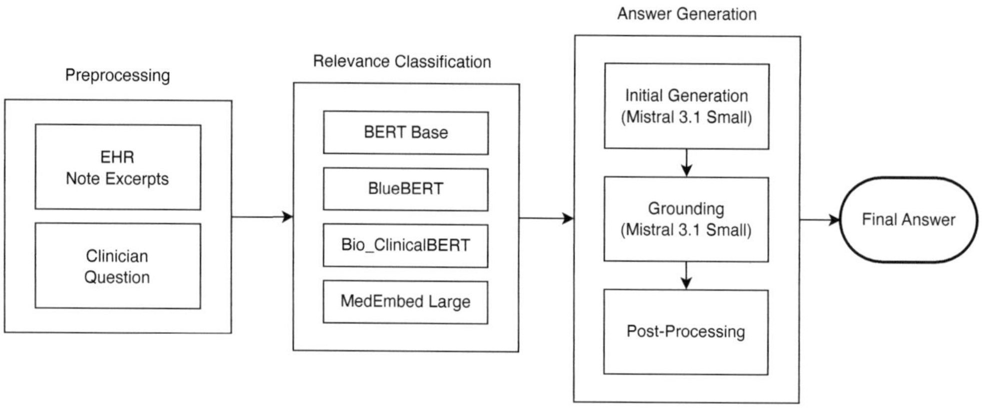

Figure 1: Our proposed system architecture, structured as a modular pipeline involving preprocessing, relevance classification, and answer generation. The entire system is designed to operate within a 24GB VRAM limit.

To address the class imbalance present in the dataset, we adopt the focal loss (Lin et al., 2020) as the objective function during training. Originally developed for dense object detection in computer vision, focal loss adjusts the contribution of each sample to the overall loss based on its classification difficulty. Specifically, it down-weights the contribution of well-classified samples and focuses more on harder-to-classify examples, mitigating the impact of the overrepresented non-relevant class in our dataset.

The training setup employs a learning rate of 5×10^{-5}, with a linear warm-up schedule comprising 10% of the total training steps. The training process is conducted over seven epochs, with a batch size of 16 samples per step. Model checkpoints are saved at each epoch, with the best model based on validation F1 score selected as the final model.

Upon completing individual model training, the predictions from each model are aggregated to form the ensemble output. For each input sentence, the relevance label is determined by majority voting, wherein the label receiving the highest number of votes across all models is selected as the final prediction. This ensemble strategy leverages the strengths of each model, reducing the impact of individual model biases and enhancing overall classification robustness.

The output of the relevance classification step serves as the input to the subsequent answer generation module. Only sentences that are classified as relevant to the clinician's query are retained.

3.3 Answer Generation

The generation stage begins by compiling the relevant sentences identified during the classification phase. Each sentence is formatted with its unique identifier and presented in a structured evidence list. This evidence list is then combined with the clinician's question to form a comprehensive input prompt for the generation model.

For response generation, we employ the Mistral Small 3.1 language model (Mistral AI, 2025), which is designed to handle large-scale language tasks with a compact yet powerful architecture. The model is loaded using the Ollama interface with the default parameters, which provides a seamless integration for inference and allows efficient model deployment without extensive modification of the original architecture. This integration facilitates the use of the model within the existing pipeline without exceeding the 24GB VRAM limit imposed by the RTX 4090 GPU, ensuring that the entire system remains computationally feasible.

The input prompt instructs the model to generate a concise response that addresses the clinician's query while adhering to a specified word limit. If the generated response exceeds the maximum word limit of 70 words, the generation step is repeated with a modified prompt that instructs the model to produce a more succinct version of the response. This iterative refinement process ensures that the output remains within acceptable length constraints without compromising informativeness.

Following response generation, the grounding step is employed to verify and reinforce the generated response by explicitly referencing relevant evidence from the input sentences. This step mitigates the risk of unsupported claims and enhances the factual accuracy of the output, aligning it with the context provided by the EHR data.

The final post-processing step involves correcting formatting inconsistencies, such as erroneous citations or incomplete sentences. Additionally, the post-processing script ensures that the output structure is consistent across cases, aligning with the required submission format. This step is crucial for maintaining the overall quality and coherence of the generated responses.

4 Evaluation

The evaluation is conducted on a test set of 100 medical cases, focusing on factuality and relevance. Factuality is assessed using Precision, Recall, and F1 scores by comparing generated evidence citations with the ground truth.

Factuality evaluation includes Strict and Lenient modes. Strict considers only 'essential' sentence citations, while Lenient also includes 'supplementary' sentences, allowing for more flexibility.

Relevance is evaluated using BLEU (Papineni et al., 2002), ROUGE (Lin, 2004), SARI (Xu et al., 2016), BERTScore (Zhang et al., 2020), AlignScore (Zha et al., 2023), and MEDCON (Yim et al., 2023), assessing both linguistic quality and clinical grounding.

The overall score is the average of the Strict Citation F1 score (Factuality) and a composite Relevance score, calculated by normalizing and averaging the individual metric scores.

Metric	Score	Micro	Macro
Overall Score	36.4	–	–
Factuality	44.7	–	–
Relevance	28.1	–	–
Strict Precision	–	58.7	63.6
Strict Recall	–	36.1	42.7
Strict F1	–	44.7	46.4
Lenient Precision	–	61.7	68.5
Lenient Recall	–	35.9	41.4
Lenient F1	–	45.4	47.8
BLEU	0.6	–	–
ROUGE-L	19.9	–	–
SARI	49.0	–	–
BERTScore	23.9	–	–
AlignScore	43.0	–	–
MEDCON (UMLS)	32.4	–	–

Table 1: Official evaluation results on the test dataset across overall performance, citation-based, and text generation metrics.

4.1 Evaluation Results

Table 1 presents the evaluation results of our system across various metrics, focusing on both factuality and relevance. While the system achieves consistent scores in citation-based evaluation, with strict F1 scores of 44.7 (micro) and 46.4 (macro), it underperforms in text generation metrics, particularly in BLEU (0.6) and BERTScore (23.9). This suggests that while the model effectively identifies relevant evidence, further refinement is required to enhance the fluency and linguistic alignment of generated responses.

4.2 Ablation Study

Table 2 reports the lenient F1 scores across the four test cases for each individual model, all pairwise and three-way combinations, and the complete four model ensemble. Among the single models, Bio_ClinicalBERT performs best, which is consistent with its clinical-domain pretraining. However, the standard BERT model—despite lacking biomedical specialization—proves surprisingly effective, particularly in combination with other models. In fact, BERT appears to play a stabilizing role in most ensemble variants. Its inclusion consistently improves performance, often more than one might expect given its standalone score. This suggests that general-domain representations may provide complementary context cues that specialized models overlook especially when clinical language overlaps with common phrasing. Performance improves steadily as models are added, with all three-model combinations outperforming any two-model setup. Interestingly, the top three-model combination excludes BERT, but only slightly edges out the BERT-inclusive variants›. Ultimately, the full ensemble outperforms all others, confirming that diversity in model training is meaningful to relevance prediction.

4.3 Resource Usage

The entire pipeline—including preprocessing, relevance classification, and answer generation—runs comfortably within the 24GB VRAM limit of a single NVIDIA RTX 4090 GPU. During inference, the relevance classification stage takes approximately 1 second per case on average. The answer generation stage, which uses the Mistral Small 3.1 model

Variant	F1
BERT	0.524
Bio_ClinicalBERT	0.544
BlueBERT	0.515
MedEmbed	0.507
BERT + Bio_ClinicalBERT	0.579
BERT + BlueBERT	0.563
BERT + MedEmbed	0.522
Bio_ClinicalBERT + BlueBERT	0.532
Bio_ClinicalBERT + MedEmbed	0.552
BlueBERT + MedEmbed	0.546
BERT + Bio_ClinicalBERT + BlueBERT	0.563
BERT + Bio_ClinicalBERT + MedEmbed	0.602
BERT + BlueBERT + MedEmbed	0.602
Bio_ClinicalBERT + BlueBERT + MedEmbed	0.603
Full ensemble (all 4)	**0.619**

Table 2: Ablation study: F1 scores for each single model, model combination, and the full ensemble.

via the Ollama interface, averages 15 seconds per case. Post-processing, which involves formatting corrections and citation verification, adds an additional 0.001 seconds per case on average and is performed entirely on CPU.

Altogether, the full inference pipeline processes each case in about 16 seconds end-to-end. These performance characteristics confirm the system's suitability for real-time or near-real-time deployment in clinical or low-latency environments. Additionally, the total cost for running inference over the full test set is negligible when using standard compute infrastructure, making the approach both scalable and accessible.

4.4 Error Analysis

The model exhibits false positives in sentences with clinical terms or medication instructions that are not directly relevant to the query, such as "You were started on a milrinone drip, with improvement in your heart's pump function". This suggests over-reliance on clinical terminology rather than contextual alignment. Conversely, false negatives often involve broader prognostic statements or mental health assessments, where relevance is implied across multiple sentences. This indicates a need for improved contextual understanding to handle less explicit but clinically relevant content.

5 Conclusions

This paper presents a modular pipeline for relevance classification and grounded answer generation in the ArchEHR-QA shared task, operating under constrained computational resources. The use of pre-trained models with minimal fine-tuning proved effective in leveraging both general-domain and medical-specific knowledge, resulting in consistent citation-based evaluation scores. However, lower scores in BLEU and BERTScore indicate that further refinement is necessary to improve the fluency and linguistic alignment of generated responses. Future work will explore methods for enhancing response generation, including advanced grounding techniques and multi-sentence contextual modeling.

Limitations

The reliance on a single RTX 4090 GPU constrained the computational capacity available for training and fine-tuning, limiting the scope of model experimentation and hyperparameter optimization. Additionally, the development dataset consisted of only 20 cases, restricting the diversity of clinical scenarios encountered during training and potentially impacting the system's ability to generalize effectively.

Acknowledgements

The authors acknowledge the support of project PN 23 38 01 01, "Contributions to the consolidation of emerging technologies specific to the Internet of Things and complex systems," which provided technical resources essential to this research. Ana Uban was supported by a grant of the Ministry of Research, Innovation and Digitization, CNCS - UEFISCDI, project SIROLA, number PN-IV-P1-PCE-2023-1701, within PNCDI IV, and by CCCDI - UEFISCDI, project number PN-IV-P7-7.1-PTE-2024-0046, within PNCDI IV.

References

Emily Alsentzer, John Murphy, William Boag, Wei-Hung Weng, Di Jindi, Tristan Naumann, and Matthew McDermott. 2019. Publicly available clinical BERT embeddings. In *Proceedings of the 2nd Clinical Natural Language Processing Workshop*, pages 72–78, Minneapolis, Minnesota, USA. Association for Computational Linguistics.

Abhinand Balachandran. 2024. Medembed: Medical-focused embedding models.

Jayetri Bardhan, Kirk Roberts, and Daisy Zhe Wang. 2023. Question answering for electronic health records: A scoping review of datasets and models.

Jacob Devlin, Ming-Wei Chang, Kenton Lee, and Kristina Toutanova. 2019. BERT: Pre-training of deep bidirectional transformers for language understanding. In *Proceedings of the 2019 Conference of the North American Chapter of the Association for Computational Linguistics: Human Language Technologies, Volume 1 (Long and Short Papers)*, pages 4171–4186, Minneapolis, Minnesota. Association for Computational Linguistics.

Alistair Johnson, Tom Pollard, Steven Horng, Leo Anthony Celi, and Roger Mark. 2023a. MIMIC-IV-Note: Deidentified free-text clinical notes.

Alistair Johnson, Tom Pollard, and Roger Mark. 2023b. MIMIC-III clinical database.

Sunjun Kweon, Jiyoun Kim, Heeyoung Kwak, Dongchul Cha, Hangyul Yoon, Kwanghyun Kim, Jeewon Yang, Seunghyun Won, and Edward Choi. 2024. Ehrnoteqa: An llm benchmark for real-world clinical practice using discharge summaries.

Chin-Yew Lin. 2004. ROUGE: A package for automatic evaluation of summaries. In *Text Summarization Branches Out*, pages 74–81, Barcelona, Spain. Association for Computational Linguistics.

Tsung-Yi Lin, Priya Goyal, Ross Girshick, Kaiming He, and Piotr Dollár. 2020. Focal loss for dense object detection. *IEEE Transactions on Pattern Analysis and Machine Intelligence*, 42(2):318–327.

Mistral AI. 2025. Mistral small 3.1. `https://mistral.ai/news/mistral-small-3-1`.

Kishore Papineni, Salim Roukos, Todd Ward, and Wei-Jing Zhu. 2002. Bleu: a method for automatic evaluation of machine translation. In *Proceedings of the 40th Annual Meeting of the Association for Computational Linguistics*, pages 311–318, Philadelphia, Pennsylvania, USA. Association for Computational Linguistics.

Yifan Peng, Shankai Yan, and Zhiyong Lu. 2019. Transfer learning in biomedical natural language processing: An evaluation of bert and elmo on ten benchmarking datasets. In *Proceedings of the 2019 Workshop on Biomedical Natural Language Processing (BioNLP 2019)*, pages 58–65.

Pablo Romero, Lifeng Han, and Goran Nenadic. 2025. Medication extraction and entity linking using stacked and voted ensembles on LLMs. In *Proceedings of the Second Workshop on Patient-Oriented Language Processing (CL4Health)*, pages 303–315, Albuquerque, New Mexico. Association for Computational Linguistics.

Jiwoong Sohn, Yein Park, Chanwoong Yoon, Sihyeon Park, Hyeon Hwang, Mujeen Sung, Hyunjae Kim, and Jaewoo Kang. 2024. Rationale-guided retrieval augmented generation for medical question answering. *Preprint*, arXiv:2411.00300.

Sarvesh Soni, Surabhi Datta, and Kirk Roberts. 2023. Quehry: A question answering system to query electronic health records. *Journal of the American Medical Informatics Association*, 30(6):1091–1102.

Sarvesh Soni and Dina Demner-Fushman. 2025a. A dataset for addressing patient's information needs related to clinical course of hospitalization. *arXiv preprint*.

Sarvesh Soni and Dina Demner-Fushman. 2025b. Overview of the archehr-qa 2025 shared task on grounded question answering from electronic health records. In *The 24th Workshop on Biomedical Natural Language Processing and BioNLP Shared Tasks*, Vienna, Austria. Association for Computational Linguistics.

Wei Xu, Courtney Napoles, Ellie Pavlick, Quanze Chen, and Chris Callison-Burch. 2016. Optimizing statistical machine translation for text simplification. *Transactions of the Association for Computational Linguistics*, 4:401–415.

Wen-wai Yim, Yujuan Fu, Asma Ben Abacha, Neal Snider, Thomas Lin, and Meliha Yetisgen. 2023. Acibench: a novel ambient clinical intelligence dataset for benchmarking automatic visit note generation. *Scientific Data*, 10(1):586.

Yuheng Zha, Yichi Yang, Ruichen Li, and Zhiting Hu. 2023. AlignScore: Evaluating factual consistency with a unified alignment function. In *Proceedings of the 61st Annual Meeting of the Association for Computational Linguistics (Volume 1: Long Papers)*, pages 11328–11348, Toronto, Canada. Association for Computational Linguistics.

Tianyi Zhang, Varsha Kishore, Felix Wu, Kilian Q. Weinberger, and Yoav Artzi. 2020. Bertscore: Evaluating text generation with BERT. In *8th International Conference on Learning Representations, ICLR 2020, Addis Ababa, Ethiopia, April 26-30, 2020*. OpenReview.net.

Appendix A: Prompt Templates

Generation Prompt

You are a professional clinical assistant.

Using only the provided evidence, write a concise, clinical answer to the clinician's question. Do not include any citations. The answer must be no more than 75 words.

For example: **Clinician Question:** Why did they perform the emergency salvage repair on him?

Evidence: - |1| He was transferred to the hospital on 2025-1-20 for emergent repair of his ruptured thoracoabdominal aortic aneurysm. - |2| He was immediately taken to the operating room where he underwent an emergent salvage repair of ruptured thoracoabdominal aortic aneurysm with a 34-mm Dacron tube graft using deep hypothermic circulatory arrest. - |8| Thoracoabdominal wound healing well with exception of very small open area mid wound that is @1cm around and 1/2cm deep, no surrounding erythema.

Answer: His aortic aneurysm was caused by the rupture of a thoracoabdominal aortic aneurysm, which required emergent surgical intervention |1|. He underwent a complex salvage repair using a 34-mm Dacron tube graft and deep hypothermic circulatory arrest to address the rupture |2|. The extended recovery time and hospital stay were necessary due to the severity of the rupture and the complexity of the surgery, though his wound is now healing well with only a small open area noted |8|.

Another example: **Clinician Question:** Why was ERCP recommended over a medication-based treatment for CBD sludge?

Evidence: - |1| During the ERCP a pancreatic stent was required to facilitate access to the biliary system (removed at the end of the procedure), and a common bile duct stent was placed to allow drainage of the biliary obstruction caused by stones and sludge. - |5| On hospital day 4 (post-procedure day 3) the patient returned to ERCP for re-evaluation of her biliary stent as her LFTs and bilirubin continued an upward trend. - |6| On ERCP the previous biliary stent was noted to be acutely obstructed by biliary sludge and stones. - |7| As the patient's INR was normalized to 1.2, a sphincterotomy was safely performed, with removal of several biliary stones in addition to the common bile duct stent.

Answer: Medications can sometimes help in managing bile duct sludge, but in this case, ERCP was necessary due to the severity of the obstruction and its complications. The initial ERCP revealed significant biliary obstruction caused by sludge and stones, requiring the placement of a stent to restore bile drainage |1|. However, even after this intervention, the liver function tests and bilirubin levels continued to rise, indicating that the obstruction was not fully resolved |5|. A follow-up ERCP confirmed that the stent itself had become acutely obstructed by sludge and stones, necessitating further intervention |6|. During this procedure, a sphincterotomy was performed, and several stones were physically removed, which medications alone could not have achieved |7|. These findings confirm that ERCP was essential in addressing his condition and preventing further complications.

Now, using the provided evidence, write a concise, clinical answer to the clinician's question. Do not include any citations (this will be done in a future step). The answer must be no more than 75 words.

Question: {*clinician_question*}

Evidence: {*evidence_text*}

Retry Prompt for Length Enforcement

The last answer was too long ({*word_count*} words). Please try again, but make sure the answer is no more than {*MAX_WORDS*} words.

Last answer: {*last_answer*}

Grounding Prompt

You are a clinical documentation assistant.

You will be given: - A clinical question - A professional clinical answer to the question - A list of evidence sentences with sentence IDs in the format |sentence_id|

Your task is to add appropriate citations to the clinical answer. For each statement in the answer, cite the sentence(s) from the evidence that support it using the format |sentence_id|. If multiple sentences support a statement, separate them with commas, e.g., |3,4,7|. Do not use ranges like |1-3|. Cite only at the end of the sentence (after the period), and always add a newline after the citation. Do not add a newline after the final sentence. Do not change the wording of the answer. Simply append the appropriate citation(s).

Clinician Question: {*clinician_question*}

Clinical Answer: {*answer*}

Evidence: {*evidence_text*}

KR Labs at ArchEHR-QA 2025: A Verbatim Approach for Evidence-Based Question Answering

Ádám Kovács[1], Paul Schmitt[2], Gábor Recski[1,2]

[1]KR Labs
lastname@krlabs.eu

[2]TU Wien
firstname.lastname@tuwien.ac.at

Abstract

We present a lightweight, domain-agnostic *verbatim* pipeline for evidence-grounded question answering. Our pipeline operates in two steps: first, a sentence-level extractor flags relevant note sentences using either zero-shot LLM prompts or supervised ModernBERT classifiers. Next, an LLM drafts a question-specific template, which is filled verbatim with sentences from the extraction step. This prevents hallucinations and ensures traceability. In the ArchEHR-QA 2025 shared task, our system scored 42.01%, ranking top-10 in core metrics and outperforming the organiser's 70B-parameter Llama-3.3 baseline. We publicly release our code and inference scripts under an MIT license.

1 Introduction

Modern question-answering (QA) and retrieval-augmented generation (RAG) systems play a vital role in many high-stakes domains for information extraction and generation tasks. In medicine, a typical use case involves clinicians asking questions based on a patient's electronic health record (EHR) notes, rather than manually sifting through lengthy notes, which can be time-consuming. However, in practice, RAG and QA pipelines often misalign evidence and produce incorrect information, commonly referred to as hallucinations (Ji et al., 2023; Madsen et al., 2024). We argue that a reliable QA system should guarantee complete traceability of answers. To tackle this problem, we propose a *verbatim* pipeline that clearly separates extraction and generation to mitigate hallucinations:

- **Sentence-level extraction**, using either zero-shot LLMs or supervised ModernBERT classifiers.
- **Template-constrained generation**, dynamically creating answer templates filled exclusively with verbatim sentences selected from the extraction phase.

We participated in the ArchEHR-QA 2025 shared task on grounded question answering (QA) from electronic health records (EHRs). Our approach involved (i) utilizing a zero-shot `gemma-3-27b-it`[1] LLM (Team et al., 2025) and (ii) generating synthetic data for sentence extraction from EHRs to train a compact extractor. For this purpose, we trained a Clinical ModernBERT classifier (Lee et al., 2025; Warner et al., 2024), achieving performance comparable to the LLM extractor. Both extractors were then fed into the same LLM template generator. Our solution achieved an overall score of **42.01%**, ranking in the **top 10** for core metrics, and surpassed the organizers' 70B-parameter Llama-3.3 baseline by a large margin.

Our contributions include a modular, traceable QA architecture that mitigates hallucinations, a method to generate synthetic EHR question-answer corpus and train custom models. Additionally, we are releasing all the code on GitHub[2] under the MIT License. The remainder of the paper discusses background (Section 2), method (Section 3), and evaluation (Section 4).

2 Background

2.1 Dataset

Early clinical QA datasets such as emrQA (Pampari et al., 2018) and CliCR (Šuster and Daelemans, 2018) used fill-in-the-blank methods and lacked explicit sentence-level evidence. ArchEHR-QA (Soni and Demner-Fushman, 2025b,a) addresses this by pairing clinician-authored questions with de-identified MIMIC-III (Johnson et al., 2016) notes, annotated at the sentence-level as *essential*, *supplementary*, or *irrelevant*. Answers must be concise

[1] https://huggingface.co/google/gemma-3-27b-it

[2] https://github.com/KRLabsOrg/verbatim-rag/tree/archehr

BioNLP 2025 Shared Tasks, pages 69–74
August 1, 2025 ©2025 Association for Computational Linguistics

(under 75 words) and explicitly cite relevant sentences.

2.2 Limitations of Standard RAG

Standard RAG models, despite external grounding, still frequently hallucinate unsupported or contradictory information (Ji et al., 2023). Existing approaches like post-hoc verification (Friel and Sanyal, 2023; Manakul et al., 2023) or classifiers trained on hallucination corpora such as RAGTruth (Niu et al., 2024) (e.g., RAG-HAT (Song et al., 2024), LettuceDetect (Ádám Kovács and Recski, 2025)) add extra complexity and latency. Post-hoc saliency methods (Serrano and Smith, 2019; Jain and Wallace, 2019) and LLM self-explanations (Madsen et al., 2024) have also been found unreliable. Our approach proactively prevents hallucinations through strict template-driven sentence extraction and verbatim insertion.

2.3 Synthetic Training Data

Due to limited access and annotation restrictions, obtaining sentence-level labeled clinical datasets is challenging. Recent works address this by generating synthetic data via perturbation or LLM prompting (Niu et al., 2024; Lozano et al., 2023; Frayling et al., 2024; Bai et al., 2024). We follow this approach, generating synthetic EHR snippets, clinician-style questions, and sentence relevance annotations (details in Section 3.3).

3 Method

3.1 System Overview

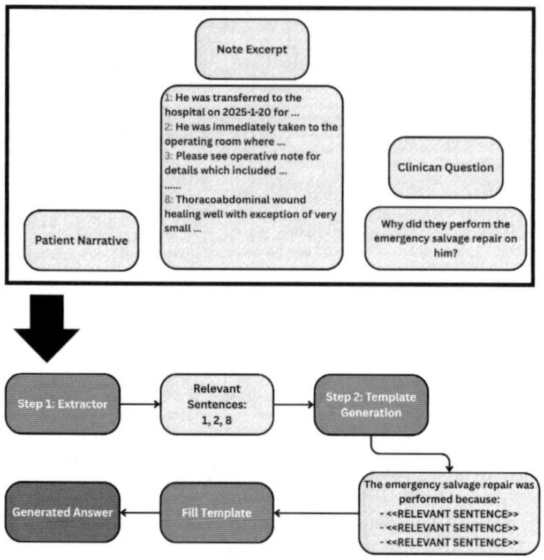

Figure 1: System overview. The pipeline first selects relevant sentences and then generates a question-specific answer using a dynamic template.

Figure 1 depicts our system architecture. First, an extraction step identifies relevant sentences from the input (patient narrative, clinician question, and note excerpt). We implemented both zero-shot and supervised models. Second, the generation step uses `gemma-3-27b-it` to dynamically draft an answer template, filled verbatim with extracted sentences. If exceeding 75 words, answers are compressed via a summarization prompt, preserving sentence-level citations.

3.2 Evidence Extraction

We evaluated two extractors: (i) We prompted `gemma-3-27b-it` to explicitly label sentences as relevant via a step-by-step process. (ii) We fine-tuned a Clinical ModernBERT classifier (Lee et al., 2025), trained on our synthetic data (Section 3.3). It independently evaluates each sentence in context (question + patient narrative). Lee et al. (2025) is a variant of ModernBERT (Warner et al., 2024) adapted specifically for biomedical and clinical text. Clinical ModernBERT supports extended input sequences (up to 8,192 tokens) and includes domain-specific vocabulary enhancements, making it particularly suitable for handling long clinical narratives. To provide additional context during classification, we included one sentence before and after the target sentence, forming a passage of up to three sentences. We chose a window size of one sentence before and after the target based on preliminary experimentation. The target sentence was explicitly marked with `[START]` and `[END]` tokens. The full input was structured using the standard BERT classification format. During fine-tuning, we merged essential and supplementary labels into a single positive class. We addressed class imbalance using weighted binary cross-entropy loss. We trained for 3 epochs (batch size 32, learning rate 2e-5), with gradient clipping and early stopping based on F1 score.

3.3 Synthetic Data Generation

Due to the scarcity of publicly available annotated data for sentence-level relevance classification, we constructed a synthetic dataset tailored specifically to the ArchEHR-QA task. Although the official development set contains labeled sentences, it is limited to 428 sentences across only 20 question–note pairs. Initial experiments using external resources like RAGBench (Friel et al., 2025) and PubMedQA-derived corpora (Jin et al., 2019) showed poor transfer performance, emphasizing the need for task-specific synthetic data.

We generated synthetic data via few-shot prompting with `gemma-3-27b-it`. Each prompt provided dynamic examples from the development set to ensure diversity. The LLM generated synthetic instances comprising de-identified clinical note excerpts, patient narratives, clinician-authored questions, and binary relevance labels. This approach yielded **3915** synthetic notes. We varied the few-shot examples across multiple runs, as static prompting resulted in repetitive outputs. This variation greatly increased lexical and semantic diversity, aligning with other work in synthetic data generation (Li et al., 2023; Tang et al., 2023; Xu et al., 2024). Ultimately, selecting each sentence with their relevance from the note excerpts, we constructed a comprehensive dataset of **58k** synthetic training examples, each labeled at the sentence level, which formed the training set for our Clinical ModernBERT classifier. Table 1 shows an illustrative training instance.

QUESTION	**Patient narrative**: My husband, a 72-year-old with a history of COPD, was admitted for worsening shortness of breath. He's been on home oxygen for years, but it wasn't helping this time. He also developed some swelling in his ankles. He seems a little confused today... **Clinician question**: What is the likely cause of the patient's ankle edema and what was done to address it?
SENTENCE	A diuretic, furosemide 40mg PO daily, was initiated to address the lower extremity edema, which was attributed to both underlying heart failure and fluid retention secondary to COPD exacerbation. **[START]** Echocardiogram revealed mild left ventricular dysfunction with an estimated ejection fraction of 45%. **[END]** Renal function was monitored closely, and remained stable throughout hospitalization.
LABEL	RELEVANT

Table 1: An example model input for our training.

3.4 Answer Generation

The answer generation module dynamically creates a template using the LLM (`gemma-3-27b-it`) based on the clinician's question, the selected evidence sentences, and the clinical note context. After the template generation step, we directly insert the extracted evidence sentences verbatim into the generated template, referencing sentence IDs explicitly. An example filled template generated by our pipeline is shown in Figure 2.

```
The emergency salvage repair was performed due to:
- He was transferred to the hospital on 2025-01-20 for emergent
repair of his ruptured thoracoabdominal aortic aneurysm. |1|
- He was immediately taken to the operating room where he underwent an
emergent salvage repair of ruptured thoracoabdominal aortic aneurysm with
a 34-mm Dacron tube graft using deep hypothermic circulatory arrest. |2|
- Thoracoabdominal wound healing well with exception of very small open
area mid-wound that is ~1 cm around and 0.5 cm deep, no surrounding
erythema. |8|
```

Figure 2: Example answer generated by our *verbatim* method, inserting evidence sentences verbatim into a dynamically generated template.

```
He was transferred to the hospital on 2025-01-20 for emergent repair of
      his ruptured thoracoabdominal aortic aneurysm |1|. He underwent an
       emergent salvage repair with a 34-mm Dacron tube graft using deep
       hypothermic circulatory arrest |2|. See also: |8|
```

Figure 3: Concise answer produced by our summarization step to comply with the 75-word limit.

If the filled answer exceeds the 75-word constraint of the task, we use an additional summarization prompt to rewrite the answer more concisely, ensuring all selected evidence remains cited and intact. An example summarization of the answer from Figure 2 is illustrated in Figure 3.

4 Evaluation

We evaluated our pipeline in the ArchEHR-QA 2025 shared task (Soni and Demner-Fushman, 2025b) using official metrics that emphasize two main aspects. Factuality is measuring alignment of the cited evidence with manually annotated sentences. Citation-level F1 scores are computed under strict (essential sentences only) and lenient (essential and supplementary sentences) conditions. Relevance is evaluating how closely generated answers match ground-truth answers through BLEU (Papineni et al., 2002), ROUGE (Lin, 2004), BERTScore (Zhang et al., 2020), and MEDCON (Yim et al., 2023) metrics.

Our best system (zero-shot LLM based on `gemma-3-27b-it`) scored **42.01%** overall, placing within the top 10 in multiple core metrics and significantly outperforming the organizers' baseline (a 70B-parameter Llama-3.3 model) across most metrics. Table 2 summarizes these metrics. A strong point of our system is factuality recall (56.8% strict, 56.6% lenient), approximately 5 points above the leaderboard average. This indicates strong capability for reliably retrieving relevant clinical evidence. Precision was more moderate (48.1% strict, 50.7% lenient), suggesting that our methods are more recall oriented in the extraction phase. In terms of

Table 2: ArchEHR-QA 2025 test-set scores: our zero-shot LLM system (*KR-Labs* versus the organizer baseline, Llama-3.3-70B).

Team	Ov.	Fact.	Rel.	Strict μ			Len. μ			Strict M			Len. M			BLEU	R-L	SARI	BERT	Align	MEDC
				P	R	F1	P	R	F1	P	R	F1	P	R	F1						
KR-Labs	42.0	52.1	31.9	48.1	56.8	52.1	50.7	56.6	53.5	55.8	62.3	54.3	60.4	60.6	56.2	2.0	21.4	57.9	26.3	49.0	35.2
Organizer baseline	30.7	33.6	27.8	71.6	21.9	33.6	77.0	22.3	34.6	77.4	31.5	39.0	83.0	30.8	39.9	0.1	15.2	47.8	20.5	57.7	25.6

relevance, our system achieved competitive Align-Score (49.0%) and MEDCON (35.2%).

We compared our zero-shot LLM extractor and the fine-tuned Clinical ModernBERT extractor on the development set, these findings can be seen in Table 3. The comparison highlights a clear trade-off: the LLM-based extractor provides higher precision and balanced F1, while the Clinical ModernBERT demonstrates strong recall, capturing nearly all relevant information at the expense of precision. Our final submission employed the LLM extractor for its balanced performance.

Table 3: Sentence-level extraction on the development set.

Extractor	Precision	Recall	F1
LLM (gemma-3-27b-it)	0.56	0.73	**0.63**
Clinical ModernBERT	0.46	**0.91**	0.61

Interestingly, final test scores were closely matched between our extractors: the LLM-based model scored 42.01%, while Clinical ModernBERT achieved a near-identical 41.85%. This underscores the effectiveness of our synthetic data training methodology, enabling a lightweight model to achieve comparable performance to a larger LLM.

Overall, our results demonstrate that lean methods can achieve competitive performance in EHR QA, highlighting the value of synthetic data generation. We show that even smaller LLMs, when used for data creation, can enable the training of lightweight models that rival larger systems—while requiring significantly fewer computational resources.

5 Ethical Considerations

Our experiments were conducted exclusively on a secure, private A100 GPU server. This ensured that we adhered to all data licensing requirements and maintained confidentiality throughout the project lifecycle, making the data inaccessible externally. Our work relies on de-identified clinical text and the generation of synthetic data. However, it is important to note that clinical AI systems can perpetuate harmful biases (Bender et al., 2021; Obermeyer et al., 2019). In any deployment setting, we recommend implementing a human-in-the-loop review process, maintaining strict provenance tracking of cited evidence, and conducting thorough bias audits to ensure patient safety and fairness.

6 Limitations

Our *verbatim* RAG pipeline explicitly cites source sentences to mitigate hallucinations; however, several practical limitations remain. Due to the task's strict 75-word limit, our approach often required summarization after the initial verbatim insertion step, meaning the purely verbatim property was not consistently maintained across all answers. Additionally, although extracted sentences were cited exactly, the dynamically generated templates themselves were produced by an LLM, potentially introducing subtle hallucinations or inaccuracies at the framing level. Future work should include explicit checks on template factuality. Finally, user studies and clinician feedback are essential to confirm whether our structured, template-based answers effectively address real-world clinician information needs.

7 Conclusion

In this paper we presented a lightweight and transparent *verbatim* pipeline for grounded question answering from clinical texts. Our method separates sentence-level extraction from template-based generation, significantly reducing hallucinations and maintaining traceable evidence. Participating in the ArchEHR-QA 2025 shared task, our system ranked among the top-10 submissions on key metrics and significantly outperformed a substantially larger baseline (70B-parameter Llama-3.3). We also demonstrated the effectiveness of synthetic training data generated by smaller LLMs for developing competitive, resource-efficient models.

References

Fan Bai, Keith Harrigian, Joel Stremmel, Hamid Hassanzadeh, Ardavan Saeedi, and Mark Dredze. 2024. Give me some hard questions: Synthetic data generation for clinical qa. *Preprint*, arXiv:2412.04573.

Emily M. Bender, Timnit Gebru, Angelina McMillan-Major, and Shmuel Shmitchell. 2021. On the dangers of stochastic parrots: Can language models be too big? In *Proceedings of the 2021 ACM Conference on Fairness, Accountability, and Transparency (FAccT)*, pages 610–623.

Erlend Frayling, Jake Lever, and Graham McDonald. 2024. Zero-shot and few-shot generation strategies for artificial clinical records. *arXiv preprint arXiv:2403.08664*.

Robert Friel, Masha Belyi, and Atindriyo Sanyal. 2025. Ragbench: Explainable benchmark for retrieval-augmented generation systems. *arXiv preprint*, arXiv:2407.11005.

Robert Friel and Atindriyo Sanyal. 2023. Chainpoll: A high efficacy method for llm hallucination detection. *Preprint*, arXiv:2310.18344.

Sarthak Jain and Byron C. Wallace. 2019. Attention is not Explanation. In *Proceedings of the 2019 Conference of the North American Chapter of the Association for Computational Linguistics: Human Language Technologies, Volume 1 (Long and Short Papers)*, pages 3543–3556, Minneapolis, Minnesota. Association for Computational Linguistics.

Ziwei Ji, Nayeon Lee, Rita Frieske, Tiezheng Yu, Dan Su, Yan Xu, Etsuko Ishii, Ye Jin Bang, Andrea Madotto, and Pascale Fung. 2023. Survey of hallucination in natural language generation. *ACM Comput. Surv.*, 55(12).

Qiao Jin, Bhuwan Dhingra, Zhengping Liu, William W. Cohen, and Xinghua Lu. 2019. Pubmedqa: A dataset for biomedical research question answering. *arXiv preprint*, arXiv:1909.06146.

Alistair EW Johnson, Tom J Pollard, Lu Shen, Li-wei H Lehman, Mengling Feng, Mohammad Ghassemi, Benjamin Moody, Peter Szolovits, Leo Anthony Celi, and Roger G Mark. 2016. Mimic-iii, a freely accessible critical care database. *Scientific data*, 3:160035.

Simon A. Lee, Anthony Wu, and Jeffrey N. Chiang. 2025. Clinical modernbert: An efficient and long context encoder for biomedical text. *Preprint*, arXiv:2504.03964.

Rumeng Li, Xun Wang, and Hong Yu. 2023. Two directions for clinical data generation with large language models: Data-to-label and label-to-data. In *Findings of the Association for Computational Linguistics: EMNLP 2023*, pages 7129–7143.

Chin-Yew Lin. 2004. ROUGE: A package for automatic evaluation of summaries. In *Text Summarization Branches Out*, pages 74–81, Barcelona, Spain. Association for Computational Linguistics.

Alejandro Lozano, Scott L Fleming, Chia-Chun Chiang, and Nigam Shah. 2023. Clinfo.ai: An open-source retrieval-augmented large language model system for answering medical questions using scientific literature. *Preprint*, arXiv:2310.16146.

Andreas Madsen, Sarath Chandar, and Siva Reddy. 2024. Are self-explanations from large language models faithful? In *Findings of the Association for Computational Linguistics: ACL 2024*, pages 295–337, Bangkok, Thailand. Association for Computational Linguistics.

Potsawee Manakul, Adian Liusie, and Mark Gales. 2023. SelfCheckGPT: Zero-resource black-box hallucination detection for generative large language models. In *Proceedings of the 2023 Conference on Empirical Methods in Natural Language Processing*, pages 9004–9017, Singapore. Association for Computational Linguistics.

Cheng Niu, Yuanhao Wu, Juno Zhu, Siliang Xu, KaShun Shum, Randy Zhong, Juntong Song, and Tong Zhang. 2024. RAGTruth: A hallucination corpus for developing trustworthy retrieval-augmented language models. In *Proceedings of the 62nd Annual Meeting of the Association for Computational Linguistics (Volume 1: Long Papers)*, pages 10862–10878, Bangkok, Thailand. Association for Computational Linguistics.

Ziad Obermeyer, Brian Powers, Christine Vogeli, and Sendhil Mullainathan. 2019. Dissecting racial bias in an algorithm used to manage the health of populations. *Science*, 366(6464):447–453.

Anusri Pampari, Preethi Raghavan, Jennifer Liang, and Jian Peng. 2018. emrQA: A large corpus for question answering on electronic medical records. In *Proceedings of the 2018 Conference on Empirical Methods in Natural Language Processing*, pages 2357–2368, Brussels, Belgium. Association for Computational Linguistics.

Kishore Papineni, Salim Roukos, Todd Ward, and Wei-Jing Zhu. 2002. Bleu: a method for automatic evaluation of machine translation. In *Proceedings of the 40th Annual Meeting of the Association for Computational Linguistics*, pages 311–318, Philadelphia, Pennsylvania, USA. Association for Computational Linguistics.

Sofia Serrano and Noah A. Smith. 2019. Is Attention Interpretable? In *Proceedings of the 57th Annual Meeting of the Association for Computational Linguistics*, pages 2931–2951, Florence, Italy. Association for Computational Linguistics.

Juntong Song, Xingguang Wang, Juno Zhu, Yuanhao Wu, Xuxin Cheng, Randy Zhong, and Cheng Niu. 2024. RAG-HAT: A hallucination-aware tuning pipeline for LLM in retrieval-augmented generation. In *Proceedings of the 2024 Conference on Empirical Methods in Natural Language Processing: Industry Track*, pages 1548–1558, Miami, Florida, US. Association for Computational Linguistics.

Sarvesh Soni and Dina Demner-Fushman. 2025a. A dataset for addressing patient's information needs related to clinical course of hospitalization. *arXiv preprint*.

Sarvesh Soni and Dina Demner-Fushman. 2025b. Overview of the archehr-qa 2025 shared task on grounded question answering from electronic health records. In *The 24th Workshop on Biomedical Natural Language Processing and BioNLP Shared Tasks*, Vienna, Austria. Association for Computational Linguistics.

Simon Šuster and Walter Daelemans. 2018. CliCR: a dataset of clinical case reports for machine reading comprehension. In *Proceedings of the 2018 Conference of the North American Chapter of the Association for Computational Linguistics: Human Language Technologies, Volume 1 (Long Papers)*, pages 1551–1563, New Orleans, Louisiana. Association for Computational Linguistics.

Ruixiang Tang, Xiaotian Han, Xiaoqian Jiang, and Xia Hu. 2023. Does synthetic data generation of llms help clinical text mining? *arXiv preprint arXiv:2303.04360*.

Gemma Team, Aishwarya Kamath, Johan Ferret, Shreya Pathak, Nino Vieillard, Ramona Merhej, Sarah Perrin, Tatiana Matejovicova, Alexandre Ramé, Morgane Rivière, Louis Rouillard, Thomas Mesnard, Geoffrey Cideron, Jean bastien Grill, Sabela Ramos, Edouard Yvinec, Michelle Casbon, Etienne Pot, Ivo Penchev, and 197 others. 2025. Gemma 3 technical report. *Preprint*, arXiv:2503.19786.

Benjamin Warner, Antoine Chaffin, Benjamin Clavié, Orion Weller, Oskar Hallström, Said Taghadouini, Alexis Gallagher, Raja Biswas, Faisal Ladhak, Tom Aarsen, Nathan Cooper, Griffin Adams, Jeremy Howard, and Iacopo Poli. 2024. Smarter, better, faster, longer: A modern bidirectional encoder for fast, memory efficient, and long context finetuning and inference. *Preprint*, arXiv:2412.13663.

Ran Xu, Wenqi Shi, Yue Yu, Yuchen Zhuang, Yanqiao Zhu, May D. Wang, Joyce C. Ho, Chao Zhang, and Carl Yang. 2024. Bmretriever: Tuning large language models as better biomedical text retrievers. *arXiv preprint arXiv:2404.18443*.

Wen-wai Yim, Yujuan Fu, Asma Ben Abacha, Neal Snider, Thomas Lin, Meliha Yetisgen, and *et al.* 2023. Aci-bench: a novel ambient clinical intelligence dataset for benchmarking automatic visit note generation. *Scientific Data*, 10(1):586.

Tianyi Zhang, Varsha Kishore, Felix Wu, Kilian Q. Weinberger, and Yoav Artzi. 2020. Bertscore: Evaluating text generation with bert. *Preprint*, arXiv:1904.09675.

Ádám Kovács and Gábor Recski. 2025. LettuceDetect: A hallucination detection framework for RAG applications. *Preprint*, arXiv:2502.17125.

LAILab at ArchEHR-QA 2025: Test-time scaling for evidence selection in grounded question answering from electronic health records

Tuan-Dung Le[1,2], Thanh Duong[1,2], Shohreh Haddadan[1], Behzad Jazayeri[1], Brandon Manley[1], Thanh Q. Thieu[1,2]
[1]Moffitt Cancer Center and Research Institute, USA
[2]University of South Florida, USA

{tuandung.le , thanh.duong, shohreh.haddadan, behzad.jazayeri, brandon.manley, thanh.thieu}@moffitt.org

Abstract

This paper presents our approach to the ArchEHR shared task on generating answers to real-world patient questions grounded in evidence from electronic health records (EHRs). We investigate the zero-shot capabilities of general-purpose, domain-agnostic large language models (LLMs) in two key aspects: identifying essential supporting evidence and producing concise, coherent answers. To this aim, we propose a two-stage pipeline: (1) evidence identification via test-time scaling (TTS) and (2) generating the final answer conditioned on selected evidences from the previous stage. Our approach leverages high-temperature sampling to generate multiple outputs during the evidence selection phase. This TTS-based approach effectively explores more potential evidences which results in significant improvement of the factuality score of the answers.

1 Introduction

Large language models (LLMs) tuned with reinforcement learning from human feedback (RLHF), have transformed automatic question answering (QA) systems, leading to their widespread adoption in various domains. In clinical settings, QA systems have been used to answer health-related inquiries (Demner-Fushman et al., 2020) which require medical domain knowledge. Patient-specific QA, more critically, require grounding responses in evidence extracted from electronic health records (EHRs) to ensure factual accuracy and reliability. Training and fine-tuning of clinical-specific LLMs have been shown to outperform general models on NLP tasks, including patient-specific QA (Lehman et al., 2023). However, this approach faces several significant challenges. First, task-specific clinical data is often scarce and difficult to obtain due to strict privacy regulations and patient safety concerns. Second, manual expert annotation of such data is prohibitively expensive. Most critically, even when clinical datasets are de-identified, there remains a non-trivial risk of inadvertently disclosing protected health information (PHI) through model training and deployment (Das et al., 2025) specifically in real-world applications where models are accessible externally such as patient portals. These constraints, coupled with the increasing zero-shot capabilities of LLMs, motivate an alternative paradigm: leveraging general-purpose domain-agnostic LLMs and elicit their domain-specific knowledge and reasoning abilities at inference time. This approach known as test-time scaling (TTS) offers a promising path toward mitigating data scarcity, reducing annotation costs, improving robustness to input variability, and minimizing privacy risks in clinical NLP applications in real-world settings (Zhang et al., 2025).

In this paper, we present a TTS-based solution to the ArchEHR Shared Task (Soni and Demner-Fushman, 2025b). We argue that TTS is particularly well-suited for this task due to limited availability of annotated training data and the method's practicality in real-world deployment scenarios, such as integration into patient portals. We propose a two-stage pipeline methodology consisting of evidence identification followed by answer generation. In the first stage, we employ a parallel TTS strategy by generating multiple outputs at a high temperature and selecting frequently predicted sentences as essential evidence. In the second stage, we prompt the model to generate concise and grounded answers conditioned on the selected evidence, using different prompting strategies to optimize response quality.

2 Task Description

The **ArchEHR-QA 2025** shared task aims at automatically providing answers to real-world patient questions grounded in evidence from EHRs (Soni and Demner-Fushman, 2025b). The dataset con-

BioNLP 2025 Shared Tasks, pages 75–80
August 1, 2025 ©2025 Association for Computational Linguistics

sists of 20 cases in the development set and 100 in the test set (Soni and Demner-Fushman, 2025a). Each case includes patient question, clinician-rewritten version, and excerpts from patients' clinical notes. Each sentence from the excerpt is manually labeled as *essential*, *supplementary*, or *not relevant*, indicating the relevance of the sentence to the answer. Systems are evaluated on two criteria: factuality and relevance. Overall factuality is assessed using strict micro F1, where only essential evidence sentences are considered relevant, with manual annotations as reference labels. Automated relevance is measured by comparing generated answers to reference texts, which include patient narrative, clinician question, and ground-truth evidence sentences. Relevance metrics are BLEU (Papineni et al., 2002), ROUGE (Lin, 2004), SARI (Xu et al., 2016), BERTScore (Zhang et al., 2020), AlignScore (Zha et al., 2023), and MEDCON (Yim et al., 2023). The final leaderboard score averages the overall factuality score and the normalized average of all automated relevance metrics. The organizers also conduct additional post-challenge evaluations, including relevance comparisons to clinician-written answers and manual assessments, offering a more comprehensive view of system performance (Soni and Demner-Fushman, 2025b).

3 Approach

3.1 Overview

To address the challenges posed in low-resource settings given only 20 cases in development set, we leverage the strong zero-shot capabilities of LLMs. Our preliminary experiments in prompting LLMs to directly generate answers using corresponding citations result in high variability across runs and inconsistent sets of cited evidence generated at each run by the same prompt. This method also often leads to low overall factuality scores. These initial findings align with the baseline scores reported by the organizers using a similar strategy.

To address this limitation towards a more reliable patient-specific QA system grounded in evidences from note excerpts, we propose a two-stage prompting strategy. In the first stage, we apply parallel test-time scaling to identify a broader set of potentially essential evidence sentences. In the second stage, we generate the final answer conditioned on the evidence selected during the first stage.

3.2 Stage 1: Evidence identification

The goal of this stage is to identify essential sentences from the note excerpt to serve as evidence to answer the patient's question. Given a clinical note consisting of sentences s_i for $i = 1, 2, \ldots, N_{sent}$ where N_{sent} is the total number of sentences in the note, we prompt a LLM to generate a list of relevant sentence indices i. We apply a zero-shot chain-of-thought prompting strategy (Wei et al., 2022), using the following prompt:

```
Given a clinical note and a patient's question,
identify the sentence indices that provide evidence to
answer the question. Each sentence in the clinical
note is indexed. Return only the relevant sentence
indices as a comma-separated list.

Clinical note: ...
Patient question: ...

Think step by step before finalizing your answer.
Provide your final answer within \boxed{{}}.
```

We extract a list of sentence indices from each model-generated output, representing the sentences identified as essential. To encourage diversity in evidence selection, we sample multiple candidate outputs by varying the decoding temperature. A lower temperature (e.g., 0) results in more deterministic outputs, while a higher temperature (e.g., 0.6 or 1.0) increases randomness, allowing the model to explore more candidate solutions (Renze, 2024). We prompt the model once using temperature 0 (greedy decoding), 64 times with a temperature of 0.6, and either 128 or 256 times with temperature a of 1 to encourage diverse output generation. Let c_i denote the number of times sentence s_i is predicted as essential across all runs. A sentence is included in the final evidence set if $c_i \geq t$, where t is a threshold in the range $[1, N_{gen}]$ and N_{gen} is the total number of generations.

For this stage, we employ two open instruction-tuned LLMs: Qwen2.5-32B-Instruct (Yang et al., 2024) and LLaMA-3.3-70B-Instruct (Grattafiori et al., 2024). We investigate the effectiveness of three question variations provided in the dataset: patient narratives, patient questions, and clinician questions. Results on the development set indicate that prompts with solely patient narratives as input consistently achieve the highest performance. Accordingly, all prompts in our experiments use only patient narrative as input.

3.3 Stage 2: Answer generation

We prompt the LLM to generate the final answer using the essential sentence indices identified in

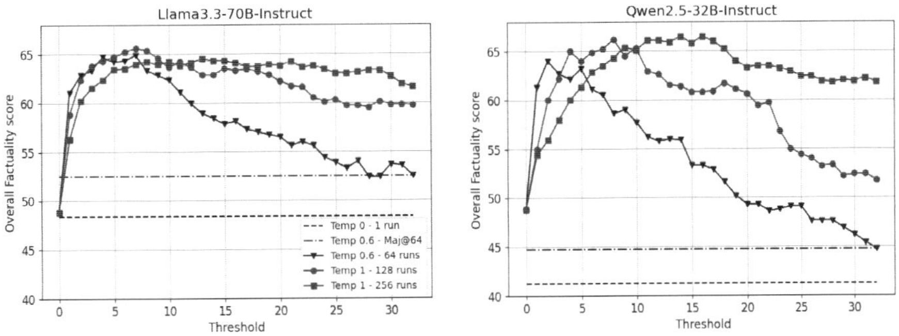

Figure 1: Test-time scaling improves factuality score on development set.

the previous stage with the following prompt:

```
Given a clinical note, a patient's question and a list
of sentence indices that represent the essential
supporting evidence, write a 5-sentence (fewer than
100 words) answer that addresses the patient's
concern. Each sentence must end with the evidence
indices immediately after the period, in this format:
"The treatment was successful.|1,2|\n"

You must cite all essential indices in the answer. Do
not introduce any information that is not grounded in
the clinical note. To ensure high-quality answer,
reuse as much phrasing and sentence structure from the
clinical note as possible.

Clinical note: ...
Patient question: ...
Essential sentences: <list of sentence indices from
stage 1>
```

We conduct an ablation study by varying the instruction components to evaluate their impact on the overall score. Specifically, we experiment with constraints such as allowing free-form generation, limiting the answer length to a fixed number of sentences or words, and encouraging the model to reuse phrasing, sentence structure, or exact evidence sentences from the clinical note.

In this stage, we experiment with Gemini-2.0-flash(Google, 2024) and Gemini-2.5-pro-preview(Google, 2025)[1], as these models more reliably follow instructions and consistently generate answers in the required submission format, whereas the open-source LLMs used in Stage 1 occasionally fail to meet these criteria.

4 Results and Discussions

4.1 Dev performance

Figure 1 shows the performance of the evidence identification stage. These results indicate that generating multiple outputs with higher temperatures and setting lower selection thresholds consistently improves factuality scores. With a temperature of 1.0, Llama-3.3-70B-Instruct achieves a factuality score of 65.4 using a threshold of 7 over 128 runs, while Qwen2.5-32B-Instruct achieves the highest score of 66.5 with thresholds of 14 and 16 over 256 runs. This approach outperforms both single-pass greedy decoding and self-consistency with majority voting (Wang et al., 2022).

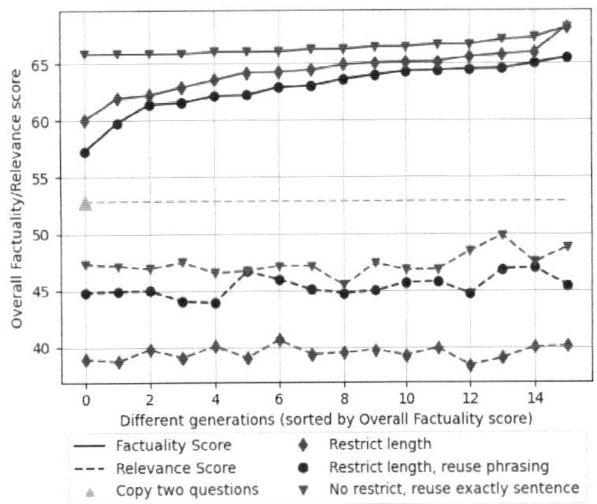

Figure 2: Factuality and relevance scores of answer generation strategies on the development set, evaluated over 16 runs using Gemini-2.0-Flash at temperature 0.6.

For each case, we prompt the model 16 times with different configurations using the best evidence set identified in Stage 1. Figure 2 presents the performance of our answer generation strategies across the 16 runs. When the model is restricted to generate answers with a maximum of 5 sentences and fewer than 100 words, the model achieves an overall relevance score of approximately 40, with an average output length of approximately 80 words on the development set.

It's important to note that despite explicitly in-

[1]These models are accessed via Vertex AI, the platform recommended by PhysioNet for responsible MIMIC data use.

#ID	Stage 1	Stage 2	Leaderboard			Post-challenge re-evaluation			
	t	Answer generation settings	Ovr.	Fact.	Auto Rel.	Human Ovr.	Fact.	Auto Rel.	Human Rel.
1	16	Gemini-2.5-pro-preview 5 sentences, $\leq$ 100 words reuse phrasing and sentence structure	48.2	59.2	37.3	**43.1**	**53.8**	38.0	**32.4**
2	16	Gemini-2.0-flash 5 sentences, $\leq$ 100 words add unused citations to last sentence reuse exact sentence when possible	49.7	59.1	40.3	42.6	53.5	41.5	31.7
3	14	Gemini-2.0-flash no limit reuse exact sentence when possible	**51.0**	**60.4**	**41.6**	41.5	53.3	**42.0**	29.6

Table 1: Details of our three submissions on the test set. Leaderboard scores are based on initial relevance labels and concatenated evidence sentences from the clinical notes, while post-challenge re-evaluation scores use reconciled relevance labels and clinician-written reference answers.

structing the model to include all essential sentences from stage 1, LLMs often omit or introduce citations outside the provided list, leading to variability across runs. The factuality score varies by up to 6 points, while the relevance score remains relatively stable. Removing length constraints improves citation consistency, with the model more reliably preserving the majority of the evidence sentences identified in the previous stage.

We observe that automated relevance metrics favor answers that closely align with the reference, which integrates information from the patient narrative, clinician questions, and ground-truth essential sentences. Prompting the model to reuse phrasing or directly incorporating sentences from the clinical note consistently boosts relevance scores to the 45–47 range. Further improvements are achieved by directly copying sentences from the identified evidence and ordering them based on importance or model confidence to prioritize key information within the first 75 words of the generated response. Moreover, using the patient narrative and clinician question directly as the answer (or appending them to the beginning of the answer) yields a relevance score of 52.9, significantly improving all automated relevance scores, except for SARI score due to copying questions. However, we refrain from adopting these direct copy strategies in our final submission, as they diverge from the objective of the challenge, which emphasize generating coherent responses.

A medical expert at our institute provides answers for the development set based on the annotated essential sentences. Their responses yield an average relevance score of 27.2 with an average length of 54 words, excluding case 16, where our expert notes that the clinical note lacks relevant evidence to answer the patient's question.

4.2 Test submissions

Details of our three test submissions are shown in Table 1. We run Qwen2.5-32B-Instruct 256 times and select essential sentences using thresholds of 14 or 16, chosen based on development set performance. For the first submission, we use Gemini-2.5-pro-preview, which includes all essential sentences within 5 sentences likely due to its stronger reasoning capabilities. The other two use Gemini-2.0-flash to boost automated relevance scores.

Post-challenge re-evaluation based on reconciled relevance labels results in factuality scores dropping by up to 7.1 points, while automated relevance scores varies only slightly, increasing by at most 1.2 points. This aligns with our development set observations and highlights the limitations of automated relevance metrics. Mitigating the limitations of automated relevance scores, the organizers evaluated human relevance by comparing our answers with clinician-written reference answers. Interestingly, human relevance scores often diverged from automated ones, favoring shorter responses with less verbatim replication of the evidence sentences.

5 Related Work

Extractive question answering—a task closely related to grounded question answering—aims to extract patient-specific answer spans from clinical notes in response to clinical queries. Recent approaches have leveraged large language models (LLMs) to address this challenge through a variety of techniques. Fine-tuning language models such as ClinicalBert for sequence generation (Moon et al., 2023) and sequence labeling(Yue et al., 2021) tasks was used for extractive QA from unstructured EHR notes. Hamidi and Roberts (2023) experiment prompting ChatGPT 3.5 and Claude and report a manual evaluation of accuracy, relevance, comprehensiveness, and coherence on a set of patient-specific questions. Lehman et al. (2023) evaluate the performance of various clinical domain specific LLMs with different sizes ranging from 220M to 175B parameters, and use in context learning (ICL) for extractive QA on a dataset on radiology reports (Soni et al., 2022). Their results demonstrate that fine-tuning clinical domain specific models outperform ICL methods on extractive QA.

6 Conclusion

Zero-shot prompting of large language models for patient-specific question answering—grounded in clinical notes—results in inconsistent evidence selection, leading to lower factuality scores. Parallel scaling strategy at test-time mitigates this problem in a low-resource setting. We experiment with generating multiple outputs at higher temperatures and selecting frequently predicted sentences as essential evidence which improves factuality score of evidence identification. We then generate answers conditioned on the selected evidence, and further enhance relevance by engineering the prompt to align the answer to the question while preserving coherence.

Limitations

Our proposed approach has several limitations. First, applying TTS by generating multiple outputs increases computational cost and latency. We run the Qwen2.5-32B-Instruct model 256 times on 4 H100 GPUs to identify evidence, averaging 4 seconds per case, followed by answer generation with Gemini-2.0-Flash via API, which takes an additional 1 second. Due to the cost, we avoid using API-based models for evidence selection and instead rely solely on open-source instruction-tuned LLMs. Exploring more efficient TTS methods with recent open-weight reasoning models such as DeepSeek-R1(Guo et al., 2025) and Qwen3(Yang et al., 2025) is a promising direction for future work. Second, the frequency-based evidence selection is tuned on a small development set of 20 examples, which may not generalize well to unseen cases. Third, while the use of API-based models for answer generation is acceptable for this shared task, it may not be feasible or allowed in real-world clinical settings due to privacy and regulatory constraints. Finally, the answer quality is sensitive to prompt design in the second stage, with minor phrasing changes often leading to significant output variability.

References

Badhan Chandra Das, M. Hadi Amini, and Yanzhao Wu. 2025. Security and privacy challenges of large language models: A survey. *ACM Comput. Surv.*, 57(6).

Dina Demner-Fushman, Yassine Mrabet, and Asma Ben Abacha. 2020. Consumer health information and question answering: helping consumers find answers to their health-related information needs. *Journal of the American Medical Informatics Association*, 27(2):194–201.

Google. 2024. Gemini-2.0-flash-001. https://console.cloud.google.com/vertex-ai/publishers/google/model-garden/gemini-2.0-flash-001.

Google. 2025. Gemini-2.5-pro-preview-03-25. https://console.cloud.google.com/vertex-ai/publishers/google/model-garden/gemini-2.5-pro-preview-03-25.

Aaron Grattafiori, Abhimanyu Dubey, Abhinav Jauhri, Abhinav Pandey, Abhishek Kadian, Ahmad Al-Dahle, Aiesha Letman, Akhil Mathur, Alan Schelten, Alex Vaughan, and 1 others. 2024. The llama 3 herd of models. *arXiv preprint arXiv:2407.21783*.

Daya Guo, Dejian Yang, Haowei Zhang, Junxiao Song, Ruoyu Zhang, Runxin Xu, Qihao Zhu, Shirong Ma, Peiyi Wang, Xiao Bi, and 1 others. 2025. Deepseek-r1: Incentivizing reasoning capability in llms via reinforcement learning. *arXiv preprint arXiv:2501.12948*.

Alaleh Hamidi and Kirk Roberts. 2023. Evaluation of ai chatbots for patient-specific ehr questions. *arXiv preprint arXiv:2306.02549*.

Eric Lehman, Evan Hernandez, Diwakar Mahajan, Jonas Wulff, Micah J Smith, Zachary Ziegler, Daniel Nadler, Peter Szolovits, Alistair Johnson, and Emily Alsentzer. 2023. Do we still need clinical language models? In *Conference on health, inference, and learning*, pages 578–597. PMLR.

Chin-Yew Lin. 2004. Rouge: A package for automatic evaluation of summaries. In *Text summarization branches out*, pages 74–81.

Sungrim Moon, Huan He, Heling Jia, Hongfang Liu, Jungwei Wilfred Fan, and 1 others. 2023. Extractive clinical question-answering with multianswer and multifocus questions: data set development and evaluation study. *JMIR AI*, 2(1):e41818.

Kishore Papineni, Salim Roukos, Todd Ward, and Wei-Jing Zhu. 2002. Bleu: a method for automatic evaluation of machine translation. In *Proceedings of the 40th annual meeting of the Association for Computational Linguistics*, pages 311–318.

Matthew Renze. 2024. The effect of sampling temperature on problem solving in large language models. In *Findings of the Association for Computational Linguistics: EMNLP 2024*, pages 7346–7356, Miami, Florida, USA. Association for Computational Linguistics.

Sarvesh Soni and Dina Demner-Fushman. 2025a. A dataset for addressing patient's information needs related to clinical course of hospitalization. *arXiv preprint*.

Sarvesh Soni and Dina Demner-Fushman. 2025b. Overview of the archehr-qa 2025 shared task on grounded question answering from electronic health records. In *The 24th Workshop on Biomedical Natural Language Processing and BioNLP Shared Tasks*, Vienna, Austria. Association for Computational Linguistics.

Sarvesh Soni, Meghana Gudala, Atieh Pajouhi, and Kirk Roberts. 2022. Radqa: A question answering dataset to improve comprehension of radiology reports. In *Proceedings of the thirteenth language resources and evaluation conference*, pages 6250–6259.

Xuezhi Wang, Jason Wei, Dale Schuurmans, Quoc Le, Ed Chi, Sharan Narang, Aakanksha Chowdhery, and Denny Zhou. 2022. Self-consistency improves chain of thought reasoning in language models. *arXiv preprint arXiv:2203.11171*.

Jason Wei, Xuezhi Wang, Dale Schuurmans, Maarten Bosma, Fei Xia, Ed Chi, Quoc V Le, Denny Zhou, and 1 others. 2022. Chain-of-thought prompting elicits reasoning in large language models. *Advances in neural information processing systems*, 35:24824–24837.

Wei Xu, Courtney Napoles, Ellie Pavlick, Quanze Chen, and Chris Callison-Burch. 2016. Optimizing statistical machine translation for text simplification. *Transactions of the Association for Computational Linguistics*, 4:401–415.

An Yang, Anfeng Li, Baosong Yang, Beichen Zhang, Binyuan Hui, Bo Zheng, Bowen Yu, Chang Gao, Chengen Huang, Chenxu Lv, and 1 others. 2025. Qwen3 technical report. *arXiv preprint arXiv:2505.09388*.

An Yang, Baosong Yang, Beichen Zhang, Binyuan Hui, Bo Zheng, Bowen Yu, Chengyuan Li, Dayiheng Liu, Fei Huang, Haoran Wei, and 1 others. 2024. Qwen2. 5 technical report. *arXiv preprint arXiv:2412.15115*.

Wen-wai Yim, Yujuan Fu, Asma Ben Abacha, Neal Snider, Thomas Lin, and Meliha Yetisgen. 2023. Aci-bench: a novel ambient clinical intelligence dataset for benchmarking automatic visit note generation. *Scientific data*, 10(1):586.

Xiang Yue, Xinliang Frederick Zhang, Ziyu Yao, Simon Lin, and Huan Sun. 2021. Cliniqg4qa: Generating diverse questions for domain adaptation of clinical question answering. In *2021 IEEE International Conference on Bioinformatics and Biomedicine (BIBM)*, pages 580–587. IEEE.

Yuheng Zha, Yichi Yang, Ruichen Li, and Zhiting Hu. 2023. AlignScore: Evaluating factual consistency with a unified alignment function. In *Proceedings of the 61st Annual Meeting of the Association for Computational Linguistics (Volume 1: Long Papers)*, pages 11328–11348, Toronto, Canada. Association for Computational Linguistics.

Qiyuan Zhang, Fuyuan Lyu, Zexu Sun, Lei Wang, Weixu Zhang, Zhihan Guo, Yufei Wang, Irwin King, Xue Liu, and Chen Ma. 2025. What, how, where, and how well? a survey on test-time scaling in large language models. *arXiv preprint arXiv:2503.24235*.

Tianyi Zhang, Varsha Kishore, Felix Wu, Kilian Q. Weinberger, and Yoav Artzi. 2020. Bertscore: Evaluating text generation with BERT. In *8th International Conference on Learning Representations, ICLR 2020, Addis Ababa, Ethiopia, April 26-30, 2020*. OpenReview.net.

UTSA-NLP at ArchEHR-QA 2025: Improving EHR Question Answering via Self-Consistency Prompting

Sara Shields-Menard*, **Zach Reimers***, **Joshua Gardner**, **David Perry, and Anthony Rios**
The University of Texas at San Antonio
{Sara.Shields-Menard2, Zachary.Reimers2, Joshua.Gardner, David.Perry}@my.utsa.edu
{Anthony.Rios}@utsa.edu

Abstract

We describe our system for the ArchEHR-QA Shared Task on answering clinical questions using electronic health records (EHRs). Our approach uses large language models in two steps: first, to find sentences in the EHR relevant to a clinician's question, and second, to generate a short, citation-supported response based on those sentences. We use few-shot prompting, self-consistency, and thresholding to improve the sentence classification step to decide which sentences are essential. We compare several models and find that a smaller 8B model performs better than a larger 70B model for identifying relevant information. Our results show that accurate sentence selection is critical for generating high-quality responses and that self-consistency with thresholding helps make these decisions more reliable.

1 Introduction

The prevalence of electronic health records (EHRs) has prompted increased digital communication between patients and clinicians. Patient portal use has been linked to better health outcomes for those with chronic conditions, especially by improving self-management and treatment (Brands et al., 2022) and doctor-patient relationships (Carini et al., 2021). While patients benefit from EHRs and patient portals, clinicians now face one extra hour of work per day, much outside of work hours, due to the high volume of patient-initiated messages and results (Akbar et al., 2021). The result is increased clinician burden and burnout. This work focuses on developing methods that could potentially answer patient portal questions.

Large language models are in the early stages of adoption in medical systems and have begun to be implemented in clinical decision support (CDS), medical question-answering systems, and medical documentation (Liu et al., 2023a). Clinical decision support (CDS) has been used to help reduce clinician burnout by sending rule-based alerts to clinicians and patients based on their EHRs. The clinician either canceled or ignored these alerts due to improper timing or alert fatigue. Large Language Models (LLMs) such as ChatGPT have shown beneficial use to assist the clinician in alert logic (Liu et al., 2023b). ChatGPT has also been used as a chatbot assistant to respond to patient questions from a social media forum (Ayers et al., 2023). Overall, studies have found that AI-generated responses are longer than those written by physicians (Ayers et al., 2023; Liu et al., 2023b). Some models are rated as more empathetic and of higher quality (Ayers et al., 2023), while reviewers note others to have a noticeably artificial tone (Li et al., 2023). Recent efforts such as Med-PaLM and Med-PaLM 2 have advanced long-form medical QA by incorporating improved prompting strategies, fine-tuning on medical datasets, and human-centric evaluation frameworks (Singhal, 2023, 2025). These models demonstrate strong performance on USMLE-style questions and improved physician-rated safety and factuality in long-form answers (Pfohl, 2024; Callahan, 2021; Ayers, 2023).

While promising results have been shown using LLMs with Question-Answer queries for specific medical topics like cancer, hepatic disease, and Obstetrics and gynecology, the use of LLMs for patient-specific question-answer responses is limited (Liu et al., 2023b). Even with a medical chat model that can respond to patient questions using online medical content (Li et al., 2023), the patient's question is not directly answered with their own EHR information, which may provide critical references to drug interactions, surgery recoveries, or lab results. One limitation of previous work is that the data to train these models is outdated by the time the model is ready to be deployed, so the

*These two authors contributed equally to this work.

BioNLP 2025 Shared Tasks, pages 81–90
August 1, 2025 ©2025 Association for Computational Linguistics

use of current patient EHRs is essential for maximum clinician and patient benefit. Further, most LLMs used in prior work are not grounded in real-time, patient-specific data, which has been shown to impact accuracy and safety in recent benchmark evaluations (Singhal, 2025).

The ArchEHR-QA shared task provides a wide range of real patient EHRs with clinician and patient questions to generate concise responses to patient/clinician questions grounded in the patient EHRs (Soni and Demner-Fushman, 2025b). In this study, we designed a prompting pipeline for LLMs to identify and compile relevant patient notes in response to a clinician's question. We specifically used few-shot prompting combined with self-consistency and thresholding to select the EHR sentences most relevant for response generation.

Our work contributes the following findings to the solution for the ArchEHR-QA: BioNLP at ACL 2025 Shared Task on Grounded Electronic Health Record Question Answering:

- A method for sentence classification using few-shot prompting with self-consistency and thresholding.
- An analysis showing that sentence selection quality is the primary driver of overall performance, with errors concentrated in the *supplementary* class.

2 Methodology

Overall, this we aim to generate clinically grounded answers to patient (or clinician) questions using evidence from electronic health records (Soni and Demner-Fushman, 2025a). We provide an overview of our approach in Figure 1. A more complete overview is shown in the Appendix in Figure 2. Our approach to the task was two-fold: **(1.)** we develop an "relevant sentence identifier" using few-shot prompting and self-consistency with thresholding. Specifically, the first objective was to classify each sentence in the note excerpt as *essential,supplementary*, or *not relevant* to answer the clinician's question[1]. **(2.)** we generate final answers to the question using zero-shot prompting that transforms a list of relevant sentences into a 75-word response with citations indicating the sentence used.

[1]In initial experiments we found limited difference in performance between using the clinician, patient, or even a combination of clinician an patient questions.

Formally, given a natural language question q and a set of context sentences from an electronic health record $C = \{s_1, s_2, \ldots, s_n\}$, the goal is to generate an answer a consisting of at most 75 words, grounded in a subset of the sentences in C. Our two-part approach uses large language models (LLMs) for: **(1)** sentence selection, where we learn a function $f_{\text{rel}}(q, C) \rightarrow C' \subseteq C$ that identifies relevant sentences, and **(2)** answer generation, where we use a function $f_{\text{gen}}(q, C') \rightarrow a$ to compose a fluent, grounded answer. Sentence citations from C' are retained in a to support factual consistency. We provide an overview of our approach in the Appendix.

2.1 Relevant Sentence Identification.

To identify which sentences in the clinical note are useful for answering the question, we use few-shot prompting with a large language model. Each sentence in the note is independently classified as *essential*, *supplementary*, or *not relevant* to the clinician-formulated question. We construct our few-shot prompts using labeled examples from the development set, where each example includes a question and a sentence with its gold relevance label. A balanced set of 30 such examples (sentences) is randomly sampled and inserted into the prompt for each test case.

2.2 Self-Consistency and Thresholds.

Overall, we find that the model has a hard time detecting essential (relevant) sentences, and will often default to not-relevant. To mitigate class imbalance issues at inference time, we adopt self-consistency decoding with thresholding. Despite using a pre-trained model, the majority class of *not relevant* may be predicted most of the time. Thresholding is used to ensure other classes, especially *essential*, are predicted and can be used for response generation. For each sentence, we sample 20 independent predictions using the LLaMA 8B model with temperature set to 1.0. Thresholds are applied to determine the final label: a sentence is labeled *essential* if it appears in at least 2 out of 20 predictions; if not, it is labeled *supplementary* if it appears at least once; otherwise, it is labeled *not relevant.* This strategy biases the classification toward relevant categories. Increasing the number of samples beyond 20 (e.g., to 111 or 200) yielded limited gains in F1 score on the development set. Prior thresholding attempts used the data description statistics (both median and mean) of

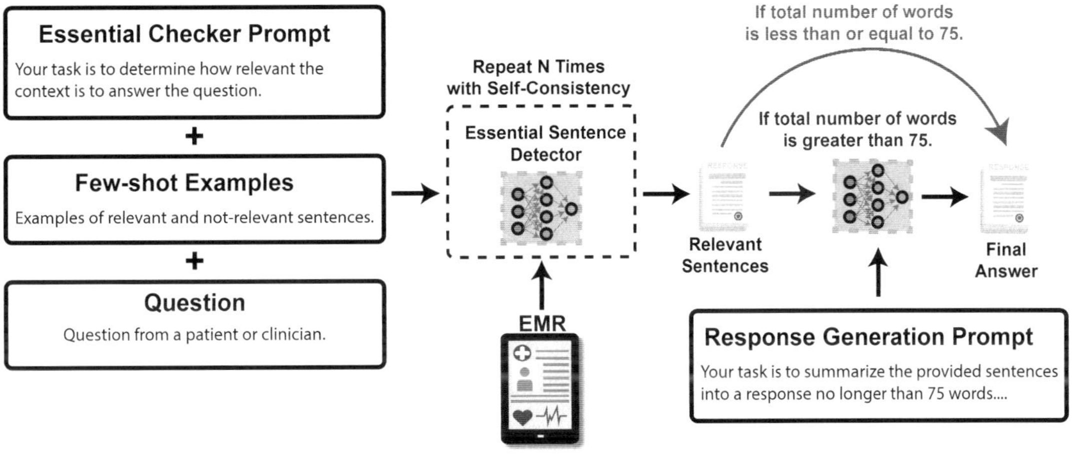

Figure 1: Overview of our multi-step approach for generating clinically grounded answers from electronic health records. An essential sentence detector uses a prompt and few-shot examples to classify sentences from the EMR with self-consistency sampling. Relevant sentences are passed to the response generation step. If the total word count is within the 75-word limit, the sentences are used directly. Otherwise, the sentences are summarized using a language model prompted to stay within the word constraint.

the distribution of each classification of sentences per case. For example, the median number of *essential* sentences is 5.5 out of 21 total sentences per case (Soni and Demner-Fushman, 2025a). The threshold was set at 26 percent, meaning that of the 20 self-consistency samples, at least 5 would have to be labeled *essential* for the final label to be *essential*. We repeated several attempts with these median and mean thresholds, but these high thresholds yielded F1 scores with several false negatives for the *essential* label.

2.3 Final Response Generation.

We use zero-shot prompting for response generation using the Llama 70B quantized model without any in-context examples. Instead, a system prompt (see Appendix) was provided to guide the structure and formatting of the output. For lenient evaluation, both *essential* and *supplementary* sentences were used as relevant sentences to compose the response, while strict evaluation included only sentences labeled as *essential*. Sentences marked as *not relevant* were excluded from all responses. In cases where no *essential* sentences were identified, a placeholder response ("No citations found") was generated along with a randomly sampled citation (pipe-delimited from 1–10) to satisfy format requirements that sentences must have at least one citation. If the extracted content was under the 75-word limit, the sentence set was used directly without further processing. We did not attempt to pass these sentences of less than 75 words to the model to make the response even more concise because we were concerned that further reduction might negatively impact ROUGE and BERT scores by decreasing n-gram overlap. Only when the combined sentence content exceeded the limit, we used the Llama 70B model to summarize the selected sentences into a coherent answer under the constraint with the zero-shot prompt. If citations were removed by the Llama model, citations for any *essential* or *supplementary* sentences excluded due to length were appended to the final sentence of the response.

2.4 Models.

We evaluated both LLaMA 3.1 8B and 3.1 70B (w416b quantization) models (Grattafiori et al., 2024) on the development set for sentence classification and answer generation.[2] Based on development performance, we selected two configurations for test set evaluation: **(1)** a entirely 70B pipeline for both sentence selection and answer generation (denoted 70B-LENIENT and 70B-STRICT), and **(2)** a hybrid configuration using LLaMA 8B for essential sentence identification and LLaMA 70B for response generation called 8B-LENIENT and 8B-STRICT.

[2]We also evaluated UltraMedical (Zhang et al., 2024), but it did not outperform the LLaMA models in initial experiments. Although UltraMedical showed some promise, its responses were consistently too lengthy and overly elaborate, making it unsuitable for the 75-word constraint on the test set.

Configuration	Ovr	Rel.	Fact.	SMP	SMR	SMF1	SmP	SmR	SmF1	LMP	LMR	LMF1	LmP	LmR	LmF1	ROU	MED	BE	AS	BL	SA
8B-LENIENT	42.4	32.1	52.7	44.9	75.0	51.4	41.6	71.7	52.7	61.9	73.1	60.4	54.2	71.7	60.4	20.7	32.7	27.3	30.6	1.8	59.6
8B-STRICT	36.1	25.9	46.4	42.5	44.0	38.0	48.8	44.2	46.4	63.1	78.5	40.9	59.2	39.2	47.1	14.8	28.1	15.6	40.5	1.3	56.9
SHAREDTASK-BASELINE	35.9	28.7	43.1	70.3	47.1	49.4	63.4	32.6	43.1	78.5	38.9	46.5	71.8	27.0	39.2	18.7	29.4	24.2	52.1	0.2	48.0
GT-LENIENT	61.90	39.39	84.40	78.25	100	85.52	73.02	100	84.40	100	100	100	100	100	100	28.52	40.62	35.43	59.21	6.61	65.95
GT-STRICT	74.58	49.16	100	100	100	100	100	100	100	100	78.25	85.52	100	73.02	84.40	36.42	48.64	42.56	82.25	11.18	73.89
8B-LENIENT (W/O THRESH.)	29.3	24.5	34.1	47.4	29.1	30.3	49.3	26.1	34.1	58.0	28.8	31.9	57.5	22.2	32.1	15.4	22.2	13.1	36.6	2.7	57.0
8B-LENIENT (W/O THRESH. AND SC)	26.3	22.3	30.3	34.7	25.1	24.0	43.8	23.2	30.3	42.8	25.2	25.6	50.7	19.6	28.2	10.0	17.4	11.5	30.8	2.7	57.8

Table 1: Development set performance across configurations. **Ovr** is the overall score (mean of factuality and relevance). **Rel.** and **Fact.** are overall relevance and factuality. **SMP/SMR/SMF1** and **LMP/LMR/LMF1** are strict micro/macro precision, recall, and F1 ("essential" only). **SmP/SmR/SmF1** and **LmP/LmR/LmF1** are lenient versions ("essential" + "supplementary"). **ROU**, **BL**, **SA**, **BE**, **AS**, and **MED** denote ROUGE-L, BLEU, SARI, BERTScore, AlignScore, and MEDCON.

3 Results

Here, we present our results from several attempts using the development data set and our three submissions to the competition using the test data set.

3.1 Evaluation Metrics

The ArchEHR scoring script was used to evaluate all attempts for classification accuracy (strict and lenient F1 scores) and response quality, including fluency, relevance, and medical accuracy, using metrics such as BLEU (Papineni et al., 2002), ROUGE (Lin, 2004), SARI (Xu et al., 2016), BERTScore (Zhang et al., 2019), MedCon (Yim et al., 2023), and AlignScore (Zha et al., 2023). The scoring script parsed the responses for citations and then, if needed, truncated the response to 75 words. Only the 75 words and the pipe-delimited citations were evaluated.

3.2 Validation Results.

We report the results on the dev set using leave-one-out cross-validation in Table 1. The best performance on the development set was achieved using the LLAMA3-8B model to identify both *essential* and *supplementary* sentences, followed by response generation using the LLAMA3-70B model (denoted LENIENT (8B)). This configuration outperformed all others, including strict variants that relied solely on *essential* sentences.

To assess the potential upper bound of the task, we evaluate ground truth (GT) configurations that assume perfect classification. GT-STRICT uses only the gold *essential* sentences, while GT-LENIENT includes both gold *essential* and *supplementary* sentences as input to the response generation model. These oracle settings achieved substantially higher scores, 74.58 and 61.90, respectively, demonstrating the significant headroom remaining for improving sentence selection models. This comparison also highlights the performance of our lenient model, which achieves 42.37 overall.

Finally, we conducted ablations to examine the effect of self-consistency and thresholding in the classification process. Removing thresholding alone reduced overall performance to 26.34, while removing both thresholding and self-consistency resulted in a score of 29.31 (Table 1). These declines were primarily driven by reduced F1 scores for sentence classification, underscoring the importance of threshold-based calibration in achieving stable, high-quality predictions.

3.3 Competition Results.

The 8B model was used to classify sentences using lenient evaluation metrics, and the 70B model was then used to generate responses based on those classifications. This combination ("8B-lenient") outperformed the 70B model when it was used alone to both classify sentences and generate responses ("70B-lenient" and "70B-strict"; Table 4 in the Appendix). For classification tasks, the 8B model had better consistency in label prediction and produced more factually correct and relevant answers than the 70B quantized models. Furthermore, the balanced recall and precision scores indicate that the thresholds were well-established as the model was able to identify most of the essential sentences. The increased performance in sentence classification led to higher-quality response generation and improved the response generation metrics. Responses had better alignment, quality, and included more medical-specific content. Despite having fewer parameters, the 8B model outperformed the 70B quantized model across almost every metric, especially in classification, which showed to be a key point in generating high-quality responses.

3.4 Error Analysis.

While the model shows the ability to differentiate between classes, performance was negatively affected by class imbalance. The overwhelming

Class	TP	FP	FN	TN
Essential	64	67	74	223
Supplementary	18	89	33	288
Not-relevant	130	60	109	129

Table 2: Confusion Matrix for "Strict" Results.

Class	TP	FP	FN	TN
Essential	129	109	60	130
Not-relevant	130	60	109	129

Table 3: Confusion matrix for "Lenient" Results.

number of *not relevant* sentences and the relatively small number of *supplementary* sentences led to label mismatches and reduced classification accuracy.

In the strict classification setting, the model was expected to predict three distinct classes. As shown in Table 2, the *supplementary* class proved particularly difficult to identify, with only 18 out of 51 instances correctly predicted. High false negative rates for both *essential* and *not relevant* sentences suggest that important information was often missed, and irrelevant content was not reliably excluded.

In the lenient setting, where *essential* and *supplementary* sentences were grouped into a single class, the task was reduced to binary classification (Table 3). This improved recall for relevant content, and the model successfully identified a larger number of relevant sentences. However, the distinction between *essential* and *supplementary* information introduced ambiguity. While the lenient setup benefited answer generation on the development set, it also produced a high number of false positives, likely due to the low thresholding strategy that aimed to capture as many relevant sentences as possible.

To better understand these trends, we conducted a manual error analysis on development set predictions. One common error involved *not relevant* sentences being misclassified as *essential* or *supplementary*. For example, in response to the clinician question "Why was a procedure used instead of a medication?", two sentences containing only the acronym of the procedure (which also appeared in the question) were incorrectly labeled *essential*. Although these sentences referenced the procedure, they did not explain the reasoning behind it. This suggests that the model may rely too heavily on lexical overlap without considering the deeper intent of the question.

We also observed the opposite error, where sentences labeled as *essential* were misclassified as *not relevant*. In one case, the clinician question concerned a patient's oxygen flow, and a relevant sentence referenced "hypoxia" and "respiratory failure". These terms are clinically important for evaluating oxygen status, yet the model failed to recognize the connection. This misclassification may be due to the model's reliance on surface features rather than contextual relationships.

Ambiguity in the *supplementary* label also introduced challenges. In one example, a two-part clinician question asked about the lasting effects of poisoning and the patient's confusion. The model often misclassified *essential* sentences as *supplementary* or vice versa, suggesting it struggled to distinguish between past clinician explanations and future clinical concerns. Additionally, a sentence mentioning psychiatry was misclassified as *not-relevant* instead of *supplementary*, likely because the model failed to connect psychiatric care with the patient's mental state in the question.

4 Conclusion

Our approach to the ArchEHR-QA Shared Task showed that sentence classification is essential for generating high-quality, grounded responses from electronic health records. Using few-shot prompting with self-consistency and thresholding improved performance, and the smaller LLAMA3.1-8B model outperformed the larger 70B model in identifying relevant sentences. However, distinguishing supplementary content remained difficult due to label imbalance.

Future work should explore incorporating sentence context and document structure to improve classification, along with adaptive thresholding based on model confidence. Fine-tuning with clinician feedback and expanding evaluation to include human judgments will be important for improving real-world reliability and clinical applicability.

Acknowledgments

This material is based upon work supported by the National Science Foundation (NSF) under Grant No. 2145357.

Limitations

This work has several limitations. First, the dataset was relatively small, consisting of only 20 development cases and 100 test cases, which may limit the generalizability of the results. Additionally, the evaluation relied solely on quantitative metrics, without manual review of patient context and medical accuracy. It also lacked evaluation of personable aspects such as empathy and professionalism. Finally, the imposed word limit on responses introduced a scoring bias, particularly disadvantaging longer or complex patient cases that required more nuanced explanations.

Another limitation lies in the reliance on self-consistency thresholding as a heuristic rather than a learned calibration method. Although it improved performance, the threshold values were tuned manually and may not generalize well across datasets with different distributions of relevance labels. Future work could explore adaptive or data-driven methods to calibrate sentence selection confidence.

Additionally, while the 8B model outperformed the 70B model in sentence classification, this may reflect the effects of quantization, prompt format sensitivity, or differences in instruction tuning. These variables were not systematically controlled or analyzed. Further investigation is needed to isolate whether smaller models offer consistent advantages or whether specific tuning strategies are responsible for the performance gains.

The current approach treats each sentence independently during classification, ignoring the surrounding context that may be critical in understanding clinical relevance. Sentences referring to previous or subsequent medical events could be misclassified due to this lack of discourse awareness. Integrating document-level context or sequential modeling could help mitigate this issue.

References

Fatema Akbar, Gloria Mark, E. Margaret Warton, Mary E. Reed, Stephanie Prausnitz, Jeffrey A. East, Mark F. Moeller, and Tracy A. Lieu. 2021. Physicians' electronic inbox work patterns and factors associated with high inbox work duration. *Journal of the American Medical Informatics Association*, 28(5):923–930.

John W. Ayers, Adam Poliak, Mark Dredze, and et al. 2023. Comparing physician and artificial intelligence chatbot responses to patient questions posted to a public social media forum. *JAMA Internal Medicine*, 183(6):589–596. Published by the American Medical Association.

John W. et al. Ayers. 2023. Comparing physician and artificial intelligence chatbot responses to patient questions posted to a public social media forum. *JAMA Internal Medicine*, 183(6):589–596.

M. Brands, S. Gouw, M. Beestrum, R. Cronin, K. Fijnvandraat, and S. Badawy. 2022. Patient-centered digital health records and their effects on health outcomes: Systematic review. *Journal of Medical Internet Research*, 24(12):e43086.

Alexander et al. Callahan. 2021. Using aggregate patient data at the bedside via an on-demand consultation service. *NEJM Catalyst Innovations in Care Delivery*, 2.

E. Carini, L. Villani, A. M. Pezzullo, A. Gentili, A. Barbara, W. Ricciardi, and S. Boccia. 2021. The impact of digital patient portals on health outcomes, system efficiency, and patient attitudes: Updated systematic literature review. *Journal of Medical Internet Research*, 23(9):e26189.

Aaron Grattafiori, Abhimanyu Dubey, Abhinav Jauhri, Abhinav Pandey, Abhishek Kadian, Ahmad Al-Dahle, Aiesha Letman, Akhil Mathur, Alan Schelten, Alex Vaughan, and 1 others. 2024. The llama 3 herd of models. *arXiv preprint arXiv:2407.21783*.

Yunxiang Li, Zihan Li, Kai Zhang, Ruilong Dan, Steve Jiang, and You Zhang. 2023. Chatdoctor: A medical chat model fine-tuned on a large language model meta-ai (llama) using medical domain knowledge. *Cureus*, 15(6).

Chin-Yew Lin. 2004. Rouge: A package for automatic evaluation of summaries. In *Text summarization branches out*, pages 74–81.

Siru Liu, Aileen P. Wright, Barron L. Patterson, Jonathan P. Wanderer, Robert W. Turer, Scott D. Nelson, Allison B. McCoy, Dean F. Sittig, and Adam Wright. 2023a. Using ai-generated suggestions from chatgpt to optimize clinical decision support. *Journal of the American Medical Informatics Association*, 30(7):1237–1245.

Siru Liu, Aileen P. Wright, Barron L. Patterson, Jonathan P. Wanderer, Robert W. Turer, Scott D. Nelson, Allison B. McCoy, Dean F. Sittig, and Adam Wright. 2023b. Using ai-generated suggestions from chatgpt to optimize clinical decision support. *Journal of the American Medical Informatics Association*, 30(7):1237–1245.

Kishore Papineni, Salim Roukos, Todd Ward, and Wei-Jing Zhu. 2002. Bleu: a method for automatic evaluation of machine translation. In *Proceedings of the 40th annual meeting of the Association for Computational Linguistics*, pages 311–318.

Stephen R. et al. Pfohl. 2024. A toolbox for surfacing health equity harms and biases in large language models. *Nature Medicine*.

Karan et al. Singhal. 2023. Large language models encode clinical knowledge. *Nature*, 620:172–180.

Karan et al. Singhal. 2025. Toward expert-level medical question answering with large language models. *Nature Medicine*, 31:943–950.

Sarvesh Soni and Dina Demner-Fushman. 2025a. A dataset for addressing patient's information needs related to clinical course of hospitalization. *arXiv preprint*.

Sarvesh Soni and Dina Demner-Fushman. 2025b. Overview of the archehr-qa 2025 shared task on grounded question answering from electronic health records. In *The 24th Workshop on Biomedical Natural Language Processing and BioNLP Shared Tasks*, Vienna, Austria. Association for Computational Linguistics.

Wei Xu, Courtney Napoles, Ellie Pavlick, Quanze Chen, and Chris Callison-Burch. 2016. Optimizing statistical machine translation for text simplification. *Transactions of the Association for Computational Linguistics*, 4:401–415.

Wen-wai Yim, Yujuan Fu, Asma Ben Abacha, Neal Snider, Thomas Lin, and Meliha Yetisgen. 2023. Aci-bench: a novel ambient clinical intelligence dataset for benchmarking automatic visit note generation. *Scientific data*, 10(1):586.

Yuheng Zha, Yichi Yang, Ruichen Li, and Zhiting Hu. 2023. Alignscore: Evaluating factual consistency with a unified alignment function. *arXiv preprint arXiv:2305.16739*.

Kaiyan Zhang, Sihang Zeng, Ermo Hua, Ning Ding, Zhang-Ren Chen, Zhiyuan Ma, Haoxin Li, Ganqu Cui, Biqing Qi, Xuekai Zhu, and 1 others. 2024. Ultramedical: Building specialized generalists in biomedicine. *Advances in Neural Information Processing Systems*, 37:26045–26081.

Tianyi Zhang, Varsha Kishore, Felix Wu, Kilian Q Weinberger, and Yoav Artzi. 2019. Bertscore: Evaluating text generation with bert. *arXiv preprint arXiv:1904.09675*.

A Appendix

A. Prompt for Sentence Relevance Classification

We use a single, structured prompt to classify the relevance of EHR sentences with respect to a clinical question. For clarity, we display the full prompt below in a formatted box, broken into the system instruction and examples.

Prompt for "Essential Checker"

Your task is to determine how relevant the Context is to answering the Question.

Assign one of the following labels:
- essential: The Context provides critical information needed to answer the Question.
- supplementary: The Context provides useful but non-essential information related to the Question.
- not-relevant: The Context does not provide useful information for answering the Question.

Important: Output only the label — "essential", "supplementary", or "not-relevant". Do not include any other text. *"You will be given a Question and a Context. The Context is a sentence excerpted from a patient's electronic health record.*

Your task is to determine how relevant the Context is to answering the Question.

Assign one of the following labels:
- essential: The Context provides critical information needed to answer the Question.
- supplementary: The Context provides useful but non-essential information related to the Question.
- not-relevant: The Context does not provide useful information for answering the Question.

Important: Output only the label — "essential", "supplementary", or "not-relevant". Do not include any other text.

Examples:[3]

[3]Examples have been changed to ensure anonyminity of data.

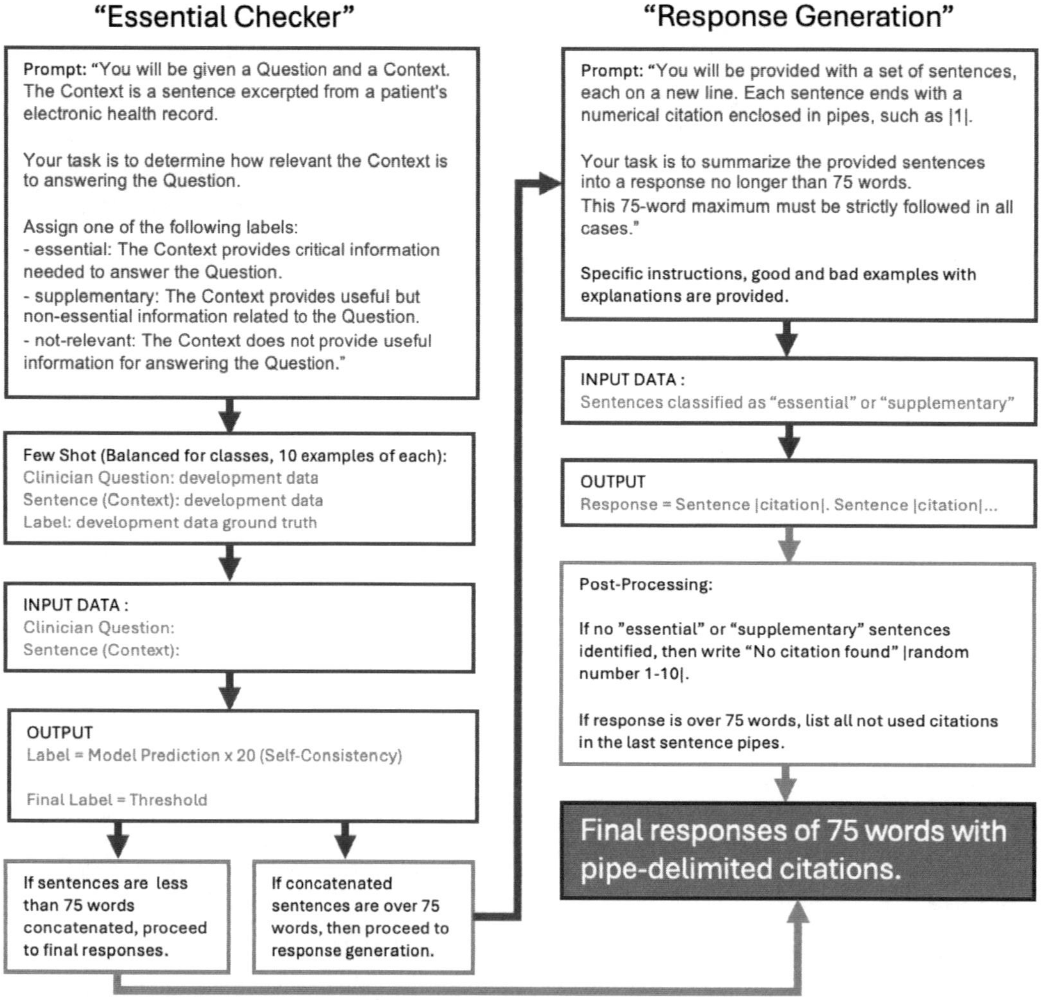

Figure 2: Overall Method Figure.

Configuration	Overall	Rel.	Fact.	SMP	SMR	SMF1	SmP	SmR	SmF1	LMP	LMR	LMF1	LmP	LmR	LmF1	R	Med	BE	AS	BL	SA
8B Results	40.45	27.92	52.97	45.06	77.3	52.59	43.71	67.22	52.97	49.6	77.43	56.65	47.02	68.39	55.73	11.11	17.75	29.4	56.58	0.72	22.7
Organizers Baseline	30.7	27.8	33.6	77.4	31.5	39	71.6	21.9	33.6	83	30.8	39.9	77	22.3	34.6	15.2	25.6	20.5	57.7	0.1	47.8
70B Lenient Results	29.93	18.59	41.28	44.4	51.06	41.7	38.67	44.26	41.28	47.41	50.71	43.64	41.42	44.83	43.06	8.83	12.56	10.78	54.81	1.42	12.86
70B Strict Results	28.05	18.92	37.18	43.77	44.53	38.3	36.27	38.15	37.18	46.88	44.4	39.97	38.64	38.44	38.54	8.77	12.32	11.33	54.96	1.3	12.71

Table 4: Overall Score Comparison of Model Configurations for Test Data. **Ovr** is the overall score (mean of factuality and relevance). **Rel.** and **Fact.** are overall relevance and factuality. **SMP/SMR/SMF1** and **LMP/LMR/LMF1** are strict micro/macro precision, recall, and F1 ("essential" only). **SmP/SmR/SmF1** and **LmP/LmR/LmF1** are lenient versions ("essential" + "supplementary"). **ROU**, **BL**, **SA**, **BE**, **AS**, and **MED** denote ROUGE-L, BLEU, SARI, BERTScore, AlignScore, and MEDCON.

Question: What medications is the patient currently taking?
Context: The patient is currently prescribed metformin and lisinopril.
Label: essential

Question: Has the patient experienced any recent falls?
Context: The patient reports no falls over the past six months.
Label: essential

Question: What is the patient's preferred pharmacy?
Context: The patient prefers CVS Pharmacy on Main Street.
Label: essential

Question: What medications is the patient currently taking?
Context: The patient lives with their daughter and two grandchildren.
Label: not-relevant

Question: Has the patient experienced any recent falls?
Context: The patient has a history of osteoarthritis in the knees.
Label: supplementary

Question: What is the patient's preferred pharmacy?
Context: The patient reports good control of their blood sugar levels.
Label: not-relevant

Question: What medications is the patient currently taking?
Context: The patient reports an allergy to penicillin.
Label: supplementary

Question: Has the patient experienced any recent falls?
Context: Patient noted to have unsteady gait and occasional dizziness.
Label: supplementary

Question: What is the patient's preferred pharmacy?
Context: The patient was discharged home with follow-up scheduled in two weeks.
Label: not-relevant

Question: What medications is the patient currently taking?
Context: At discharge, the patient was advised to continue taking atorvastatin daily.
Label: essential

Question: Has the patient experienced any recent falls?
Context: The patient was admitted after slipping on ice and fracturing their wrist last month.
Label: essential

Additional Examples:

B. Prompt for Answer Generation

We use a single zero-shot prompt to guide answer generation. The model receives a list of preselected sentences with citations and is asked to generate a 75-word summary with citation formatting preserved.

Prompt for "Response Generator"

Your task is to summarize the provided sentences into a response no longer than 75 words.
This 75-word maximum must be strictly followed in all cases.

Do not add any notes, comments, or additional text after the summary.
This will result in the response exceeding the 75-word limit.

Each sentence in your output should start on a new line.
Each sentence must have one or more citations at the end, formatted as integers inside pipes (e.g., |2| or |3,5,7|).

When combining multiple original sentences into one, list all relevant citations in order, separated by commas inside a single pair of pipes (e.g., |2,4,5|).
If multiple sequential citations are combined, list them individually, not as a range (e.g., |7,8,9,10|, not |7-10|).

If any sentences from the input are omitted completely from your summary, their citations must still be preserved by adding them to the final sentence's citation list.

Only output the summarized response. Do not include any commentary, labels, or additional text."
"You will be provided with a set of sentences, each on a new line. Each sentence ends with a numerical citation enclosed in pipes, such as |1|.

Your task is to summarize the provided sentences into a response no longer than 75 words.
This 75-word maximum must be strictly followed in all cases.

Do not add any notes, comments, or additional text after the summary.
This will result in the response exceeding the 75-word limit.

Each sentence in your output should start on a new line.
Each sentence must have one or more citations at the end, formatted as integers inside pipes (e.g., |2| or |3,5,7|).

When combining multiple original sentences into one, list all relevant citations in order, separated by commas inside a single pair of pipes (e.g., |2,4,5|).
If multiple sequential citations are combined, list them individually, not as a range (e.g., |7,8,9,10|, not |7-10|).

If any sentences from the input are omitted completely from your summary, their citations must still be preserved by adding them to the final sentence's citation list.

Only output the summarized response. Do not include any commentary, labels, or additional text."

Examples:

Example 1 — Input and Output

Input:
The company launched a new product in April |1|.
Sales exceeded expectations within the first month |2|.
Customer feedback highlighted a few technical issues |3|.
The technical team promised a software update to address concerns |4|.

Output:
The company launched a new product in April, and sales exceeded expectations in the first month |1,2|.
Customer feedback highlighted technical issues, and the technical team promised a software update to address them |3,4|.

Example 2 — Output

A new downtown cafe offering organic food received praise for its atmosphere but some criticism for high prices |1,2,3,4|.
It plans to expand to a second location next year |5|.

Example 3 — Output

The software update brought a redesigned interface and improved navigation |1,2|.
Although users reported new bugs, a patch issued two weeks later resolved major issues but caused minor compatibility problems on older devices |3,4,5|.

Bad Example (What **Not** to Do)

Input:
The research team published their findings in a leading journal |1|.
They discovered a new species of bacteria in the Arctic |2|.
The bacteria showed resistance to extreme cold temperatures |3|.
Further studies are needed to understand its potential applications |4|.

Output (Incorrect):
The research team published their findings about a new cold-resistant bacteria discovered in the Arctic |1-3|.
Further studies are needed to understand its applications |4|.

Issues:

- Incorrect citation format: |1-3| is a range, but it should be |1,2,3|.
- Word count and sentence coverage are fine, but citation formatting makes this output invalid.

UTSamuel at ArchEHR-QA 2025: A Clinical Question Answering System for Responding to Patient Portal Messages Using Generative AI

Samuel M. Reason, Liwei Wang, Hongfang Liu, Ming Huang*
McWilliams School of Biomedical Informatics,
University of Texas Health Science Center at Houston
***Corresponding author (Ming.Huang@uth.tmc.edu)**

Abstract

Responding to patient portal messages places a substantial burden on clinicians. To mitigate this, automatically generating answers to patient questions by considering their medical records is a critical solution. In this study, we proposed a clinical question answering system for the BioNLP 2025 Shared Task on Grounded Electronic Health Record Question Answering. The system processed each patient message case by selecting relevant sentences as evidences from the associated clinical notes and generating a concise, medically accurate answer to the patient's question. A generative AI model from OpenAI (GPT-4o) was leveraged to assist with sentence selection and answer generation. Each response is grounded in source text, limited to 75 words, and includes sentence-level citations. The system was evaluated on 100 test cases using alignment, citation, and summarization metrics. Our results indicate the significant potential of the clinical question answering system based on generative AI models to streamline communication between patients and healthcare providers by automatically generating responses to patient messages.

1 Introduction

Patient portal messaging has become a critical communication channel between patients and healthcare providers, extending interaction beyond scheduled visits (Huang, Fan et al. 2022, Huang, Khurana et al. 2023). This platform enables dynamic exchanges on complex issues like new symptoms, disease follow-ups, medication concerns, and other medical inquiries (De, Huang et al. 2021, Huang, Wen et al. 2022).

With the increasing adoption of digital technologies by healthcare organizations to foster patient engagement and care, patient portals have become more prevalent, leading to a substantial surge in portal message volume (Huang, Khurana et al. 2022, Zhou, Arriaga et al. 2022). While this increased communication holds the promise of improved patient care and satisfaction, it has also created challenges in terms of efficient management and timely responses. Consequently, secure messaging has contributed to a heavier workload and burnout among clinicians by increasing patient-clinician interactions between in-person visits. For instance, primary care physicians commonly spend 1.5 hours daily processing around 150 inbox messages, often extending their work beyond regular clinic hours (Akbar, Mark et al. 2021). This constant influx of patient messages has become a significant stressor in clinical settings, particularly for primary care physicians, exacerbating burnout. Thus, the development of a clinical question answering system that can automatically generate answers to patient questions derived from their messages is essential to aid clinicians in responding effectively to patient portal communications (Ren, Wu et al. 2023, Ren, Wu et al. 2024).

The BioNLP 2025 shared task on grounded question answering (QA) from electronic health records (EHRs) focuses on automatically generating answers to patients' health-related questions that are grounded in the evidence from patients' clinical notes (Soni and Demner-Fushman 2025a). This QA task emphasizes direct citation of supporting evidence and grounding within the relevant clinical notes of patients. The need for accurate, transparent, and reproducible QA methods is especially important in clinical settings, where misinterpretation or hallucination can lead to critical errors.

This paper presents a clinical QA system developed leveraging generative AI models. The system selects sentences relevant to the clinical

BioNLP 2025 Shared Tasks, pages 91–95
August 1, 2025 ©2025 Association for Computational Linguistics

question and uses them to generate a plain-language response. No training data, external models, or automation was used. The emphasis throughout development was on traceability, consistency, and alignment with the shared task format.

2 Methods

2.1 Dataset

The dataset for this task includes patient questions (based on real patient queries) and associated EHR data (from MIMIC-III) containing vital clinical evidence (Soni and Demner-Fushman 2025b). Each question-note combination is a "case." Clinical note excerpts are provided with pre-assigned sentence numbers, which systems must use for citing evidence. Additionally, each sentence is manually annotated with a "relevance" label ("essential," "supplementary," or "not-relevant") indicating its role in answering the question. The development set of 20 cases provides these relevance labels. The test set contains 100 cases without the relevance labels.

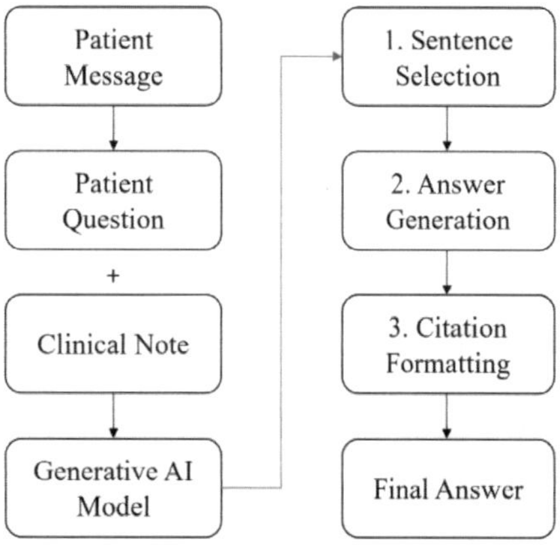

Figure 1: Overview of the clinical QA system for the BioNLP 2025 shared task

2.2 System Design

As shown in Figure 1, the clinical QA system for automatically generating answers to patient questions was implemented as a three-step pipeline applied for each patient message:

1. Sentence Selection – identifying essential and supplementary sentences from the clinical note

2. Answer Generation – using a structured prompt to compose a response with Generative AI models

3. Citation Formatting – ensuring each sentence is properly cited using its unique sentence ID

All work was done directly in an interactive session of ChatGPT (GPT-4o) (Hurst, Lerer et al. 2024) through HIPAA compliant Azure OpenAI Studio, without the use of application programming interface (APIs) and model fine-tuning.

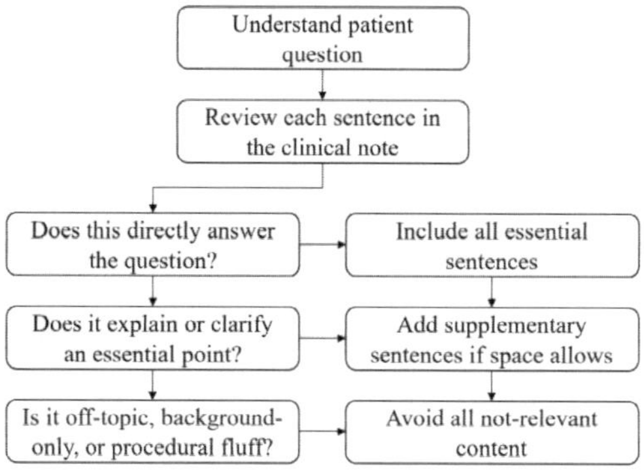

Figure 2: Sentence selection logic

2.3 Sentence Selection Strategy

Each sentence from the clinical note excerpt was reviewed and categorized as one of the following three categories:

1. Essential – directly answers the clinician's question (e.g., diagnosis, treatment, hospital course)
2. Supplementary – adds clinical context (e.g., medications, labs, background)
3. Not relevant – unrelated or duplicative information

Sentences were selected based on clinical reasoning and their alignment with the question's intent. If essential information was distributed across multiple sentences or incomplete without context, supplementary sentences were added for clarity.

To refine the sentence selection process, we used the 20 development cases provided with gold-standard relevance annotations. These cases included sentence-level labels (essential, supplementary, or not relevant), which allowed us to evaluate how well different selection strategies automatically designed by generative AI aligned with human annotations. Several iterations of sentence triage logic were tested and revised based on comparisons to these keys. The final sentence selection logic is illustrated in Figure 2. The process to develop the final sentence selection logic is detailed in Section 2.3.1. Once finalized, this selection logic was applied to the 100 test cases, which were processed without access to gold relevance labels.

2.3.1 Self-evaluation Approach for Sentence Selection

To improve sentence selection, we used ChatGPT to perform a self-evaluation analysis based on the 20-case development set. After ChatGPT generated answers without access to these labels, we uploaded the annotations to assess its performance by providing the following prompt:

"I'm going to upload a file which, for each case, shows which sentences are essential, supplementary, or not relevant. I want you to analyze how you did on using essential sentences for your answer—how many did you use, how many did you miss, etc.? Do the same for supplementary and not-relevant sentences."

Based on this self-evaluation analysis, ChatGPT recommended several logic changes, which we adopted in the final system: (1) Read the clinician question to determine its clinical focus (e.g., diagnosis, treatment, prognosis). (2) Classify note sentences as essential, supplementary, or not relevant based on their alignment with the question. (3) Generate the answer by including essential sentences first, then supplementary ones if needed. This refinement process—enabled by prompting ChatGPT to self-assess—improved the completeness of generated answers, particularly in aligning with the information explicitly required by the question.

2.4 Answer Generation and Citation Formatting

The answer to the patient question was composed using a structured prompt. The prompt included the question, the selected sentences, and explicit instructions. A typical prompt was as follows:

"Write a medically accurate answer to the question below using only the sentences provided. Limit the answer to 75 words. Keep the language clear and professional. At the end of each sentence in your answer, cite the original sentence ID in this format: |ID|."

Additionally, the prompt strategy enables all generated answers to: (1) remain under 75 words (2) cite each supporting sentence using its ID (|sentence_id|) (3) use only content from the provided note excerpt (4) be written in medically appropriate, clear language. This constraint-based format ensured that responses were traceable and aligned with the evidence selection.

2.5 Evaluation

The generated answers will be evaluated on two key aspects: Factuality (how well they are grounded in clinical evidence) and Relevance (how well they answer the question). Factuality is measured using Precision, Recall, and F1 scores by comparing the evidence sentences cited in the generated answer against a manually annotated ground truth set of essential and supplementary sentences. Two F1 scores are calculated: a strict score considering only 'essential' sentences as correct evidence, and a lenient score including both 'essential' and 'supplementary' sentences. Relevance is assessed by comparing the generated answer text to the ground truth 'essential' sentences and the original question using metrics like BLEU (Papineni, Roukos et al. 2002), ROUGE (Lin 2004), SARI (Xu, Napoles et al. 2016), BERTScore (Zhang, Kishore et al. 2019), AlignScore (Zha, Yang et al. 2023), and MEDCON (Yim, Fu et al. 2023). The overall score for ranking will be the average of the strict Factuality F1 score

	Metric	Min	Max	Mean	Median	Score
Overall	**Overall**	19.3	53.7	39.8	39.2	37.8
	Factuality	13.2	60.5	47.7	45.3	47.8
	Relevance	25.2	48.8	31.8	33.1	27.8
Factuality	**Strict F1(i)**	13.2	60.5	47.7	45.3	47.8
	Strict F1(a)	18.7	62.6	51.4	48.5	49.0
	Lenient F1(i)	13.5	62.7	48.8	46.4	49.7
	Lenient F1(a)	18.6	64.8	52.6	50.0	51.8
Relevance	**BLEU**	0.1	14.3	1.7	2.6	0.6
	ROUGE	15.2	46.5	22.7	24.3	20.0
	SARI	36.7	73.1	54.4	55.5	56.7
	BERTScore	19.9	53.9	26.3	28.3	24.2
	AlignScore	35.2	92.4	52.9	54.2	35.4
	MEDCON	23.2	49.3	32.9	33.8	29.6

*F1(i) and F(a) denote F1 (micro) and F1 (macro), respectively.

Table 1: Analysis of key performance metrics

and a combined score derived from the normalized Relevance metrics.

3 Results

The clinical QA system was evaluated on 100 test cases using the official metrics provided by the shared task organizers. Its key performance metrics among the 30 participants are listed in Table 1.

Among the overall metrics (Factuality and Relevance), Factuality performance was relatively strong at 47.8, exceeding both the mean (47.7) and median (45.3). This indicates a consistent use of relevant evidence sentences. The strict and lenient micro F1 scores (47.8 and 49.7, respectively) were also higher than their respective means and medians.

For Relevance, the system scored 27.8, slightly lower than mean (31.8) and median (33.1). The score of SARI (56.7) is higher than mean (54.4) and median (55.5), suggesting the answers were readable and cleanly edited. However, metrics like ROUGE-Lsum (20.0), BLEU (0.6), BERTScore (24.2), and MEDCON (29.6) were slightly lower than mean and median because the system focused on giving short, evidence-backed answers rather than exact matches to the reference summaries.

4 Discussion

This study explored using a single generative AI model (GPT-4o) through OpenAI's interact session and prompts to generate answers to patient questions with evidence from their medical records. Our goal focused on the straightforward application of readily accessible generative AI models via the interact session, rather than developing complex clinical QA models. This approach leverages the easy deployment of generative AI, which bypasses the need for in-depth model development expertise such as API calls, fine-tuning, and knowledge injection.

The performance of the clinical QA system was comparable to the mean and median, indicating the feasibility of using a single generative AI for answering patient questions via direct interaction. Its stronger performance in Factuality compared to the mean and median highlights the effectiveness of the designed sentence selection logic in consistently utilizing relevant evidence for answer generation.

5 Limitations

The system was developed under tight time constraints. While the current system only used a single generative AI model and straightforward interactive workflow, our plans included experimentation with multiple strategies involving different generative AI models for a hybrid system, collaborative learning, and advanced evidence sentence classification. These extensions were not explored due to lack of time.

The generative AI model (ChatGPT) was accessed through the web interface for simplicity. Although the interactive session allows the ease use of generative AI models, this limited reproducibility and scalability. The interactive nature of the workflow made it difficult to test multiple prompting strategies at scale or implement programmatic validation. Use of the API could have enabled more consistent experimentation and versioning.

6 Conclusion

We present a clinical QA system developed through an interactive workflow with a generative AI model. The system selects relevant sentences and uses them to construct a short, evidence-grounded answer with sentence-level citations. No model fine-tuning or APIs were required. Our findings show the feasibility of the strategy to develop a clinical QA system for generating answers to patient questions in portal messages. Future work may explore multi-model workflows, collaborative learning, and more structured evaluation pipelines.

Acknowledgments

Research reported in this publication was supported by the National Library of Medicine of the National Institutes of Health under award numbers R01LM011934, the National Human Genome Research Institute under award number R01HG012748, the National Institute of Aging under award number R01AG072799, and the Cancer Prevention Institute of Texas (CPRIT) under award number RR230020. The content is solely the responsibility of the authors and does not necessarily represent the official views of the National Library of Medicine, the National Human Genome Research Institute, the National Institutes of Health, or the State of Texas.

References

Akbar, F., G. Mark, E. M. Warton, M. E. Reed, S. Prausnitz, J. A. East, M. F. Moeller and T. A. Lieu (2021). "Physicians' electronic inbox work patterns and factors associated with high inbox work duration." Journal of the American Medical Informatics Association 28(5): 923-930.

De, A., M. Huang, T. Feng, X. Yue and L. Yao (2021). "Analyzing patient secure messages using a fast health care interoperability resources (FIHR)–based data model: development and topic modeling study." Journal of medical Internet research 23(7): e26770.

Huang, M., J. Fan, J. Prigge, N. D. Shah, B. A. Costello and L. Yao (2022). "Characterizing patient-clinician communication in secure medical messages: retrospective study." Journal of Medical Internet Research 24(1): e17273.

Huang, M., A. Khurana, G. Mastorakos, A. Wen, H. He, L. Wang, S. Liu, Y. Wang, N. Zong and J. Prigge (2022). "Patient portal messaging for asynchronous virtual care during the COVID-19 pandemic: retrospective analysis." JMIR Human Factors 9(2): e35187.

Huang, M., A. Khurana, G. Mastorakos, J. Zhou, N. Zong, Y. Yu, J. E. Prigge, C. A. Patten, H. Liu and B. A. Costello (2023). Characterizing the Users of Patient Portal Messaging: A Single Institutional Cohort Study. 2023 IEEE 11th International Conference on Healthcare Informatics (ICHI).

Huang, M., A. Wen, H. He, L. Wang, S. Liu, Y. Wang, N. Zong, Y. Yu, J. E. Prigge and B. A. Costello (2022). "Midwest rural - urban disparities in use of patient online services for COVID - 19." The Journal of Rural Health 38(4): 908-915.

Hurst, A., A. Lerer, A. P. Goucher, A. Perelman, A. Ramesh, A. Clark, A. Ostrow, A. Welihinda, A. Hayes and A. Radford (2024). "Gpt-4o system card." arXiv preprint arXiv:2410.21276.

Lin, C.-Y. (2004). Rouge: A package for automatic evaluation of summaries. Text summarization branches out.

Papineni, K., S. Roukos, T. Ward and W.-J. Zhu (2002). Bleu: a method for automatic evaluation of machine translation. Proceedings of the 40th annual meeting of the Association for Computational Linguistics.

Ren, Y., D. Wu, A. Khurana, G. Mastorakos, S. Fu, N. Zong, J. Fan, H. Liu and M. Huang (2023). Classification of Patient Portal Messages with BERT-based Language Models. 2023 IEEE 11th International Conference on Healthcare Informatics (ICHI).

Ren, Y., Y. Wu, J. W. Fan, A. Khurana, S. Fu, D. Wu, H. Liu and M. Huang (2024). "Automatic uncovering of patient primary concerns in portal messages using a fusion framework of pretrained language models." Journal of the American Medical Informatics Association 31(8): 1714-1724.

Soni, S. and D. Demner-Fushman (2025a). Overview of the ArchEHR-QA 2025 Shared Task on Grounded Question Answering from Electronic Health Records. The 24th Workshop on Biomedical Natural Language Processing and BioNLP Shared Tasks. Vienna, Austria, Association for Computational Linguistics.

Soni, S. and D. Demner-Fushman (2025b). "A Dataset for Addressing Patient's Information Needs related to Clinical Course of Hospitalization." arXiv preprint.

Xu, W., C. Napoles, E. Pavlick, Q. Chen and C. Callison-Burch (2016). "Optimizing statistical machine translation for text simplification." Transactions of the Association for Computational Linguistics 4: 401-415.

Yim, W.-w., Y. Fu, A. Ben Abacha, N. Snider, T. Lin and M. Yetisgen (2023). "Aci-bench: a novel ambient clinical intelligence dataset for benchmarking automatic visit note generation." Scientific data 10(1): 586.

Zha, Y., Y. Yang, R. Li and Z. Hu (2023). "AlignScore: Evaluating factual consistency with a unified alignment function." arXiv preprint arXiv:2305.16739.

Zhang, T., V. Kishore, F. Wu, K. Q. Weinberger and Y. Artzi (2019). "Bertscore: Evaluating text generation with bert." arXiv preprint arXiv:1904.09675.

Zhou, J., R. I. Arriaga, H. Liu and M. Huang (2022). A Tale of Two Perspectives: Harvesting System Views and User Views to Understand Patient Portal Engagement. 2022 IEEE 10th International Conference on Healthcare Informatics (ICHI).

LAMAR at ArchEHR-QA 2025: Clinically Aligned LLM-Generated Few-Shot Learning for EHR-Grounded Patient Question Answering

Seksan Yoadsanit[1,2], **Nopporn Lekuthai**[1,2], **Watcharitpol Sermsrisuwan**[1,2], **Titipat Achakulvisut**[1]

[1] Department of Biomedical Engineering, Faculty of Engineering, Mahidol University, Nakhon Pathom, Thailand
[2] Faculty of Medicine Ramathibodi Hospital, Mahidol University, Bangkok, Thailand
Correspondence: titipat.ach@mahidol.ac.th

Abstract

This paper presents an approach to answering patient-specific medical questions using electronic health record (EHR) grounding with ArchEHR-QA 2025 datasets. We address medical question answering as an alignment problem, focusing on generating responses factually consistent with patient-specific clinical notes through in-context learning techniques. We show that LLM-generated responses, used as few-shot examples with GPT-4.1 and Gemini-2.5-Pro, significantly outperform baseline approaches (overall score = 49.1), achieving strict precision, recall, and F1-micro scores of 60.6, 53.6, and 56.9, respectively, on the ArchEHR-QA 2025 test leaderboard. It achieves textual similarity between answers and essential evidence using BLEU, ROUGE, SARI, BERTScore, AlignScore, and MEDCON scores of 6.0, 32.1, 65.8, 36.4, 64.3, and 43.6, respectively. Our findings highlight the effectiveness of combining EHR grounding with few-shot examples for personalized medical question answering, establishing a promising approach for developing accurate and personalized medical question answering systems. We release our code at https://github.com/biodatlab/archehr-qa-lamar.

1 Introduction

Large language models (LLMs) have significantly influenced medical question-answering systems by generating clinically relevant content grounded in electronic health records (EHRs) for more personalized and context-aware patient care (Yang et al., 2022). Clinical-related questions are among the most frequently asked topics, reflecting the public's natural curiosity about their health literacy and the rising healthcare costs in many countries, which drive individuals to seek alternative sources of information (Savery et al., 2020). Despite some hallucinations, recent frontier models typically maintain reasonable factual accuracy. We theorized that aligning with human expectations on answering style, citation practices, and information structuring is the main challenge. Thus, we formulate our approach to align the model response to human expectation with the limited data provided in this shared task.

While fine-tuning LLMs on medical records or textbooks can improve alignment, it demands extensive datasets, limiting scalability (Singhal et al., 2023). Few-shot learning offers a promising alternative by guiding models with representative examples that demonstrate task-specific reasoning patterns without requiring fine-tuning, though designing optimal examples remains challenging (Brown et al., 2020). Similarly, Retrieval-augmented generation (RAG) enables LLMs to access external knowledge sources such as structured medical databases and clinical literature, providing accurate, up-to-date answers by incorporating the medical knowledge without retraining (Alkhalaf et al., 2024; Lewis et al., 2020). However, questions remain about how effectively retrieved information is integrated and grounded in the model's final output, particularly in clinical contexts where alignment with human preferences is crucial.

In this paper, we present an approach to answering patient-specific medical questions using electronic health record (EHR) grounding with the ArchEHR-QA 2025 dataset (Soni and Demner-Fushman, 2025b). We address medical question answering as an alignment problem, focusing on generating responses factually consistent with patient-specific clinical notes through in-context learning techniques. Our system leverages LLM-generated responses as few-shot examples with GPT-4.1 and Gemini-2.5-Pro, achieving strict precision, recall, and F1-micro scores of 60.6, 53.6, and 56.9, respectively, on the test leaderboard.

BioNLP 2025 Shared Tasks, pages 96–103
August 1, 2025 ©2025 Association for Computational Linguistics

2 Related work

Large Language Models (LLMs) have demonstrated significant potential across diverse medical question-answering applications. Initial research focused on general medical knowledge retrieval (Shi et al., 2024), while subsequent work has expanded into specialized domains including USMLE-style multiple-choice questions (Lucas et al., 2024), clinical decision support (Benary et al., 2023), medical exam preparation (Artsi et al., 2024), and patient-facing information systems (Goodwin et al., 2022). Despite these advances, LLMs continue to face challenges. Hallucinations remain a key concern in medical settings (Agarwal et al., 2024), and newer models have made progress in reducing them (Kim et al., 2025). However, real-world EHRs introduce an even bigger hurdle: clinical data are often messy, incomplete, and inconsistent (Holmes et al., 2021). Issues such as outdated knowledge and inconsistent reasoning also persist and demand ongoing attention (Ji et al., 2023).

In-context learning (ICL) provides an efficient alternative to model fine-tuning, enabling LLMs to learn from demonstrations embedded directly in prompts without requiring parameter adjustments. Dong et al. (2022) demonstrate that ICL leverages pre-trained capabilities to recognize task patterns from limited examples, reducing dependency on supervised datasets (Dong et al., 2024). Few-shot prompting, popularized by Brown et al. (2020) with GPT-3, showed that LLMs can achieve competitive performance across diverse tasks, including medical question answering, by conditioning on carefully selected examples. This approach significantly reduces barriers to adapting LLMs for specialized applications like clinical reasoning without requiring domain-specific retraining or extensive annotated data (Brown et al., 2020).

Alkhalaf et al. demonstrated that combining generative AI with Retrieval-Augmented Generation (RAG) significantly improves clinical information extraction from EHRs, achieving 99.25% accuracy using LLaMA 2 13B with zero-shot prompting (Alkhalaf et al., 2024). Beyond methodology, RAG component quality is critical for performance, as highlighted by research using the MEDRAG toolkit across 41 configurations with varying models, retrievers, and knowledge corpora. This comprehensive analysis revealed that properly implemented RAG systems can boost accuracy by up to 18% compared to chain-of-thought prompting across multiple medical question-answering tasks, emphasizing the importance of appropriate language model selection, retrieval strategy, and knowledge corpus construction (Xiong et al., 2024).

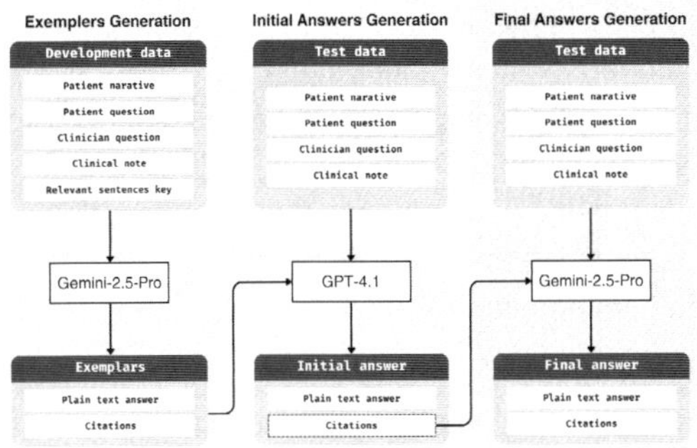

Figure 1: The multistage few-shot prompting pipeline of our system submitted to ArchEHR-QA 2025.

3 Datasets

We utilize the ArchEHR-QA 2025 dataset (Soni and Demner-Fushman, 2025a) , which comprises 20 development cases and 100 testing question-note pairs. The dataset includes a patient question, a clinician question, and a clinical note. The development set features ground-truth annotations for evidence sentences, while the test set requires natural language answers accompanied by cited sentence numbers. The patient's question is inspired by real patient inquiries. Clinical note excerpts are derived from the MIMIC-III database (Johnson et al., 2016). Answers consist of sentences referenced by the ID from the clinical note.

4 Methodology

We viewed the problem as an alignment issue. We aimed to generate an answer that was correctly cited and factually aligned with the clinical note. To help generate aligned answers, we explored zero-shot, few-shot prompting, and retrieval augmented generation (RAG) with external sources including MedlinePlus and Merck Manual.

4.1 Baseline

We applied zero-shot and chain-of-thought (CoT) prompting conditions as our baselines. We select non-thinking models, including GPT-4.1, Gemini-2.0-Flash, or Claude-3.7-Sonnet (non-thinking), due to their significant computational and financial overhead. Each model was prompted to reason step by step before generating a final answer, providing a reference for measuring the impact of few-shot examples and retrieval-augmented generation (RAG).

4.2 In-Context Learning through Few-shot Prompting

We explored several few-shot prompting strategies, including:

- **Basic Few-shot.** We selected two examples from the ArchEHR website as few-shot.
- **LLM-Generated Exemplars as Few-shot.** Since the relevant sentence labeling can only be found in the development dataset, we used Gemini-2.5-Pro to generate answers from the development set. These answers, paired with their corresponding clinical notes and questions, were used as few-shot examples in subsequent prompts.
- **LLM-Generated Exemplars with Reasoning.** We want to see if examples with reasoning can help improve the answer. Here, we used Gemini-2.5-Pro to generate both reasoning steps and final answers. These reasoning-annotated examples were included in prompts to simulate clinical thinking.

4.3 Retrieval-Augmented Generation (RAG)

We tested external context enhancement using 10,232 MedlinePlus (National Library of Medicine (US), 2025) and 2,927 Merck Manual articles (Merck & Co., 2025). This experiment aimed to determine whether external medical knowledge could improve answer accuracy over a few-shot exemplar. Articles were embedded using MedCPT (Jin et al., 2023) and indexed for retrieval. We compared 3 retrieval approaches:

- **Full-Text Clinical Articles.** We retrieved the complete texts of relevant clinical publications identified by our RAG pipeline. We input the entire article to provide the model with comprehensive contextual information for answer generation.
- **Concise Article Summaries.** We prompted Gemini-2.5-Pro to distill each full-text article into a one-paragraph summary to reduce prompt length and boost information density.
- **Synthetic Clinical Cases.** We prompted Gemini-2.5-Pro with a few-shot examples from the ArchEHR page to transform and format the retrieved articles into a realistic patient scenario featuring patient narrative, patient question, clinical question, clinical notes, and answer to mimic the ArchEHR dataset.

5 Evaluation

Each answer includes sentences and their references to the clinical note. Generated sentences are evaluated on factuality and relevance. Factuality compares cited evidence to ground truth using precision, recall, and F1 scores, with both strict (essential sentences only) and lenient (essential and supplementary sentences). Relevance measures textual similarity between answers and essential evidence using BLEU (Papineni et al., 2002), ROUGE (Lin, 2004), SARI (Xu et al., 2016), BERTScore (Zhang et al., 2019), AlignScore (Zha et al., 2023), and MEDCON (Yim et al., 2023). The final score averages normalized Strict Citation F1 and composite Relevance.

6 Experimental Setup

We use zero-shot and chain-of-thought prompting with GPT-4.1, Gemini-2.0 Flash, and Claude-3.7-Sonnet as our baseline. In few-shot prompting, we use Gemini-2.5-Pro to generate 19 exemplar answers for each item in the development set, excluding the item itself. Generated answers are sent to Gemini-2.5-Pro to trim and summarize answers to a 75-word limit. We set up top-k=5 in retrieval for all RAG experiments.

7 Results and discussions

We evaluated various prompting and retrieval strategies on the development dataset to assess their impact on citation accuracy, factuality, and relevance across multiple language models. These experiments informed our final multi-stage pipeline design for the test dataset. The following sections present key results and their implications for our system.

7.1 Development dataset observation

7.1.1 Zero-shot Baselines

On the development set, GPT-4.1, Gemini-2.0-Flash, and Claude-3.7-Sonnet achieved overall zero-shot scores of 47.9, 45.0, and 48.1, respectively, with Claude-3.7-Sonnet performing best (Ta-

Approach	Model	Development dataset																				
		Overall	Factual.	Relev.	SP_μ	SR_μ	SF_μ	LP_μ	LR_μ	LF_μ	SP_M	SR_M	SF_M	LP_M	LR_M	LF_M	BLEU	ROUGE-L	SARI	BERTScore	AlignScore	MEDCON
Baseline	GPT-4.1	47.9	53.8	42.1	54.0	53.6	53.8	68.6	49.7	57.7	60.4	64.1	57.1	74.2	59.4	62.2	6.9	34.1	69.5	37.3	61.8	42.7
	Gemini-2.0-Flash	45.0	49.2	40.8	55.5	44.2	49.2	70.0	40.7	51.5	60.2	53.7	53.5	73.0	50.7	56.2	6.3	30.6	65.4	35.9	65.4	41.0
	Claude-3.7-Sonnet	48.1	57.1	39.2	51.1	64.5	57.1	65.5	60.3	62.8	52.6	72.5	57.5	66.9	68.0	64.1	5.7	31.6	64.2	35.6	56.6	41.2
Few-shot: basic	GPT-4.1	48.6	57.8	39.4	**60.8**	55.1	57.8	72.8	48.1	58.0	**65.5**	64.5	59.2	**79.0**	59.6	62.5	5.5	31.4	65.5	35.6	57.1	41.3
	Gemini-2.0-Flash	47.5	56.3	38.8	54.0	58.7	56.3	66.0	52.4	58.4	56.4	66.4	56.2	70.5	62.3	61.3	5.2	30.2	65.5	33.8	56.3	41.9
	Claude-3.7-Sonnet	48.1	55.6	40.6	49.7	63.0	55.6	62.3	57.7	59.9	55.1	73.9	57.4	67.8	70.1	63.5	7.0	32.8	66.3	35.2	60.7	41.5
Few-shot: LLM-generated exemplars	GPT-4.1	**51.5**	**61.1**	41.8	56.4	66.7	**61.1**	71.8	61.9	**66.5**	60.3	**77.2**	**63.6**	76.9	74.0	**71.4**	6.3	32.7	66.5	36.4	64.0	44.9
	Gemini-2.0-Flash	50.5	59.6	41.5	54.9	65.2	59.6	65.9	57.1	61.2	56.9	75.9	59.8	68.8	69.5	63.6	7.5	32.1	66.7	36.2	60.9	45.6
	Claude-3.7-Sonnet	49.6	58.4	40.7	51.1	**68.1**	58.4	65.2	63.5	64.3	53.4	76.3	58.4	69.1	72.0	65.8	7.1	31.7	65.8	35.2	61.3	43.2
Few-shot: exemplars with reasoning	GPT-4.1	47.3	55.6	39.0	52.2	59.4	55.6	67.5	56.1	61.3	56.6	70.9	58.2	75.1	70.8	67.3	6.1	31.6	65.8	36.5	54.6	39.6
	Gemini-2.0-Flash	51.0	58.1	**43.9**	54.4	62.3	58.1	65.8	55.0	59.9	57.2	74.2	59.3	68.5	68.9	64.0	**9.8**	**35.5**	70.6	**40.0**	60.3	**47.4**
	Claude-3.7-Sonnet	49.3	55.5	43.1	58.4	52.9	55.5	68.0	45.0	54.1	61.7	60.5	57.3	73.6	55.2	58.2	9.1	34.6	**70.8**	38.2	64.1	41.6
RAG: articles	GPT-4.1	46.4	53.3	39.5	47.0	61.6	53.3	65.2	62.4	63.8	53.1	71.1	55.9	70.2	70.2	66.4	6.8	32.4	64.4	36.5	54.0	43.0
	Gemini-2.0-Flash	45.7	50.0	41.5	56.4	44.9	50.0	**73.6**	42.9	54.2	61.0	57.2	53.2	76.8	55.0	59.4	6.9	32.8	67.2	35.6	66.4	40.1
	Claude-3.7-Sonnet	47.4	55.3	39.5	51.2	60.1	55.3	66.0	56.6	61.0	56.2	69.8	56.4	70.1	65.5	62.6	6.9	31.6	67.0	37.3	58.0	36.4
RAG: article summaries	GPT-4.1	46.7	52.8	40.5	49.1	57.2	52.8	62.7	53.4	57.7	54.9	66.6	55.4	68.8	63.0	61.1	7.2	33.4	67.4	36.8	55.4	42.9
	Gemini-2.0-Flash	45.7	49.0	42.3	52.0	46.4	49.0	60.2	39.2	47.4	57.1	55.1	50.0	66.7	49.9	51.1	6.4	33.0	66.8	38.7	**67.1**	41.7
	Claude-3.7-Sonnet	46.9	54.7	39.1	47.1	65.2	54.7	66.0	**66.7**	66.3	50.6	76.5	56.4	68.0	**74.7**	67.0	5.3	31.4	64.2	34.4	59.1	40.4
RAG: synthetic cases	GPT-4.1	47.1	56.3	37.9	50.0	64.5	56.3	62.4	58.7	60.5	55.6	74.2	59.8	68.3	68.3	65.1	4.8	29.2	64.0	34.6	54.7	40.3
	Gemini-2.0-Flash	48.9	58.2	39.7	55.2	61.6	58.2	66.9	54.5	60.1	59.7	68.6	58.0	72.9	65.1	62.8	4.8	30.9	66.7	35.7	59.4	40.5
	Claude-3.7-Sonnet	47.8	55.3	40.3	51.2	60.1	55.3	66.0	56.6	61.0	56.2	69.8	56.4	70.1	65.5	62.6	6.8	32.3	69.3	37.2	61.0	35.2

Submission	Test dataset																				
	Overall	Factual.	Relev.	SP_μ	SR_μ	SF_μ	LP_μ	LR_μ	LF_μ	SP_M	SR_M	SF_M	LP_M	LR_M	LF_M	BLEU	ROUGE-L	SARI	BERTScore	AlignScore	MEDCON
Exemplars as few-shot with Gemini-2.0-Flash	48.5	54.6	**42.5**	**62.6**	48.4	54.6	**65.9**	48.2	55.6	**67.6**	62.7	58.7	**71.4**	60.2	59.7	**6.3**	31.9	**67.7**	**37.0**	**68.7**	43.3
Exemplars as few-shot with GPT-4.1	48.6	**57.3**	39.8	61.4	**53.8**	**57.3**	64.7	**53.7**	**58.7**	65.7	**64.2**	**60.4**	70.2	**62.3**	**62.0**	4.2	29.6	64.6	33.8	63.7	43.1
Multistage few-shot prompting (Figure 1)	**49.1**	56.9	41.4	60.6	53.6	56.9	64.0	53.5	58.3	65.4	64.0	60.2	70.0	62.2	61.8	6.0	**32.1**	65.8	36.4	64.3	**43.6**

Table 1: Results on development and test sets. SP = Strict Precision, SR = Strict Recall, SF = Strict F1, LP = Lenient Precision, LR = Lenient Recall, LF = Lenient F1. Subscripts μ and M denote micro and macro respectively.

ble 1). Across all models, we observed consistently high macro recall but low micro recall. This suggests that while models can identify relevant evidence across different cases, they often fail to capture all necessary citations in longer notes with many sentences, indicating challenges in evidence completeness for long and complex cases.

7.1.2 Few shots outcome

GPT-4.1 with LLM-generated exemplars achieved the highest overall score of 51.5 on the development set, outperforming the reasoning-based few-shot approach. These exemplars notably improved factual recall, increasing the overall factuality score from 53.8 (baseline) to 61.1 without relying on external data. This highlights that well-structured, relevant examples can enhance the model's ability to cite appropriate evidence. In contrast, the reasoning-based few-shot setup achieved a lower overall score of 47.3, compared to 51.5 for few-shot prompting without reasoning. This suggests that explicitly including reasoning steps may not yield additional benefit in this task, and that the model may perform implicit reasoning more effectively when guided by concise, LLM-curated exemplars.

7.1.3 RAG: Full text vs. Article summary vs. Synthetic clinical cases

Among RAG variants with Gemini-2.0-Flash, top-5 synthetic cases yielded the best performance, achieving an overall score of 48.9 and improving factuality from 49.2 to 58.2 compared to the baseline. This suggests structured, case-like inputs better support clinical reasoning than unstructured text. RAG using full-text articles and summaries produced lower factuality scores (50.0 and 49.0, respectively). Although converting articles into cases improved alignment, these formats remained less effective than LLM-crafted exemplars, likely due to misalignment between retrieved content and the target question. Overall, the RAG approaches performed worse than the best few-shot LLM-generated exemplars. We hypothesized that the quality and relevance of in-context examples may be more important than retrieved knowledge.

7.2 Test dataset results

Based on the development set results, GPT-4.1 with LLM-generated exemplars as few-shot achieved the best overall performance. On the test set, GPT-4.1 demonstrated strong factuality (overall factuality = 57.3), while Gemini-2.5-Pro performed better in terms of relevance (overall relevance = 42.5). We leveraged both models by developing a multistage few-shot prompting pipeline without external data for our final submission, achieving an overall score of 49.1 (Table 1). This pipeline uses Gemini-2.5-Pro to generate 20 exemplar answers with citations from the development dataset. These 20 exemplars are used as in-context examples for GPT-4.1's initial answer generation on the test dataset. We then extract references from these initial answers. In the final stage, we input the test dataset and its corresponding retrieved references into Gemini-2.5-Pro to generate the final grounded answers (Figure 1).

8 Conclusion and Future Work

Our study demonstrates that few-shot learning with LLM-generated examples significantly improves EHR-grounded medical question answering. We achieved performance gains on the ArchEHR-QA 2025 benchmark without requiring model retraining or external knowledge sources. Models can leverage implicit patterns when guided by in-context learning demonstrations. Future work may explore example selection for ICL or demonstration strategy (Zhang et al., 2024; Huang et al., 2023), which can help improve the model's alignment with the ground truth. We can also improve the reference of clinical notes to achieve better recall.

Limitations

LLM-generated few-shot examples may incorporate subtle biases or inaccuracies that propagate through the system. Our implementation relies on underlying EHR data quality, which may vary in completeness and structure across clinical settings. In practice, real-world EMR heterogeneity amplifies these challenges: clinicians document information across free-text notes, scanned documents, and copied entries that vary widely in format, often include redundant or contradictory details, and fragment critical data. Moreover, we rely on proprietary model APIs with 19–20-shot prompts, which drive up computation time and latency and limit scalability in resource-constrained settings.

Despite strong benchmark performance, real-world deployment would require the validation of our prompting strategies on unstructured production EHR systems, incorporating robust NLP preprocessing (entity normalization, de-duplication) alongside human oversight to ensure clinical safety, data privacy, and appropriateness. We also need to prune exemplars, distill models, and conduct cost–benefit analyses to reduce inference time and API costs, all while upholding data privacy and regulatory compliance.

References

Vibhor Agarwal, Yiqiao Jin, Mohit Chandra, Munmun De Choudhury, Srijan Kumar, and Nishanth Sastry. 2024. Medhalu: Hallucinations in responses to healthcare queries by large language models. *Preprint*, arXiv:2409.19492.

Mohammad Alkhalaf, Ping Yu, Mengyang Yin, and Chao Deng. 2024. Applying generative ai with retrieval augmented generation to summarize and extract key clinical information from electronic health records. *Journal of Biomedical Informatics*, 156:104662.

Yaara Artsi, Vera Sorin, Eli Konen, Benjamin S Glicksberg, Girish Nadkarni, and Eyal Klang. 2024. Large language models for generating medical examinations: systematic review. *BMC Medical Education*, 24(1):354.

Manuela Benary, Xing David Wang, Max Schmidt, Dominik Soll, Georg Hilfenhaus, Mani Nassir, Christian Sigler, Maren Knödler, Ulrich Keller, Dieter Beule, Ulrich Keilholz, Ulf Leser, and Damian T. Rieke. 2023. Leveraging large language models for decision support in personalized oncology. *JAMA Network Open*, 6(11):e2343689–e2343689.

Tom Brown, Benjamin Mann, Nick Ryder, Melanie Subbiah, Jared D. Kaplan, Prafulla Dhariwal, Arvind Neelakantan, Pranav Shyam, Girish Sastry, Amanda Askell, Sandhini Agarwal, Ariel Herbert-Voss, Gretchen Krueger, Tom Henighan, Rewon Child, Aditya Ramesh, Daniel Ziegler, Jeffrey Wu, Clemens Winter, and 12 others. 2020. Language models are few-shot learners. In *Advances in Neural Information Processing Systems*, volume 33, pages 1877–1901.

Qingxiu Dong, Lei Li, Damai Dai, Ce Zheng, Jingyuan Ma, Rui Li, Heming Xia, Jingjing Xu, Zhiyong Wu, Baobao Chang, Xu Sun, Lei Li, and Zhifang Sui. 2024. A survey on in-context learning. In *Proceedings of the 2024 Conference on Empirical Methods in Natural Language Processing*, pages 1107–1128, Miami, Florida, USA. Association for Computational Linguistics.

Travis R. Goodwin, Dina Demner-Fushman, Kyle Lo, Lucy Lu Wang, Hoa T. Dang, and Ian M. Soboroff. 2022. Automatic question answering for multiple stakeholders, the epidemic question answering dataset. *Scientific Data*, 9:432.

John Holmes, James Beinlich, Mary Regina Boland, Kathryn Bowles, Yong Chen, Tessa Cook, George Demiris, Michael Draugelis, Laura Fluharty, Peter Gabriel, Robert Grundmeier, Clarence Hanson, Daniel Herman, Blanca Himes, Rebecca Hubbard, Charles Kahn, Jr, Dokyoon Kim, Ross Koppel, Qi Long, and Jason Moore. 2021. Why is the electronic health record so challenging for research and clinical care? *Methods of information in medicine*, 60.

Ziniu Huang, Jing Zhou, Guoxin Xiao, and Gong Cheng. 2023. Enhancing in-context learning with answer feedback for multi-span question answering. *arXiv preprint arXiv:2306.04508*.

Ziwei Ji, Tiezheng Yu, Yan Xu, Nayeon Lee, Etsuko Ishii, and Pascale Fung. 2023. Towards mitigating llm hallucination via self reflection. In *Findings of the Association for Computational Linguistics: EMNLP 2023*, pages 1827–1843, Singapore. Association for Computational Linguistics.

Qiao Jin, Won Kim, Qingyu Chen, Donald C Comeau, Lana Yeganova, W John Wilbur, and Zhiyong Lu. 2023. Medcpt: Contrastive pre-trained transformers with large-scale pubmed search logs for zero-shot biomedical information retrieval. *Bioinformatics*, 39(11):btad651. Published: 01 November 2023.

Alistair E. W. Johnson, Tom J. Pollard, Lu Shen, Liwei H. Lehman, Mengling Feng, Mohammad Ghassemi, Benjamin Moody, Peter Szolovits, Leo Anthony Celi, and Roger G. Mark. 2016. Mimic-iii, a freely accessible critical care database. *Scientific Data*, 3(1):160035.

Yubin Kim, Hyewon Jeong, Shan Chen, Shuyue Stella Li, Mingyu Lu, Kumail Alhamoud, Jimin Mun, Cristina Grau, Minseok Jung, Rodrigo Gameiro, Lizhou Fan, Eugene Park, Tristan Lin, Joonsik Yoon, Wonjin Yoon, Maarten Sap, Yulia Tsvetkov, Paul Liang, Xuhai Xu, and 6 others. 2025. Medical hallucination in foundation models and their impact on healthcare. *medRxiv*.

Patrick Lewis, Ethan Perez, Aleksandra Piktus, Fabio Petroni, Vladimir Karpukhin, Naman Goyal, Heinrich Küttler, Mike Lewis, Wen-tau Yih, Tim Rocktäschel, Sebastian Riedel, and Douwe Kiela. 2020. Retrieval-augmented generation for knowledge-intensive nlp tasks. In *Advances in Neural Information Processing Systems*, volume 33, pages 9459–9474.

Chin-Yew Lin. 2004. Rouge: A package for automatic evaluation of summaries. In *Text Summarization Branches Out*, pages 74–81, Barcelona, Spain. Association for Computational Linguistics.

Mary M Lucas, Justin Yang, Jon K Pomeroy, and Christopher C Yang. 2024. Reasoning with large language models for medical question answering. *Journal of the American Medical Informatics Association*, 31(9):1964–1975.

Inc. Merck & Co. 2025. Msd manual consumer version. Retrieved May 3, 2025.

National Library of Medicine (US). 2025. Medlineplus. Bethesda (MD): National Library of Medicine; cited 2025 May 3.

Kishore Papineni, Salim Roukos, Todd Ward, and Wei-Jing Zhu. 2002. Bleu: a method for automatic evaluation of machine translation. In *Proceedings of the 40th Annual Meeting of the Association for Computational Linguistics*, pages 311–318, Philadelphia, Pennsylvania, USA. Association for Computational Linguistics.

Max Savery, Asma Ben Abacha, Soumya Gayen, and Dina Demner-Fushman. 2020. Question-driven summarization of answers to consumer health questions. *Scientific Data*, 7:322.

Yucheng Shi, Shaochen Xu, Tianze Yang, Zhengliang Liu, Tianming Liu, Quanzheng Li, Xiang Li, and Ninghao Liu. 2024. Mkrag: Medical knowledge retrieval augmented generation for medical question answering. In *Proceedings of the 2024 American Medical Informatics Association Annual Symposium (AMIA)*. Distinguished Paper Award.

Karan Singhal, Shekoofeh Azizi, Tao Tu, S. Sara Mahdavi, Jason Wei, and 1 others. 2023. Toward expert-level medical question answering with large language models. *Nature Medicine*, 29:1–9.

Sarvesh Soni and Dina Demner-Fushman. 2025a. A dataset for addressing patient's information needs related to clinical course of hospitalization. *arXiv preprint*.

Sarvesh Soni and Dina Demner-Fushman. 2025b. Overview of the archehr-qa 2025 shared task on grounded question answering from electronic health records. In *The 24th Workshop on Biomedical Natural Language Processing and BioNLP Shared Tasks*, Vienna, Austria. Association for Computational Linguistics.

Guangzhi Xiong, Qiao Jin, Zhiyong Lu, and Aidong Zhang. 2024. Benchmarking retrieval-augmented generation for medicine. In *Findings of the Association for Computational Linguistics: ACL 2024*, pages 6233–6251, Bangkok, Thailand. Association for Computational Linguistics.

Wei Xu, Courtney Napoles, Ellie Pavlick, Quanze Chen, and Chris Callison-Burch. 2016. Optimizing statistical machine translation for text simplification. *Transactions of the Association for Computational Linguistics*, 4:401–415.

Xi Yang, Aokun Chen, Nima PourNejatian, Hoo Chang Shin, Kaleb E. Smith, Christopher Parisien, Colin Compas, Cheryl Martin, Anthony B. Costa, Mona G. Flores, Ying Zhang, Tanja Magoc, Christopher A. Harle, Gloria Lipori, Duane A. Mitchell, William R. Hogan, Elizabeth A. Shenkman, Jiang Bian, and Yonghui Wu. 2022. A large language model for electronic health records. *npj Digital Medicine*, 5:194.

Wen-wai Yim, Yujuan Fu, Asma Ben Abacha, Neal Snider, Thomas Lin, and Meliha Yetisgen. 2023. Aci-bench: a novel ambient clinical intelligence dataset for benchmarking automatic visit note generation. *Scientific Data*, 10(1):586.

Yuheng Zha, Yichi Yang, Ruichen Li, and Zhiting Hu. 2023. Alignscore: Evaluating factual consistency with a unified alignment function. In *Proceedings of the 61st Annual Meeting of the Association for Computational Linguistics (Volume 1: Long Papers)*, pages 11328–11348, Toronto, Canada. Association for Computational Linguistics.

Tianyi Zhang, Varsha Kishore, Felix Wu, Kilian Q. Weinberger, and Yoav Artzi. 2019. Bertscore: Evaluating text generation with bert. In *International Conference on Learning Representations*.

Yuwei Zhang, Shi Feng, and Chenhao Tan. 2024. Demonstration selection for in-context learning via reinforcement learning. *arXiv preprint arXiv:2412.03966*.

A Appendix

This appendix provides documentation of the prompts used in our multistage few-shot pipeline (Figure 1). The following sections describe the exact prompts, design rationale, and implementation considerations used throughout the system.

A.1 Prompt for Exemplar Generation

Figure B illustrates the prompt employed for generating exemplars. This stage utilizes the development dataset, which includes the patient narrative, patient question, clinician question, clinical note, and relevant sentence key. Gemini-2.5-Pro generates concise, citation-grounded answers in plain text, which are subsequently used as few-shot examples for downstream prompting.

A.2 Prompt for Initial Answers Generation

We use the prompt in Figure A to generate initial answers from the test dataset. Input components include the patient narrative, patient question, clinician question, and clinical note. The exemplars derived from the development data (as described in Section A.1) are incorporated into the prompt for initial responses with GPT-4.1.

LLM-Generated Exemplars as Few-shot

 # Examples

`{exemplars}`

 # To answer

Patient Narrative: `{patient_narrative}`
Patient Question: `{patient_question}`
Clinician Question: `{clinical_question}`
Clinical Note: `{clinical_note}`

Return your response in the format below strictly.

<answer>
Your answer based on the things you have seen in the Example Patient Narrative, Example Patient Question, Example Clinician Question, Example Clinical Note and the Example Answer. Please do not use a hyphen ('-') in the citation. List all the citations.
</answer>

Figure A: **Prompt used for LLM-generated exemplars as few-shot for final answers generation**

Exemplars and Final Answers Generation

You are a medical expert tasked with providing clear, accurate answers to medical questions based on relevant sentences from the clinical notes. Your response should be detailed, evidence-based, and reference specific points from the relevant sentences using the numbered citations. You are only allowed to use the relevant sentences to answer the question.

Example Patient Narrative:
I had severe abdomen pain and was hospitalised for 15 days in ICU, diagnosed with CBD sludge. Doctor advised for ERCP. My question is if the sludge was there does not any medication help in flushing it out? Whether ERCP was the only cure?

Example Patient Question:
My question is if the sludge was there does not any medication help in flushing it out? Whether ERCP was the only cure?

Example Clinician Question:
Why was ERCP recommended over a medication-based treatment for CBD sludge?

Example Clinical Note:
1: During the ERCP a pancreatic stent was . . .
2: However, due to the patient's elevated INR . . .
3: Frank pus was noted . . .
4: The Vancomycin was discontinued.
5: On hospital day 4 . . .
6: On ERCP the previous biliary stent . . .
7: As the patient's INR was normalized . . .
8: At the conclusion of the procedure . . .

Example Relevant Sentences: [1, 5, 6, 7]

Example Answer:
Medications can sometimes help in managing bile duct sludge, but in this case, ERCP was necessary... |1|... |5|... |6|... |7|.

Now, please provide a similar detailed answer for the following case:

Patient Narrative: `{patient_narrative}`
Patient Question: `{patient_question}`
Clinician Question: `{clinical_question}`
Clinical Note: `{clinical_note}`
Relevant Sentences: {relevant_sentences}

Answer Format:

<answer>
Use ALL of the relevant sentences to answer the question. Make sure to answer the question based on the relevant sentences. See the example answer for the format (use the |sentence number| to reference).
</answer>

Note: Think about the question and relevant sentences carefully. You may reshuffle the sentences, but should not include any other content.

Figure B: **Prompt used for exemplars and final answers generation**

A.3 Prompt for Final Answers Generation

For the final answer generation stage, we reutilize the prompt shown in Figure B. However, instead of using the development sentence key, we provide the model with retrieved sentences cited in the initial answers. This configuration enables Gemini-2.5-Pro to generate a grounded, citation-supported response based on the test data and previously extracted evidence.

Neural at ArchEHR-QA 2025: Agentic Prompt Optimization for Evidence-Grounded Clinical Question Answering

Sai Prasanna Teja Reddy[1], Abrar Majeedi, Viswanatha Reddy Gajjala, Zhuoyan Xu, Siddhant Rai, Vaishnav Potlapalli

[1]University of Chicago
bogireddyteja@uchicago.edu

Abstract

Automated question answering (QA) over electronic health records (EHRs) can bridge critical information gaps for clinicians and patients, yet it demands both precise evidence retrieval and faithful answer generation under limited supervision. In this work, we present *Neural*, the runner-up in the BioNLP 2025 ArchEHR-QA shared task on evidence-grounded clinical QA. Our proposed method decouples the task into (1) sentence-level evidence identification and (2) answer synthesis with explicit citations. For each stage, we automatically explore the prompt space with DSPy's MIPROv2 optimizer, jointly tuning instructions and few-shot demonstrations on the development set. A self-consistency voting scheme further improves evidence recall without sacrificing precision. On the hidden test set, our method attains an overall score of 51.5, placing second stage while outperforming standard zero-shot and few-shot prompting by over 20 and 10 points, respectively. These results indicate that data-driven prompt optimization is a cost-effective alternative to model fine-tuning for high-stakes clinical QA, advancing the reliability of AI assistants in healthcare.

1 Introduction

Automatically generating answers to patients' medical questions using information from their electronic health records (EHRs) poses significant challenges, but also offers substantial potential for improving clinical communication and patient engagement (Soni and Demner-Fushman, 2025b). The ArchEHR-QA 2025 shared task directly targets this problem by providing patient questions alongside excerpts from clinicians' notes, and requiring systems to generate grounded responses that explicitly cite the supporting sentences.

Recent advances in Large Language Models (LLMs) have shown promising results in the domain of answering clinical questions based on unstructured patient notes (Singhal et al., 2025). However, fine-tuning LLMs for answering clinical questions based on unstructured patient notes is constrained by the limited availability of supervised clinical data, which increases the risk of overfitting. Consequently, prompt-based methods offer a practical and cost effective solution. Despite the advantages, prompt engineering comes with its own set of challenges (Karayanni et al., 2024). Crafting effective prompts for complex tasks often requires expert effort and iterative refinement. This difficulty is amplified in the clinical domain, where identifying the correct evidence from lengthy medical notes is critical for accurate answers. Prior studies have explored manual prompt designs and chain-of-thought cues for medical QA (Tai and Tannier, 2025), yet these ad-hoc methods may not yield optimal performance. Automated prompt optimization techniques (Wang et al., 2023) offer a systematic alternative, but often treat each task holistically and may not incorporate domain expertise effectively.

In this work, we introduce a two-stage LLM pipeline for clinical question answering that explicitly separates evidence identification and answer generation. In each stage, prompts are automatically optimized using the MIPROv2 optimizer from DSPy (Khattab et al., 2021, 2024). The first stage is dedicated to identifying the relevant information within the clinical note, while the second stage leverages this information to generate a precise and contextually appropriate answer. By decomposing the task into these two distinct stages, it becomes possible to define clear, stage-specific evaluation objectives, namely, F1 score for evidence retrieval performance and the mean of word limit score, citation format score, BLEU, ROUGE, SARI, BERT, ALIGN, and MEDCON scores for answer quality metrics. This decomposition also enables the use of optimization algorithms to systematically search for prompts that maximize these objectives. To further improve reliability, we integrate a self-

BioNLP 2025 Shared Tasks, pages 104–109

consistency (Wang et al., 2022) approach in the evidence retrieval stage: the model is run multiple times, and a majority vote over the outputs determines the final cited sentences, reducing variability and errors.

In summary, our contributions are:

- **Decomposed Prompt Optimization Framework:** We propose a two-stage pipeline that modularizes clinical QA, enabling distinct and targeted prompt optimization for evidence retrieval and answer generation, a paradigm shift from monolithic optimization approaches.

- **Systematic Instruction Space Exploration:** We demonstrate the efficacy of leveraging advanced optimizers like MIPROv2 to discover high-performing, task-specific prompt configurations from limited development data, enhancing both performance and reproducibility.

- We perform a rigorous evaluation on an expert-annotated clinical QA dataset, demonstrating that our prompt-optimized pipeline yields significant improvements in factual accuracy and answer relevance compared to established baselines, underscoring its effectiveness for reliable clinical-QA.

2 Related Work

Clinical QA: Developing QA systems for clinical data has long been an interest in biomedical NLP. Earlier datasets like emrQA (Pampari et al., 2018) generated large-scale QA pairs from electronic medical records by repurposing annotations, but these often contained synthetic questions or required mapping to structured outputs. Recent research has shown that large LLMs can achieve near-expert performance on medical QA benchmarks (Singhal et al., 2025).

Prompt Optimization: There is a growing interest in automated prompt search or optimization. More recently, methods such as APE (Zhou et al., 2022) and OPRO (Yang et al., 2023) treat prompt design as a black-box optimization problem, iteratively refining prompts by evaluating model outputs. MIPRO (Opsahl-Ong et al., 2024) extends this idea to multi-stage LLM programs, jointly optimizing the instructions and demonstration examples of each module in a pipeline. Our work leverages the latest optimizer, MIPROv2 (Opsahl-Ong et al., 2024), which uses a combination of prompt proposal and Bayesian search to find high-performing prompts efficiently.

Self-Consistency: Large LLMs can produce variable outputs given the same prompt, especially under chain-of-thought reasoning. The self-consistency decoding strategy (Wang et al., 2022) addresses this by sampling multiple outputs and choosing the result that is most consistent across samples.

3 Methodology

Our method draws on a human-inspired decoupling strategy, separating evidence gathering from solution formulation. In Stage 1, we identify relevant resources analogous to conducting a web search or literature review by retrieving key sentences. In Stage 2, we frame the final solution by synthesizing insights from the retrieved facts. We operationalize this intuition as a modular, two-stage pipeline tailored to clinical QA.

Consider each clinical note excerpt is segmented into individual sentences $s_1, s_2, \ldots, s_n$, and each sentence s_i is annotated with a label $y_i \in$ (essential, not-relevant, supplementary. The label indicates whether s_i contains information essential for answering a given patient/clinician question q. This sentence-level annotation forms the basis of Stage-1, while Stage-2 uses the content of the essential sentences (post consistency testing) to produce the final answer a_{gen}.

3.1 Sentence-Level Essentiality Classification

For a question–note pair let

$$Y^+ = \{i \mid y_i = 1\}, \text{and } \hat{Y}^+ = \{i \mid \hat{y}_i = 1\},$$

denote, respectively, the indices of *gold-standard essential* sentences and the indices predicted essential by the model. We begin with a manually crafted prompt that presents the question q and the sentence sequence $\{s_1, \ldots, s_n\}$ and requests a binary relevance label for every sentence in addressing the q.

Prompt-Optimization Objective (Stage-1): We invoke the **MIPROv2** to optimize the prompt. Treating the instruction text (and any embedded demonstrations) as discrete parameters $P \in \mathcal{P}$, MIPROv2 iteratively: (i) proposes a candidate prompt P, (ii) applies the fixed LLM to the training set, and (iii) updates P so as to *maximize* the sentence-level $F_1(Y^+, \hat{Y}^+)$. By searching this space of instructions and few-shot exemplars, the optimizer converges on a prompt P^* that elicits labels with markedly higher precision and recall, thereby yielding a more reliable evidence set for Stage 2.

Self-Consistency Voting: To improve the reliability of Stage 1, we apply a *self-consistency voting* scheme: the classifier is executed $R = 5$ times on the same $(q, \{s_i\})$ input, each run differing only in its stochastic seed. Let $\hat{y}_i^{(r)} \in \{0, 1\}$ be the binary prediction for sentence s_i in run r ($1 =$ essential). The final label is obtained by majority vote,

$$v_i = \sum_{r=1}^{R} \hat{y}_i^{(r)}, \quad \hat{y}_i = \begin{cases} 1 & \text{if } v_i \geq \tau = \lceil R/2 \rceil, \\ 0 & \text{otherwise,} \end{cases}$$

This aggregation suppresses spurious single-run errors and retains sentences identified as essential by at least three of the five passes, thereby reducing variance and boosting the expected F_1 of the evidence selection step.

3.2 Answer Generation from Essential Sentences

Let q be the input question and let $E = \{ s_i \mid \hat{y}_i = 1 \}$ denote the set of sentences that Stage 1 predicted as *essential*. Given the pair (q, E), Stage 2 must produce a concise natural-language answer a_{gen} that (i) directly addresses q, (ii) contains at most 75 words, and (iii) cites the supporting sentences in E using the required parenthetical notation. We initialise Stage 2 with a hand-written prompt template and then invoke **MIPROv2** to optimize this template. Let P denote a prompt parameterised by its instruction wording and any embedded demonstrations, and let $g_\theta(\cdot\,; P)$ be the fixed LLM generator. Given an input pair (q, E) the model outputs $a_{\text{gen}} = g_\theta((q, E); P)$.

Prompt-Optimization Objective (Stage-2): The goal is to maximise the composite reward

$$\mathcal{R}(a_{gen}, a^*, E) = \underbrace{\mathbf{1}\big[|a_{gen}| \leq 75\big]}_{\text{length}} + \underbrace{\mathbf{1}\big[\text{format}(a_{gen}, E)\big]}_{\text{citations}} + \underbrace{\tfrac{1}{6} \sum_{m \in \mathcal{M}} m(a_{\text{gen}}, a^*)}_{\text{surface \& semantic quality}}$$

where a^* is the reference answer, $|\cdot|$ counts words, and

$$\mathcal{M} = \{\text{BLEU, ROUGE, SARI, BERT, Align, MEDCON}\}.$$

The indicator terms enforce hard constraints on length and citation format, while the mean of the six metrics rewards lexical overlap, semantic fidelity, factual consistency, and medical–concept coverage.

Search Procedure: MIPROv2 explores the discrete prompt space $\mathcal{P}$ by iteratively proposing candidate prompts, evaluating them on a validation set, and selecting

$$P^* = \underset{P \in \mathcal{P}}{\text{argmax}}\ \mathbb{E}_{(q,E,a^*)} \Big[\mathcal{R}(g_\theta((q, E); P), a^*, E) \Big]$$

The optimal prompt P^* consistently elicits answers that are succinct, properly cited, and of high quality according to all surface–level and semantic metrics, thus satisfying the Stage 2 requirements.

4 Experimental Setup

Dataset: We evaluated our system on the ArchEHR-QA 2025 dataset (Soni and Demner-Fushman, 2025a). This dataset contains 120 question-note cases derived from MIMIC-III/IV clinical notes. Each case includes a patient question (often a layperson's phrasing) and a clinician-rewritten question focusing on the key medical query, along with a relevant excerpt from the patient's EHR notes. The notes are annotated with sentence numbers and labels indicating relevance ("essential," "supplementary," "not relevant") to the question. The official split provides 20 cases as a development set and 100 cases as a test set . We used the 20 development cases for prompt optimization and for all ablations. Final results on the test set were obtained via the Codabench submission system.

Evaluation Metrics: Following the official ArchEHRQA shared task protocol, we evaluate each submission along two complementary axes: *Factuality* and *Relevance*, which help capture evidence faithfulness and response quality. *Factuality* is quantified by matching the set of note sentences cited by the model against expert-annotated evidence and computing precision, recall, and F1. We report a **strict** variant that counts only *essential*

Table 1: Evaluation of participants on factuality and relevance metrics. **Bold** indicates the best performance in each column, underlined the second best. Here P^S, R^S, F_1^S denote micro-averaged *strict* precision, recall and F1; P^L, R^L, F_1^L denote micro-averaged *lenient* precision, recall and F1; AVG_{fact} and $\text{AVG}_{\text{relev}}$ are the official "Overall Factuality" and "Overall Relevance" scores, and "Overall" is the combined score. Abbreviations: R.L.= ROUGE-Lsum, B.S. = BERTScore, A.S. = AlignScore, M.C. = MEDCON .

Model	Factuality							Relevance							Overall
	P^S	R^S	F_1^S	P^L	R^L	F_1^L	AVG_{fact}	BLEU	R.L.	SARI	B.S.	A.S.	M.C.	$\text{AVG}_{\text{relev}}$	
DMISLab	57.9	59.3	58.6	61.2	59.2	60.2	58.6	**14.3**	**46.5**	36.7	**53.9**	**92.4**	**49.3**	**48.8**	**53.7**
Ours	55.4	63.8	59.3	58.4	63.7	60.9	59.3	8.5	34.1	**73.1**	39.1	67.3	40.0	43.7	51.5
LAILab	56.0	**65.5**	**60.4**	59.7	**66.0**	**62.7**	**60.4**	6.5	32.7	69.2	37.4	65.3	38.4	41.6	51.0
LAMAR	60.6	53.6	56.9	64.0	53.5	58.3	56.9	6.0	32.1	65.8	36.4	64.3	43.6	41.4	49.1
ssagarwal	68.8	36.2	47.5	71.7	35.6	47.6	47.5	4.7	31.1	70.0	36.9	74.9	38.0	42.6	45.0
Few-Shot	71.2	38.2	49.8	74.5	37.8	50.2	49.8	1.7	25.5	53.9	28.7	54.5	39.7	34.0	41.9
Zero-Shot	**71.6**	21.9	33.6	**77.0**	22.3	34.6	33.6	0.1	15.2	47.8	20.5	57.7	25.6	27.8	30.7

citations and a **lenient** variant that also accepts *supplementary* evidence, following the task guidelines. *Relevance* is evaluated as the arithmetic mean of complementary surface and semantic level metrics: BLEU (Papineni et al., 2002), ROGUE (Lin, 2004), SARI (Xu et al., 2016), BERTScore (Zhang et al., 2020), AlignScore (Zha et al., 2023), and MEDCON (Yim et al., 2023).

Baselines: To gauge the performance of our prompt optimization approach, we compare it against two baselines:

- **Zero-Shot Prompting:** A single, succinct instruction per stage. This reflects the common practice of "plug-and-play" prompting without any exemplars.

- **Few-Shot Prompting:** Adds two manually selected demonstrations to each stage's prompt but preserves the terse directive style. This isolates the value of exemplars alone, without optimization.

5 Results

Table 1 presents the comparative performance of our system alongside competing submissions on the ArchEHR-QA 2025 test set. Our approach ranked **second overall**, achieving a combined score of **51.5**, with individual scores of **59.3** for factuality and **43.7** for relevance. Crucially, our system maintained **consistently high performance across all evaluation axes**, in contrast to other systems that exhibited strong performance on isolated metrics but lacked robustness overall. We observe a **substantial margin of improvement over baseline prompting strategies**: our method outperforms the **zero-shot** and **few-shot** variants by approximately **20 and 10 points**, respectively, on the overall score. These gains underscore the effectiveness of *automated prompt optimization*, which systematically discovers high-performing instructions and demonstrations tailored to each stage of the QA pipeline. Moreover, our system's relative stability across metrics—including both surface-level (BLEU, ROUGE, SARI) and semantic (BERTScore, AlignScore, MEDCON) relevance measures—suggests that **prompt optimization not only improves individual metrics but also contributes to the holistic quality and trustworthiness of generated answers**. These findings affirm our central claim: that prompt optimization is not merely a heuristic tuning step, but a principled and impactful method for enhancing LLM-based clinical QA systems.

6 Conclusion

We propose a two-stage approach for clinical question answering on medical notes, leveraging DSPy's MIPROv2 optimizer to autonomously fine-tune prompts for each stage. In Stage 1, the method extracts essential evidence from the notes by optimizing the prompt to maximize the evidence F1 score. In Stage 2, the system generates answers by optimizing a prompt based on a composite metric incorporating several metric (word limit score, citation format score, BLEU, ROUGE, etc.), yielding concise, structured, and clinically reliable response. This prompt-optimized pipeline demonstrates substantial improvements over baselines, highlighting the efficacy of prompt optimization within a modular LLM framework. The results suggest that prompt engineering can transit from heuristic practice to data-driven optimization process, identifying high-performing prompts tailored to specific tasks. For medical question answering systems, this advancement enhances both evidence retrieval and answer trustworthiness, representing a significant step toward the development of reliable AI assistants for clinicians and patients.

Future research directions include integrating web search agents to retrieve external medical knowledge absent from clinical notes, further enriching the capabilities and completeness of automated clinical QA systems.

7 Limitations

Despite strong performance on the ArchEHR-QA benchmark, our two-stage prompt-optimized framework faces limitations rooted in both data and model design. The curated and annotated EHR excerpts used for evaluation do not reflect the messiness of real-world clinical notes, which often suffer from incompleteness, inconsistency, and institutional variability; this makes generalization across healthcare settings difficult, especially given the lack of standardization and privacy restrictions on accessing realistic data. Furthermore, the model has not been domain-adapted and relies on a generic tokenizer, potentially missing specialized medical vocabulary crucial for understanding nuanced queries. The modular two-step process, while flexible, introduces latency and risk of compounding errors, especially as the size of the candidate space in MIPROv2 grows. This reranker also depends heavily on metrics like BLEU, which can reward surface-level similarity over true semantic alignment and are sensitive to the distribution of training data. Together, these factors raise concerns about both scalability and the quality of alignment, even when evaluation scores appear strong.

8 LLM Settings

In both stages of our pipeline—sentence-level evidence identification and answer synthesis—we employ the GPT-4.1 model accessed via the OpenAI API. To accommodate the extensive clinical context and few-shot demonstrations during prompt optimization, we allocate a maximum context window of 10,000 tokens. All prompt-optimization experiments (i.e., MIPROv2's evaluation of candidate prompt templates and few-shot exemplars) are conducted with a low-variance decoding strategy, setting the temperature to 0.3. This relatively "cold" sampling regime promotes determinism, ensuring that our optimizer receives consistent feedback on prompt efficacy as measured by evidence-retrieval F1 or composite relevance metrics.

For the self-consistency mechanism in Stage 1, we leverage stochastic sampling to capture the model's latent uncertainty. Specifically, we issue R = 5 independent generations per question–note pair, each sampled at temperature 0.7. A majority-vote over these five runs determines the final label for each sentence, suppressing spurious outliers while preserving genuinely informative evidence. All other decoding parameters (e.g., top-p, frequency and presence penalties) are held at their API defaults, isolating temperature and context length as the principal levers in our experimental configuration.

9 Prompts and Code Availability

To promote transparency and reproducibility, we release all manual and optimized prompt templates, together with our full pipeline implementation at our GitHub repository[1].

References

Nader Karayanni, Aya Awwad, Chein-Lien Hsiao, and Surish P Shanmugam. 2024. Keeping experts in the loop: Expert-guided optimization for clinical data classification using large language models. *arXiv preprint arXiv:2412.02173*.

Omar Khattab, Christopher Potts, and Matei Zaharia. 2021. Relevance-guided supervision for OpenQA with ColBERT. *Transactions of the Association for Computational Linguistics*, 9:929–944.

Omar Khattab, Arnav Singhvi, Paridhi Maheshwari, Zhiyuan Zhang, Keshav Santhanam, Sri Vardhamanan, Saiful Haq, Ashutosh Sharma, Thomas T. Joshi, Hanna Moazam, Heather Miller, Matei Zaharia, and Christopher Potts. 2024. Dspy: Compiling declarative language model calls into self-improving pipelines.

Chin-Yew Lin. 2004. ROUGE: A package for automatic evaluation of summaries. In *Text Summarization Branches Out*, pages 74–81, Barcelona, Spain. Association for Computational Linguistics.

Krista Opsahl-Ong, Michael J Ryan, Josh Purtell, David Broman, Christopher Potts, Matei Zaharia, and Omar Khattab. 2024. Optimizing instructions and demonstrations for multi-stage language model programs. *arXiv preprint arXiv:2406.11695*.

Anusri Pampari, Preethi Raghavan, Jennifer Liang, and Jian Peng. 2018. emrqa: A large corpus for question answering on electronic medical records. *arXiv preprint arXiv:1809.00732*.

[1] https://github.com/ViswanathaReddyGajjala/ArchEHR-QA-Neural

Kishore Papineni, Salim Roukos, Todd Ward, and Wei-Jing Zhu. 2002. Bleu: a method for automatic evaluation of machine translation. In *Proceedings of the 40th Annual Meeting on Association for Computational Linguistics*, ACL '02, page 311–318, USA. Association for Computational Linguistics.

Karan Singhal, Tao Tu, Juraj Gottweis, Rory Sayres, Ellery Wulczyn, Mohamed Amin, Le Hou, Kevin Clark, Stephen R Pfohl, Heather Cole-Lewis, et al. 2025. Toward expert-level medical question answering with large language models. *Nature Medicine*, pages 1–8.

Sarvesh Soni and Dina Demner-Fushman. 2025a. A dataset for addressing patient's information needs related to clinical course of hospitalization. *arXiv preprint*.

Sarvesh Soni and Dina Demner-Fushman. 2025b. Overview of the archehr-qa 2025 shared task on grounded question answering from electronic health records. In *The 24th Workshop on Biomedical Natural Language Processing and BioNLP Shared Tasks*, Vienna, Austria. Association for Computational Linguistics.

Chi-en Amy Tai and Xavier Tannier. 2025. Clinical trial cohort selection using large language models on n2c2 challenges. *arXiv preprint arXiv:2501.11114*.

Xinyuan Wang, Chenxi Li, Zhen Wang, Fan Bai, Haotian Luo, Jiayou Zhang, Nebojsa Jojic, Eric P Xing, and Zhiting Hu. 2023. Promptagent: Strategic planning with language models enables expert-level prompt optimization. *arXiv preprint arXiv:2310.16427*.

Xuezhi Wang, Jason Wei, Dale Schuurmans, Quoc Le, Ed Chi, Sharan Narang, Aakanksha Chowdhery, and Denny Zhou. 2022. Self-consistency improves chain of thought reasoning in language models. *arXiv preprint arXiv:2203.11171*.

Wei Xu, Courtney Napoles, Ellie Pavlick, Quanze Chen, and Chris Callison-Burch. 2016. Optimizing statistical machine translation for text simplification. *Transactions of the Association for Computational Linguistics*, 4:401–415.

Chengrun Yang, Xuezhi Wang, Yifeng Lu, Hanxiao Liu, Quoc V Le, Denny Zhou, and Xinyun Chen. 2023. Large language models as optimizers. *arXiv preprint arXiv:2309.03409*.

Wen-wai Yim, Yujuan Fu, Asma Ben Abacha, Neal Snider, Thomas Lin, and Meliha Yetisgen. 2023. Aci-bench: a novel ambient clinical intelligence dataset for benchmarking automatic visit note generation. *Scientific data*, 10(1):586.

Yuheng Zha, Yichi Yang, Ruichen Li, and Zhiting Hu. 2023. Alignscore: Evaluating factual consistency with a unified alignment function. *arXiv preprint arXiv:2305.16739*.

Tianyi Zhang, Varsha Kishore, Felix Wu, Kilian Q. Weinberger, and Yoav Artzi. 2020. Bertscore: Evaluating text generation with bert.

Yongchao Zhou, Andrei Ioan Muresanu, Ziwen Han, Keiran Paster, Silviu Pitis, Harris Chan, and Jimmy Ba. 2022. Large language models are human-level prompt engineers. In *The Eleventh International Conference on Learning Representations*.

UIC at ArchEHR-QA 2025: Tri-Step Pipeline for Reliable Grounded Medical Question Answering

Mohammad Arvan[1] **Anuj Gautam**[2] **Mohan Zalake**[1] **Karl M. Kochendorfer**[1]
{marvan3, zalake, kkoche1}@uic.edu, agautam@uillinois.edu

[1]University of Illinois at Chicago
[2]University of Illinois

Abstract

Automated response generation from electronic health records (EHRs) holds potential to reduce clinician workload, but it introduces important challenges related to factual accuracy and reliable grounding in clinical evidence. We present a structured three-step pipeline that uses large language models (LLMs) for evidence classification, guided response generation, and iterative quality control. To enable rigorous evaluation, our framework combines traditional reference-based metrics with a claim-level "LLM-as-a-Judge" methodology. On the ArchEHR-QA benchmark, our system achieves 82.0 percent claim-level evidence faithfulness and 51.6 percent citation-level factuality, demonstrating strong performance in generating clinically grounded responses. These findings highlight the utility of structured LLM pipelines in healthcare applications, while also underscoring the importance of transparent evaluation and continued refinement. All code, prompt templates, and evaluation tools are publicly available.

1 Introduction

Artificial intelligence (AI) holds transformative potential for healthcare, particularly in automating routine clinical tasks. A significant challenge in contemporary clinical practice is managing patient messages efficiently, a process that often requires clinicians to synthesize information from electronic health records (EHRs) and compose personalized, accurate responses. This time-consuming task imposes substantial cognitive and emotional burdens on medical professionals, contributing to burnout and potentially diminishing the quality of patient care (Shanafelt et al., 2022).

The ArchEHR challenge addresses this critical need by focusing on automated clinical response generation from EHRs. This process presents two primary technical challenges: the accurate extraction of relevant information from patient historical records, and the generation of factual, faithful, context-appropriate responses suitable for patient communication. Large language models (LLMs) have shown promising capabilities in medical question answering, with some studies reporting that they match or exceed clinicians in empathy and communication quality (Ayers et al., 2023). However, their real-world deployment remains constrained by risks of factual errors, hallucinations (i.e., the generation of incorrect or fabricated information), and misunderstandings of medical context.

A fundamental challenge in advancing this field lies in the evaluation of AI-generated responses. Traditional text similarity metrics such as BLEU (Papineni et al., 2002) and ROUGE (Lin, 2004) have demonstrated poor correlation with human judgments across various tasks and scenarios (Liu et al., 2016; Lowe et al., 2017; Xu et al., 2023; Fabbri et al., 2021; Ernst et al., 2023). This limitation necessitates novel approaches to ensure the reliability and safety of automated clinical communication systems.

To address these challenges, we present a grounded medical question-answering system specifically designed for the ArchEHR challenge. Our approach innovates by treating evidence classification as a multiple-choice task, where an LLM selects among predefined clinical evidence categories. This classification then informs a structured clinical response generation process, followed by automated quality control and iterative revision to enhance response adherence to the required format and citation standards.

The remainder of this paper is organized as follows: Section 2 reviews related work on medical question answering, the spectrum of LLM usage, and evaluation methodologies for natural language generation. Section 3 details our methodology, including the LLM-based classification system, response generation process, and evaluation

BioNLP 2025 Shared Tasks, pages 110–117

framework. Section 4 presents experimental results on the ArchEHR-QA dataset. We then discuss the implications and limitations of our findings in Section 5. Finally, we conclude with a summary and directions for future work. Our source code, prompts, and evaluation scripts are available at `https://github.com/mo-arvan/grounded-medical-question-answering`.

2 Related Work

Our work intersects three fundamental areas: medical question answering (QA) using large language models (LLMs), the spectrum of LLM usage strategies, and evaluation methods for natural language generation (NLG). Together, these domains support the development of a reliable medical QA system. In this section, we summarize recent research in each area to contextualize our contributions.

Medical QA with LLMs Recent advances in large language models have significantly transformed medical QA, demonstrating strong performance in few-shot and zero-shot settings (Kung et al., 2023; Nori et al., 2023; Brin et al., 2023; Singhal et al., 2022). Despite their strengths, these models continue to face critical challenges. Chief among these are hallucinations, referring to generated statements that are not supported by underlying medical evidence or knowledge sources (Zhang et al., 2023; Yang et al., 2024), and difficulties in maintaining accurate, up-to-date clinical knowledge (Zhou et al., 2023; Gao et al., 2023). Our work addresses these limitations through a combination of targeted constraints and comprehensive evaluation protocols designed to ensure response faithfulness.

Spectrum of LLM Usage The complexity of medical queries has prompted the adoption of distinct modeling strategies aimed at improving reasoning and accuracy. One widely used approach involves task decomposition, in which a complex problem is reformulated into smaller, sequential reasoning tasks. These are often structured as chains or directed acyclic graphs (DAGs) of intermediate steps (Wei et al., 2022; Shen et al., 2023). Although effective, these structures are typically defined in advance and lack adaptability. Alternatively, AI agents offer a more dynamic approach. These systems autonomously generate and execute plans informed by contextual cues (Kim et al., 2024). However, such flexibility introduces increased system complexity and requires more rigorous evaluation to verify reliability (Anthropic, 2025). Our framework adopts a pipeline strategy that decomposes responses into interpretable stages. This approach balances control and transparency with adaptability across diverse query types. The exploration of more autonomous agent-based approaches is deferred to future work.

Evaluation of Natural Language Generation Evaluation of generated medical text involves multiple complementary methodologies. Traditional reference-based metrics, including BLEU (Papineni et al., 2002) and ROUGE (Lin, 2004), assess surface-level lexical overlap between system outputs and gold-standard references. However, such metrics often correlate poorly with human judgments of quality and relevance (Liu et al., 2016; Lowe et al., 2017; Xu et al., 2023; Fabbri et al., 2021; Ernst et al., 2023). More recent semantic-oriented metrics, such as BERTScore (Zhang et al., 2020) and AlignScore (Zha et al., 2023), use contextual embeddings to better capture semantic equivalence, offering improved sensitivity beyond surface similarity.

LLM-based judgment frameworks, particularly those employing the "LLM-as-a-Judge" paradigm, have demonstrated greater alignment with human evaluators (Zheng et al., 2023; Ashktorab et al., 2024; Hong et al., 2024; Ru et al., 2024; Gilardi et al., 2023). These techniques often break the evaluation process into finer-grained subtasks such as claim extraction and factual verification (Ru et al., 2024). Although promising, concerns remain about evaluator bias and model inconsistency (Schroeder and Wood-Doughty, 2024; Thakur et al., 2024). Our evaluation framework integrates both reference-based and LLM-based methods for a more comprehensive analysis of text quality and reliability.

These three strands of prior work collectively inform our methodology for building a reliable and interpretable medical QA system. By integrating structured decomposition strategies, constraint-driven generation, and multi-method evaluation, we tackle key challenges in producing trustworthy, clinically relevant outputs. This approach also supports future adaptability as techniques in each domain continue to evolve.

3 Methodology

To achieve rigorous and clinically reliable automation of message generation in healthcare, we present a methodology encompassing three sequential stages: (1) evidence classification using Large Language Models (LLMs), (2) generation of clinician-facing responses with iterative quality control, and (3) comprehensive evaluation across diverse medical datasets. This structured pipeline ensures transparency through principled processing and systematic validation. It ultimately supports robust clinical decision-making.

Prompt Templates To standardize and guide LLM behavior across each stage, we employ a suite of carefully designed prompt templates publicly available at GitHub[1]. These templates include:

- **Evidence Classification**: Categorizing relevant evidence segments from EHRs.
- **Grounded Question Answering**: Generating clinician responses grounded in classified evidence.
- **Answer Revision**: Refining responses through iterative feedback.

Evidence Classification We formulate evidence classification as a multiple-choice task, wherein the LLM assigns EHR evidence segments to one of three classes: *relevant*, *supplementary*, or *not relevant*. To ensure consistency in the output, categorical labels are constrained using Enum types (Willard and Louf, 2023). Additionally, to improve interpretability and encourage faithful predictions, the model is prompted to provide a rationale before selecting its final label (Wei et al., 2022).

Response Generation and Quality Control The LLM generates responses designed for clinicians that emphasize clarity, coherence, and professional tone after identifying relevant evidence. These outputs undergo a systematic quality assurance process based on metrics such as structural consistency, citation accuracy, and length. When deficiencies are detected, iterative feedback prompts the LLM to revise and improve outputs. This feedback loop enforces adherence to clinical communication standards.

[1]https://github.com/mo-arvan/grounded-medical-question-answering/tree/master/prompts

Evaluation Strategy Our evaluation strategy includes two phases: benchmarking foundational medical reasoning and assessing the full clinical message pipeline.

The first phase evaluates the LLM's performance using multiple-choice datasets closely aligned with our evidence classification framework: MMLU-Pro-Med, MedQA-US, MedMCQA, and PubMedQA (Wang et al., 2024; Jin et al., 2021; Pal et al., 2022; Jin et al., 2019). These datasets collectively measure domain-specific competency.

The second phase involves a comprehensive evaluation of the pipeline. This includes evidence classification, response generation and automated quality control applied to the ArchEHR-QA dataset (Soni and Demner-Fushman, 2025b,a), which is sourced from real-world EHR scenarios. Performance is assessed across two major dimensions:

Factuality is evaluated using Precision, Recall, and F1 Scores that compare the evidence cited in the generated responses to manually annotated ground-truth evidence. A "strict" Citation F1 considers only essential evidence, whereas a "lenient" variant also incorporates supplementary evidence.

Relevance is measured by comparing generated answers to essential EHR sentences and the original clinical question. Metrics employed include BLEU, ROUGE, SARI, BERTScore, AlignScore, and MEDCON.

Faithfulness Verification via Claim-Level Triple Extraction We introduce a custom, interpretable faithfulness metric grounded in claim-level triple extraction to evaluate factual consistency. Faithfulness, defined as the extent to which generated outputs accurately reflect source evidence, is a critical factor for clinical dependability (Ru et al., 2024). However, it is often difficult to measure due to incomplete references and the resource-intensive nature of expert reviews.

Our approach extracts atomic subject-predicate-object triples from both generated responses and their supporting EHR evidence. In a fungal infection case, the following triples, for example, are identified:

> ("Yeast", "was seen with", "bacteria on initial sputum gram stain"),
> ("Torulopsis glabrata", "was identified in", "blood/fungal culture"), and
> ("Antifungal therapy", "was started after", "fungal findings were confirmed").

In this example, the first two triples are supported by evidence, while the third lacks grounding. Each claim's support is verified by a separate LLM. Faithfulness is then quantified as the proportion of claims backed by evidence—in this case, 66.7%. This metric provides scalable and explainable factuality assessment.

Summary In summary, our methodology integrates structured prompt-guided classification, coherent response generation with iterative quality checks, and a rigorous evaluation framework. These include domain-specific benchmarks and interpretable factuality metrics. This design creates reliable, transparent, and extensible automation for generating clinical messages grounded in EHR data.

4 Results

This section presents our evaluation of model performance on two complementary tasks: general medical knowledge assessment and grounded clinical question answering. We first measure accuracy on standard multiple-choice benchmarks to assess general medical knowledge competence. We then evaluate the ability of the models to generate factually grounded and contextually relevant answers to clinical questions using the ArchEHR-QA dataset.

4.1 General Medical Knowledge Assessment

Table 1 summarizes accuracy scores across four established medical knowledge benchmarks. GPT-4o consistently outperforms both GPT-4o-mini and the baseline GPT-4[†] across all datasets.

In particular, GPT-4o achieves 77.67% on MedMCQA, marking an 8 percentage point improvement over GPT-4. On MedQA, it attains 88.69%, surpassing GPT-4 by 5 points. For MMLU-Pro-Med, GPT-4o sets a new state of the art with 81.56% accuracy. Although performance on PubMedQA is lower at 45.80%, this is expected due to the dataset's reliance on detailed comprehension of specialized biomedical literature. The lack of retrieval capabilities particularly challenges models in this setting.

4.2 Grounded Medical Question Answering

We next evaluate models on the ArchEHR-QA dataset, which benchmarks clinical question answering grounded in patient electronic health records. To ensure comparability with prior work, we use the official evaluation scripts provided by the challenge organizers.

Table 2 reports factuality and relevance scores for GPT-4o and GPT-4o-mini on both the development and test sets. GPT-4o achieves factuality scores of 51.85% (dev) and 51.59% (test), along with relevance scores of 29.96% and 33.33%, respectively. GPT-4o-mini scores 27.27% for factuality and 29.21% for relevance on the development set. As only GPT-4o was submitted to the challenge, test set outcomes for GPT-4o-mini are unavailable.

In addition to factuality and relevance, we assess response faithfulness. As shown in Table 3, GPT-4o attains 76.1% on the development set and 82.0% on the test set. GPT-4o-mini achieves a lower score of 65.6% on the development set.

These results collectively indicate that GPT-4o not only generates responses that are more accurate and pertinent but also maintains a strong alignment with provided clinical evidence.

5 Discussion

Our findings show that large generative models, such as GPT-4o, demonstrate superior performance on medical question-answering tasks, excelling across both knowledge-based and clinically grounded queries. Furthermore, the model is maintaining a high degree of factual consistency in evidence-grounded outputs.

Despite these advances, key trade-offs emerge between extractive and generative approaches. Evaluation metrics employed by ArchEHR-QA emphasize lexical overlap with reference texts, thereby favoring extractive methods. Generative models, by contrast, tend to produce more fluent and coherent responses but may not replicate the precise phrasing found in reference answers. To better capture the factual accuracy of generative outputs, we adopted a structured evaluation using the LLM-as-a-Judge framework. This approach enables scalable verification by assessing whether individual assertions in a generated response are supported by underlying evidence.

However, assessing factual consistency alone does not guarantee citation-level reliability. Recent studies highlight that large language models can incorrectly attribute statements to references that do not actually support them, introducing risks in high-stakes domains like healthcare. Notably, prior evidence shows that up to 30% of model-generated

Model	MedMCQA	MedQA	MMLU-Pro-Med	PubMedQA
GPT-4o	**77.67**	**88.69**	**81.56**	**45.80**
GPT-4o-mini	68.13	74.39	74.07	44.80
GPT-4†	69.88	83.97	-	39.60

Table 1: Performance comparison (accuracy %) across medical knowledge datasets. Results marked with † are baseline results from Xiong et al. (2024).

Model	Set	Factuality	Relevance
GPT-4o	Dev	51.85	29.96
GPT-4o	Test	51.59	33.33
GPT-4o-mini	Dev	27.27	29.21

Table 2: Factuality and relevance scores for GPT-4o and GPT-4o-mini on development and test sets of ArchEHR.

Model	Set	Faithfulness
GPT-4o	Dev	76.1
GPT-4o	Test	82.0
GPT-4o-mini	Dev	65.6

Table 3: Faithfulness scores for GPT-4o and GPT-4o-mini on development and test sets.

statements may contain unsupported reference citations (Wu et al., 2025). Ensuring that all cited sources genuinely substantiate the content remains a critical challenge.

Given these limitations, the safe and responsible deployment of LLMs in clinical environments requires comprehensive validation and routine monitoring using scalable methods like those employed in this study. Importantly, expert human review remains essential, particularly in scenarios where accuracy and reliability are paramount for patient safety.

A practical advantage in the clinical setting is that real-time response generation is often not required. This relaxed time constraint allows the system to proactively generate multiple candidate questions and corresponding answers for each incoming patient message in advance. Consequently, clinicians are not burdened with crafting questions themselves and can instead select from a curated list of contextually appropriate Q&A pairs. This workflow-integrated approach streamlines clinical decision-making and promotes more efficient patient communication.

Looking ahead, integrating external knowledge retrieval with interactive clinical tools presents a promising avenue to enhance both model performance and usability. Future research should also examine the impact of such systems on key outcomes, including clinician workload, as existing evidence in this area remains mixed (Garcia et al., 2024). In addition, comprehensive human preference studies comparing outputs from extractive and generative systems will be essential to align evaluation frameworks with the practical expectations and needs of clinicians.

6 Conclusion

Our work shows that generative models such as GPT-4o perform well across a range of clinical question answering tasks. These models also demonstrate strong factual alignment with source evidence when evaluated using structured, claim-level assessment methods.

However, several important challenges remain. Distinguishing between claim faithfulness, which assesses whether individual assertions align with evidence, and citation faithfulness, which considers whether referenced sources support the claims, continues to be difficult. In addition, label consistency and the design of evaluation frameworks require further improvement to ensure more reliable assessments.

Addressing these challenges, together with incorporating direct feedback from clinicians, is essential for enabling trustworthy and effective deployment of these models in real-world biomedical settings.

Limitations

One important challenge identified during Phase 1 of our evaluation involved testing models on general medical knowledge benchmarks, including MMLU-Pro-Med (Wang et al., 2024), MedQA-US (Jin et al., 2021), MedMCQA (Pal et al., 2022), and PubMedQA (Jin et al., 2019). A central limitation in this context is the lack of transparency surrounding the training data used by commercial large language models. Without clear documentation of training corpora, there is a significant risk of data leakage, where benchmark content may inadvertently overlap with training inputs. This overlap can lead to inflated performance metrics, which misrepresent a model's generalizability and complicate direct comparisons between models. Because these benchmarks aim to assess a broad range of medical knowledge and reasoning skills, even partial contamination reduces the credibility of conclusions drawn from model performance. Although commercial LLMs exhibit strong capabilities, the opacity of their training data sources remains a fundamental barrier to reproducible and trustworthy evaluation. This limitation underscores the need for greater dataset transparency or the development of evaluation strategies that explicitly control for training-evaluation separation.

In addition, this study did not include expert validation of the model-generated responses. Due to time constraints, we were unable to engage licensed medical professionals in a systematic review process. While our structured framework incorporates LLM-as-a-Judge assessments, the absence of expert oversight limits our ability to confirm the clinical accuracy and safety of model outputs. Future work should incorporate formal expert evaluation to ensure that responses meet professional standards and are suitable for use in healthcare settings.

Ethical Considerations

The system was developed using Azure OpenAI Services in accordance with PhysioNet's responsible use guidelines [2]. We avoided using any protected health information during development.

References

Anthropic. 2025. Building effective agents. Accessed: 2025-02-18.

Zahra Ashktorab, Michael Desmond, Qian Pan, James M. Johnson, Martin Santillan Cooper, Elizabeth M. Daly, Rahul Nair, Tejaswini Pedapati, Swapnaja Achintalwar, and Werner Geyer. 2024. Aligning human and LLM judgments: Insights from evalassist on task-specific evaluations and ai-assisted assessment strategy preferences. *CoRR*, abs/2410.00873.

JW Ayers, A Poliak, M Dredze, EC Leas, Z Zhu, JB Kelley, DJ Faix, AM Goodman, CA Longhurst, M Hogarth, and 1 others. 2023. Comparing physician and artificial intelligence chatbot responses to patient questions posted to a public social media forum. jama intern med 183 (6): 589–596.

Dana Brin, Vera Sorin, Akhil Vaid, Ali Soroush, Benjamin S Glicksberg, Alexander W Charney, Girish Nadkarni, and Eyal Klang. 2023. Comparing chatgpt and gpt-4 performance in usmle soft skill assessments. *Scientific Reports*, 13(1):16492.

Ori Ernst, Ori Shapira, Ido Dagan, and Ran Levy. 2023. Re-examining summarization evaluation across multiple quality criteria. In *Findings of the Association for Computational Linguistics: EMNLP 2023*, pages 13829–13838, Singapore. Association for Computational Linguistics.

Alexander R. Fabbri, Wojciech Kryscinski, Bryan McCann, Caiming Xiong, Richard Socher, and Dragomir R. Radev. 2021. Summeval: Re-evaluating summarization evaluation. *Trans. Assoc. Comput. Linguistics*, 9:391–409.

Yunfan Gao, Yun Xiong, Xinyu Gao, Kangxiang Jia, Jinliu Pan, Yuxi Bi, Yi Dai, Jiawei Sun, Qianyu Guo, Meng Wang, and Haofen Wang. 2023. Retrieval-augmented generation for large language models: A survey. *CoRR*, abs/2312.10997.

Patricia Garcia, Stephen P Ma, Shreya Shah, Margaret Smith, Yejin Jeong, Anna Devon-Sand, Ming Tai-Seale, Kevin Takazawa, Danyelle Clutter, Kyle Vogt, and 1 others. 2024. Artificial intelligence–generated draft replies to patient inbox messages. *JAMA Network Open*, 7(3):e243201–e243201.

Fabrizio Gilardi, Meysam Alizadeh, and Maël Kubli. 2023. Chatgpt outperforms crowd-workers for text-annotation tasks. *CoRR*, abs/2303.15056.

Giwon Hong, Aryo Pradipta Gema, Rohit Saxena, Xiaotang Du, Ping Nie, Yu Zhao, Laura Perez-Beltrachini, Max Ryabinin, Xuanli He, Clémentine Fourrier, and Pasquale Minervini. 2024. The hallucinations leaderboard - an open effort to measure hallucinations in large language models. *CoRR*, abs/2404.05904.

Di Jin, Eileen Pan, Nassim Oufattole, Wei-Hung Weng, Hanyi Fang, and Peter Szolovits. 2021. What disease does this patient have? a large-scale open domain question answering dataset from medical exams. *Applied Sciences*, 11(14):6421.

Qiao Jin, Bhuwan Dhingra, Zhengping Liu, William Cohen, and Xinghua Lu. 2019. PubMedQA: A dataset for biomedical research question answering. In *Proceedings of the 2019 Conference on Empirical Methods in Natural Language Processing and the 9th International Joint Conference on Natural Language Processing (EMNLP-IJCNLP)*, pages 2567–2577, Hong Kong, China. Association for Computational Linguistics.

[2]`https://physionet.org/news/post/gpt-responsible-use`

Yubin Kim, Chanwoo Park, Hyewon Jeong, Yik Siu Chan, Xuhai Xu, Daniel McDuff, Hyeonhoon Lee, Marzyeh Ghassemi, Cynthia Breazeal, and Hae Won Park. 2024. Mdagents: An adaptive collaboration of llms for medical decision-making. In *Advances in Neural Information Processing Systems 38: Annual Conference on Neural Information Processing Systems 2024, NeurIPS 2024, Vancouver, BC, Canada, December 10 - 15, 2024*.

Tiffany H Kung, Morgan Cheatham, Arielle Medenilla, Czarina Sillos, Lorie De Leon, Camille Elepaño, Maria Madriaga, Rimel Aggabao, Giezel Diaz-Candido, James Maningo, and 1 others. 2023. Performance of chatgpt on usmle: potential for ai-assisted medical education using large language models. *PLoS digital health*, 2(2):e0000198.

Chin-Yew Lin. 2004. ROUGE: A package for automatic evaluation of summaries. In *Text Summarization Branches Out*, pages 74–81, Barcelona, Spain. Association for Computational Linguistics.

Chia-Wei Liu, Ryan Lowe, Iulian Serban, Michael Noseworthy, Laurent Charlin, and Joelle Pineau. 2016. How NOT to evaluate your dialogue system: An empirical study of unsupervised evaluation metrics for dialogue response generation. In *Proceedings of the 2016 Conference on Empirical Methods in Natural Language Processing, EMNLP 2016, Austin, Texas, USA, November 1-4, 2016*, pages 2122–2132. The Association for Computational Linguistics.

Ryan Lowe, Michael Noseworthy, Iulian Vlad Serban, Nicolas Angelard-Gontier, Yoshua Bengio, and Joelle Pineau. 2017. Towards an automatic Turing test: Learning to evaluate dialogue responses. In *Proceedings of the 55th Annual Meeting of the Association for Computational Linguistics (Volume 1: Long Papers)*, pages 1116–1126, Vancouver, Canada. Association for Computational Linguistics.

Harsha Nori, Nicholas King, Scott Mayer McKinney, Dean Carignan, and Eric Horvitz. 2023. Capabilities of GPT-4 on medical challenge problems. *CoRR*, abs/2303.13375.

Ankit Pal, Logesh Kumar Umapathi, and Malaikannan Sankarasubbu. 2022. Medmcqa: A large-scale multi-subject multi-choice dataset for medical domain question answering. In *Proceedings of the Conference on Health, Inference, and Learning*, volume 174 of *Proceedings of Machine Learning Research*, pages 248–260. PMLR.

Kishore Papineni, Salim Roukos, Todd Ward, and Wei-Jing Zhu. 2002. Bleu: a method for automatic evaluation of machine translation. In *Proceedings of the 40th Annual Meeting of the Association for Computational Linguistics, July 6-12, 2002, Philadelphia, PA, USA*, pages 311–318. ACL.

Dongyu Ru, Lin Qiu, Xiangkun Hu, Tianhang Zhang, Peng Shi, Shuaichen Chang, Cheng Jiayang, Cunxiang Wang, Shichao Sun, Huanyu Li, Zizhao Zhang, Binjie Wang, Jiarong Jiang, Tong He, Zhiguo Wang, Pengfei Liu, Yue Zhang, and Zheng Zhang. 2024. Ragchecker: A fine-grained framework for diagnosing retrieval-augmented generation. In *Advances in Neural Information Processing Systems 38: Annual Conference on Neural Information Processing Systems 2024, NeurIPS 2024, Vancouver, BC, Canada, December 10 - 15, 2024*.

Kayla Schroeder and Zach Wood-Doughty. 2024. Can you trust LLM judgments? reliability of llm-as-a-judge. *CoRR*, abs/2412.12509.

Tait D Shanafelt, Colin P West, Lotte N Dyrbye, Mickey Trockel, Michael Tutty, Hanhan Wang, Lindsey E Carlasare, and Christine Sinsky. 2022. Changes in burnout and satisfaction with work-life integration in physicians during the first 2 years of the covid-19 pandemic. In *Mayo Clinic Proceedings*, volume 97, pages 2248–2258. Elsevier.

Yongliang Shen, Kaitao Song, Xu Tan, Dongsheng Li, Weiming Lu, and Yueting Zhuang. 2023. Hugginggpt: Solving AI tasks with chatgpt and its friends in hugging face. In *Advances in Neural Information Processing Systems 36: Annual Conference on Neural Information Processing Systems 2023, NeurIPS 2023, New Orleans, LA, USA, December 10 - 16, 2023*.

Karan Singhal, Shekoofeh Azizi, Tao Tu, S. Sara Mahdavi, Jason Wei, Hyung Won Chung, Nathan Scales, Ajay Kumar Tanwani, Heather Cole-Lewis, Stephen Pfohl, Perry Payne, Martin Seneviratne, Paul Gamble, Chris Kelly, Nathaneal Schärli, Aakanksha Chowdhery, Philip Andrew Mansfield, Blaise Agüera y Arcas, Dale R. Webster, and 11 others. 2022. Large language models encode clinical knowledge. *CoRR*, abs/2212.13138.

Sarvesh Soni and Dina Demner-Fushman. 2025a. A dataset for addressing patient's information needs related to clinical course of hospitalization. *arXiv preprint*.

Sarvesh Soni and Dina Demner-Fushman. 2025b. Overview of the archehr-qa 2025 shared task on grounded question answering from electronic health records. In *The 24th Workshop on Biomedical Natural Language Processing and BioNLP Shared Tasks*, Vienna, Austria. Association for Computational Linguistics.

Aman Singh Thakur, Kartik Choudhary, Venkat Srinik Ramayapally, Sankaran Vaidyanathan, and Dieuwke Hupkes. 2024. Judging the judges: Evaluating alignment and vulnerabilities in llms-as-judges. *CoRR*, abs/2406.12624.

Yubo Wang, Xueguang Ma, Ge Zhang, Yuansheng Ni, Abhranil Chandra, Shiguang Guo, Weiming Ren, Aaran Arulraj, Xuan He, Ziyan Jiang, Tianle Li, Max Ku, Kai Wang, Alex Zhuang, Rongqi Fan, Xiang Yue, and Wenhu Chen. 2024. Mmlu-pro: A more robust and challenging multi-task language understanding benchmark. In *Advances in Neural Information Processing Systems 38: Annual Conference on Neural Information Processing Systems 2024, NeurIPS 2024, Vancouver, BC, Canada, December 10 - 15, 2024*.

Jason Wei, Xuezhi Wang, Dale Schuurmans, Maarten Bosma, Brian Ichter, Fei Xia, Ed H. Chi, Quoc V. Le, and Denny Zhou. 2022. Chain-of-thought prompting elicits reasoning in large language models. In *Advances in Neural Information Processing Systems 35: Annual Conference on Neural Information Processing Systems 2022, NeurIPS 2022, New Orleans, LA, USA, November 28 - December 9, 2022*.

Brandon T Willard and Rémi Louf. 2023. Efficient guided generation for llms. *arXiv preprint arXiv:2307.09702*, 6.

Kevin Wu, Eric Wu, Kevin Wei, Angela Zhang, Allison Casasola, Teresa Nguyen, Sith Riantawan, Patricia Shi, Daniel Ho, and James Zou. 2025. An automated framework for assessing how well llms cite relevant medical references. *Nature Communications*, 16(1):3615.

Guangzhi Xiong, Qiao Jin, Zhiyong Lu, and Aidong Zhang. 2024. Benchmarking retrieval-augmented generation for medicine. In *Findings of the Association for Computational Linguistics: ACL 2024*, pages 6233–6251, Bangkok, Thailand. Association for Computational Linguistics.

Fangyuan Xu, Yixiao Song, Mohit Iyyer, and Eunsol Choi. 2023. A critical evaluation of evaluations for long-form question answering. In *Proceedings of the 61st Annual Meeting of the Association for Computational Linguistics (Volume 1: Long Papers), ACL 2023, Toronto, Canada, July 9-14, 2023*, pages 3225–3245. Association for Computational Linguistics.

Yifan Yang, Qiao Jin, Qingqing Zhu, Zhizheng Wang, Francisco Erramuspe Álvarez, Nicholas Wan, Benjamin Hou, and Zhiyong Lu. 2024. Beyond multiple-choice accuracy: Real-world challenges of implementing large language models in healthcare. *CoRR*, abs/2410.18460.

Yuheng Zha, Yichi Yang, Ruichen Li, and Zhiting Hu. 2023. Alignscore: Evaluating factual consistency with A unified alignment function. In *Proceedings of the 61st Annual Meeting of the Association for Computational Linguistics (Volume 1: Long Papers), ACL 2023, Toronto, Canada, July 9-14, 2023*, pages 11328–11348. Association for Computational Linguistics.

Tianyi Zhang, Varsha Kishore, Felix Wu, Kilian Q. Weinberger, and Yoav Artzi. 2020. Bertscore: Evaluating text generation with BERT. In *8th International Conference on Learning Representations, ICLR 2020, Addis Ababa, Ethiopia, April 26-30, 2020*. OpenReview.net.

Yue Zhang, Yafu Li, Leyang Cui, Deng Cai, Lemao Liu, Tingchen Fu, Xinting Huang, Enbo Zhao, Yu Zhang, Yulong Chen, Longyue Wang, Anh Tuan Luu, Wei Bi, Freda Shi, and Shuming Shi. 2023. Siren's song in the AI ocean: A survey on hallucination in large language models. *CoRR*, abs/2309.01219.

Lianmin Zheng, Wei-Lin Chiang, Ying Sheng, Siyuan Zhuang, Zhanghao Wu, Yonghao Zhuang, Zi Lin, Zhuohan Li, Dacheng Li, Eric P. Xing, Hao Zhang, Joseph E. Gonzalez, and Ion Stoica. 2023. Judging llm-as-a-judge with mt-bench and chatbot arena. In *Advances in Neural Information Processing Systems 36: Annual Conference on Neural Information Processing Systems 2023, NeurIPS 2023, New Orleans, LA, USA, December 10 - 16, 2023*.

Hongjian Zhou, Boyang Gu, Xinyu Zou, Yiru Li, Sam S. Chen, Peilin Zhou, Junling Liu, Yining Hua, Chengfeng Mao, Xian Wu, Zheng Li, and Fenglin Liu. 2023. A survey of large language models in medicine: Progress, application, and challenge. *CoRR*, abs/2311.05112.

DMIS Lab at ArchEHR-QA 2025: Evidence-Grounded Answer Generation for EHR-based QA via a Multi-Agent Framework

Hyeon Hwang[1*] Hyeongsoon Hwang[1*] Jongmyung Jung[1] Jaehoon Yoon[2,4]
Minju Song[1] Yein Park[1] Dain Kim[1] Taewhoo Lee[1,4] Jiwoong Sohn[1,3]
Chanwoong Yoon[1] Sihyeon Park[1] Jiwoo Lee[1] Heechul Yang[1] Jaewoo Kang[1,4†]
[1]Korea University [2]Hanyang University [3]ETH Zürich [4]AIGEN Sciences
{hyeon-hwang, hhs8746, kangj}@korea.ac.kr

Abstract

The increasing utilization of patient portals has amplified clinicians' workloads, primarily due to the necessity of addressing detailed patient inquiries related to their health concerns. The ArchEHR-QA 2025 shared task aims to alleviate this burden by automatically generating accurate, evidence-grounded responses to patients' questions based on their Electronic Health Records (EHRs). This paper presents a six-stage multi-agent framework specifically developed to identify essential clinical sentences for answering patient questions, leveraging large language models (LLMs). Our approach begins with OpenAI's o3 model generating focused medical context to guide downstream reasoning. In the subsequent stages, GPT-4.1-based agents assess the relevance of individual sentences, recruit domain experts, and consolidate their judgments to identify essential information for constructing coherent, evidence-grounded responses. Our framework achieved an Overall Factuality score of 62.0 and an Overall Relevance Score of 52.9 on the development set, and corresponding scores of 58.6 and 48.8, respectively, on the test set.

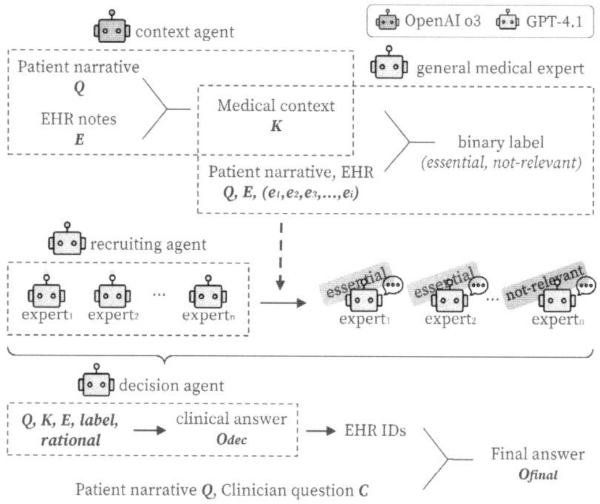

Figure 1: Overview of our six-stage *multi-agent* pipeline for evidence-grounded EHR question answering. The *Context Agent* generates a medical context K from the patient question Q and EHR sentences E. A *General-Medical Expert* labels each sentence as *essential* or *not-relevant* with a brief rationale. The *Recruiting Agent* selects domain-specific *Expert Agents*. The *Decision Agent* integrates all information, generate the answer O_{dec} that cites sentence IDs. The patient narrative, clinician question, and chosen ehr sentences are then assembled into the final reply O_{final}.

1 Introduction

The increased use of patient portals has significantly increased clinicians' workload, especially concerning responding to patients' inbox messages. These messages frequently include detailed questions regarding patients' medical conditions, treatments, and healthcare procedures. Addressing these inquiries manually by clinicians is not only time-consuming but can also delay patient care. To mitigate this burden, the ArchEHR-QA 2025 shared task (Soni and Demner-Fushman, 2025b) focuses on automatically generating accurate and clinically-grounded responses to patients' health-related questions by leveraging information contained within their Electronic Health Records (EHRs).

In this paper, we introduce a six-stage multi-agent framework specifically designed to select appropriate EHR sentences for effectively answering patients' questions in the ArchEHR-QA 2025 shared task. Figure 1 briefly shows how our framework generates evidence-grounded answer based on multiple LLM agents. Our approach begins with context generation using OpenAI's advanced o3 model (OpenAI, 2025b), selected for its demonstrated superior reasoning capabilities in medical contexts. Subsequent stages employ specialized GPT-4.1 (OpenAI, 2025a)-based agents to evaluate the relevance of clinical note sentences individually and collectively, recruit domain-specific

*Equal contribution †Corresponding author

BioNLP 2025 Shared Tasks, pages 118–125
August 1, 2025 ©2025 Association for Computational Linguistics

experts dynamically based on the patient's narrative, and integrate diverse expert perspectives into a consensus-driven decision process. The final stage involves synthesizing the identified essential clinical evidence to produce a comprehensive, clinically grounded answer. On the development set, our framework achieved an Overall Factuality score of 62.0 and an Overall Relevance Score of 52.9. On the test set, it attained an Overall Factuality score of 58.6 and an Overall Relevance Score of 48.8.

2 Related Works

Recent advances in large language model (LLM)-based multi-agent systems (Xi et al., 2025; Wang et al., 2023; Guo et al., 2024) have demonstrated significant promise in complex reasoning tasks. Such systems have particularly shown effectiveness in medical domains, with successful implementations like MDAgents (Kim et al., 2024) and MedAgent (Tang et al., 2024) illustrating that deploying specialized agents, each tailored for distinct analytical functions, facilitates robust clinical decision-making processes and precise information extraction. Given the complexity and the sensitive nature of clinical data processing, utilizing a multi-agent framework is particularly well-suited for the ArchEHR-QA 2025 shared task.

3 Method

We propose a six-stage multi-agent framework that automatically extracts the subset of sentences essential for answering patient and clinician questions from a patient's electronic health record (EHR) corpus. The context generation stage employs OpenAI o3 (OpenAI, 2025b) selected for its superior reasoning performance to generate focused and clinically relevant medical context, whereas the remaining stages rely on specialised agents based on GPT-4.1 (OpenAI, 2025a) that provide complementary analytic perspectives and collectively converge on a final consensus.

3.1 Problem Definition

Throughout this section, Q denotes the patient narrative, $E = \{e_1, \ldots, e_n\}$ the set of EHR sentences, and $S \subseteq E$ the subset labeled essential. An agent is defined as a function of the form:

$$A_{\text{role}}(\text{M}, \text{I}) = \text{O},$$

- ***model*** **M** is an instantiated LLM (e.g. OpenAI o3 or GPT-4.1).
- ***input*** **I** is a role-specific set of inputs, comprising prompts, auxiliary context, and intermediate metadata.
- ***output*** **O** is the structured result expected from that role (e.g. a context paragraph, a binary relevance label with rationale.)

ContextAgent Example

```
### Patient Question
ICU 15 days for severe abdominal pain; diagnosed with
common-bile-duct (CBD) sludge and started Udiliv,
but doctor still advises ERCP.
Can medication alone clear the sludge?

### Generated Medical Context
1. UDCA may dissolve microscopic gallbladder sludge
but not obstructive CBD sludge, especially
when infection or jaundice is present...

2. Despite of ICU care and ongoing Udiliv,
the sludge has persisted strong enough evidence
that medication has not yet relieved the obstruction.
...
```

Figure 2: An example of medical contexts generated by the *ContextAgent*, with long explanations truncated for brevity.

3.2 Multi-Agent Framework

Context Generation. The context agent A_{ctx} uses the high-performance model M_{ctx} (OpenAI o3) to generate a medical context K to address the patient question Q, guiding downstream reasoning.

$$K = A_{\text{ctx}}\big(M_{\text{ctx}}, (Q, E)\big)$$

General Medical Expert Relevance Screening. The general medical expert agent A_{gen} evaluates the essentiality of each individual EHR sentence. The agent is provided with the patient's question Q, the generated medical context K, the full set of EHR notes E, and the specific sentence e_i under evaluation. The agent outputs a binary label $\ell_i \in \{\text{essential}, \text{not-relevant}\}$, indicating the essentiality of e_i, along with a rationale r_i. This process is repeated independently for each sentence.

$$(\ell_i^{\text{gen}}, r_i^{\text{gen}}) = A_{\text{gen}}\big(M_{\text{gen}}, (Q, K, E, e_i)\big)$$

Experts Recruitment. The recruiting agent A_{rec} synthesizes the patient narrative Q, medical context K, and the full set of EHR notes E to assemble an expert panel, denoted as

$$\text{Experts} = \{A_{\text{exp}}^{(1)}, \ldots, A_{\text{exp}}^{(m)}\} = A_{\text{rec}}(Q, K, E).$$

Domain-Specific Assessment. Each expert agent $A_{\text{exp}}^{(j)} \in$ Experts then receives the (Q, K, E) to perform a domain-specific evaluation. Based on this comprehensive input, the expert evaluates the essentiality of all sentences in the EHR collectively, leveraging their specialized medical knowledge to make sentence-level judgments. The relevance label and rationale set produced by the j-th expert agent are given by

$$(\boldsymbol{L}^{(j)}, \boldsymbol{R}^{(j)}) = A_{\text{exp}}^{(j)}\big(M_{\text{exp}}^{(j)}, (Q, K, E)\big),$$

where $\boldsymbol{L}^{(j)} = (\ell_1^{(j)}, \ldots, \ell_n^{(j)})$ represents the set of sentence-level labels, and $\boldsymbol{R}^{(j)} = (r_1^{(j)}, \ldots, r_n^{(j)})$ represents the corresponding rationales provided by the expert for each sentence $e_i \in E$. Each label $\ell_i^{(j)} \in \{\text{essential, not-relevant}\}$ encodes the expert's judgment regarding the essentiality of sentence e_i, and each rationale $r_i^{(j)}$ provides the justification for that judgment.

Consensus Integration. The aggregated package

$$I_{\text{dec}} = \Big(Q,\ K,\ E,\ (\ell_i^{\text{gen}}, r_i^{\text{gen}})_{i=1}^{n},\ \{(\mathbf{L}^{(j)}, \mathbf{R}^{(j)})\}_{j=1}^{m}\Big)$$

is forwarded to the decision agent A_{dec}. This agent consolidates the upstream judgments to determine the definitive essential-sentence set S, and uses S to craft a comprehensive, evidence-grounded clinical answer O_{dec} to the patient's question Q.

$$O_{dec} = A_{\text{dec}}\big(M_{\text{dec}}, I_{\text{dec}}\big),$$

Final Answer Generation. Finally, we extract the IDs of the essential notes identified in the O_{dec}, and then concatenate the patient narrative, clinician question, and the selected essential notes to generate a comprehensive response O_{final}.

4 Experimental settings

4.1 Dataset

To evaluate our framework, we utilize the benchmark dataset (Soni and Demner-Fushman, 2025a) provided by ArchEHR-QA 2025. This dataset consists of case-based collections, each comprising a patient narrative, a patient question, a clinician question, and associated EHR data intended to support answering the question. The EHR data for each case is composed of multiple sentences, each annotated with a unique sentence ID. The dataset consists of a development set and a test set. Among these, only the development set provides sentence-level relevance labels (categorized as essential, supplementary, or not relevant) for evaluating the performance of answer generation.

Method	Factuality (Strict Micro) Precision	Recall	F1
Multiclass classification			
w/ Experts	**64.8**	52.1	57.8
w/ Context K	61.0	52.1	56.2
w/ Experts + Context K	64.2	52.1	57.6
Binary classification			
w/ Experts	50.0	69.5	58.1
w/ Context K	52.1	69.5	59.6
w/ Experts + Context K	53.4	**73.9**	**62.0**

Table 1: Factuality score comparison for multiclass models (*essential / supplementary / not-relevant*) and binary models (*essential / not-relevant*) using *w/ Experts, w/ Context K*, and *w/ Experts + Context K*.

4.2 Metrics

We adopt three evaluation metrics in accordance with the official scoring criteria of ArchEHR-QA 2025: *Overall Factuality Score*, *Overall Relevance Score*, and *Overall Score*.

Overall Factuality Score measures the F1 score between the set of sentence IDs cited in the final answer and those cited in the gold answer. This score is computed based on the counts of true positives, false positives, and false negatives aggregated across each case.

Overall Relevance Score evaluates the semantic and lexical similarity between the final and gold answers using a combination of BLUE (Papineni et al., 2002), ROUGE (Lin, 2004), SARI (Xu et al., 2016), BERTScore (Zhang* et al., 2020), AlignScore (Zha et al., 2023), and MEDCON (Yim et al., 2023) metrics. The final score is obtained by combining the normalized scores of these individual metrics. The gold answer in this context is constructed by concatenating the patient narrative, clinician question, and essential EHR sentences provided for each case.

Overall Score serves as the primary evaluation metric for this challenge. It is defined as the average of the *Overall Factuality Score* and the *Overall Relevance Score*.

4.3 Results

To verify the quality of our framework, this section presents (i) a sentence-level factuality analysis on the development set, (ii) a multi-metric relevance

Method	Relevance						
	BLEU	ROUGE-L	SARI	B.S.	A.S.	MEDCON	Overall
Questions + Predicted sentences	**19.2**	**53.6**	34.8	**58.2**	**97.6**	**54.1**	**52.9**
Summary of predicted sentences	2.0	25.1	**55.7**	27.6	42.6	37.4	31.7

Table 2: Comparison of relevance scores between two answer generation methods: using full question with predicted essential sentences vs. using a summary of the predicted essential sentences. Abbreviations: ROUGE-L= ROUGE-Lsum, B.S. = BERTScore, A.S. = AlignScore.

Team	Overall	Factuality (Strict Micro)			Relevance						
		Precision	Recall	F1	BLEU	ROUGE-L	SARI	B.S.	A.S.	MEDCON	Overall
DMIS Lab (Ours)	**53.7**	57.9	59.3	58.6	**14.3**	**46.5**	36.7	**53.9**	**92.4**	**49.3**	**48.8**
Neural	51.5	55.4	63.8	59.3	8.5	34.1	**73.1**	39.1	67.3	40.0	43.7
LAILab	51.0	56.0	**65.5**	**60.4**	6.5	32.7	69.2	37.4	65.3	38.4	41.6
LAMAR	49.1	60.6	53.6	56.9	6.0	32.1	65.8	36.4	64.3	43.6	41.4
ssagarwal	45.0	**68.8**	36.2	47.5	4.7	31.1	70.0	36.9	74.9	38.0	42.6

Table 3: Official results of the leaderboard (Top 5) on ArchEHR-QA 2025 dataset. The teams are ranked based on Overall score. Abbreviations: ROUGE-L = ROUGE-Lsum, B.S. = BERTScore, A.S. = AlignScore.

analysis, and (iii) a comparison of test-set scores on the official ArchEHR-QA 2025 leaderboard.

4.3.1 Factuality analysis

The sentence–level evaluation, summarised in Table 1, reveals a clear benefit from contextual conditioning. The *multiclass* (essential / supplementary / not-relevant) variant achieves the highest precision (62.2%) but simultaneously records the lowest recall (51.4%), resulting in an F1 of 56.3. Conversely, the *binary* (essential / not-relevant) classifier attains the greatest recall (69.5%) at the expense of precision (50.0%), yielding an F1 of 58.1. When the identical binary classification approach is prefixed with the automatically generated medical context K, recall increases further to 73.9% while precision recovers to 53.4%, producing the best strict-micro F1 of **62.0**. These results indicate that (i) finer-grained labels do not compensate for the recall penalty inherent in multiclass formulations, and (ii) domain-aware context provides the disambiguating cues necessary to recover clinically critical sentences, thereby maximising overall factuality.

4.3.2 Relevance analysis

Table 2 compares two answer-construction strategies. Passing the generator the *question* concatenated with the sentences predicted *essential* achieves higher scores across most relevance metrics: BLEU increases from 2.0 to 19.2, ROUGE-Lsum from 25.1 to 53.6, AlignScore from 42.6 to 97.6, and MEDCON from 37.4 to 54.1, resulting in an overall relevance score of **52.9**. By contrast, generating a free-form *summarised answer* results in an overall relevance score of 31.7. Based on this result, we adopted the *questions + essential sentences* strategy for our final test submission.

4.3.3 Official Leaderboard

Table 3 summarises official test-set results. Our system (DMIS Lab) ranks first with an Overall score of 53.7, balancing a factuality F1 of 58.6 and a relevance overall of 48.8. The consistency between development and test splits underscores the effectiveness of the proposed multi-agent architecture.

5 Conclusion

In this paper, we presented a multi-agent framework for answering patients' health-related questions using their EHRs. Our method decomposes the task into distinct stages: context generation, relevance assessment, expert recruitment, and consensus integration. Each stage is handled by specialized LLM-based agents. This structured, modular approach enables robust identification of essential clinical sentences and the generation of coherent, evidence-grounded responses. Our framework achieved strong performance on both the development and test sets in terms of factuality. These results highlight the potential of LLM-based multi-agent systems in clinical question answering and suggest promising directions for future work in automating patient-clinician communication based on real EHR data.

Acknowledgements

This research was supported by (1) the National Research Foundation of Korea (NRF-2023R1A2C3004176, RS-2023-00262002), (2) the Ministry of Health & Welfare, Republic of Korea (HR20C002103), and (3) ICT Creative Consilience Program through the Institute of Information & Communications Technology Planning & Evaluation(IITP) grant funded by the Korea government(MSIT)(IITP-2025-RS-2020-II201819).

Limitations

De-identification Assumptions. ArchEHR-QA 2025 provides de-identified notes, but real clinical systems often contain partially identifiable information. Our framework does not include additional privacy-preserving mechanisms and would need adaptation before deployment on raw, identifiable EHR data.

Dependence on Closed source LLMs. Our framework relies on OpenAI's o3 and GPT-4.1 models. Although these models currently provide state-of-the-art reasoning, they are proprietary, incur nontrivial inference costs, and can change without notice. Reproducing or extending our results with fully open-source alternatives may require prompt and hyper-parameter retuning.

Latency and Cost. The framework's inference time and computational cost remain substantial, posing challenges for real-time deployment in high-volume patient-portal environments. These resource demands may limit its practical scalability without further optimization.

References

Taicheng Guo, Xiuying Chen, Yaqi Wang, Ruidi Chang, Shichao Pei, Nitesh V. Chawla, Olaf Wiest, and Xiangliang Zhang. 2024. Large language model based multi-agents: A survey of progress and challenges. *Preprint*, arXiv:2402.01680.

Yubin Kim, Chanwoo Park, Hyewon Jeong, Yik Siu Chan, Xuhai Xu, Daniel McDuff, Hyeonhoon Lee, Marzyeh Ghassemi, Cynthia Breazeal, and Hae Won Park. 2024. MDAgents: An adaptive collaboration of LLMs for medical decision-making. In *The Thirty-eighth Annual Conference on Neural Information Processing Systems*.

Chin-Yew Lin. 2004. ROUGE: A package for automatic evaluation of summaries. In *Text Summarization Branches Out*, pages 74–81, Barcelona, Spain. Association for Computational Linguistics.

OpenAI. 2025a. Introducing gpt-4.1 in the api. `https://openai.com/index/gpt-4-1/`. Accessed: 2025-05-09.

OpenAI. 2025b. o3 and o4-mini system card. `https://cdn.openai.com/pdf/2221c875-02dc-4789-800b-e7758f3722c1/o3-and-o4-mini-system-card.pdf`. Accessed May 9, 2025.

Kishore Papineni, Salim Roukos, Todd Ward, and Wei-Jing Zhu. 2002. Bleu: a method for automatic evaluation of machine translation. In *Proceedings of the 40th Annual Meeting on Association for Computational Linguistics*, ACL '02, page 311–318, USA. Association for Computational Linguistics.

Sarvesh Soni and Dina Demner-Fushman. 2025a. A dataset for addressing patient's information needs related to clinical course of hospitalization. *arXiv preprint*.

Sarvesh Soni and Dina Demner-Fushman. 2025b. Overview of the archehr-qa 2025 shared task on grounded question answering from electronic health records. In *The 24th Workshop on Biomedical Natural Language Processing and BioNLP Shared Tasks*, Vienna, Austria. Association for Computational Linguistics.

Xiangru Tang, Anni Zou, Zhuosheng Zhang, Ziming Li, Yilun Zhao, Xingyao Zhang, Arman Cohan, and Mark Gerstein. 2024. MedAgents: Large language models as collaborators for zero-shot medical reasoning. In *Findings of the Association for Computational Linguistics: ACL 2024*, pages 599–621, Bangkok, Thailand. Association for Computational Linguistics.

Zhenhailong Wang, Shaoguang Mao, Wenshan Wu, Tao Ge, Furu Wei, and Heng Ji. 2023. Unleashing the emergent cognitive synergy in large language models: A task-solving agent through multi-persona self-collaboration. *arXiv preprint arXiv:2307.05300*.

Zhiheng Xi, Wenxiang Chen, Xin Guo, Wei He, Yiwen Ding, Boyang Hong, Ming Zhang, Junzhe Wang, Senjie Jin, Enyu Zhou, and 1 others. 2025. The rise and potential of large language model based agents: A survey. *Science China Information Sciences*, 68(2):121101.

Wei Xu, Courtney Napoles, Ellie Pavlick, Quanze Chen, and Chris Callison-Burch. 2016. Optimizing statistical machine translation for text simplification. *Transactions of the Association for Computational Linguistics*, 4:401–415.

Wen-wai Yim, Yujuan Fu, Asma Ben Abacha, Neal Snider, Thomas Lin, and Meliha Yetisgen. 2023. Aci-bench: a novel ambient clinical intelligence dataset for benchmarking automatic visit note generation. *Scientific Data*, 10(1):586.

Yuheng Zha, Yichi Yang, Ruichen Li, and Zhiting Hu. 2023. AlignScore: Evaluating factual consistency with a unified alignment function. In *Proceedings of the 61st Annual Meeting of the Association for Computational Linguistics (Volume 1: Long Papers)*, pages 11328–11348, Toronto, Canada. Association for Computational Linguistics.

Tianyi Zhang*, Varsha Kishore*, Felix Wu*, Kilian Q. Weinberger, and Yoav Artzi. 2020. Bertscore: Evaluating text generation with bert. In *International Conference on Learning Representations*.

A Appendix

A.1 Prompt example for Each Agent

ContextAgent system_message

```
- You are a ContextBuildingAgent.

- Your role is to carefully read the patient's question and generate a relevant medical context
    that would help answer it accurately.

Guidelines:
- Focus only on medically meaningful information that would assist in answering the patient's
    question.
- Include important background knowledge, clinical reasoning steps, diagnostic considerations,
    and treatment options that are directly relevant to the question.
- Do not fabricate information unrelated to the patient's case.
- The context should be clear, concise, medically accurate, and structured to support clinical
    decision-making.

Output:
- Plain text only.
- Write in a factual and professional tone as if you are preparing supporting information for a
    medical expert.
```

RecruitingAgent prompt

```
- You are a recruiting agent.

- Given a patient question and EHR note, your task is to identify the most relevant medical
    experts.

- Return a JSON object with a key called 'experts' whose value is a list of strings.
- Example: {"experts": ["cardiology", "gastroenterology"]}

- Do not include any explanation or additional text. Only return the JSON object.
```

AnalysisAgent prompt

```
- You are a medical reviewer. Your task is to evaluate whether each individual sentence
in a clinical note is relevant to answering the patient's question.

- Each sentence is identified by its ID. For every sentence, return:
Whether the sentence is 'essential' or 'not-relevant', and A brief justification for your
judgment, explaining why the sentence does or does not contribute to answering the question.

- A sentence is considered **essential** if it directly or indirectly helps answer the question
through evidence, explanation, clarification, or medically meaningful context.
- Avoid marking sentences as essential if they only provide background or loosely related
    information.
```

ExpertAgent prompt

```
- You are a board-certified clinical expert in {expertise}.

- You are evaluating each sentence in an EHR note from the unique clinical
perspective of your own specialty ({expertise}).

- Your role is to assess whether the sentence meaningfully contributes to
answering the patient's question, based on your specialty's reasoning
principles, typical clinical decision-making, and domain-specific interpretation.

- If the sentence contains medically meaningful evidence, logic, or interpretation
```

```
  that a {expertise} specialist would find critical to answer the question,
  label it "ESSENTIAL".

- If the sentence contains no valuable insight or decision-making relevance from
  your specialty perspective, label it "NOT-RELEVANT".

- Avoid generic reasoning. Always ground your decision in your expert role.
```

DecisionAgent prompt

```
- You are a skilled medical expert. Your task is to provide an accurate and evidence-based answer
     to a patient's question using the provided EHR note.

- Your answer must be medically sound and supported by evidence extracted from the provided EHR
    note sentences.

- When composing your answer, you **must** include citation IDs (enclosed in pipe symbols |, for
    example, |3,4|) only for the parts of your answer that are directly supported by evidence
    from the EHR note.

- Each sentence in your answer should be on a separate line.

- **Before writing your answer, carefully verify whether the EHR note includes any sentences that
     are truly relevant to answering the patient's question.**
```

A.2 Basic Structure of Agent

Listing A.2: Agent Class Definition

```python
class Agent:
    def __init__(self, agent_name,
                 model='model',
                 temperature=0,
                 system_message='You are a helpful assistant.'):
        self.agent_name  = agent_name
        self.model       = model
        self.temperature = temperature
        self.system_message = system_message
        self.client = openai.OpenAI()

    def generate_response(self, user_msg: str) -> str:
        rsp = self.client.chat.completions.create(
            model=self.model,
            messages=[{"role": "system",
                        "content": self.system_message},
                       {"role": "user",
                        "content": user_msg}],
            temperature=self.temperature,
            max_tokens=2048,
        )
        return rsp.choices[0].message.content.strip()
```

CogStack-KCL-UCL at ArchEHR-QA 2025: Investigating Hybrid LLM Approaches for Grounded Clinical Question Answering

Shubham Agarwal[1], Thomas Searle[1,2,3], Kawsar Noor[2], Richard Dobson[1,2,3]

[1]Department of Biostatistics & Health Informatics, King's College London, London, U.K.
[2]Institute of Health Informatics, University College London, London, U.K.
[3]CogStack Limited, London, U.K.

Correspondence: shubham.agarwal@kcl.ac.uk

Abstract

We present our system for the ArchEHR shared task, which focuses on answering clinical and patient-facing questions grounded in real-world EHR data. Our core contribution is a 2-Stage prompting pipeline that separates evidence selection from answer generation while employing in-context learning strategies. Our experimentation leveraged the open-weight Gemma-v3 family of models, with our best submission using the Gemma-12B model securing 5th place overall on the unseen test set. Through systematic experimentation, we demonstrate the effectiveness of task decomposition in improving both factual accuracy and answer relevance in grounded clinical question answering.

1 Introduction

As the adoption of digital systems in healthcare become ubiquitous, patients will expect to be able pose questions of their recent experiences. Responding to these questions in a rapid, thorough and most importantly safe way will ensure patients are more involved on their care and receive overall improved care.

Effective communication between patients and their healthcare providers is a cornerstone of quality care as it plays a critical role in treatment adherence, recovery, and overall health outcomes (Zolnierek and DiMatteo, 2009). Patient portals have emerged as a key tool for facilitating this communication, providing individuals with direct access to their health information and enabling ongoing interaction with their care teams (Irizarry et al., 2015). Modern patient portals go beyond simple data access—they support secure messaging, prescription refill requests, and delivery of tailored educational materials (Lyles et al., 2020).

A growing body of research highlights that patient engagement through these digital platforms is associated with improved health literacy, better understanding, increased medication adherence, and greater satisfaction with care (Han et al., 2019; Otte-Trojel et al., 2014; Carini et al., 2021; Dendere et al., 2019). Portals enabling record review and follow-up questions have been shown to foster better self-management and reduce conflict in decision-making (Najafi et al., 2022; Shay and Lafata, 2015).

Beyond empowerment, these digital systems help reduce medical errors, improve communication of complex information, and foster trust between patients and providers (Bell et al., 2017; DesRoches et al., 2020). Integrating robust and responsive question-answering capabilities into patient portals offers a promising direction for advancing truly patient-centered care. The potential of conversational agents to further enhance communication and engagement is increasingly recognized, with recent studies showing early but promising results in clinical contexts (Laranjo et al., 2018).

2 Background

2.1 Retrieval Augmented Generation

A widely adopted framework for building question-answering systems is Retrieval-Augmented Generation (RAG) (Lewis et al., 2020), which uses a Causal Large Language Models (LLMs) to generate answers. In this framework, a retriever first selects relevant passages from a knowledge source, which are then passed as context to the LLM that leverages it to generate answers. In the medical domain, RAG has been applied to tasks such as clinical decision support (Zakka et al., 2024), medical literature retrieval (Tian et al., 2024), and patient education (Xiong et al., 2024), generating patient-friendly explanations of medical conditions and procedures (Yang et al., 2025).

2.2 ArchEHR shared task

Responding to patients' queries on portals offers numerous benefits, as discussed in Section 1, but it

BioNLP 2025 Shared Tasks, pages 126–135

has become a contributor to clinician burden. Automatically generating accurate, safe, and context-aware responses to patient questions using data from EHRs can help alleviate this pressure.
Although RAG offers a compelling framework for this task, it faces several limitations, especially in the medical domain. Generated responses can be incorrect, particularly when retrieved documents are ambiguous or conflicting, leading to hallucinations (Wong et al., 2025). The complexity of RAG systems can make it challenging to trace the reasoning behind generated answers, which is especially critical in medical contexts (Yang et al., 2025).

The ArchEHR shared task (Soni and Demner-Fushman, 2025a), hosted on PhysioNet (Goldberger et al., 2000), proposes a benchmark specifically designed to evaluate grounded question answering in the clinical domain. The task focuses on answering patient-facing questions using evidence from EHR notes, with a strong emphasis on two core criteria: **factuality**, which measures whether the generated answer is supported by cited evidence, and **relevance**, which assesses how well the response aligns with the patient's original query (Soni and Demner-Fushman, 2025c). This aims at addressing the above discussed limitations and thereby advancing safe and trustworthy patient-centered clinical QA systems.

2.2.1 Dataset

The ArchEHR dataset (Soni and Demner-Fushman, 2025b) is constructed using the MIMIC-III database (Johnson et al., 2016), a large, publicly available resource of de-identified ICU records, ensuring realistic clinical language and complexity. Each instance in the dataset contains a patient-posed question, a clinician-refined rewrite, a set of evidence sentences from real clinical notes, and a gold standard answer.

The task requires systems to generate responses grounded in the provided evidence, with citations to specific supporting sentences. The task does not enforce answers to use either or only one of the patient or clinician focused questions.

3 Methodology

We present two groups of approaches for the task:

- Firstly, we prompt LLMs either in 2-Stages or a combined 1-Stage approach, as described in Section 3.1.1 and 3.1.2 respectively.
- Secondly, we experiment with a *classical* sentence embeddings, fine-tuning classifiers for classification of relevant evidence to inform the generation step.

These approaches are similar to a general RAG process, but importantly our retriever step is constrained to only sentences, and to three distinct classes of informativeness for the generated summary, i.e. essential, supplementary or not-relevant sentence classes.

3.1 Approaches

3.1.1 2-Stage Prompt Approach

This approach consists of two stages, each targeting a specific subtask: **Stage 1** – Sentence Classification and Retrieval and **Stage 2** – Generation. For both the stages, an LLM is prompted to perform the specified subtask. Below is a description of the stages:

- **Stage 1:** Given a query, Stage 1 focuses on identifying the most relevant sentences from the clinical notes. By passing focused context to the generation stage, it improves performance as the generation stage focuses on clinically meaningful evidence, leading to more precise and context-aware responses.
- **Stage 2:** Stage 2 performs generation using the filtered context, allowing the model can leverage all its capability effectively to produce accurate, clinically relevant answers. This ensures that the final response is not only coherent but also grounded in the relevant evidence, minimizing the risk of hallucinations or misinformation.

The 2-Stage prompt is the proposed approach in this work. Its ability to break down the task into manageable stages improves the clarity and performance of each step, resulting in a higher performing pipeline. Further details are discussed in Section 5.

3.1.2 1-Stage Prompt and 2-Stage Fine-tuned Classifier Approach

The 1-Stage Prompt combines sentence classification and answer generation into a single prompt to the LLM, requiring the model to both identify relevant evidence and generate a response at once. This approach simplifies and aims to streamline the process by tackling the task as a single coherent objective. This approach also utilises the prompting techniques mentioned in Section 3.2.

The 2-Stage fine-tuned classifier approach follows the same 2 stage structure as the 2-Stage prompting method, but uses a fine-tuned classifier to perform the sentence classification in-place of an LLM. Specifically, we use Sentence-BERT embeddings (Reimers and Gurevych, 2019) to encode sentences and train a classifier to perform the task using the dev test. This approach allows for greater control over the sentence filtering stage and enables fine-tuning on the task-specific data.

3.2 Few Shot Learning

To guide the model through both stages, we leverage in-context learning via few-shot prompting to ensure consistent and contextually accurate outputs. Carefully designed prompts which include a small number of examples, help the model understand the task, distinguish relevant from irrelevant information, and structure its responses appropriately.

Prompt design was iteratively refined based on empirical performance during sentence classification and answer generation phases. Our approach integrates task-specific examples that included varied clinical scenarios to better guide the model, allowing it to also grasp the clinical nuances of the tasks. Appendix A.2 provides our prompt templates.

3.3 Output Guardrails and Format Enforcement

To ensure consistency and adherence to format requirements across both stages, we implement guardrails across both stages of the pipeline. An output parser validates the model's responses in both stages with the expected format criteria. In cases where the initial output fails to adhere to the required format, we utilize an additional parser that leverages an LLM to reattempt answering and formatting. This lowers the probability of a response being discarded by allowing it to be reformatted correctly.

3.4 Pre-Trained Models

In our pipeline, we use the Gemma family of models (Gemma Team et al., 2025), specifically the instruction tuned models. These models are openly available and based off the closed source Google Gemini models. We selected Gemma models due to their strong performance on instruction-following tasks and their demonstrated reasoning capabilities with more manageable parameter sizes. Notably, the Gemma v3 models outperform their predecessors across multiple reasoning tasks (Gemma Team et al., 2025), making them suitable for complex clinical question answering. Our initial experiments also included Mistral 7B v0.2 instruct model (Jiang et al., 2023).
Experimentation utilized a shared university resource machine with 3 Nvidia A100 GPUs via KCL CREATE (King's College London e-Research team, 2025). We also utilised LLama-cpp and GGML / GGUF quantized models for directly running models on locally available hardware.

We attempted to use the Gemma 27B with initial experiments for 1-Stage prompting but found the model refused to consistently return results on the dev set. We did not continue experimenting with this model and do not report results. Similarly, we attempted to use the Qwen 2.5 7B instruct model (Qwen Team, 2024). We did not report the results for it as the performance was poor for all approaches.

3.5 Evaluation

The ArchEHR task is evaluated through cited evidence classification performance representing *Factuality* and the quality of the generated responses using the cited evidence representing *Relevance*.

Factuality is measured through precision, recall and F1 of prediction of each source sentence representing of one of three classes 'essential', 'supplementary', 'not-relevant'. Scoring is *strict* if only 'essential' labels are included or *lenient* if both 'essential' and 'supplementary' sentences are counted towards final calculations.

Relevance uses a collection of n-gram based automated evaluation metrics BLUE (Papineni et al., 2002), ROUGE (Lin, 2004), SARI (Xu et al., 2016) and model based metrics BERTScore (Zhang et al., 2020), AlignScore (Zha et al., 2023) and MEDCON (wai Yim et al., 2023). Scoring generated text for relevance to a provided question can be subjective, but aggregating a range of scores provides some means to automatically evaluate system performance at scale.

While open-domain metrics can give a broad indication of fluency and semantic similarity, MEDCON directly assesses the preservation of medical relevance, offering a more trustworthy signal in safety-critical clinical question answering.

External knowledge was permitted during this

task, and a real system would likely include the integration of external knowledge supplementing existing knowledge within the LLM or model approach. For example, a local, regional or national clinical guideline could be referenced by an LLM during a generation if a question involved why a course of action was taken.

Our approach did not use any external knowledge, or external clinical knowledge base such as UMLS (Bodenreider, 2004) or SNOMED CT (Stearns et al., 2001). This is further mentioned in Section 6.

4 Results

4.1 2-Stage Prompt Approach

Our approach was evaluated on the dev set and test set using the specified metrics, and the results demonstrate promising performance for clinical question-answering tasks. As shown in the results table 1, the Gemma 12B model outperformed the other models across all metrics, achieving an overall score of **47.03**. This suggests that the larger models are better equipped to follow instructions and capture the complex relationships and context within clinical data*. While the larger models consistently outperformed the smaller ones*, the smaller models exhibited a strong ability to handle complex clinical data.

4.2 1-Stage Prompt and 2-Stage Fine-tuned Classifier Approach

As shown in Appendix A.1, both the 1-Stage prompt and 2-Stage fine-tuned classifier approaches underperform relative to the 2-Stage prompt approach, especially for the Gemma 12B model which shows a performance decrease of 34.7% and 47.1% respectively. The 1-Stage Prompt approach lags in both factuality and relevance, except for a slight gain in relevance for the Gemma 4B model.

Similarly, the 2-Stage fine-tuned classifier approach is subpar overall except Factuality for Gemma 4B model. Notably, this approach achieves high precision scores (for strict and lenient), with lenient macro precision score of 86.25.

5 Discussion

The development of our approach for the ArchEHR task evolved through several iterations, each building on previous insights. The 1-Stage prompt approach exposed the limitations of a monolithic design, as the LLM struggled with handling both classification and generation simultaneously. To address this, we introduced a 2-Stage fine-tuned classifier approach, which showed promise and achieved high factuality and precision but was constrained by limited data for effective training. With these insights, we adopted the 2-Stage prompt approach, which retained the advantages of task separation without requiring fine-tuning. This approach outperformed the others, delivering stronger results in both factuality and relevance.

This approach mimics the Chain-of-Thought reasoning process (Wei et al., 2022), whereby breaking down the task into smaller, sequential subtasks encourages more structured reasoning, improves factual alignment, and reduces cognitive load on the model, enabling it to perform each step more reliably and accurately. It also provides a more interpretable pipeline where each stage can be independently evaluated, enhancing overall system transparency.

While the proposed approach achieves strong results, it depends heavily on prompt design and the inherent capabilities of the underlying LLM. We further discuss the limitations and future work in the below sections.

6 Conclusions & Future Work

Our work presents a 2-Stage few-shot prompting approach to grounded clinical QA from real-world EHR data. Leveraging the Gemma-v3-12B model, our best approach secures 5th place overall on the unseen test set, demonstrating a good balance between *factuality*, recognising the correct sentences that should be used in the generated answer, and *relevance* the quality of the generated text from the cited evidence. This systematic task decomposition enhances performance along with providing a more transparent method, crucial for sensitive healthcare contexts.

Our future work involves integration of external world knowledge into system responses, either as 'guardrails' or to directly improve system responses. An example of such world knowledge could be clinical guideline that informed or impacted a course of action, but is not directly referenced in the source EHR notes. Secondly, we aim to explore fine-tuning a Casual Large Language Model on a more expansive and curated dataset for sentence classification. This would enhance

*Except Gemma 27B model as discussed in Section 3.4

Table 1: Pipeline performance for 2-stage prompting approach

Model	Factuality					Relevance			Overall score
	Strict		*Lenient*		Overall	BLEU	SARI	Overall	
	Macro F1	Micro F1	Macro F1	Micro F1	factuality			relevance	
Dev set performance									
Mistral 7B	41.89	38.65	43.38	42.21	38.65	3.44	57.4	35.3	36.98
Gemma 1B	25.41	23.38	29.91	28.76	23.38	3.2	62.99	32.54	27.96
Gemma 4B	36.4	31.9	38.2	37.1	31.9	4.1	65.5	38.5	35.2
Gemma 12B	**51.35**	**49.81**	**51.59**	**48.92**	**49.82**	**8.99**	**71.84**	**44.2**	**47.03**
Test set performance									
Gemma 12B	51.4	47.5	52.1	47.6	47.5	4.7	70.0	42.6	45.0

the quality and consistency of context filtering, thereby improving downstream answer quality and reducing reliance on prompt-based reasoning by the LLM.
We look to integrate the development and testing of these methods as we actively pursue safe and reliable clinical QA over EHRs.

Limitations

Our work is presented as a solution to the ArchEHR shared task, and provides results on a small development and unseen larger test set. Our best method generalises well to the unseen test demonstrating the suitability of our method to the task.

However, the proposed system is limited in a number of ways. Firstly, the task and proposed system assumes that entire sentences are either wholly relevant or useful to a response, representing a form of *extractive* summarisation, whereas it is likely an optimal response will likely be helped to *abstractively* summarise from across one or more partial sentences to generate a response.

Secondly, the dataset is small and only representative of a single provider USA based ICU. Further work could expand evaluation of such systems across health systems and geographies.

Usage of our proposed system in a 'production' environment will likely require extensive use of hardware resources, namely GPU compute. Due to the sensitivity of patient EHR data, clinical providers will likely require patient QA systems that leverage LLM technology to be secure and isolated from other systems alongside adhering to regulatory standard such as HIPPA or GDPR. In deployment of clinical informatics systems it is especially important to balance availability of hardware with model and system performance.

Acknowledgments

This work was supported by Health Data Research UK, an initiative funded by UK Research and Innovation, Department of Health and Social Care (England) and the devolved administrations, and leading medical research charities. SA, TS, RD are part-funded by the National Institute for Health Research (NIHR) Biomedical Research Centre at South London and Maudsley NHS Foundation Trust and King's College London. RD is also supported by The National Institute for Health Research University College London Hospitals Biomedical Research Centre. This paper represents independent research part funded by the National Institute for Health Research (NIHR) Biomedical Research Centre at South London and Maudsley NHS Foundation Trust and King's College London. The views expressed are those of the authors and not necessarily those of the NHS, the NIHR or the Department of Health and Social Care. The funders had no role in study design, data collection and analysis, decision to publish, or preparation of the manuscript.

References

Sigall K Bell, Tom Delbanco, Joann G Elmore, Paul S Fitzgerald, Alan Fossa, Katharine Harcourt, and Suzanne G Leveille. 2017. Frequency and types of patient-reported errors in electronic health record ambulatory care notes. *JAMA Network Open*, 320(18):1867–1878.

Olivier Bodenreider. 2004. The unified medical language system (umls): integrating biomedical terminology. *Nucleic acids research*, 32(suppl_1):D267–D270.

Elena Carini, Luca Villani, Anna Maria Pezzullo, Antonio Gentili, Antonella Barbara, Walter Ricciardi, and Stefania Boccia. 2021. The impact of digital patient portals on health outcomes, system efficiency, and patient attitudes: Updated systematic review. *Journal of Medical Internet Research*, 23(9):e26189.

Ronald Dendere, Christine Slade, Andrew Burton-Jones, Clair Sullivan, Andrew Staib, and Monika Janda. 2019. Patient portals facilitating engagement with inpatient electronic medical records: A systematic review. *Journal of Medical Internet Research*, 21(4):e12779.

Catherine M DesRoches, Sigall K Bell, Zhaohui Dong, Joann G Elmore, Paul Fitzgerald, Katharine Harcourt, and Suzanne G Leveille. 2020. Patients managing medications and reading their visit notes: A survey of opennotes participants. *Annals of Internal Medicine*, 172(1):35–38.

Gemma Team, Aishwarya Kamath, Johan Ferret, Shreya Pathak, Nino Vieillard, Ramona Merhej, Sarah Perrin, Tatiana Matejovicova, Alexandre Ramé, Morgane Rivière, Louis Rouillard, Thomas Mesnard, Geoffrey Cideron, Jean-Bastien Grill, Sabela Ramos, Edouard Yvinec, Michelle Casbon, Etienne Pot, Ivo Penchev, and 197 others. 2025. Gemma 3 technical report. *arXiv [cs.CL]*.

A L Goldberger, L A Amaral, L Glass, J M Hausdorff, P C Ivanov, R G Mark, J E Mietus, G B Moody, C K Peng, and H E Stanley. 2000. PhysioBank, PhysioToolkit, and PhysioNet: components of a new research resource for complex physiologic signals. *Circulation*, 101(23):E215–20.

Haixia Han, Kelly Gleason, Ruopeng Sun, Heather Miller, Huong Kang, Yan Gai, and David Rosenthal. 2019. Using patient portals to improve patient outcomes: Systematic review. *JMIR Human Factors*, 6(4):e15038.

Taya Irizarry, Annette DeVito Dabbs, and Christine R Curran. 2015. Patient portals and patient engagement: a state of the science review. *Journal of medical Internet research*, 17(6):e148.

Albert Q Jiang, Alexandre Sablayrolles, Arthur Mensch, Chris Bamford, Devendra Singh Chaplot, Diego de las Casas, Florian Bressand, Gianna Lengyel, Guillaume Lample, Lucile Saulnier, and 1 others. 2023. Mistral 7b. *arXiv preprint arXiv:2310.06825*.

Alistair E W Johnson, Tom J Pollard, Lu Shen, Li-Wei H Lehman, Mengling Feng, Mohammad Ghassemi, Benjamin Moody, Peter Szolovits, Leo Anthony Celi, and Roger G Mark. 2016. MIMIC-III, a freely accessible critical care database. *Sci Data*, 3:160035.

King's College London e-Research team. 2025. King's computational research, engineering and technology environment (CREATE).

Liliana Laranjo, Adam G Dunn, Helen L Tong, Ahmet Baki Kocaballi, Jing Chen, Rifat Bashir, Didi Surian, Blanca Gallego, Farah Magrabi, and Enrico Coiera. 2018. Conversational agents in healthcare: a systematic review. *Journal of the American Medical Informatics Association*, 25(9):1248–1258.

Patrick Lewis, Ethan Perez, Aleksandra Piktus, Fabio Petroni, Vladimir Karpukhin, Naman Goyal, Heinrich Küttler, Mike Lewis, Wen-tau Yih, Tim Rocktäschel, and 1 others. 2020. Retrieval-augmented generation for knowledge-intensive nlp tasks. *Advances in neural information processing systems*, 33:9459–9474.

Chin-Yew Lin. 2004. Rouge: A package for automatic evaluation of summaries. In *Text Summarization Branches Out*, pages 74–81. Association for Computational Linguistics.

Courtney R Lyles, Eugene C Nelson, Susan Frampton, Patricia C Dykes, Anupama G Cemballi, and Urmimala Sarkar. 2020. Using electronic health record portals to improve patient engagement: research priorities and best practices. *Annals of internal medicine*, 172(11_Supplement):S123–S129.

Faezeh Najafi, Pirhossein Shojaei, Saeed Shojaee Moghaddam, Mehdi Jafari, and Arash Rashidian. 2022. Impact of patient engagement on healthcare quality: A scoping review. *Annals of Global Health*, 88(1):78.

Tine Otte-Trojel, Antoinette de Bont, Joris van de Klundert, and Thomas G Rundall. 2014. How outcomes are achieved through patient portals: a realist review. *Journal of the American Medical Informatics Association*, 21(4):751–757.

Kishore Papineni, Salim Roukos, Todd Ward, and Wei-Jing Zhu. 2002. Bleu: a method for automatic evaluation of machine translation. In *Proceedings of the 40th Annual Meeting of the Association for Computational Linguistics*, pages 311–318. Association for Computational Linguistics.

Qwen Team. 2024. Qwen2.5: A party of foundation models.

Nils Reimers and Iryna Gurevych. 2019. Sentence-bert: Sentence embeddings using siamese bert-networks. *arXiv preprint arXiv:1908.10084*.

L Aubree Shay and Jennifer Elston Lafata. 2015. Where is the evidence? a systematic review of shared decision making and patient outcomes. *Medical decision making*, 35(1):114–131.

Sarvesh Soni and Dina Demner-Fushman. 2025a. ArchEHR-QA: BioNLP at ACL 2025 shared task on grounded electronic health record question answering.

Sarvesh Soni and Dina Demner-Fushman. 2025b. A dataset for addressing patient's information needs related to clinical course of hospitalization. *arXiv preprint*.

Sarvesh Soni and Dina Demner-Fushman. 2025c. Overview of the ArchEHR-QA 2025 shared task on grounded question answering from electronic health records. In *The 24th Workshop on Biomedical Natural Language Processing and BioNLP Shared Tasks*, Vienna, Austria. Association for Computational Linguistics.

M Q Stearns, C Price, K A Spackman, and A Y Wang. 2001. SNOMED clinical terms: overview of the development process and project status. *Proc. AMIA Symp.*, pages 662–666.

Shubo Tian, Qiao Jin, Lana Yeganova, Po-Ting Lai, Qingqing Zhu, Xiuying Chen, Yifan Yang, Qingyu Chen, Won Kim, Donald C Comeau, and 1 others. 2024. Opportunities and challenges for chatgpt and large language models in biomedicine and health. *Briefings in Bioinformatics*, 25(1):bbad493.

Wen wai Yim, Yujuan Fu, Asma Ben Abacha, Neal Snider, Thomas Lin, and Meliha Yetisgen. 2023. Aci-bench: a novel ambient clinical intelligence dataset for benchmarking automatic visit note generation. *Scientific Data*, 10(1):586.

Jason Wei, Xuezhi Wang, Dale Schuurmans, Maarten Bosma, Fei Xia, Ed Chi, Quoc V Le, Denny Zhou, and 1 others. 2022. Chain-of-thought prompting elicits reasoning in large language models. *Advances in neural information processing systems*, 35:24824–24837.

Lionel Wong, Ayman Ali, Raymond Xiong, Shannon Zeijang Shen, Yoon Kim, and Monica Agrawal. 2025. Retrieval-augmented systems can be dangerous medical communicators. *arXiv preprint arXiv:2502.14898*.

Guangzhi Xiong, Qiao Jin, Zhiyong Lu, and Aidong Zhang. 2024. Benchmarking retrieval-augmented generation for medicine. In *Findings of the Association for Computational Linguistics ACL 2024*, pages 6233–6251.

Wei Xu, Courtney Napoles, Ellie Pavlick, Quanze Chen, and Chris Callison-Burch. 2016. Optimizing statistical machine translation for text simplification. *Transactions of the Association for Computational Linguistics*, 4:401–415.

Rui Yang, Yilin Ning, Emilia Keppo, Mingxuan Liu, Chuan Hong, Danielle S Bitterman, Jasmine Chiat Ling Ong, Daniel Shu Wei Ting, and Nan Liu. 2025. Retrieval-augmented generation for generative artificial intelligence in health care. *npj Health Systems*, 2(1):2.

Cyril Zakka, Rohan Shad, Akash Chaurasia, Alex R Dalal, Jennifer L Kim, Michael Moor, Robyn Fong, Curran Phillips, Kevin Alexander, Euan Ashley, and 1 others. 2024. Almanac—retrieval-augmented language models for clinical medicine. *Nejm ai*, 1(2):AIoa2300068.

Yuheng Zha, Yichi Yang, Ruichen Li, and Zhiting Hu. 2023. Alignscore: Evaluating factual consistency with a unified alignment function. In *Proceedings of the 61st Annual Meeting of the Association for Computational Linguistics*, pages 11300–11316. Association for Computational Linguistics.

Tianyi Zhang, Varsha Kishore, Felix Wu, Kilian Q Weinberger, and Yoav Artzi. 2020. Bertscore: Evaluating text generation with bert. In *International Conference on Learning Representations (ICLR)*.

Kelly B Haskard Zolnierek and M Robin DiMatteo. 2009. Physician communication and patient adherence to treatment: a meta-analysis. *Medical care*, 47(8):826–834.

A Appendix

A.1 Results for all approaches

Table 2: Pipeline performance for 1-Stage prompt approach

Dev set Performance

Model	Factuality					Relevance			Overall score
	Strict		*Lenient*		Overall factuality	BLEU	SARI	Overall relevance	
	Macro F1	Micro F1	Macro F1	Micro F1					
Gemma 4B	24.66	25.25	24.65	24.89	25.25	3.57	68.63	41.79	33.52
Gemma 12B	21.10	22.56	21.03	21.13	22.56	1.36	65.70	38.84	30.70

Table 3: Pipeline performance for 2-Stage approach with fine-tuned classifier

Dev set Performance

Model	Factuality					Relevance			Overall score
	Strict				Overall factuality	BLEU	SARI	Overall relevance	
	Macro Precision	Macro F1	Micro Precision	Micro F1					
Gemma 4B	72.08	35.0	71.42	37.43	37.43	1.32	56.38	27.932	32.67
Gemma 12B	70.0	22.93	69.56	19.8	19.8	2.99	62.29	29.89	24.88

A.2 Prompt for 2-Stage pipeline

A.2.1 For Stage 1

```
"""
<bos><start_of_turn>user You are a clinical assistant. Use the context below to perform the given
    task. Your response must be a JSON list of citations.

Answer using the given context to help, if you don't know the answer, just say that you don't know,
    don't try to make up an answer.
Format of Context:
ID: <chunk ID> ; text : <text>
ID: <chunk ID> ; text : <text>
...

The context contains all sentences from note excerpts. These sentences have two categories: relevant
    and not relevant.
Your task is to using reasoning and filter out the ones that are relevant to the question, and
    respond with their ID. Ensure to pick all the relevant ones, prioritise higher recall over
    precision.
Include all chunks that are directly relevant and reasonably connected to answering the question.
    Only exclude chunks that are clearly unrelated.

Output format:
Your output must be a list of structured objects with:
   - 'citation': the chunk ID (e.g., '1')

   - 'citation': the chunk ID (e.g., '2')

   - 'citation': the chunk ID (e.g., '4')

DO NOT add explanations, only the above output.
NOTE: The sentences may not be directly relevant, you will have to infer it.

Examples:
```

```
# Example 1:
**Context:**
ID: 1 ; Text: "The patient complained of frequent urination and excessive thirst. Laboratory tests revealed elevated blood glucose levels."
ID: 2 ; Text: "The patient was diagnosed with type 2 diabetes mellitus."
ID: 3 ; Text: "Dietary counseling was initiated to help manage blood sugar levels."
ID: 4 ; Text: "The patient also reported occasional headaches over the past month."

**Question:** What is the patient's diagnosis?
**Answer:**
[{{"citation": "2"}},
{{"citation": "1"}},
{{"citation": "3"}}]
**Reasoning:
ID 2 gives the direct diagnosis (must include).
ID 1 gives symptoms and test results leading to diagnosis (should include).
ID 3 mentions management for blood sugar slightly grey, but include as it supports the context of the diagnosis.
ID 4 about headaches is unrelated (exclude).

# Example 2:
**Context:**
ID: 1 ; Text: "The patient sustained a fractured right femur after a fall from a ladder."
ID: 2 ; Text: "An open reduction and internal fixation (ORIF) surgery was performed to stabilize the fracture."
ID: 3 ; Text: "The patient was prescribed physical therapy after hospital discharge."
ID: 4 ; Text: "The patient's blood pressure was also found to be elevated during admission."

**Question:** What treatment did the patient receive for the femur fracture?
**Answer:**
[{{"citation": "2"}},
{{"citation": "3"}},
{{"citation": "1"}}]
**Reasoning:
ID 2 describes surgical treatment (must include).
ID 3 is post-surgical physical therapy (treatment-related; include).
ID 1 gives context about the fracture itself include because it's important background to understand the treatment.
ID 4 about blood pressure is unrelated (exclude).

Context: {context}

Question: {query}

DO NOT add explanations, only the mentioned output <end_of_turn>
<start_of_turn>model
"""
```

A.2.2 Prompt for Stage 2

```
"""
<bos><start_of_turn>user You are a clinical assistant. Use all of the context below to answer the question. Your response must be a JSON list of sentence-grounding pairs.
Answer the question using the given context to help, if you don't know the answer, just say that you don't know, don't try to make up an answer.

Format of Context:
ID: <chunk ID> ; text : <text>
ID: <chunk ID> ; text : <text>
...

Output format:
Your output must be a list of structured objects with:
    - 'statement': part of the response
    - 'citation': the chunk ID (e.g., '1') it came from to ground it in evidence

    - 'statement': part of the response
    - 'citation': the chunk ID (e.g., '2') it came from to ground it in evidence
```

```
    - 'statement': part of the response
    - 'citation': the chunk ID (e.g., '4') it came from to ground it in evidence

DO NOT add explanations, only the above output.
NOTE: Use all of the sources and cite all sources, do not omit any one, all are relevant.

Examples

# Example 1:
**Context:**
ID: 1 ; Text: "The patient complained of frequent urination and excessive thirst. Laboratory tests
    revealed elevated blood glucose levels."
ID: 2 ; Text: "The patient was diagnosed with type 2 diabetes mellitus."

**Question:** What is the patient's diagnosis?
**Answer:**
[{{"statement": "The patient was diagnosed with type 2 diabetes mellitus.", "citation": "2"}},
{{"statement": "Laboratory tests revealed elevated blood glucose levels.", "citation": "1"}}]

# Example 2:
**Context:**
ID: 1 ; Text: "An open reduction and internal fixation (ORIF) surgery was performed to stabilize the
    fracture."
ID: 2 ; Text: "The patient was prescribed physical therapy after hospital discharge."

**Question:** What treatment did the patient receive for the femur fracture?
**Answer:**
[{{"statement": "The patient underwent open reduction and internal fixation (ORIF) surgery to
    stabilize the femur fracture.", "citation": "1"}},
{{"statement": "The patient was prescribed physical therapy after hospital discharge.", "citation":
    "2"}}]

Context: {context}
Question: {query}

You can combine the sentences too, there is a word limit , so be succinct.
DO NOT add explanations, only the mentioned output.
USE ALL SOURCES, ALL OF THEM ARE IMPORTANT. <end_of_turn>
<start_of_turn>model
"""
```

SzegedAI at ArchEHR-QA 2025: Combining LLMs with traditional methods for grounded question answering

Soma Bálint Nagy, Bálint Nyerges, Zsombor Mátyás Kispéter, Gábor Tóth, András Tamás Szlúka, Gábor Kőrösi, Zsolt Szántó, Richárd Farkas,

Institute of Informatics, University of Szeged
2. Árpád tér, Szeged, Hungary
{nagysoma,tothg,korosig,szantozs,rfarkas}@inf.u-szeged.hu

Abstract

In this paper, we present the SzegedAI team's submissions to the ArchEHR-QA 2025 shared task. Our approaches include multiple prompting techniques for large language models (LLMs), sentence similarity methods, and traditional feature engineering. We are aiming to explore both modern and traditional solutions to the task. To combine the strengths of these diverse methods, we employed different ensembling strategies.

1 Introduction

The ArchEHR-QA 2025 shared task (Soni and Demner-Fushman, 2025b) aimed to help reduce the workload of clinicians by automatically generating answers to patients' questions. These answers were based on information from patients' electronic health records (EHRs) (Soni and Demner-Fushman, 2025a). The goal was to ensure that the answers were grounded in the clinical notes, with clear references to the specific sentences in the records. The task focused on two main evaluation criteria: factuality, which checks if the references are correct, and relevancy, which evaluates the quality of the answers.

In our solution, we combined strategies based on large language models (LLMs) with classical NLP techniques, such as the bag-of-words representation of overlapping terms between the question and the sentences. Our results include a comparison of different LLMs, such as Gemini (Team et al., 2024), Gemma 3 (Team et al., 2025), LLama (Grattafiori et al., 2024) and its medical fine-tuned versions (Ankit Pal, 2024; Christophe et al., 2024; Kim et al., 2025). We applied prompting strategies that either directly generate answers with references or select relevant sentences and generate responses from them. Additionally, we combined the outputs of the models using a voting mechanism, along with feature-rich classification techniques trained on the development set.

2 System Overview

We developed two main approaches:

1. **Pipeline Approach**: A two-step process that first identifies essential sentences in the clinical notes and then generates an answer based on these sentences.

2. **End-to-End Approach**: A single-step process that directly generates responses with appropriate citations using an LLM.

Our primary focus was on the pipeline approach, where we experimented with different methods for both essential sentence identification and answer generation. For essential sentence identification, we looked at the problem from both classical machine learning and LLM-based perspectives. The ML approach utilized feature engineering with lexical and semantic similarity metrics between questions and clinical note sentences, and other textual features. While the LLM-based approaches employed various prompting strategies to identify essential sentences through direct citation, two-agent interaction, and pairwise question-sentence evaluation.

We also explored ensemble techniques for essential sentence identification that combined the strengths of our various approaches through voting mechanisms and feature-rich classification. These ensemble models incorporated predictions from previous methods to improve overall performance.

For answer generation in our pipeline approach, we developed methods that used the identified essential sentences as input to craft concise, coherent responses that answered the question while properly citing the source sentences.

In our end-to-end approach, we prompted LLMs with carefully designed instructions to simultaneously identify relevant clinical evidence and generate coherent answers with citations in a single step.

BioNLP 2025 Shared Tasks, pages 136–149
August 1, 2025 ©2025 Association for Computational Linguistics

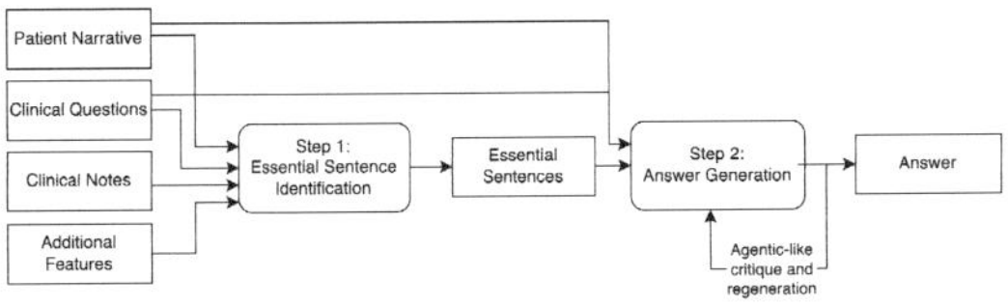

Figure 1: Overview of our pipeline system architecture for the ArchEHR-QA task, showing the two-step process of essential sentence identification followed by answer generation.

Both approaches were enhanced with an agentic reflection loop where initial responses were programmatically validated against task requirements (citation format, answer length, coverage of essential information) and iteratively refined based on specific feedback.

3 Methods

Sections 3.1 and 3.2 detail the various techniques we tested for the pipeline approach, while section 3.3 summarizes our procedure for the end-to-end approach.

3.1 Essential Sentence Identification

The first step of our pipeline approach was the identification of the essential sentences. We treated this problem as a binary classification task (essential or not-relevant), without considering "supplementary" as a separate category. We explored three main approaches:

3.1.1 Supervised Machine Learning-based Classification

We implemented a traditional machine learning approach using a LightGBM classifier (Ke et al., 2017), treating each sentence as a separate training or test instance. The following feature templates were used:

- Bag-of-words representations and overlap between question and sentence
- Semantic embeddings and cosine similarities (between question-sentence and between adjacent sentences)
- Length features (question, sentence, and their difference)
- Positional indicators (first/last sentence in note)

3.1.2 LLM-based Classification

We leveraged the contextual awareness and potential domain knowledge of LLMs through various prompting strategies (the prompts are available in Appendix A):

Answer with References This approach prompted the LLM to generate answers with citations to relevant clinical note sentences, which were labeled as essential. Unlike our pipeline's answer generation step, it omitted validation for length limits and formatting requirements. We also developed a two-stage variation (v2) that first identified the key sentence answering the question, then found supporting context sentences.

Agentic This method used two LLM instances: one generated an uncited answer, while the second identified supporting sentences from the clinical notes, which were labeled as essential.

References Only This approach focused solely on identifying essential sentences without generating a complete answer. The LLM was prompted to analyze the question and clinical notes, then output the numbers of sentences containing essential information. We used chain-of-thought reasoning and tested both zero-shot and one-shot variants.

Question-Sentence Compare This strategy evaluated individual question-sentence pairs rather than full cases, with the LLM classifying each sentence as essential or not. For reliability, we applied majority voting across three separate evaluations of each sentence.

3.1.3 Ensemblers

We developed two distinct ensemble approaches for essential sentence identification:

Supervised Ensembler This approach combined traditional machine learning features with the predictions from our various LLM-based methods as additional input features. This hybrid method leveraged both the structured learning of traditional classifiers and the contextual understanding provided by LLMs.

Answer with references - voting We created five variations of our "Answer with References" prompt with slight modifications. Sentences that were marked as essential by at least three of the five generated answers were considered essential in the final output, creating a majority-voting ensemble.

3.2 Answer Generation

The second step of our pipeline approach is the answer generation. Here our prompts contained the patient narrative, clinician question, and the full list of sentences identified as essential by our classification methods. We developed an iterative prompting strategy with an agentic reflection loop where each generated answer was programmatically validated against several key requirements: proper citation formatting, answer length constraints, comprehensive coverage of all essential information, and proper citation of all identified essential sentences.

When an answer failed to meet any of these criteria, we provided the LLM with the original prompt, the unsatisfactory answer, and specific feedback identifying the shortcomings. This initiated an iterative refinement process where the model would revise its response based on the targeted feedback, continuing until all quality requirements were satisfied.

3.3 End-to-End Approach

In contrast to our pipeline approach, we also explored an end-to-end method that directly generated answers with appropriate citations in a single step. For this approach, we provided the LLM with all sentences from the clinical notes rather than prefiltering for essential ones. The prompt explicitly specified that not all sentences contained relevant information and that the model should only cite sentences that directly underpinned its answer.

The end-to-end prompts instructed the model to generate a coherent answer using the clinical notes, include proper citations, address key aspects of the question concisely, and adhere to formatting requirements—all in a single step.

This approach was also enhanced with an agentic reflection loop, though with a different set of validation criteria. Since no separate sentence identification step existed, validation focused primarily on formatting correctness, citation syntax, and answer length constraints.

4 Results

In this section, we present the results of our methods. We begin by showing the performance of our models on the development set, followed by the performance of our submissions on the test set.

4.1 Experimental setup

On the development set, we focus on factuality (essential sentence identification) as the primary criteria.

The supervised machine learning-based classifier was trained on the development set with 100 estimators, gradient boosting decision trees, a fixed random seed of 42, and a minimum of 10 data points in each leaf. The model was validated using k-fold cross-validation, where k was 5. To calculate semantic representation we used LaBSE (Feng et al., 2020).

In our experiments, we compared various LLMs to find the best for the shared task[1]. Besides our baseline models, LLama 3.3 70B and Gemma 3 27B, we utilized fine-tuned models for different biomedical goals. Llama3-OpenBioLLM-70B model fine-tuned for biomedical tasks using DPO and a curated medical instruction dataset. Llama3-Med42-70B is optimized for medical question answering and clinical knowledge with instruction tuning. Llama-3-Meerkat-70B (Kim et al., 2025) is built for medical reasoning, trained with synthetic CoT data and diverse instruction datasets. Along with the open source models, we also used Gemini 1.5 Flash model.

4.2 Essential Sentence Identification

First, we evaluated our systems on the development set, which is shown in the Table 1.

Supervised classification Despite the limited number of training examples, our `supervised machine learning-based classification` model that mainly applies bag-of-words and semantic similarity-based features performed comparably to many prompt-based solutions. It achieved better results than 9 out of 13 LLM-based approaches.

LLMs Among the tested LLMs, the Gemini 1.5 Flash outperformed both the original and biomedical `LLaMA 70Bs` and `Gemma 3 27B` by a large margin. In the challenge of 70B LLaMa variants, 2 out of 3 fine-tuned models preceded the original model, where the `Llama-3-Meerkat-70B` was the best. Interestingly, the smaller Gemma model, which was not fine-tuned on medical data, achieved comparable results to the best LLaMA model.

Prompting strategies When comparing prompting strategies, the best results were obtained

[1]We used 4 A100 GPU for the open sourced LLMs.

	LLM	strict-micro			strict-macro		
		P	R	F1	P	R	F1
Competition baseline	Llama 3.3 70B	0.634	0.326	0.431	**0.703**	0.471	0.494
Supervised classifier	-	0.521	0.529	0.525	0.510	0.514	0.499
Answer with references	Gemini	0.566	0.558	0.562	0.608	**0.638**	0.578
Answer with references	Llama 3.3 70B	0.397	0.362	0.379	0.416	0.349	0.357
Answer with references	Llama3-Med42-70B	0.341	0.341	0.341	0.358	0.400	0.340
Answer with references	Llama3-OpenBioLLM-70B	0.333	0.275	0.406	0.309	0.308	0.289
Answer with references	Llama-3-Meerkat-70B	0.336	0.406	0.385	0.360	0.434	0.362
Answer with references	Gemma 27B	0.400	0.406	0.403	0.419	0.428	0.398
Answer with references v2	Gemini	0.631	0.384	0.477	0.651	0.443	0.477
Agentic	Gemini	0.500	0.442	0.469	0.583	0.530	0.495
References only - zero shot	Gemini	0.657	0.500	0.568	0.659	0.568	0.574
References only - 1 shot	Gemini	0.699	0.522	0.598	0.662	0.591	0.583
Question - sentence compare	Gemini	0.477	0.536	0.505	0.481	0.519	0.457
Question - sentence compare	Gemini	0.503	0.536	0.519	0.517	0.518	0.462
End-to-end	Gemini	0.693	0.507	0.587	0.534	0.438	0.473
Answer with references - voting	Gemini	0.514	0.398	0.449	0.538	0.490	0.454
Supervised ensembler	Gemini	**0.750**	**0.608**	**0.672**	0.685	0.586	**0.616**

Table 1: Factuality results of the independent systems on the development set. The `Competition baseline` used the LLaMA 3.3 70B model in a zero-shot setting prompting it to generate cited answers; if responses were invalid, they retried up to five times to get a valid one. Detailed descriptions of the `Supervised classifier` method can be found in Section 3.1.1; `Answer with references (V2)`, `Agentic`, `References only` and `Question - sentence compare` are in Section 3.1.2; `End-to-end` in 3.3; and `Answer with references - voting` and `Supervised ensembler` are in 3.1.3.

with the `References only` and `Answer with references` approaches for sentence identification, but the `End-to-end` approch also acheived similarly high score.

Ensemblers The voting method over the `Answer with references` can't improve the performance. Instead of the `Supervised ensembler` that applies all of the Gemini-based system's output as features besides the features of the `Supervised classifier`, achieved the highest score on the development set.

4.3 Submissions

We selected three distinct models as submissions to reflect the variety of approaches we had previously evaluated on the development set, results presented in the Table 2. The first model, `Supervised classifier (SC)`, aimed to evaluate the performance of traditional machine learning methods on the shared task. The `End-to-end (E2E)` model was one of the most purely prompt-based solutions, and we uploaded our best system from the development set, the `Supervised ensembler (SE)`.

The `SC` model performed notably worse on the test set than on the development set. Since we did not use the development set for hyperparameter tuning during cross-validation, we suspect that the limited amount of training data failed to generalize well to the test set. A similar pattern was observed with our `SE` model. But in this case, the factuality score is matched with the `E2E` model, which was in third place on the development set. In the case of relevance, the `SE` model, which generates answers based on selected essential sentences, outperformed the `E2E` model. Consequently, the `SE` also achieved a higher score on the overall metric, so we selected this model as our official submission.

	SC	E2E	SE
Overall	0.321	0.407	**0.427**
Overall Factuality	0.317	0.470	**0.472**
Strict F1 (micro)	0.317	0.470	**0.472**
Strict F1 (macro)	0.309	**0.523**	0.514
Overall Relevance	0.325	0.344	**0.382**
BLEU	0.018	0.008	**0.032**
ROUGELsum	0.227	0.211	**0.292**
SARI	0.558	0.597	**0.642**
BERTScore	**0.288**	0.275	0.191
AlignScore	0.272	**0.631**	0.195
MEDCON (UMLS)	**0.586**	0.344	0.278

Table 2: Official scores of our systems on the test set.

5 Conclusion

In this paper, we presented the SzegedAI team's submissions to the ArchEHR-QA 2025 shared task. Our models combined traditional machine learning techniques with LLM-based predictions. We explored a range of models and prompting strategies, and integrated their outputs using a feature-rich classification framework to identify the most relevant information from clinical notes in response to patient questions. Our submission achieved 11th place in the automatic evaluation of the shared task.

Limitations

In this paper, we relied heavily on the development set for evaluations, but the small size of this dataset limits the accurate comparison of the different methods.

Most of our LLM-based methods were limited to one prompt per question, except the `Agentic`, `End-to-end`, and `Answer generation` methods, which were limited to five cycles, and the `Question-sentence compare` applied an LLM call for each sentence in a clinical note.

While our supervised machine learning-based systems performed well on the development set, their performance dropped on the test set, suggesting potential overfitting and limited generalization due to the small training size. Increasing the amount of training data would likely improve results, but the `Supervised classifier` is inherently less generalizable than LLMs.

Our evaluation focused on factuality metrics, with less emphasis on the relevance of the answer, which plays a critical role in real-life applications.

Acknowledgments

The research is supported by the European Union project RRF-2.3.1-21-2022-00004 within the framework of the Artificial Intelligence National Laboratory.

We acknowledge the Digital Government Development and Project Management Ltd. for awarding us access to the Komondor HPC facility based in Hungary.

References

Malaikannan Sankarasubbu Ankit Pal. 2024. Openbiollms: Advancing open-source large language models for healthcare and life sciences. https://huggingface.co/aaditya/OpenBioLLM-Llama3-70B.

Clément Christophe, Praveen K Kanithi, Tathagata Raha, Shadab Khan, and Marco AF Pimentel. 2024. Med42-v2: A suite of clinical llms.

Fangxiaoyu Feng, Yinfei Yang, Daniel Cer, Naveen Arivazhagan, and Wei Wang. 2020. Language-agnostic bert sentence embedding. *arXiv preprint arXiv:2007.01852*.

Aaron Grattafiori, Abhimanyu Dubey, Abhinav Jauhri, Abhinav Pandey, Abhishek Kadian, Ahmad Al-Dahle, Aiesha Letman, Akhil Mathur, Alan Schelten, Alex Vaughan, and 1 others. 2024. The llama 3 herd of models. *arXiv preprint arXiv:2407.21783*.

Guolin Ke, Qi Meng, Thomas Finley, Taifeng Wang, Wei Chen, Weidong Ma, Qiwei Ye, and Tie-Yan Liu. 2017. Lightgbm: A highly efficient gradient boosting decision tree. *Advances in neural information processing systems*, 30.

Hyunjae Kim, Hyeon Hwang, Jiwoo Lee, Sihyeon Park, Dain Kim, Taewhoo Lee, Chanwoong Yoon, Jiwoong Sohn, Jungwoo Park, Olga Reykhart, Thomas Fetherston, Donghee Choi, Soo Heon Kwak, Qingyu Chen, and Jaewoo Kang. 2025. Small language models learn enhanced reasoning skills from medical textbooks. *npj Digital Medicine*, 8(1):240.

Sarvesh Soni and Dina Demner-Fushman. 2025a. A dataset for addressing patient's information needs related to clinical course of hospitalization. *arXiv preprint*.

Sarvesh Soni and Dina Demner-Fushman. 2025b. Overview of the archehr-qa 2025 shared task on grounded question answering from electronic health records. In *The 24th Workshop on Biomedical Natural Language Processing and BioNLP Shared Tasks*, Vienna, Austria. Association for Computational Linguistics.

Gemini Team, Petko Georgiev, Ving Ian Lei, Ryan Burnell, Libin Bai, Anmol Gulati, Garrett Tanzer, Damien Vincent, Zhufeng Pan, Shibo Wang, and 1 others. 2024. Gemini 1.5: Unlocking multimodal understanding across millions of tokens of context. *arXiv preprint arXiv:2403.05530*.

Gemma Team, Aishwarya Kamath, Johan Ferret, Shreya Pathak, Nino Vieillard, Ramona Merhej, Sarah Perrin, Tatiana Matejovicova, Alexandre Ramé, Morgane Rivière, and 1 others. 2025. Gemma 3 technical report. *arXiv preprint arXiv:2503.19786*.

A Prompts for essential sentence identification

This section shows the prompts that were applied to the results of the paper.

A.1 Prompt for "Answer with References"

```
Task: Generate a concise, helpful answer to a patient's health question using only information from the clinical note. Each statement in your answer must be grounded in specific sentences from the note.
Example:
Patient's Narrative: Took my 59 yo father to ER ultrasound discovered he had an aortic aneurysm. He had a salvage repair (tube graft). Long surgery / recovery for couple hours then removed packs. why did they do this surgery????? After this time he spent 1 month in hospital now sent home.
Patient's Question: why did they do this surgery?????
Clinician's Rephrased Question: Why did they perform the emergency salvage repair on him?
Clinical Note (numbered sentences):
1: He was transferred to the hospital on 2025-1-20 for emergent repair of his ruptured thoracoabdominal aortic aneurysm.
2: He was immediately taken to the operating room where he underwent an emergent salvage repair of ruptured thoracoabdominal aortic aneurysm with a 34-mm Dacron tube graft using deep hypothermic circulatory arrest.
3: Please see operative note for details which included cardiac arrest x2.
4: Postoperatively he was taken to the intensive care unit for monitoring with an open chest.
5: He remained intubated and sedated on pressors and inotropes.
6: On 2025-1-22, he returned to the operating room where he underwent exploration and chest closure.
7: On 1-25 he returned to the OR for abd closure JP/ drain placement/ feeding jejunostomy placed at that time for nutritional support.
8: Thoracoabdominal wound healing well with exception of very small open area mid wound that is @1cm around and 1/2cm deep, no surrounding erythema.
9: Packed with dry gauze and covered w/DSD.
Example Answer:
His aortic aneurysm was caused by the rupture of a thoracoabdominal aortic aneurysm, which required emergent surgical intervention (1). He underwent a complex salvage repair using a 34-mm Dacron tube graft and deep hypothermic circulatory arrest to address the rupture (2). The extended recovery time and hospital stay were necessary due to the severity of the rupture and the complexity of the surgery, though his wound is now healing well with only a small open area noted (8).
Now, please generate an answer for the following case:
Patient's Narrative: {patient_narrative}
Patient's Question: {patient_question}
Clinician's Rephrased Question: {clinician_question}
Clinical Note (numbered sentences):
{numbered_note}
Instructions:

First, carefully identify which sentences are ESSENTIAL to answering the clinician's rephrased question. Focus on sentences that directly explain the medical reasoning, procedures performed, and clinical findings.
When writing your answer, ONLY include information from these essential sentences. Each statement in your answer MUST be supported by at least one citation.
For each statement in your answer, cite the specific sentence number(s) that support it using parentheses, e.g., "The procedure was successful (3, 5)."
Be very precise with your citations - only cite sentences that directly support each specific claim you make.

Your Answer:
```

A.2 Prompts "Agentic LLM classification"

A.2.1 Stage 1: Answer generation prompt

```
Task: Generate a helpful, concise answer to a patient's health question using only information from the clinical note.
Example:
Patient's Narrative: Took my 59 yo father to ER ultrasound discovered he had an aortic aneurysm. He had a salvage repair (tube graft). Long surgery / recovery for couple hours then removed packs. why did they do this surgery????? After this time he spent 1 month in hospital now sent home.
Patient's Question: why did they do this surgery?????
Clinician's Rephrased Question: Why did they perform the emergency salvage repair on him?
Clinical Note (numbered sentences):
1: He was transferred to the hospital on 2025-1-20 for emergent repair of his ruptured thoracoabdominal aortic aneurysm.
2: He was immediately taken to the operating room where he underwent an emergent salvage repair of ruptured thoracoabdominal aortic aneurysm with a 34-mm Dacron tube graft using deep hypothermic circulatory arrest.
3: Please see operative note for details which included cardiac arrest x2.
4: Postoperatively he was taken to the intensive care unit for monitoring with an open chest.
5: He remained intubated and sedated on pressors and inotropes.
6: On 2025-1-22, he returned to the operating room where he underwent exploration and chest closure.
7: On 1-25 he returned to the OR for abd closure JP/ drain placement/ feeding jejunostomy placed at that time for nutritional support.
8: Thoracoabdominal wound healing well with exception of very small open area mid wound that is @1cm around and 1/2cm deep, no surrounding erythema.
9: Packed with dry gauze and covered w/DSD.
Example Answer:
His aortic aneurysm was caused by the rupture of a thoracoabdominal aortic aneurysm, which required emergent surgical intervention. He underwent a complex salvage repair using a 34-mm Dacron tube graft and deep hypothermic circulatory arrest to address the rupture. The extended recovery time and hospital stay were necessary due to the severity of the rupture and the complexity of the surgery, though his wound is now healing well with only a small open area noted.
Now, please generate an answer for the following case:
Patient's Narrative: {patient_narrative}
Patient's Question: {patient_question}
Clinician's Rephrased Question: {clinician_question}
Clinical Note (numbered sentences):
```

```
{numbered_note}
Instructions:

Answer the clinician's rephrased question directly and clearly.
Use only information found in the clinical note.

Your Answer:
```

A.2.2 Stage 2: Source identification prompt

```
Task: Identify which sentences from the clinical note support statements in the patient answer.

Example:
Clinical Note (numbered sentences):
1: He was transferred to the hospital on 2025-1-20 for emergent repair of his ruptured thoracoabdominal aortic aneurysm.
2: He was immediately taken to the operating room where he underwent an emergent salvage repair of ruptured thoracoabdominal aortic aneurysm with a 34-mm Dacron tube graft using deep hypothermic circulatory arrest.
3: Please see operative note for details which included cardiac arrest x2.
4: Postoperatively he was taken to the intensive care unit for monitoring with an open chest.
5: He remained intubated and sedated on pressors and inotropes.
6: On 2025-1-22, he returned to the operating room where he underwent exploration and chest closure.
7: On 1-25 he returned to the OR for abd closure JP/ drain placement/ feeding jejunostomy placed at that time for nutritional support.
8: Thoracoabdominal wound healing well with exception of very small open area mid wound that is @1cm around and 1/2cm deep, no surrounding erythema.
9: Packed with dry gauze and covered w/DSD.
Example input text:
His aortic aneurysm was caused by the rupture of a thoracoabdominal aortic aneurysm, which required emergent surgical intervention. He underwent a complex salvage repair using a 34-mm Dacron tube graft and deep hypothermic circulatory arrest to address the rupture. The extended recovery time and hospital stay were necessary due to the severity of the rupture and the complexity of the surgery, though his wound is now healing well with only a small open area noted.
Essential Sentences: 1, 2, 8

Now, please generate an answer for the following case:
Clinical Note (numbered sentences):
{numbered_note}
Input text:
{generated_answer}
Instructions:

Carefully analyze the answer and identify ALL sentences from the clinical note that directly support information in the answer.
Do not include sentences that contain information not referenced in the text.
List ONLY the sentence numbers (without any additional text) in a comma-separated format.
Your response should follow this format exactly:

Essential Sentences: [list of numbers]
For example: "Essential Sentences: 1, 3, 5, 7"
```

A.3 Prompts for "Answer with references v2"

A.3.1 Stage 1: Best sentence identification prompt

```
Task: Identify the SINGLE BEST sentence from the clinical note that directly answers the clinician's question.

Patient's Narrative: {patient_narrative}
Patient's Question: {patient_question}
Clinician's Rephrased Question: {clinician_question}

Clinical Note (numbered sentences):
{numbered_note}

Instructions:

    Analyze each sentence in the clinical note carefully.
    Identify the ONE sentence that most directly answers the clinician's question about why a procedure was performed, what caused a condition, how something was treated, or other clinical reasoning.
    Choose the sentence that contains the core explanation, not just related information.
    Provide ONLY the sentence number in your response, with no additional text.

Few-Shot Examples:

Example 1:
Patient Question: "My question is if the sludge was there does not the medication help in flushing it out? Whether ERCP was the only cure?"
Clinician Question: "Why was ERCP recommended to him over continuing a medication-based treatment?"
Clinical Note:

    Brief Hospital Course:
    During the ERCP a pancreatic stent was required to facilitate access to the biliary system (removed at the end of the procedure), and a common bile duct stent was placed to allow drainage of the biliary obstruction caused by stones and sludge.
    However, due to the patient's elevated INR, no sphincterotomy or stone removal was performed.
    Frank pus was noted to be draining from the common bile duct, and post-ERCP it was recommended that the patient remain on IV Zosyn for at least a week.
```

```
    The Vancomycin was discontinued.
    On hospital day 4 (post-procedure day 3) the patient returned to ERCP for re-evaluation of her biliary stent as her LFTs and bilirubin continued an upward trend.
    On ERCP the previous biliary stent was noted to be acutely obstructed by biliary sludge and stones.
    As the patient's INR was normalized to 1.2, a sphincterotomy was safely performed, with removal of several biliary stones in addition to the common bile duct stent.
    At the conclusion of the procedure, retrograde cholangiogram was negative for filling defects.

Best Sentence Answer: 2

Reasoning: Sentence 2 is the best single sentence because it directly explains why ERCP was necessary - it reveals that stones and sludge were causing a biliary obstruction that required stent placement to allow drainage. This is the core reason why medication alone wouldn't be sufficient - there was a physical blockage that needed mechanical intervention.

Example 2:
Patient Question: "I overdosed October 4th on trihexyphenidyl, thorazine, and cocaine. I have had chest pain in my left upper quadrant ever since. Any ideas?"
Clinician Question: "Is the pain connected to the overdose or something else?"
Clinical Note:

    Brief Hospital Course:
    Bipolar d/o, PTSD, schizophrenia: Psychiatry consult recommended that all psych medications be held until they could be re-prescribed by pt's outpatient psychiatrist.
    During hospital course, thorazine was restarted but discontinued soon after because pt became tachycardic; pt remained asymptomatic during these episodes of tachycardia.
    Tachycardia resolved with discontinuation of thorazine, IV hydration, and small dose of IV benzodiazepene x 1.
    Social work consult was obtained because pt did not have a PCP nor did he have a psychiatrist.
    He could not see his former psychiatrist due to insurance reasons.
    With the help of social work, pt was set up with a PCP who would be able to refer him to a new psychiatrist in a timely fashion.
    He was instructed to follow-up with his new psychiatrist to restart his psychiatric medications.
    Chest pain: Pt complained of chest pain during hospital course that appeared musculoskeletal as it was reproducible with palpation and pt reported more pain with movement.
    EKG showed no ischemic changes and troponins were flat x 4. CK was elevated, peaking at 1405 but downtrended without any intervention.
    TTE was obtained due to history of cocaine use to rule out cardiac events.
    EF was >55%; TTE was unremarkable.
    He was monitored on telemetry without significant events.
    Discharge Instructions:
    It was a pleasure taking care of you at the hospital.
    You were admitted with confusion that was likely due to a combination of the medications you were taking and the street drugs that you may have also been used.
    Your heart rhythm was monitored because many of these drugs can affect your heart.
    Your EKG and blood tests showed that you likely did not have a heart attack.
    An ultrasound of your heart was also normal.
    Your confusion cleared during your hospital stay.
    You were seen by our psychiatry team who recommended holding all of your medications while you were in the hospital.
    It is very important that you follow-up with a primary care doctor who can refer you to a psychiatrist.
    This psychiatrist can then prescribe to you the medications you were normally taking.

Best Sentence Answer: 9

Reasoning: Sentence 9 is the best choice because it directly addresses the nature of the chest pain, identifying it as musculoskeletal based on clinical examination (reproducible with palpation and worsening with movement). This directly answers whether the pain is connected to the overdose or something else by suggesting a musculoskeletal cause.

Your Answer:
```

A.3.2 Stage 2: Context sentences identification prompt

```
Task: Identify additional sentences from the clinical note that provide necessary context for understanding the answer to the clinician's question.

Patient's Narrative: {patient_narrative}
Patient's Question: {patient_question}
Clinician's Rephrased Question: {clinician_question}

Clinical Note (numbered sentences):
{numbered_note}

The MAIN sentence that answers the question is:
Sentence {best_sentence_num}: {best_sentence_text}

Instructions:

    Analyze the clinical note to identify any OTHER sentences that provide necessary context to fully understand the answer.
    Include sentences that:
        Explain medical terminology used in the main answer
        Provide evidence supporting the main answer
        Show treatment outcomes that validate the answer
        Describe test results that confirm the diagnosis or treatment decision
        Explain why alternative treatments were not chosen
    Exclude sentences that:
        Repeat information already in the main sentence
        Contain general information not directly related to the question
        Focus on administrative details rather than clinical reasoning
```

```
    List ONLY the sentence numbers in your response, separated by commas (e.g., "3, 5, 9").
    If no additional context sentences are needed, respond with "None".
    Limit your selection to the most relevant sentences (typically 2-5 sentences).

Few-Shot Examples:

Example 1:
Patient Question: "My question is if the sludge was there does not the medication help in flushing it out? Whether ERCP was the only cure?"
Clinician Question: "Why was ERCP recommended to him over continuing a medication-based treatment?"
Clinical Note:

    Brief Hospital Course:
    During the ERCP a pancreatic stent was required to facilitate access to the biliary system (removed at the end of the procedure), and a common bile duct stent was placed to allow drainage of the biliary obstruction caused by stones and sludge.
    However, due to the patient's elevated INR, no sphincterotomy or stone removal was performed.
    Frank pus was noted to be draining from the common bile duct, and post-ERCP it was recommended that the patient remain on IV Zosyn for at least a week.
    The Vancomycin was discontinued.
    On hospital day 4 (post-procedure day 3) the patient returned to ERCP for re-evaluation of her biliary stent as her LFTs and bilirubin continued an upward trend.
    On ERCP the previous biliary stent was noted to be acutely obstructed by biliary sludge and stones.
    As the patient's INR was normalized to 1.2, a sphincterotomy was safely performed, with removal of several biliary stones in addition to the common bile duct stent.
    At the conclusion of the procedure, retrograde cholangiogram was negative for filling defects.

Main sentence that answers the question is:
Sentence 2: During the ERCP a pancreatic stent was required to facilitate access to the biliary system (removed at the end of the procedure), and a common bile duct stent was placed to allow drainage of the biliary obstruction caused by stones and sludge.

Context Sentences Answer: 6, 7, 8

Reasoning for including these context sentences:

    Sentence 6 shows that even after initial treatment, the patient's liver function tests continued to worsen, indicating that medication alone was not sufficient
    Sentence 7 demonstrates that the biliary stent became obstructed again by sludge and stones, further proving that physical removal was necessary
    Sentence 8 shows that once conditions allowed (normalized INR), a sphincterotomy was performed to physically remove the stones, which medication alone couldn't accomplish

Reasoning for NOT including other potential sentences:

    Sentence 3 mentions elevated INR preventing sphincterotomy, but doesn't directly address why medication wouldn't work
    Sentence 4 mentions pus and antibiotics, which is related to infection treatment but not directly about sludge removal
    Sentence 9 only provides procedural outcome information without explaining why ERCP was necessary over medication

Example 2:
Patient Question: "I overdosed October 4th on trihexyphenidyl, thorazine, and cocaine. I have had chest pain in my left upper quadrant ever since. Any ideas?"
Clinician Question: "Is the pain connected to the overdose or something else?"
Clinical Note:

    Brief Hospital Course:
    Bipolar d/o, PTSD, schizophrenia: Psychiatry consult recommended that all psych medications be held until they could be re-prescribed by pt's outpatient psychiatrist.
    During hospital course, thorazine was restarted but discontinued soon after because pt became tachycardic; pt remained asymptomatic during these episodes of tachycardia.
    Tachycardia resolved with discontinuation of thorazine, IV hydration, and small dose of IV benzodiazepene x 1.
    Social work consult was obtained because pt did not have a PCP nor did he have a psychiatrist.
    He could not see his former psychiatrist due to insurance reasons.
    With the help of social work, pt was set up with a PCP who would be able to refer him to a new psychiatrist in a timely fashion.
    He was instructed to follow-up with his new psychiatrist to restart his psychiatric medications.
    Chest pain: Pt complained of chest pain during hospital course that appeared musculoskeletal as it was reproducible with palpation and pt reported more pain with movement.
    EKG showed no ischemic changes and troponins were flat x 4. CK was elevated, peaking at 1405 but downtrended without any intervention.
    TTE was obtained due to history of cocaine use to rule out cardiac events.
    EF was >55%; TTE was unremarkable.
    He was monitored on telemetry without significant events.
    Discharge Instructions:
    It was a pleasure taking care of you at the hospital.
    You were admitted with confusion that was likely due to a combination of the medications you were taking and the street drugs that you may have also been used.
    Your heart rhythm was monitored because many of these drugs can affect your heart.
    Your EKG and blood tests showed that you likely did not have a heart attack.
    An ultrasound of your heart was also normal.
    Your confusion cleared during your hospital stay.
    You were seen by our psychiatry team who recommended holding all of your medications while you were in the hospital.
    It is very important that you follow-up with a primary care doctor who can refer you to a psychiatrist.
    This psychiatrist can then prescribe to you the medications you were normally taking.

Main sentence that answers the question is:
Sentence 9: # Chest pain: Pt complained of chest pain during hospital course that appeared musculoskeletal as it was reproducible with palpation and pt reported more pain with movement.
```

```
Context Sentences Answer: 3, 10, 11, 12, 13

Reasoning for including these context sentences:

    Sentence 3 provides information about the thorazine (one of the overdosed medications) causing tachycardia, which could be related to the chest discomfort
    Sentence 10 rules out cardiac ischemia through EKG and troponin tests, while noting elevated CK (which can indicate muscle damage)
    Sentence 11 mentions additional cardiac testing due to history of cocaine use
    Sentence 12 shows normal heart function on ultrasound
    Sentence 13 confirms no cardiac events were detected during monitoring

Reasoning for NOT including other potential sentences:

    Sentences 16-19 from the discharge instructions contain similar information to sentences 10-13 but are written for the patient rather than providing additional clinical details
    Sentence 2 discusses psychiatric management but doesn't address the chest pain question
    Sentences 4-8 focus on medication management and discharge planning rather than explaining the chest pain

Your Answer:
```

A.4 Prompts for "References Only"

In this prompt, the question and the clinical note are given in a user prompt.

The system prompt:

```
Your task is to find essential sentences in a clinical note to answer a clinical question.
The clinical notes contain the history of a patient and details of a clinical event, you can select sentences from each category if neccessary.
There are always at least 3 essential sentences in the clinical note.
Try to find all of the relevant sentences in the clinical note to answer the question.

You can think step by step,
step 1: Analyze the question and the clinical note.
step 2: Find the essential sentences in the clinical note to answer the question. Write the reason why each sentence is essential or not.
step 3: To a separated last line list the ids of the essential sentences in the clinical note, in the following format:
1, 2, 3{example if example else ""}
```

The user prompt:

```
# Question
{patient_narrative}

# Clinical note
{json.dumps(clinical_note, indent=2)}
```

A.5 Prompt for "Question-Sentence Compare"

```
You are a medical expert. You will be given a question relating to a patient and a sentence which may or may not contain relevant information to answering the question. Your job is to tell wether the information is relevant or not-relevant.
This is the question of the patient:
{narrative}
{clinical_question}

The sentence is:
{sentence}

Does the sentence contain relevant information? Think carefully before you answer and end your answer with a definitive yes or no answer:
```

A.6 Prompts for "Answer with references - voting"

A.6.1 Prompt variation 1

```
Task: Generate a concise, helpful answer to a patient's health question using only information from the clinical note. Each statement in your answer must be grounded in specific sentences from the note.

Example:
Patient's Narrative: Took my 59 yo father to ER ultrasound discovered he had an aortic aneurysm. He had a salvage repair (tube graft). Long surgery / recovery for couple hours then removed packs. why did they do this surgery????? After this time he spent 1 month in hospital now sent home.

Patient's Question: why did they do this surgery?????

Clinician's Rephrased Question: Why did they perform the emergency salvage repair on him?
```

```

Clinical Note (numbered sentences):
**1:** He was transferred to the hospital on 2025-1-20 for emergent repair of his ruptured thoracoabdominal aortic aneurysm.
**2:** He was immediately taken to the operating room where he underwent an emergent salvage repair of ruptured thoracoabdominal aortic aneurysm with a 34-mm Dacron tube graft using deep hypothermic circulatory arrest.
**3:** Please see operative note for details which included cardiac arrest x2.
**4:** Postoperatively he was taken to the intensive care unit for monitoring with an open chest.
**5:** He remained intubated and sedated on pressors and inotropes.
**6:** On 2025-1-22, he returned to the operating room where he underwent exploration and chest closure.
**7:** On 1-25 he returned to the OR for abd closure JP/ drain placement/ feeding jejunostomy placed at that time for nutritional support.
**8:** Thoracoabdominal wound healing well with exception of very small open area mid wound that is @1cm around and 1/2cm deep, no surrounding erythema.
**9:** Packed with dry gauze and covered w/DSD.

Example Answer:
His aortic aneurysm was caused by the rupture of a thoracoabdominal aortic aneurysm, which required emergent surgical intervention (1). He underwent a complex salvage repair using a 34-mm Dacron tube graft and deep hypothermic circulatory arrest to address the rupture (2). The extended recovery time and hospital stay were necessary due to the severity of the rupture and the complexity of the surgery, though his wound is now healing well with only a small open area noted (8).

Now, please generate an answer for the following case:

Patient's Narrative: {patient_narrative}

Patient's Question: {patient_question}

Clinician's Rephrased Question: {clinician_question}

Clinical Note (numbered sentences):
{numbered_note}

Instructions:
1. First, carefully identify which sentences are ESSENTIAL to answering the clinician's rephrased question. Focus on sentences that directly explain the medical reasoning, procedures performed, and clinical findings.

2. When writing your answer, ONLY include information from these essential sentences. Each statement in your answer MUST be supported by at least one citation.

3. For each statement in your answer, cite the specific sentence number(s) that support it using parentheses, e.g., "The procedure was successful (3, 5)."

4. Be very precise with your citations - only cite sentences that directly support each specific claim you make.

Your Answer:
```

A.6.2 Prompt variation 2

```
Task: Answer a medical question based solely on the provided clinical note. Cite sentence numbers for each claim.

Example:
Patient's Narrative: Took my 59 yo father to ER ultrasound discovered he had an aortic aneurysm. He had a salvage repair (tube graft). Long surgery / recovery for couple hours then removed packs. why did they do this surgery????? After this time he spent 1 month in hospital now sent home.

Patient's Question: why did they do this surgery?????

Clinician's Rephrased Question: Why did they perform the emergency salvage repair on him?

Clinical Note (numbered sentences):
**1:** He was transferred to the hospital on 2025-1-20 for emergent repair of his ruptured thoracoabdominal aortic aneurysm.
**2:** He was immediately taken to the operating room where he underwent an emergent salvage repair of ruptured thoracoabdominal aortic aneurysm with a 34-mm Dacron tube graft using deep hypothermic circulatory arrest.
**3:** Please see operative note for details which included cardiac arrest x2.
**4:** Postoperatively he was taken to the intensive care unit for monitoring with an open chest.
**5:** He remained intubated and sedated on pressors and inotropes.
**6:** On 2025-1-22, he returned to the operating room where he underwent exploration and chest closure.
**7:** On 1-25 he returned to the OR for abd closure JP/ drain placement/ feeding jejunostomy placed at that time for nutritional support.
**8:** Thoracoabdominal wound healing well with exception of very small open area mid wound that is @1cm around and 1/2cm deep, no surrounding erythema.
**9:** Packed with dry gauze and covered w/DSD.

Example Answer:
His aortic aneurysm was caused by the rupture of a thoracoabdominal aortic aneurysm, which required emergent surgical intervention (1). He underwent a complex salvage repair using a 34-mm Dacron tube graft and deep hypothermic circulatory arrest to address the rupture (2). The extended recovery time and hospital stay were necessary due to the severity of the rupture and the complexity of the surgery, though his wound is now healing well with only a small open area noted (8).

Now answer this question:

Question: {clinician_question}

Clinical Note:
{numbered_note}

```

```
Instructions:
- Only use information directly from the note
- Each claim must have a citation in parentheses (e.g., "The surgery was successful (3)")
- Be concise and precise
- Only cite the most relevant sentences that directly answer the question

Your Answer:
```

A.6.3 Prompt variation 3

```
Task: Help a patient understand their medical situation by answering their question using information from their clinical note.

Example:
Patient's Narrative: Took my 59 yo father to ER ultrasound discovered he had an aortic aneurysm. He had a salvage repair (tube graft). Long surgery / recovery for couple hours then removed packs. why did they do this surgery????? After this time he spent 1 month in hospital now sent home.

Patient's Question: why did they do this surgery?????

Clinician's Rephrased Question: Why did they perform the emergency salvage repair on him?

Clinical Note (numbered sentences):
**1:** He was transferred to the hospital on 2025-1-20 for emergent repair of his ruptured thoracoabdominal aortic aneurysm.
**2:** He was immediately taken to the operating room where he underwent an emergent salvage repair of ruptured thoracoabdominal aortic aneurysm with a 34-mm Dacron tube graft using deep hypothermic circulatory arrest.
**3:** Please see operative note for details which included cardiac arrest x2.
**4:** Postoperatively he was taken to the intensive care unit for monitoring with an open chest.
**5:** He remained intubated and sedated on pressors and inotropes.
**6:** On 2025-1-22, he returned to the operating room where he underwent exploration and chest closure.
**7:** On 1-25 he returned to the OR for abd closure JP/ drain placement/ feeding jejunostomy placed at that time for nutritional support.
**8:** Thoracoabdominal wound healing well with exception of very small open area mid wound that is @1cm around and 1/2cm deep, no surrounding erythema.
**9:** Packed with dry gauze and covered w/DSD.

Example Answer:
His aortic aneurysm was caused by the rupture of a thoracoabdominal aortic aneurysm, which required emergent surgical intervention (1). He underwent a complex salvage repair using a 34-mm Dacron tube graft and deep hypothermic circulatory arrest to address the rupture (2). The extended recovery time and hospital stay were necessary due to the severity of the rupture and the complexity of the surgery, though his wound is now healing well with only a small open area noted (8).

Patient's Question: {patient_question}

Clinical Note:
{numbered_note}

Instructions:
1. Analyze which sentences in the note directly address the patient's question
2. Write a clear, concise answer citing only the most important sentences
3. Each statement must include sentence numbers in parentheses: (1) or (2, 3)
4. Be factual and only use information from the note

Your Answer:
```

A.6.4 Prompt variation 4

```
Task: Perform a structured medical note analysis to answer a clinical question.

Example:
Patient's Narrative: Took my 59 yo father to ER ultrasound discovered he had an aortic aneurysm. He had a salvage repair (tube graft). Long surgery / recovery for couple hours then removed packs. why did they do this surgery????? After this time he spent 1 month in hospital now sent home.

Patient's Question: why did they do this surgery?????

Clinician's Rephrased Question: Why did they perform the emergency salvage repair on him?

Clinical Note (numbered sentences):
**1:** He was transferred to the hospital on 2025-1-20 for emergent repair of his ruptured thoracoabdominal aortic aneurysm.
**2:** He was immediately taken to the operating room where he underwent an emergent salvage repair of ruptured thoracoabdominal aortic aneurysm with a 34-mm Dacron tube graft using deep hypothermic circulatory arrest.
**3:** Please see operative note for details which included cardiac arrest x2.
**4:** Postoperatively he was taken to the intensive care unit for monitoring with an open chest.
**5:** He remained intubated and sedated on pressors and inotropes.
**6:** On 2025-1-22, he returned to the operating room where he underwent exploration and chest closure.
**7:** On 1-25 he returned to the OR for abd closure JP/ drain placement/ feeding jejunostomy placed at that time for nutritional support.
**8:** Thoracoabdominal wound healing well with exception of very small open area mid wound that is @1cm around and 1/2cm deep, no surrounding erythema.
**9:** Packed with dry gauze and covered w/DSD.

Example Answer:
```

```
His aortic aneurysm was caused by the rupture of a thoracoabdominal aortic aneurysm, which required emergent surgical intervention (1). He underwent a complex salvage repair using a 34-mm Dacron tube graft and deep hypothermic circulatory arrest to address the rupture (2). The extended recovery time and hospital stay were necessary due to the severity of the rupture and the complexity of the surgery, though his wound is now healing well with only a small open area noted (8).

Clinical Question: {clinician_question}
Patient Context: {patient_narrative}

Clinical Note:
{numbered_note}

Process:
1. First, identify the 3-5 most relevant sentences that directly answer the question
2. Organize these sentences into a logical flow
3. Write a concise answer citing each sentence number in parentheses
4. Only include information that is explicitly stated in the cited sentences

Your Answer:
```

A.6.5 Prompt variation 5

```
Task: Use step-by-step reasoning to determine which sentences in a clinical note are essential to answering a medical question.

Example:
Patient's Narrative: Took my 59 yo father to ER ultrasound discovered he had an aortic aneurysm. He had a salvage repair (tube graft). Long surgery / recovery for couple hours then removed packs. why did they do this surgery????? After this time he spent 1 month in hospital now sent home.

Patient's Question: why did they do this surgery?????

Clinician's Rephrased Question: Why did they perform the emergency salvage repair on him?

Clinical Note (numbered sentences):
**1:** He was transferred to the hospital on 2025-1-20 for emergent repair of his ruptured thoracoabdominal aortic aneurysm.
**2:** He was immediately taken to the operating room where he underwent an emergent salvage repair of ruptured thoracoabdominal aortic aneurysm with a 34-mm Dacron tube graft using deep hypothermic circulatory arrest.
**3:** Please see operative note for details which included cardiac arrest x2.
**4:** Postoperatively he was taken to the intensive care unit for monitoring with an open chest.
**5:** He remained intubated and sedated on pressors and inotropes.
**6:** On 2025-1-22, he returned to the operating room where he underwent exploration and chest closure.
**7:** On 1-25 he returned to the OR for abd closure JP/ drain placement/ feeding jejunostomy placed at that time for nutritional support.
**8:** Thoracoabdominal wound healing well with exception of very small open area mid wound that is @1cm around and 1/2cm deep, no surrounding erythema.
**9:** Packed with dry gauze and covered w/DSD.

Example Answer:
His aortic aneurysm was caused by the rupture of a thoracoabdominal aortic aneurysm, which required emergent surgical intervention (1). He underwent a complex salvage repair using a 34-mm Dacron tube graft and deep hypothermic circulatory arrest to address the rupture (2). The extended recovery time and hospital stay were necessary due to the severity of the rupture and the complexity of the surgery, though his wound is now healing well with only a small open area noted (8).

Question to answer: {clinician_question}
Patient's original query: {patient_question}

Clinical Note:
{numbered_note}

Instructions:
1. First, break down what information is needed to answer the question
2. Identify only the sentences that contain this essential information
3. Write a concise answer using only these sentences
4. Include sentence numbers in parentheses after each claim: (1) or (2, 3)
5. Be precise - only cite sentences that directly support your statements

Your Answer:
```

B Answer generation prompt

```
# Medical question answering based on essential sentences

## Patient Information
**Patient narrative:** {patient_narrative}

**Clinician question:** {clinician_question}

## Numbered essential sentences from the clinical note
{essential_text}

## Task Instructions
1. Generate a mostly extractive response from the listed sentences, which serves as an answer for the question. You must maximize lexical overlap between the source sentences and the response, while providing a useful answer.
2. Each essential sentence must be cited at least once in your answer. Include the sentence numbers in parentheses after statements that use information from those sentences, e.g., (2) or (1, 3). Cite multiple sources separated by comma, when neccessary.
```

```
3. Citations must be at the end of each generated sentence.
4. Limit your answer to a maximum of {words_limit} words, but more than 50 words. (About 4-5 sentences.)

Be straight to the point with your answer to the question, avoid phrases like "Based on the sentences". Remember, you must maximize the similarity in the wording to the original sentences.

## [For Iteration i > 1] Previous Attempts
### Attempt {i-1}
**Answer:**
{previous_answer}

**Rejection Reason:** {validation_feedback}
{
  • "Too long ({word_count} words)" -> word limit exceeded
  • "Does not cite all essential sentences: {missing_citations}" -> missed citations
  • "Citations to non-essential sentences: {invalid_citations}" -> invalid citations
}

## Instructions for revision
- Review ALL previous rejection reasons
- Ensure ALL essential sentences are properly cited
- Maintain a concise response (maximum {words_limit} words)
- Make sure to address all issues from previous attempts
```

B.1 End-to-End approach prompt

```
# Medical question answering based on clinical notes
## Task
Generate an answer to a patient's health question using only information from the clinical note. Each statement in your answer must be grounded in specific sentences from the note.

1. Generate an answer to the patient's question.
2. Include information that explain medical reasoning, procedures, relevant medical history of the patient that provides a full answer to the question.
3. EVERY sentence in your answer MUST end with at least one citation in parentheses, e.g., "The procedure was performed to treat the condition (3)." or "The treatment involved multiple steps to address your condition (3, 5)."
4. Be precise with your citations - only cite sentences that support each claim.
5. Be accurate with your citations, make sure citation format is correct: (sentence_number) OR (sentence_number_1, sentence_number_2, ...)
   - Invalid citation examples to avoid: (1-3); (1-2, 5-6); (Sentence 2)
   - Valid citation examples instead: (1, 3); (1, 2, 5, 6); (2)
6. Cite at most a couple of sentences at a time, not more.
7. Keep your answer under {words_limit} words total.
8. Do not include any sentences without citations.

[Example showing format with citations...]

## Current Case
**Patient narrative:** {patient_narrative}

**Clinician question:** {clinician_question}

### Clinical Note (numbered sentences):
{numbered_note}

## [For Iteration i > 1] Previous Attempts
### Attempt {i-1}
**Answer:** {previous_answer}
**Rejection Reason:** {validation_feedback}
{
  • "Too long ({word_count} words)" -> word limit exceeded
  • "Sentence {n} doesn't end with citation" -> missing citation
  • "Poorly formatted citation" -> citation format error
  • "Invalid citation numbers" -> cited non-existent sentences
}

Review ALL previous rejection reasons, and do not repeat these mistakes
```

LIMICS at ArchEHR-QA 2025: Prompting LLMs Beats Fine-Tuned Embeddings

Adam Remaki[1] Armand Violle[1] Vikram Natraj[1] Étienne Guével [2]
Akram Redjdal[1,3]

[1]Sorbonne Université, Inserm, Université Sorbonne Paris-Nord, Laboratoire d'Informatique Médicale et d'Ingénierie des Connaissances en e-Santé, Paris, France
[2]Sorbonne Cluster for Artificial Intelligence, Paris, France
[3]Univ Gustave Eiffel, Aix-Marseille Univ, LBA, F-13016 Marseille, France

Correspondence: adam.remaki@etu.sorbonne-universite.fr, armand.violle@sorbonne-universite.fr, natrajvikram.sivabalasubramanian@sorbonne-universite.fr, etienne.guevel@sorbonne-universite.fr, akram.redjdal@esiee.fr,

Abstract

In this paper, we investigated two approaches to clinical question-answering based on patient-formulated questions, supported by their narratives and brief medical records. The first approach leverages zero- and few-shot prompt engineering techniques with GPT-based Large Language Models (LLMs), incorporating strategies such as prompt chaining and chain-of-thought reasoning to guide the models in generating answers. The second approach adopts a two-steps structure: first, a text-classification stage uses embedding-based models (e.g., BERT variants) to identify sentences within the medical record that are most relevant to the given question; then, we prompt an LLM to paraphrase them into an answer so that it is generated exclusively from these selected sentences. Our empirical results demonstrate that the first approach outperforms the classification-guided pipeline, achieving the highest score on the development set and the test set using prompt chaining. Code: `github.com/armandviolle/BioNLP-2025`

1 Introduction

The ArchEHR-QA 2025 shared task (Soni and Demner-Fushman, 2025b) focused on grounded electronic health record question answering. The goal was to design a system that could answer patients' questions based on sentences from the patient's medical notes providing evidence supporting the answer's statements.

A development dataset (Soni and Demner-Fushman, 2025a) of 20 cases was at our disposal, structured as follows in `XML` format: a *patient narrative* ($P^{narrative}$), where the patient states their situation and asks their question(s); the original patient question ($Q^{patient}$); its clinician reformulation ($Q^{clinician}$); and a *medical note* summarizing the patient's history, presented both as a whole and sentence-by-sentence. Additionally, a `JSON` file was provided which contained a label for each sentence ("essential," "supplementary," or "not relevant") with respect to the questions.

Three guidelines were set for the generated answers: 1. a maximal length of 75 words, 2. one sentence per line with, at the end of each line, the cited `id` attribute(s) of the supporting medical note sentence(s) and 3. avoiding using external data or knowledge (relaxed later on).

The answers went through a two-step evaluation based on their *factuality*, i.e. the effective citation of "essential" sentences in the answers, and their *relevance*, i.e the semantic similarity with the inputs. Consequently, we tried to design systems suiting this layered structure with a *classification* step of the medical note sentences' relevance and a *summarization* step rephrasing relevant sentences into an answer to the $Q^{patient}$. We confronted these approaches to Large Language Models (LLMs) prompting strategies which we considered as baselines.

2 Methods

2.1 Sentence relevance classification

In this section, we present a method to identify question-relevant sentences using Sentence-BERT's bi-encoder and cross-encoder architectures (Reimers and Gurevych, 2019), enabling the LLM to generate answers grounded solely in the extracted content.

2.1.1 Single-sentence classification using short-context embeddings

First, we evaluated each clinical sentence individually against $P^{narrative}$ using pairwise comparisons.

We employed the pretrained cross-encoder ms-marco-MiniLM-L12-v2, which was originally trained on the MS MARCO dataset (Bajaj et al., 2016), a large corpus of query-document

BioNLP 2025 Shared Tasks, pages 150–159
August 1, 2025 ©2025 Association for Computational Linguistics

pairs ranked by relevance, and then fine-tuned on 15 cases (5 for validation) from ArchEHR dataset (Soni and Demner-Fushman, 2025a).

We also evaluated a bi-encoder model Jina-embedding v3 (Sturua et al., 2024). Sentences with cosine similarity score ≥ 0.5 to the query were considered essential, using 0.5 as a midpoint heuristic within the range of [0, 1]. eFigure 1 in the Appendix shows the distribution of similarity scores across label categories.

2.1.2 Multi-sentence classification using long-context embeddings

In our second approach, we utilized Jina-embedding v3's 8k-token capacity to process multiple sentences in context. Unlike the single-sentence setup, each example consists of a concatenated input of the $P^{narrative}$ and candidate sentences, formatted as `[Question] </s> [Sentence 1] </s> ... [Sentence N]`. The model outputs binary labels indicating whether each sentence is *Essential* or not (*Supplementary/Not Relevant*).

2.1.3 Data augmentation for robust classification

To address data scarcity, we generated 748 synthetic question-answer pairs from i2b2 (Uzuner et al., 2011), emrQA (Pampari et al., 2018), and MIMIC-III (Johnson et al., 2016)) clinical corpora. Each instance contained: (i) a question (generated via OpenAI's gpt-o4-mini with manual prompt tuning), (ii) clinical note excerpts, and (iii) binary relevance labels. For sentence selection, we embedded text using `text-embedding-ada-002`, retrieved top-k matches via FAISS, and assigned labels (*Essential/Supplementary/Not relevant*) based on ranking position. We evaluated augmentation effectiveness by fine-tuning both a ms-marco-MiniLM-L12-v2 cross-encoder (short-context) and a Jina Embedding v3 classifier (long-context). Details on the training are available in the section A of the Appendix.

2.2 Prompting LLMs for answer generation

In this section, we present an end-to-end method that generates the answer using LLMs. To evaluate different prompting strategies, we used the OpenAI API with data sharing explicitly disabled, ensuring that no inputs, outputs, were used to train or improve OpenAI models.

2.2.1 Zero-shot prompting

Zero-shot prompting was our first approach to generate the answer, specifically to understand how effectively LLMs could tackle both classification and paraphrasing sub-tasks at once. We adapted the prompt's *instructions* and format according to the observed output and best practices found in the literature, as well as diverse combinations of input data. We tested GPT-4.1-mini (OpenAI, 2025) and Mistral Large (AI, 2024). More details on the prompts can be found in eFigure 2 and eFigure 3 of the Appendix.

2.2.2 Prompting reasoning steps with chain-of-thought

As chain-of-thought (CoT) has proven to be an efficient prompting strategy to increase model reasoning abilities, we decomposed the task in a sequence of distinct steps to help the model tackle the task. We incorporated these *reasoning steps* into the system prompt and fed it to a GPT-4.1-mini (OpenAI, 2025) model, mostly to control outputs' format, trying to force the model to autonomously check and adapt its answer to the expected format. Prompt is presented in eFigure 4 of the Appendix.

2.2.3 Few-shot prompting

In few-shot prompting, we created pairs of question-answers to add as examples in our prompts. To generate the "gold standard" answers, we prompted GPT-4.1-mini (OpenAI, 2025) to paraphrase essential sentences from the medical note, based on the available labels in the dataset, into an answer to the $Q^{patient}$. Then, for each case, we sampled randomly a subset of pairs among the other available cases that were included in the prompt as examples, before the inference case's input. Prompts are presented in eFigure 5 and eFigure 6 of the Appendix.

2.2.4 Prompt chaining: `divide-and-conquer`

We adopted a prompt chaining approach based on the divide-and-conquer principle, breaking down the overall task into a structured sequence of smaller, interdependent subtasks. Each subtask is addressed by a language model, and the intermediate outputs are passed as inputs to subsequent stages. An overview of the full pipeline is shown in Figure 1.

This pipeline comprised five steps:

(i) Free answer generation. We prompted `o4-mini-2025-04-16` to generate a detailed and

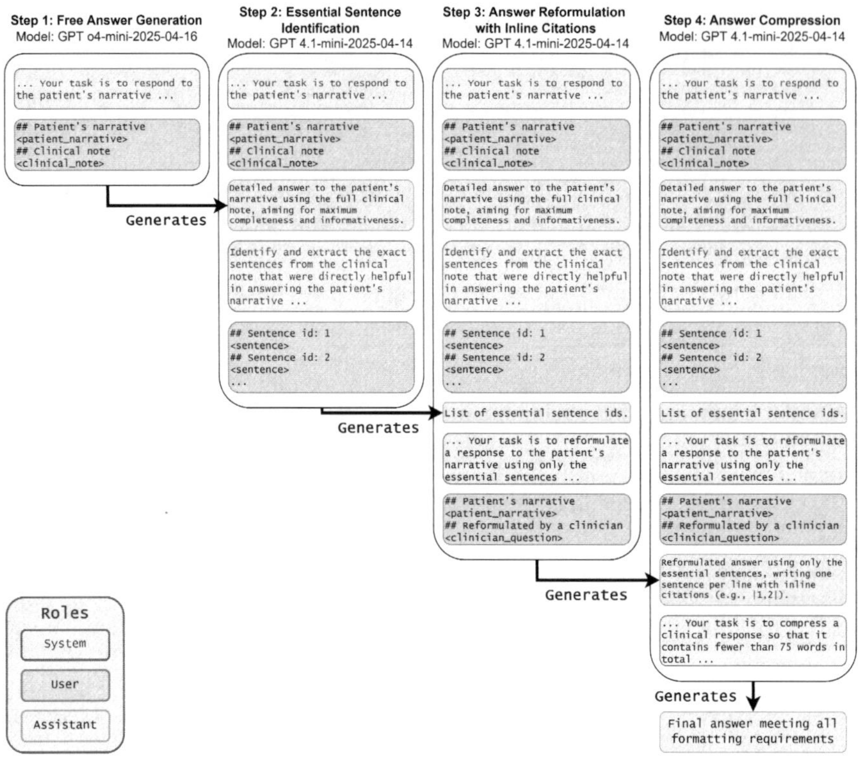

Figure 1: Overview of the prompt chaining workflow: each step refines the answer, improves grounding, and enforces formatting.

informative answer, given the $P^{narrative}$ and the associated full medical note. The prompt was designed to encourage completeness, with no formatting constraints, in order to generate as many relevant elements of the medical note as possible.
(ii) Essential sentence identification. The output from Step 1, along with the medical note (provided as a list of markdown-formatted sentences), is passed to `gpt-4.1-mini-2025-04-14`. The model was prompted to identify the minimal subset of sentences that directly support the answer.
(iii) Answer reformulation with inline citations. Using only the essential sentences from Step 2, the same model was prompted to reformulate the answer in a structured format. Each sentence appears on a new line and includes inline citations (e.g., `|3,7|`) referencing the supporting sentence IDs.
(iv) Answer compression. We prompted the same model to compress the reformulated answer into a concise version constrained to 75 words, while preserving the same inline citations.
(v) Strict answer compression (optional). If the compressed answer still exceeds 75 words, we prompt the same model again using the same compression rules, but presented in a more structured and imperative format. We allow up to three retries. If the constraint remains unmet, we restart the entire pipeline with a new seed.

The system prompts used in the pipeline are provided in eFigure 7 in the Appendix. One may note that only prompt chaining and CoT consistently produced answers within the 75-word limit. Other methods required post-processing compression, as described in Section E of the Appendix.

3 Results

3.1 Sentence relevance classification

Table 1 reports the performance of various embedding-based models in identifying essential sentences. We present precision, recall, and F1-score for each model configuration including the pretrained cross-encoder ms-marco-MiniLM-L12-v2 (with 33.4 million parameters), Jina Embedding v3 (Sturua et al., 2024) (with 572 million parameters) evaluated in both a bi-encoder (single-sentence) and a multi-sentence classification setting. The second and third columns indicate fine-tuning on the ArchEHR sample and the augmented dataset, respectively.

Model	ArchEHR FT*	Augmented FT*	Precision	Recall	F1-score
ms-marco-MiniLM			0.24 (0.20-0.28)	0.51 (0.42-0.60)	0.29 (0.25-0.34)
ms-marco-MiniLM	✓		0.37 (0.34-0.41)	0.28 (0.24-0.33)	0.29 (0.26-0.32)
ms-marco-MiniLM		✓	0.36 (0.35-0.37)	**0.90 (0.88-0.92)**	0.51 (0.50-0.52)
Jina (single-sentence)			**0.49 (0.41-0.61)**	0.55 (0.43-0.66)	**0.52 (0.44-0.59)**
Jina (multi-sentence)		✓	0.39 (0.33-0.46)	0.70 (0.59-0.82)	0.50 (0.44-0.56)

Table 1: Performance of embedding-based models for essential sentence classification on the development set. Metrics are reported as mean (95% confidence interval). *FT: fine-tuned.

3.2 Prompting LLMs for answer generation

Table 2 reports the performance of various prompting methods using large language models. The first column lists the prompting strategies. The second column presents the *factuality score*, measured as the F1-score on the essential sentence identification task. The third column shows the *relevance score*, computed as the average of several semantic similarity metrics (bleu, rouge, medcon, alignescore, bertscore, and sari) between the generated answer and the concatenation of the essential sentences, the $P^{narrative}$, and the $Q^{clinician}$.

Development Set		
Method	**Factuality**	**Relevance**
Zero-shot Mistral	51.1 (2.6)	31.1 (0.7)
Zero-shot GPT	56.6 (2.1)	32.5 (0.6)
Chain-of-thought	52.4 (1.9)	33.2 (0.5)
Few-shot	54.5 (1.9)	32.5 (0.5)
Prompt chaining	**59.3** (0.2)	**37.9** (0.3)
Test Set		
Method	**Factuality**	**Relevance**
Prompt chaining	**54.2**	**35.5**

Table 2: Comparison of methods on factuality and relevance score for the development and test sets. Results are reported as mean (standard deviation) over 10 random seeds for the development set. Test result is shown for the best-performing method.

4 Discussion

Our findings highlight several important insights regarding the classification of essential sentences in clinical narratives. First, fine-tuning on the ArchEHR dataset alone did not yield consistent performance gains. We attribute this to the dataset's limited size (only 20 annotated cases), which is insufficient for effective adaptation. Moreover, the augmented dataset significantly improved the performance of the cross-encoder model. It not only boosted F1-scores but also reduced variance across runs, suggesting that the model benefited from the synthetic data. However, fine-tuning the Jina-Embedding v3 model with augmented data and multi-sentence input did not improve performance. This may be due to the LoRA adapters being poorly suited for this fine-tuning setup, or because the model's initial performance left little room for improvement. Further investigation is needed to understand the cause.

Despite extensive experimentation with embedding-based approaches, including both single and multi-sentence configurations, we observed that LLMs outperformed them on the sentence classification task. Nevertheless, it is noteworthy that a relatively small 33M-parameter BERT cross-encoder achieved the same F1-score of 0.51 as the much larger 123B-parameter Mistral large model, highlighting a meaningful tradeoff between performance and computational cost.

Results indicate that prompting strategies isolating subtasks through sequential prompt chaining led to more accurate sentence classification, improved answer relevance, and reduced variability, with standard deviation nearly ten times smaller for the factuality score. Interestingly, zero-shot prompting outperformed both few-shot and CoT approaches. While the reason remains unclear, this may suggest that overly long system prompts were less effective for this task.

5 Conclusion

This study addressed the ArchEHR-QA challenge, where the goal is to answer patient-specific clinical questions by identifying and citing essential sentences from clinical notes. For sentence classification, augmenting the dataset with synthetic QA pairs improved performance and reduced variation. While embedding models such as bi-encoders and cross-encoders produced solid results, LLMs consistently outperformed them. For this task, prompt chaining, which isolates subtasks, gave the best result.

Limitations

The first limitations to mention are related to the LLMs we used for prompting strategies. Indeed OpenAI's GPT models and Mistral AI's models are proprietary and thus lack transparency on their training process (e.g data corpora used) and some functionality (e.g "determinism not guaranteed" when fixing seed parameter). In research, it is a major drawback as it is hard to truthfully build upon undisclosed features. Moreover, these models are pay-as-you-go, so we stuck to smaller, cheaper models that enabled us to run multiple experiments (we spent almost 100$ worth of OpenAI tokens for the challenge). Scaling up to models such as GPT-4.5, o1 or o3 may have improved performances-but it comes at a cost.

One limitation of our synthetic dataset is that the complexity of the sentence classification task often requires domain-specific medical knowledge. As a result, the generated data may not fully capture the nuances present in real clinical scenarios. Incorporating validation and annotation by medical experts could help ensure the reliability and clinical relevance of the synthetic data, thereby increasing its impact for downstream tasks.

To conclude, we reflect on the evaluation methodology, particularly the suitability of BLEU (Papineni et al., 2002) for assessing the *relevance* metric. BLEU includes a brevity factor that can disproportionately penalize predicted answers that differ in length from the reference. In our case, relatively short predicted answers (with a maximum expected length of 75 words) were evaluated against much longer references composed of concatenated $P^{narrative}$, $Q^{clinician}$, and essential sentences. This mismatch in length likely contributed to the uniformly low BLEU scores observed across the leader board.

Acknowledgments

This work was performed using HPC resources from GENCI–IDRIS (Grant 2025-AD011015342R1). Some computations were also performed using the GPU cluster resources of the Sorbonne Center for Artificial Intelligence (SCAI) at Sorbonne University. The authors thank Xavier Tannier and Stéphane Dohayon for their valuable advice on the design.

References

Mistral AI. 2024. Au Large.

Payal Bajaj, Daniel Campos, Nick Craswell, Li Deng, Jianfeng Gao, Xiaodong Liu, Rangan Majumder, Andrew McNamara, Bhaskar Mitra, Tri Nguyen, and 1 others. 2016. Ms marco: A human generated machine reading comprehension dataset. *arXiv preprint arXiv:1611.09268*.

Alistair EW Johnson, Tom J Pollard, Lu Shen, Li-wei H Lehman, Mengling Feng, Mohammad Ghassemi, Benjamin Moody, Peter Szolovits, Leo Anthony Celi, and Roger G Mark. 2016. Mimic-iii, a freely accessible critical care database. *Scientific data*, 3(1):1–9.

Zehan Li, Xin Zhang, Yanzhao Zhang, Dingkun Long, Pengjun Xie, and Meishan Zhang. 2023. Towards general text embeddings with multi-stage contrastive learning. *arXiv preprint arXiv:2308.03281*.

OpenAI. 2025. Introducing GPT-4.1 in the API.

Anand Pampari, Pradeep Raghavan, Jinfeng Liang, and Jian Peng. 2018. emrqa: A large corpus for question answering on electronic medical records. In *Proceedings of the 2018 Conference on Empirical Methods in Natural Language Processing*, pages 2357–2368.

Kishore Papineni, Salim Roukos, Todd Ward, and Wei-Jing Zhu. 2002. Bleu: a method for automatic evaluation of machine translation. In *Proceedings of the 40th annual meeting of the Association for Computational Linguistics*, pages 311–318.

Nils Reimers and Iryna Gurevych. 2019. Sentence-BERT: Sentence embeddings using Siamese BERT-networks. In *Proceedings of the 2019 Conference on Empirical Methods in Natural Language Processing and the 9th International Joint Conference on Natural Language Processing (EMNLP-IJCNLP)*, pages 3982–3992, Hong Kong, China. Association for Computational Linguistics.

Sarvesh Soni and Dina Demner-Fushman. 2025a. A dataset for addressing patient's information needs related to clinical course of hospitalization. *arXiv preprint*.

Sarvesh Soni and Dina Demner-Fushman. 2025b. Overview of the archehr-qa 2025 shared task on grounded question answering from electronic health records. In *The 24th Workshop on Biomedical Natural Language Processing and BioNLP Shared Tasks*, Vienna, Austria. Association for Computational Linguistics.

Saba Sturua, Isabelle Mohr, Mohammad Kalim Akram, Michael Günther, Bo Wang, Markus Krimmel, Feng Wang, Georgios Mastrapas, Andreas Koukounas, Andreas Koukounas, Nan Wang, and Han Xiao. 2024. jina-embeddings-v3: Multilingual embeddings with task lora. *Preprint*, arXiv:2409.10173.

Özlem Uzuner, Brett R South, Shuying Shen, and Scott L DuVall. 2011. 2010 i2b2/va challenge on concepts, assertions, and relations in clinical text. *Journal of the American Medical Informatics Association*, 18(5):552–556.

Appendix

A Training details for sentence classification

A.1 Cross-encoder finetuning

We trained the cross-encoder on both the augmented dataset and archEHR sample using identical hyperparameters: binary cross-entropy loss, AdamW optimizer (learning rate 2×10^{-5}), and batch size of 64. Training proceeded for 10 epochs.

A.2 Multi-sentence classification

We fine-tuned the LoRA adapter specifically designed for classification in the Jina-Embedding v3 model, which includes five task-specific LoRA adapters in total. These adapters are integrated into the embedding and linear layers of the multi-head attention mechanism, with a rank of 4 and $\alpha = 1$. We fine-tuned the classification adapter on our synthetic QA dataset for 5 epochs using the AdamW optimizer (learning rate: 2×10^{-5}). Due to the long input sequences, we used a batch size of 1. Class imbalance was addressed using a weighted `BCEWithLogitsLoss`, and mixed-precision training (`bfloat16`) was enabled via `torch.cuda.amp`. Inputs followed the format: `[Question] </s> [Sentence 1] </s> ... [Sentence N]` The final prediction was produced by a linear head applied to sentence embeddings extracted at the `</s>` token positions.

B Bi-encoder classification

The bi-encoder approach using Jina Embedding v3 demonstrated significantly higher cosine similarity scores between patient questions and sentences labeled as "Essential" (mean = 0.62) compared to other categories (mean = 0.41, t-test $p < 1 \times 10^{-10}$). eFigure 1 shows the distribution of similarity scores across label categories, revealing clear separation between essential and non-essential phrases.

C Classification with a large encoder

Here we report an evaluation of gte-Qwen2-7B-instruct (Li et al., 2023). When using the prompt presented in eFigure 8, the model ended up overfitting on the training set while failing to generalize the information on the validation set . For the accuracy it reached 0.98 and the f1 0.98 in training, while in validation the best metrics were : f1 0.33, recall 0.28, precision 0.41.

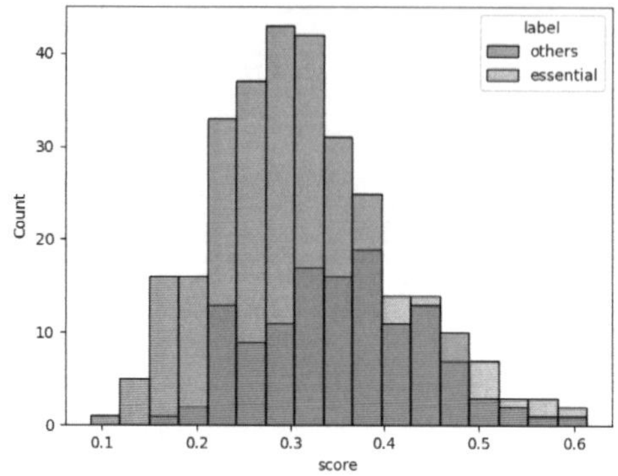

eFigure 1: Distribution of cosine similarity between question and sentence

D CoT and Few-Shot implementation details

We used the three roles offered by the `chat.completions.create` of the OpenAI API client: `system` to describe the general beahior of the model, `user` to input data and additional information helping the model to respond such as reasoning steps or examples, and `assistant` to input example responses for few-shot prompting. The `system` prompt and `user` prompts were very similar in few-shot (see Figure 6) and CoT (see Figure 4). For the `user` inference prompt, we just concatenated the selected data consisting in $P^{narrative}$, $Q^{patient}$, $Q^{clinician}$ and the sentence-by-sentence medical note excerpt in few-shot, while in CoT we first prompted the reasoning steps and then the same inputs.

For CoT, we created examples using the first 2 cases by prompting successively the reasoning steps and input data in ChatGPT. We then used the final answers as "gold standard" to provide an example for each case before inference, which resulted in the following (considering a single case):

1. We prompt the `system` role (see Figure 4 for detailed prompts).

2. We prompt the `user` role with the reasoning steps and the input data of an example case.

3. We prompt the `assistant` role with the final answer obtained with ChatGPT.

4. Finally, we prompt the actual inference case to the `user` role.

In few-shot, before generating the answers, we zero-shot a summarization task with the `system` prompt on Figure 5 and a `user` prompt containing essential sentences only and $Q^{patient}$. We used them in the few-shot strategy to provide as follows, considering a single case:

1. We randomly sampled 5 cases among the 19 other available to serve as examples.
2. We prompt the `system` role.
3. We prompt 5 times `user-assistant` roles successively, `user` prompts being the sampled cases formatted as inference prompts, and `assistant` prompts being the corresponding sampled cases' previously generated paraphrase.
4. Finally, we prompt the actual inference case to the `user` role.

For CoT, we created examples using the first 2 cases by prompting successively the reasoning steps and input data in ChatGPT. We then used the final answers as "gold standard" to provide an example for each case before inference, which resulted in the following (considering a single case):

1. We prompt the `system` role (see Figure 4 for detailed prompts).
2. We prompt the `user` role with the reasoning steps and the input data of an example case.
3. We prompt the `assistant` role with the final answer obtained with ChatGPT.
4. Finally, we prompt the actual inference case to the `user` role.

E Answer post-processing to enforce word limit

To enforce the 75-word limit required by the evaluation protocol, we apply a post-processing script to the model-generated answers. Although the summarization prompt explicitly specifies this limit, responses occasionally exceed it. The cleanup process ensures validity and evaluation compatibility through the following steps:

- **Grouped summarization:** Consecutive sentences with identical citations are grouped and summarized using `GPT-4.1-mini`, with a dynamic word limit to ensure the final output stays within the 75-word constraint.
- **Citation preservation:** Citations from the original outputs are preserved and reattached to the corresponding summarized segments to maintain factual alignment.
- **Fallback handling:** If summarization fails or exceeds the limit, a generic sentence is inserted: *"Additional supporting evidence."* with the missing citations appended.
- **Format compliance:** The evaluation script requires at least one citation line in the format `Sentence or summary. |citation_id(s)|`, but not necessarily one for every sentence.

This method prioritizes factual consistency and strict format adherence, and was found to be effective when used with a controlled summarization model such as `GPT-4.1-mini`.

F System prompts

```
You are a clinical assistant. Carefully review the patient narrative,
clinician question, and the provided clinical note sentences. Provide a
medically accurate and detailed answer to the clinician's question.

Example:

Patient Narrative:
"I had difficulty breathing and fever, and was hospitalized."

Clinical Note Sentences:
[0] Patient admitted on Wednesday evening.
[1] Patient complained of difficulty breathing.
[2] Chest X-ray showed clear infiltrates in lower lobes.
[3] White blood cell count significantly elevated, indicative of infection.
[4] Patient was discharged after five days.

Correct JSON Response:
{
  "answer": "Yes, the patient has clinical evidence of pneumonia |2,3|,
supported by X-ray infiltrates and elevated white blood cell count |2|."
}

Important Instructions:

Include ALL sentences that could partially or fully support answering
the clinician's question by mentionning them in |sentenceIDs|.
If uncertain, lean towards including the sentence.
Prioritize recall and completeness of supporting evidence.

Patient Narrative:
{patient_narrative}


Clinical Note Sentences:
{formatted_sentences}


Respond strictly in the JSON format:
{{
  "answer": "your detailed answer here cite sentences IDs between ||"
}}
```

eFigure 2: Mistral large zero-shot prompt

```
You are a clinical assistant. Your goal is to answer the patient's question
using only the sentences provided below.
- Every sentence used must be cited at the end using |sentence_id|.
- Cite all sentences that support each part of your answer.
- If multiple sentences support a point, cite all of them like |2,3|.
- Keep your total answer upto 75 words.
- Write one sentence per line.
Sentences:
{context}
Question :
{question / patient_narrative}
Answer:
```

eFigure 3: Zero-shot prompt using GPT 4.1-mini

```
# Identity
You are a helpful medical assistant answering accurately to patients' questions using evidence
from their medical records.
Your goal is to provide clinically grounded answers by highlighting relevant information from the
note excerpt while preserving its medical meaning.
Maintain a tone of light formality suitable for direct communication with patients.
You will receive detailed instructions that you MUST follow exactly.

# Instructions
- Address the patient.
- Do not produce void answers.
- Do not refer to or quote the full clinical note.
- Write the response as a series of standalone sentences, one sentence per line.
- At the end of each sentence, cite the supporting sentence ID(s) in this format: |s1|.
- If a sentence is supported by multiple note sentences, cite them like this: |s2,s3|.
- Every sentence in your response MUST be backed by one or more note_excerpt_sentences.
- Your answer must be EXACTLY between **70 and 75 words** (excluding citations). Adjust
phrasing to meet this requirement.

# Input format
You will be given input in XML format with the following elements:
- <patient_narrative>: the full narrative question from the patient.
- <patient_question>: key phrases extracted from the narrative, each within a <phrase> tag with
attributes "id" and "start_char_index".
- <clinician_question>: a rephrasing of the patient's question from a clinician's perspective.
- <note_excerpt_sentences>: sentences extracted from the patient's medical record, each within a
<sentence> tag, with attributes "id", "paragraph_id", and "start_char_index".
```

```
# Reasoning Steps
1. Identify relevant information from the note_excerpt_sentences based on the patient's question.
2. At the end of each response sentence, cite the supporting note_excerpt_sentences ID(s) like
this: |1| or |2,3| if multiple.
3. Paraphrase and summarize the relevant information.
4. Ensure the answer is between 70 and 75 words, excluding citations.

# Input
<patient_narrative>
<patient_question>
<clinician_question>
<note_excerpt_sentences>
```

eFigure 4: Prompts for `system` (top) and `user` (bottom) roles used for the CoT experiments with OpenAI API.

```
# Identity
You are a helpful medical assistant that rewrites text clearly and accurately to answer
a question.
Your goal is to paraphrase input sentences and question while preserving its medical
meaning, aiming for light formality in the tone answering to the patient.
You will be given instructions that you STRICTLY have to follow.

# Instructions
Your task is to reformulate a response to the patient's narrative using only the
essential sentences extracted from the clinical note. Follow these strict guidelines:
- Use only the provided essential sentences, patient narrative and clinician question
to generate your response.
- Do not refer to or quote the full clinical note.
- Write the response as a series of individual sentences, one sentence per line.
- At the end of each sentence, cite the supporting sentence ID(s) in this format:
|sentence_id|.
- If a sentence is supported by multiple essential sentences, cite all applicable IDs like
this: |2,3|.
- Every statement in your response must be supported by one or more essential
sentences.
- All essential sentences must be cited in your response.
- The answer should STRICTLY have **between 70 and 75 words**.

# Reasoning Steps
- Each input sentences holds valuable information to answer the patient's question.
Using every one of them should help imrpving the answer's relevance.
- Make sure that all instructions on the answer's format are followed, if not reformulate
until they are all followed.

# Output format
## Example of output format
This is the first generated sentence with cited evidence. |0|
This is another generated sentence with cited evidences. |i,j|
You can also cite multiple evidence-sentences within a response sentence. |N|

Take a deep breath and work step by step.
```

eFigure 5: System prompt used to generate essential sentences' and $P^{narrative}$' summarized paraphrase in Zero-Shot fashion.

Identity
You are a helpful medical assistant answering accurately to patients' questions considering their medical records.
Your goal is to answer highlighting the clinical evidence found in a patient's note excerpt and preserving their medical meaning, aiming for light formality in the answer to the patient.
You will be given instructions that you STRICTLY have to follow.

Instructions
- Do not refer to or quote the full clinical note.
- Write the response as a series of individual sentences, one sentence per line.
- At the end of each sentence, cite the supporting sentence ID(s) in this format: |sentence_id|.
- If a sentence is supported by multiple essential sentences, cite all applicable IDs like this: |2,3|.
- Every statement in your response must be supported by one or more essential sentences.
- The answer should STRICTLY have **between 70 and 75 words**.

Input format
Your are given an input sample containing different levels of information in XML format with the following tags:
- patient_narrative: full patient narrative question.
- patient_question: key phrases in the patient_narrative identified as the patient's question, each phrase is delimited by a 'phrase' tag along with an index "id" and its starting character in the narrative "start_char_index".
- clinician_question: rephrasing of the patient's question, posed by a clinician.
- note_excerpt_sentences: sentences extracted from the patient's medical hospital history. Each sentence is delimited by a 'sentence' tag along with "id", "paragraph_id" and "start_char_index" attributes.

Reasoning Steps
1. Answer to the patient's question using relevant information among the note_excerpt_sentences, considering the clinician question to guide the medical argumentation of your response.
2. For each sentence of the answer.
a. Identify which sentences among the note_excerpt_sentences can contain information related to this response sentence.
b. Cite its/their "id" attribute(s) enclosed in pipe symbols ("|") at the end of the sentence.
c. You have to find at least one relevant citation per response setence: none should be let without citation.
3. Try to reformulate your answer to stick more closely to the cited note_excerpt_sentences, paraphrasing them to some extent.
4. Make sure that the answer's length does not exceeds 75 words citations excluded, reformulate until this condition is met.

Ouput format
Example of output format
This is the first generated sentence with cited evidence. |0|
This is another generated sentence with cited evidences. |i,j|
You can also cite multiple evidence-sentences within a response sentence. |N|

Take a deep breath and work step by step.

eFigure 6: Prompt for `system` role used for Few-Shot experiments with OpenAI API.

Step 1: free answer generation

You are a clinical assistant. Your task is to respond to the patient's narrative using only the information found in the provided clinical note. Do not introduce any information that is not explicitly stated in the clinical note.
Your primary goal is to provide an accurate and detailed response that directly addresses the patient's narrative, strictly based on the content of the clinical note. Do not infer or assume any additional context beyond what is given.

Step 2: essential sentence identification

Identify and extract the exact sentences from the clinical note that were directly helpful in answering the patient's narrative. Only include the most relevant sentences that provide clear support for the answer. Do not include unrelated information or extra context. Return the selected sentences, followed by a list of their corresponding sentence IDs.

Step 3: answer reformulation with inline citations

You are a clinical assistant. Your task is to reformulate a response to the patient's narrative using only the essential sentences extracted from the clinical note. Follow these strict guidelines:
- Use only the provided essential sentences to generate your response.
- Include all essential sentences in your response.
- Do not refer to or quote the full clinical note.
- Write the response as a series of individual sentences-one sentence per line.
- At the end of each sentence, cite the supporting sentence ID(s) in this format: |sentence_id|.
- If a sentence is supported by multiple essential sentences, cite all applicable IDs like this: |2,3|.
- Every statement in your response must be supported by one or more essential sentences.

Step 4: answer compression

You are a clinical assistant. Your task is to compress a clinical response so that it contains fewer than 75 words in total while preserving the full set of cited sentence IDs. Follow these strict guidelines:
- Your goal is to reduce the total word count to 75 words or fewer by merging and rephrasing the original sentences.
- You must include all original sentence IDs in the final response, but you can combine them into fewer citation brackets.
- Example:
Original:
<sentence A>. |1,2|
<sentence B>. |4,8,16|
Reformulated:
<merged sentence>. |1,2,4,8,16|
- Write the response as a series of individual sentences—one sentence per line.
- Every statement in your response must be supported by one or more essential sentences.

Step 5: strict answer compression

You are a clinical assistant. Your task is to compress a clinical response so that it contains **fewer than 75 words in total** while preserving the full set of cited sentence IDs.

STRICT RULES:
- Your output must contain **less than 75 words total**. Not 75 or more. Not approximately. **Fewer than 75.**
- Merge, shorten, and rephrase aggressively, but preserve all sentence IDs. You may combine them into fewer citation brackets (e.g., |1,2,4|).
- **DO NOT exceed the word limit under any circumstance.**
- Each line must be a single sentence.
- Every statement must be supported by at least one sentence ID.

FINAL CHECK BEFORE OUTPUT:
- Each line must be a single sentence.
- Count the words in your response. If 75 or more: revise, shorten, and try again.
- The output is invalid unless it has **< 75 words**.

EXAMPLE:
Original:
<sentence A>. |1,2|
<sentence B>. |4,8,16|

Compressed:
<merged sentence>. |1,2,4,8,16|

eFigure 7: System prompts used for the prompt chaining pipeline.

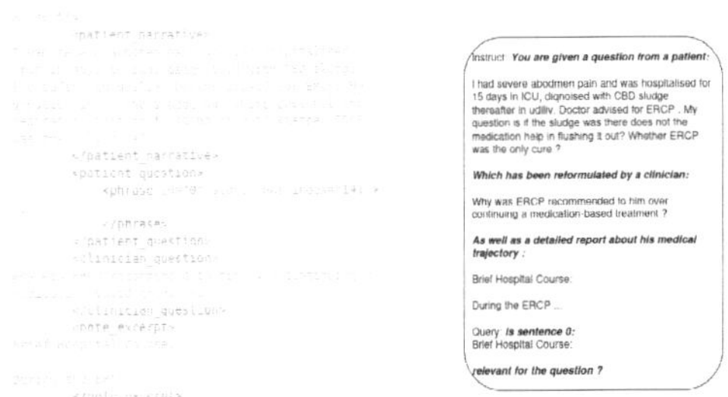

eFigure 8: Prompt for Qwen2-gte-7B-instruct

Each phrase of the excerpt makes a sample, the example shown here is for the first phrase. In bold are the added text to give context to the instruct model.

razreshili at ArchEHR-QA 2025: Contrastive Fine-Tuning for Retrieval-Augmented Biomedical QA

Arina Zemchyk
arina.zemchik@gmail.com

Abstract

We present a retrieval-augmented system for the ArchEHR-QA 2025 shared task, which focuses on generating concise, medically accurate answers to clinical questions based on a patient's electronic health record (EHR). A key challenge is following a strict citation format that references relevant sentence IDs. To improve retrieval, we fine-tuned an all-MiniLM-L6-v2 embedding model using contrastive learning on over 2,300 question–sentence triplets, with DoRA for efficient adaptation. Sentences were selected using cosine similarity thresholds and passed into a quantized Mistral-7B-Instruct model along with a structured prompt. Our system achieved similar relevance to the baseline but lower overall performance (19.3 vs. 30.7), due to issues with citation formatting and generation quality. We discuss limitations such as threshold tuning, prompt-following ability, and model size, and suggest future directions for improving structured biomedical QA.

1 Introduction

The ArchEHR-QA 2025 shared task focuses on answering medical questions based on a patient's electronic health record (EHR) (Soni and Demner-Fushman, 2025b). Each answer must be short, medically accurate, and include in-text citations using sentence IDs from the patient history (e.g., |1,2|). This makes the task challenging, especially due to the length and complexity of clinical records and the strict output formatting rules.

Our approach follows a retrieval-augmented pipeline. First, we fine-tune an embedding model to better identify relevant sentences in the patient's history. Then, we pass the selected sentences, together with the question, into a generative model (Mistral-7B[1]) that produces the answer.

Although our method did not outperform the baseline, it achieved comparable relevance score. Most of the performance gap came from formatting issues and citation errors in the generated text, which we analyze in this paper. We also discuss the challenges of tuning models with limited data and propose directions for improvement.

2 Methodology

In order to improve the accuracy and relevance of cited sentences in generated answers, the main focus of the proposed system is a domain-adapted embedding model, which can capture the nuances of a biomedical domain.

2.1 Overview

The approach consists of three main steps: (1) fine-tuning an embedding model on the development set of the shared task dataset (Soni and Demner-Fushman, 2025a), (2) selecting relevant and supplementary context sentences based on cosine similarity thresholds, and (3) generating answers using a quantized generative model (Mistral-7B) with in-context citations.

2.2 Embedding Model Fine-Tuning

To accurately retrieve relevant sentences from the patient's history, we fine-tuned the `all-MiniLM-L6-v2`[2] model with contrastive objective using DoRA (Mao et al., 2024), a parameter-efficient fine-tuning method that extends LoRA (Hu et al., 2022). DoRA improves learning capacity and training stability of LoRA, making it particularly suitable in settings with limited training data and computational resources. Additionally, parameter-efficient tuning mitigates the issue of catastrophic forgetting (Goodfellow et al., 2013), where the pretrained model loses its

[1] `https://huggingface.co/mistralai/Mistral-7B-Instruct-v0.2`

[2] `https://huggingface.co/sentence-transformers/all-MiniLM-L6-v2`

BioNLP 2025 Shared Tasks, pages 160–164

original knowledge during full fine-tuning. Details about DoRA setup can be found in Appendix A.

We constructed a dataset of 2,582 triplets from the development set (Soni and Demner-Fushman, 2025a), where each triplet consisted of:

- **Anchor:** a clinical question,
- **Positive:** a sentence labeled as "essential" or "supplementary" to a given clinical question,
- **Negative:** a sentence labeled as "not relevant" to a given clinical question.

The dataset was split into 2,341 training and 241 validation triplets to monitor performance.

We used the `Trainer`[3] from the `SentenceTransformer` library with `MultipleNegativesRankingLoss` (analogous to InfoNCE loss (Oord et al., 2018)) as the training objective. In this setup, negatives were treated as in-batch negatives, with the explicit negative in each triplet acting as a hard negative. Training was run for 50 epochs with the following key hyperparameters: batch size of 64 (train) and 128 (eval), learning rate of 1e-4, warmup ratio of 0.1, and no-duplicates batch sampling (beneficial for in-batch negative mining).

To monitor training, we evaluated embedding quality using alignment and uniformity metrics (Figure 1). Alignment is a metric that measures the closeness of positive pairs representations. Uniformity, on the other hand, depicts how well the embeddings are ditstributed on a unit hypersphere. These metrics were introduced by Wang and Isola (2020) and provide insights into how well the fine-tuned model clustered relevant sentences closer to their corresponding questions while maintaining separation from irrelevant ones.

2.3 Threshold Selection for Relevance

To define a threshold for sentence relevance, we embedded both the clinical questions and patient history sentences using the fine-tuned model and computed cosine similarity scores. Thresholds were empirically determined by testing similarity values between 0.0 and 1.0 (in increments of 0.01) on the development set, selecting the threshold that produced the highest F1 score for identifying "relevant" sentences:

- **Relevant:** cosine similarity ≥ 0.25
- **Supplementary:** $0.20 \leq$ cosine similarity < 0.25
- **Irrelevant:** cosine similarity ≤ 0.20

During answer generation on the test set, if no sentences met the "relevant" or "supplementary" criteria (i.e., all sentences were classified as "irrelevant"), the full patient history was used as context.

2.4 Generative QA Module

For answer generation, we used a quantized `Mistral-7B-Instruct-v0.2`[4] model, selected due to computational constraints. The prompt was structured into three segments:

1. **Instruction Header:** a detailed instruction block framing the task, e.g., "You are a medical assistant tasked with answering patient questions using provided case information. After each factual claim, cite supporting sentences in the format |id| or |id1, id2|. Limit the answer to 75 words."
2. **Context:** a concatenation of the retrieved relevant and supplementary sentences, each labeled with its sentence ID for proper referencing.
3. **Clinical Question:** the specific question to be answered.

The prompt also included an explicit example demonstrating correct citation style and answer formatting, to help enforce the desired output pattern. Despite these explicit instructions, we observed that the generative model frequently struggled to fully comply with strict citation formatting and word count limits, highlighting typical challenges in controlling large language models.

The full prompt template used in this work is provided in Appendix C.

2.5 Reflections

The final system's underperformance relative to the baseline may be from two main factors: (1) intrinsic weaknesses of the generative model in structured QA and (2) potentially over-restrictive relevance thresholds, which may have omitted valuable context. The small development set size also limited threshold generalizability.

[3] https://sbert.net/docs/package_reference/sentence_transformer/trainer.html

[4] https://huggingface.co/mistralai/Mistral-7B-Instruct-v0.2

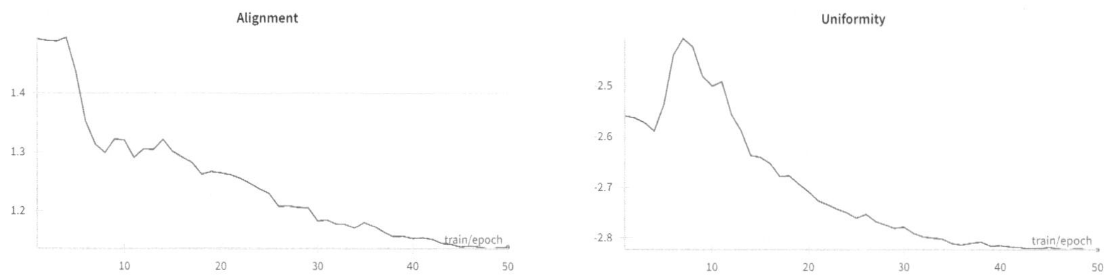

Figure 1: Training progression visualized through alignment (left) and uniformity (right) metrics on a validation set.

3 Results

We submitted one main official run for the ArchEHR-QA task using the `Mistral-7B-Instruct-v0.2`-based system. Furthermore, we experimented with another generative model `meta-lama-3-8B-Instruct`[5], but its performance was slightly lower (overall score: 19.2) and therefore it is not considered in this section. Table 1 reports the scores of our main run compared to the organizers' baseline.

Metric	Baseline	Ours
Overall	30.7	19.3
Overall Factuality	33.6	13.5
Overall Relevance	27.8	25.2
Strict Precision (micro)	71.6	36.8
Strict Recall (micro)	21.9	8.2
Strict F1 (micro)	33.6	13.5
Lenient Precision (micro)	77.0	39.7
Lenient Recall (micro)	22.3	8.4
Lenient F1 (micro)	34.6	13.9
Strict Precision (macro)	77.4	49.6
Strict Recall (macro)	31.5	14.5
Strict F1 (macro)	39.0	19.0
Lenient Precision (macro)	83.0	53.8
Lenient Recall (macro)	30.8	13.6
Lenient F1 (macro)	39.9	19.1
BLEU	0.1	0.4
ROUGE-Lsum	15.2	16.8
SARI	47.8	45.8
BERTScore	20.5	19.9
AlignScore	57.7	43.9
MEDCON (UMLS)	25.6	24.5

Table 1: Performance comparison between the baseline (organizers) and our system (razreshili) on the ArchEHR-QA test set.

Our best submission did not outperform the baseline in most official metrics, but achieved a comparable relevance score (25.2 vs. 27.8 overall relevance) and slightly higher ROUGE-Lsum (16.8 vs. 15.2) and BLEU (0.4 vs. 0.1).

3.1 Error Analysis

We conducted a detailed error analysis with the following findings:

- **Citation format errors:** Despite explicit prompt engineering, some generated answers failed to follow the required citation format (`|sent_id|`, e.g., `|1|`). This often happened when the context was complex or included many sentences. Typical mistakes included separating citations incorrectly (e.g., `|1|, |12|` instead of `|1,12|`), breaking them across lines (e.g., `|3|\n\n|2,6|` instead of `|3,2,6|`), or separating with a dot (e.g., `|6|. |1,3|` instead of `|6,1,3|`). These formatting issues might have contributed to lower scores in strict citation metrics.

- **Word limit violations:** Of 100 cases, 14 generated responses exceeded the 75-word limit. We observed that these violations were more common in truncated cases, where the context length was substantially longer: truncated cases had on average 7.4 relevant and 5.3 supplementary sentences, compared to 3.6 and 2.0 in non-truncated cases. This suggests that longer, information-rich contexts increased the likelihood of the model producing overlength answers.

4 Discussion

Our method did not outperform the baseline, but it helps show where smaller generative models struggle in biomedical question answering.

Even though adding relevant and supplementary sentences to the prompt helped us reach a similar

[5] `https://huggingface.co/meta-llama/Meta-Llama-3-8B-Instruct`

relevance score, other scores like citation accuracy and factual correctness were much lower. This means that better sentence retrieval alone is not enough—the model also needs to follow strict rules for format and content.

Smaller models like Mistral-7B and Meta-Llama-3-8B often failed to follow the required citation format or stay under the 75-word limit. In contrast, larger models like LLaMA 3.3 70B, which were used in the baseline system, are better at following instructions and producing more accurate answers. While we used a retrieval-augmented setup to shorten the context and focus the model on relevant sentences, newer models like LLaMA-3-8B or Mistral-7B support longer inputs and could process the full patient history directly. We didn't try this due to limited resources, but it could be a strong and simpler baseline for future work.

In future work, combining better retrieval with larger or more fine-tuned generative models may help improve performance on this type of task.

5 Limitations

Our approach has several limitations:

- **Small dev set:** The development set was small, which made it hard to properly adapt a sentence embedding model to a complex medical domain.
- **Strict thresholds:** The fixed similarity thresholds for selecting relevant and supplementary sentences may have removed useful context, especially for more difficult questions.
- **Generative model constraints:** We used a quantized version of Mistral-7B due to hardware limitations. While fast and memory-efficient, this model often failed to follow citation and length constraints, limiting the effectiveness of our retrieval pipeline.
- **No fine-tuning of the generator:** The generator was used as-is with prompt instructions. We didn't fine-tune it on this task, which likely hurt citation accuracy.
- **Prompt sensitivity:** Despite careful prompt design, the model often ignored citation formatting rules. This suggests that prompt-only control may be insufficient for tasks with strict output requirements.
- **No baseline for smaller embedding model:** We did not compare our fine-tuned embedding model against the original (non-adapted) version. This limits our ability to directly measure the contribution of contrastive fine-tuning to retrieval performance.

References

Ian J Goodfellow, Mehdi Mirza, Da Xiao, Aaron Courville, and Yoshua Bengio. 2013. An empirical investigation of catastrophic forgetting in gradient-based neural networks. *arXiv preprint arXiv:1312.6211*.

Edward J Hu, yelong shen, Phillip Wallis, Zeyuan Allen-Zhu, Yuanzhi Li, Shean Wang, Lu Wang, and Weizhu Chen. 2022. LoRA: Low-rank adaptation of large language models. In *International Conference on Learning Representations*.

Yulong Mao, Kaiyu Huang, Changhao Guan, Ganglin Bao, Fengran Mo, and Jinan Xu. 2024. DoRA: Enhancing parameter-efficient fine-tuning with dynamic rank distribution. In *Proceedings of the 62nd Annual Meeting of the Association for Computational Linguistics (Volume 1: Long Papers)*, pages 11662–11675, Bangkok, Thailand. Association for Computational Linguistics.

Aaron van den Oord, Yazhe Li, and Oriol Vinyals. 2018. Representation learning with contrastive predictive coding. *arXiv preprint arXiv:1807.03748*.

Sarvesh Soni and Dina Demner-Fushman. 2025a. A dataset for addressing patient's information needs related to clinical course of hospitalization. *arXiv preprint*.

Sarvesh Soni and Dina Demner-Fushman. 2025b. Overview of the archehr-qa 2025 shared task on grounded question answering from electronic health records. In *The 24th Workshop on Biomedical Natural Language Processing and BioNLP Shared Tasks*, Vienna, Austria. Association for Computational Linguistics.

Tongzhou Wang and Phillip Isola. 2020. Understanding contrastive representation learning through alignment and uniformity on the hypersphere. In *International Conference on Machine Learning (ICML)*, pages 9929–9939. PMLR.

A DoRA Fine-Tuning Configuration

```
config = LoraConfig(
    target_modules = ["value", "query"],
    use_dora=True,
    r=16,
    lora_alpha=32,
    lora_dropout=0.01,
```

```
    bias="none",
)
```

B Generation Parameters

The following parameters were used during answer generation with Mistral-7B-Instruct-v0.2:

```
generation_kwargs = {
   "pad_token_id": tokenizer.eos_token_id,
    "max_new_tokens": 512,
    "temperature": 0.2,
    "top_p": 0.95,
    "do_sample": True,
}
```

C Prompt Template

```
You are a medical assistant tasked with answering patient questions using provided
case information.
Rules:
- After every factual claim, cite the supporting sentence(s) in the format |id| or
|id1, id2|.
- Group citations if multiple sentences support the same claim (e.g., |1,2|).
- Do not create a 'References' section.
- Limit the answer to 75 words or fewer.
- Only use the provided sentences; do not hallucinate facts.
- Write clearly, medically accurately, and concisely.

Example:
Evidence:
1. The patient has alcoholic cirrhosis.
2. He has advanced hepatic encephalopathy.
3. His renal function is deteriorating.

Question:
What is the patient's prognosis?

Answer:
The patient's prognosis is poor due to alcoholic cirrhosis |1|, advanced hepatic
encephalopathy |2|, and worsening renal function |3|.

Patient Question:
{QUESTION}

Relevant Information:
- {Sentence text} |{sentence_id}|
...

Supplementary Information (less directly relevant but possibly helpful):
- {Sentence text} |{sentence_id}|

Now, based on the evidence, write your answer:
```

DKITNLP at ArchEHR-QA 2025: A Retrieval Augmented LLM Pipeline for Evidence-Based Patient Question Answering

Provia Kadusabe **Abhishek Kaushik** **Fiona Lawless**
Dundalk Institute of Technology
Regulated Software Research Centre
provia.kadusabe@dkit.ie abhishek.kaushik@dkit.ie Fiona.Lawless@dkit.ie

Abstract

This paper describes our submission for the BioNLP ACL 2025 Shared task on grounded Question Answering (QA) from Electronic Health Records (EHRs). The task aims to automatically generate answers to patients' health related questions that are grounded in the evidence from their clinical notes. We propose a two stage retrieval pipeline to identify relevant sentences to guide response generation by a Large Language Model (LLM). Specifically, our approach uses a BioBERT based bi-encoder for initial retrieval, followed by a re-ranking step using a fine-tuned cross-encoder to enhance retrieval precision. The final set of selected sentences serve as an input to Mistral 7B model which generates answers through few-shot prompting. Our approach achieves an overall score of 31.6 on the test set, outperforming a substantially larger baseline model LLaMA 3.3 70B (30.7), which demonstrates the effectiveness of retrieval-augmented generation for grounded QA.

1 Introduction

The widespread adoption of patient portals and digital health platforms has led to a growing volume of patient messages directed to healthcare providers (Martinez et al., 2024; Sieck et al., 2017). Responding to these messages in a timely, accurate, and personalized manner presents a challenge for healthcare providers often contributing to burnout (Stillman, 2023; Shanafelt et al., 2017). The ArchEHR-QA 2025 task aims to develop automated responses to patient messages that are grounded in clinical evidence from their Electronic Health Records (EHRs) (Soni and Demner-Fushman, 2025b).

Large Language Models (LLMs) have recently shown exceptional performance on general domain QA benchmarks (Singhal et al., 2025; Wang et al., 2024). However, directly applying LLMs to clinical EHR-based QA often results in models hallucinating or generating irrelevant details especially if prompted without proper grounding (Jeong et al., 2024; Elgedawy et al., 2024). The key challenge LLMs face is identifying the relevant evidence from patients' lengthy EHRs (Ahsan et al., 2024). To address this, modern QA pipelines often utilize neural retrieval models such as bi and cross-encoders (Karpukhin et al., 2020; Nogueira and Cho, 2019).

Neural retrievers typically serve as the retrieval components in Retrieval Augmented Generation (RAG) frameworks which provide LLMs with grounded document context to mitigate hallucinations and improve factuality (Lewis et al., 2020). Despite their wide adoption in open domain QA, neural retrievers are still underexplored in clinical EHR patient specific QA. A recent review found that most current QA models rely on span extraction methods which are inherently unable to generate coherent answers (Bardhan et al., 2024).

In this work, we propose a two stage retrieval pipeline as shown in figure 1. A bi-encoder first retrieves a broad set of top-K candidate sentences, these sentences are then re-ranked by a fine-tuned cross-encoder to produce top-N sentences. The top-N sentences are ultimately used as context for the LLM response generation.

2 Background & Related work

Previous research in clinical QA has primarily focused on developing datasets that map natural language queries to structured data or extract relevant spans from EHRs (Bardhan et al., 2024). A common approach involves semi-automated template-based generation of QA pairs. For instance, emrQA utilized annotations from i2b2/n2c2 clinical shared tasks to create over 1 million question answer pairs by populating templates with entities from EHRs (Pampari et al., 2018). RxWhyQA focused on extractive QA by leveraging annotated drug–reason relations to produce multi-answer and

BioNLP 2025 Shared Tasks, pages 165–170
August 1, 2025 ©2025 Association for Computational Linguistics

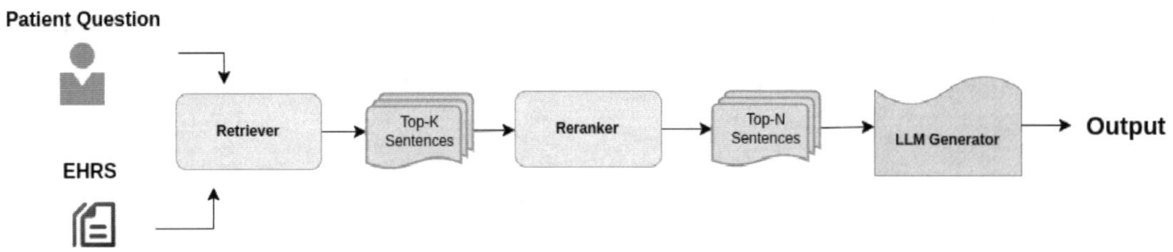

Figure 1: Our retrieval augmented pipeline for patient QA.

multi-focus questions (Moon et al., 2023). Furthermore, DrugEHRQA compiled over 70,000 medication related QA pairs from structured tables and unstructured notes, aiming to support multimodal QA systems (Bardhan et al., 2022). While these datasets have enabled development of clinical QA methods, they often rely on simple rule based or retrieval only methods that lack the capability to generate coherent and accurate answers. Although LLMs can generate coherent responses, they often struggle to extract relevant information from EHRs, which leads to irrelevant outputs (Huang et al., 2025; Maynez et al., 2020). Retrieval methods, such as RAG, have been explored to guide factual generation (Lewis et al., 2020), but existing studies mainly focus on general biomedical QA rather than patient-specific QA (Elgedawy et al., 2024; Xu et al., 2024; Chung et al., 2025; Jiang et al., 2024).

3 Methodology

In this section, we describe our proposed methodology for the task of grounded QA from EHRs.

3.1 Dataset

The dataset used in this study was provided by the organizers of the ArchEHR-QA shared task. It comprises 120 patient cases (20 development and 100 test). Each case includes a patient question, patient narrative and a clinician rewritten version of the patient question, along with the associated clinical notes with pre-annotated sentence numbers for grounding. The development set has relevance labels indicating whether each sentence is *essential*, *supplementary*, or *not-relevant* for answering the question (Soni and Demner-Fushman, 2025a).

3.2 Problem Formulation

Given a dataset $\mathcal{D}$ of patient questions and expert-annotated clinical note excerpts, the task is to classify whether a sentence $s \in \mathcal{S}$ is *essential* for answering a question $q \in \mathcal{Q}$. Each instance includes a label $y \in 0, 1$, defined as:

$$y = \begin{cases} 1 & \text{if } s \text{ is essential,} \\ 0 & \text{otherwise.} \end{cases}$$

The dataset is $\mathcal{D} = (q_i, s_i, y_i)_{i=1}^{N}$, where N is the total number of question-sentence pairs.

3.3 Model Fine-tuning

We fine-tune three BERT-based cross encoders: BERT-base (uncased) (Devlin et al., 2019), BioBERT (Lee et al., 2020) [1], and BioClinicalBERT (Alsentzer et al., 2019) [2] using the dataset described in section 3.2. For each model, the objective is to predict whether a candidate sentence s from the clinical note is *essential* to answer the patient question q.

Input Representation: Each question-sentence pair (q_i, s_i) is concatenated and tokenized as follows:

$$x_i = [\texttt{[CLS]}\ q_i\ \texttt{[SEP]}\ s_i\ \texttt{[SEP]}]$$

The resulting sequence is tokenized with a maximum length of 512 tokens and fed into the transformer encoder to produce contextualized representations:

$$h_i = \text{Transformer}(x_i)$$

The embedding corresponding to the `[CLS]` token, denoted $h_i^{\texttt{[CLS]}} \in \mathbb{R}^d$, is used as a joint representation of the question and candidate sentence.

3.3.1 Classification and Training

The joint representation is passed through a linear classification head followed by a sigmoid activation to produce a relevance score $\hat{y}_i$:

$$\hat{y}_i = \sigma(W h_i^{\texttt{[CLS]}} + b)$$

[1] `https://huggingface.co/pritamdeka/BioBERT-mnli-snli-scinli-scitail-mednli-stsb`

[2] `https://huggingface.co/emilyalsentzer/Bio_ClinicalBERT`

where $W \in \mathbb{R}^{1\times d}$ and $b \in \mathbb{R}$ are learnable parameters, and $\sigma(\cdot)$ denotes the sigmoid function. The models are optimized using binary cross-entropy loss:

$$\mathcal{L} = -\frac{1}{N}\sum_{i=1}^{N}\left[y_i \log(\hat{y}_i) + (1-y_i)\log(1-\hat{y}_i)\right]$$

3.4 Retrieval and re-ranking

3.4.1 Bi-encoder Retrieval

For initial retrieval, we adopt a bi-encoder architecture using BioBERT [3]implemented via Sentence-Transformers (Reimers and Gurevych, 2019).

Given a question q and a set of candidate sentences $\{s_j\}_{j=1}^{M}$, we first encode them independently using a bi-encoder architecture:

$$\begin{aligned} e_q &= \text{BiEncoder}(q) \\ e_{s_j} &= \text{BiEncoder}(s_j), \quad \forall j = 1,\ldots,M \end{aligned}$$

where $e_q, e_{s_j} \in \mathbb{R}^d$ are the resulting dense embeddings. Cosine similarity between the question and each candidate sentence is computed as:

$$\text{Sim}(q, s_j) = \frac{e_q \cdot e_{s_j}}{\|e_q\|\|e_{s_j}\|}$$

The top-K candidates with the highest similarity scores are selected for re-ranking:

$$\mathcal{S}_{\text{top}} = \{s_j \mid \text{rank}(\text{Sim}(q, s_j)) \le K\}$$

where rank$(\cdot)$ denotes ranking based on similarity in descending order.

3.4.2 Cross-encoder re-ranking

Each of the top-K candidates is concatenated with the question and scored for relevance using the fine-tuned cross-encoder:

$$x_j = [\texttt{[CLS]}\; q\; \texttt{[SEP]}\; s_j\; \texttt{[SEP]}]$$

$$\hat{y}_j = \sigma(W h_j^{\texttt{[CLS]}} + b)$$

where $h_j^{\texttt{[CLS]}}$ is the contextualized embedding of the input, and $\hat{y}_j \in [0,1]$ is the predicted relevance score. The top-N candidates with the highest scores are selected as evidence for generation:

$$\mathcal{S}_{\text{evidence}} = \{s_j \mid \text{rank}(\hat{y}_j) \le N\}$$

[3]https://huggingface.co/pritamdeka/BioBERT-mnli-snli-scinli-scitail-mednli-stsb

3.5 Answer Generation

For the Answer generation stage, we employ Mistral-7B-DPO [4], an instruction tuned causal language model denoted as $G(\cdot;\theta)$. This model is based on the Mistral 7B architecture (Jiang et al., 2023) and has been optimized via Direct Preference Optimization (DPO) to follow instruction and align human preferences (Rafailov et al., 2023).

Given a structured prompt P which includes the patient narrative, patient and clinician questions, and the top-N evidence sentences, the model generates free-text answers in an autoregressive manner:

$$A = G(P;\ \theta)$$

where A denotes the generated response and θ represents the pretrained model parameters. The final output consists of sentences that cite supporting evidence by including sentence identifiers inline using pipe symbols.

4 Experiments

4.1 Experimental Setup

We fine-tune the cross-encoder models on the development set using the patient question. Given the small size of the development set, we performed a fixed split over cases to separate training and validation subsets. Finetuning was conducted with a batch size of 8 for up to 10 epochs with early stopping if there is no improvement for 2 consecutive evaluations. Optimization is performed using AdamW with a weight decay of 0.01 and a learning rate of 2×10^{-5}.

For sentence retrieval, we experimented with different combinations of the number of candidates retrieved by the bi-encoder (K) and re-ranked by the cross-encoder (N). Specifically, we evaluated $(K, N) = (5, 20)$, $(7, 20)$, $(10, 25)$, $(12, 30)$, $(13, 30)$, and $(15, 35)$. The configuration $(13, 30)$ yielded the best performance and was adopted in the final retrieval pipeline.

For answer generation, we used a few-shot prompt (Brown et al., 2020) using the two examples provided in the shared task description (Soni and Demner-Fushman, 2025a). Generation was performed with a sampling temperature of 0.70, a maximum length of 200 tokens, and a target answer length of up to 75 words, as specified by the task organizers. If the model produced no output

[4]https://huggingface.co/NousResearch/Nous-Hermes-2-Mistral-7B-DPO

or generated an answer shorter than 65 words or longer than 75 words, generation was retried up to 10 times.

4.2 Evaluation

For sentence retrieval, we evaluated our models on the development set using precision, recall, and F1-score, comparing the retrieved sentences against the manually annotated ground truth. During fine-tuning, we used the same metrics on the development set to assess sentence-level classification performance. The generated responses were assessed using the official evaluation framework provided by the organizers (Soni and Demner-Fushman, 2025b), which balances two key aspects, Factuality and Relevance. Factuality was measured by calculating Precision, Recall, and F1 Scores between the cited evidence sentences in the generated answer and the manually annotated ground truth evidence sentences. Relevance, on the other hand, was assessed by comparing the generated answers to the ground truth essential note sentences and the questions using BLEU (Papineni et al., 2002), ROUGE (Lin, 2004), SARI (Xu et al., 2016), BERTScore (Zhang et al., 2019), AlignScore (Zha et al., 2023), and MEDCON (Yim et al., 2023). The overall score was computed as the mean of the Factuality and Relevance scores.

4.3 Experimental Results & Discussion

Experimental results on the development set show that among the fine-tuned models, shown in table 1, BioBERT achieved the best performance and was therefore selected as the cross-encoder re-ranker in the retrieval pipeline.

Model	Precision	Recall	F1-Score
BioClinicalBERT	41.49	72.46	52.77
BERT-base	46.06	80.43	58.58
BioBERT	**51.45**	**89.86**	**65.44**

Table 1: Performance of fine-tuned cross-encoders on the essential sentence prediction task in (%).

We also compare our system with using only few-shot prompting as shown in table 2.

Few-shot prompting achieved a slightly higher overall factuality score (47.90 vs. 45.45), however, our system outperformed it in overall relevance (35.71 vs. 31.08) and overall score (40.58 vs. 39.49). Based on these results, we selected the RAG system for testing.

Metric	RAG	Few-Shot Only
Overall Factuality Score	45.45	47.90
Overall Relevance Score	35.71	31.08
Overall Score	**40.58**	39.49

Table 2: Comparison of our system (RAG) with few-shot prompting only (no retrieval). Both methods use the Mistral 7B model.

Metric	RAG	Baseline
Overall Factuality Score	32.70	33.60
Overall Relevance Score	30.50	27.80
Overall Score	**31.6**	30.70

Table 3: Performance of our system (RAG) on the test set.

Evaluation on the test set in table 3 showed that our system achieved an overall relevance score of 30.50, outperforming the baseline score of 27.80. This suggests that our system's generated answers were more aligned to the ground-truth essential note sentences. However, it slightly underperformed in the overall factuality with a score of 32.70 compared to the baseline score of 33.60. Despite this, our system achieved a higher overall score of 31.6, surpassing the baseline score of 30.7, which was based on LLaMA 3.3 70B. While our model (Mistral 7B parameters) is significantly smaller than the LLaMa 70B model used in the baseline system, it still delivers competitive results which shows the effectiveness of retrieval augmented generation for grounded clinical question answering.

5 Conclusion & Future Work

In this work, we introduced our approach for the grounded patient QA task using EHRs. Our method uses a two stage retrieval pipeline using a BioBERT based bi-encorder for initial relevant sentence retrieval and a fine-tuned cross-encoder for re-ranking to identify the most relevant sentences for LLM (Mistral 7B) generation. Experimental results show that our proposed approach improves performance over the baseline in terms of overall score (31.6 versus 30.70).

Future work should investigate alternative model architectures and evaluate the performance of smaller LLMs on larger datasets.

6 Limitation

Our study was constrained by several factors. First, the development set used for fine-tuning was relatively small thus using a larger dataset could yield better performance. Second, our fine-tuning experiments utilized smaller pretrained language models due to resource constraints, exploring larger LLMs could further improve performance.

7 Acknowledgments

This research was funded through the CREATE-DkIT project, supported by the HEA TU-Rise program and co-financed by the Government of Ireland and the European Union through the Southern, Eastern & Midland Regional Program of the ERDF 2021-27 and the Northern & Western Regional Programme 2021–27.

References

Hiba Ahsan, Denis Jered McInerney, Jisoo Kim, Christopher Potter, Geoffrey Young, Silvio Amir, and Byron C Wallace. 2024. Retrieving evidence from ehrs with llms: possibilities and challenges. *Proceedings of machine learning research*, 248:489.

Emily Alsentzer, John R Murphy, Willie Boag, Wei-Hung Weng, Di Jin, Tristan Naumann, and Matthew McDermott. 2019. Publicly available clinical bert embeddings. *arXiv preprint arXiv:1904.03323*.

Jayetri Bardhan, Anthony Colas, Kirk Roberts, and Daisy Zhe Wang. 2022. Drugehrqa: A question answering dataset on structured and unstructured electronic health records for medicine related queries. *arXiv preprint arXiv:2205.01290*.

Jayetri Bardhan, Kirk Roberts, and Daisy Zhe Wang. 2024. Question answering for electronic health records: Scoping review of datasets and models. *Journal of Medical Internet Research*, 26:e53636.

Tom Brown, Benjamin Mann, Nick Ryder, Melanie Subbiah, Jared D Kaplan, Prafulla Dhariwal, Arvind Neelakantan, Pranav Shyam, Girish Sastry, Amanda Askell, and 1 others. 2020. Language models are few-shot learners. *Advances in neural information processing systems*, 33:1877–1901.

Philip Chung, Akshay Swaminathan, Alex J Goodell, Yeasul Kim, S Momsen Reincke, Lichy Han, Ben Deverett, Mohammad Amin Sadeghi, Abdel-Badih Ariss, Marc Ghanem, and 1 others. 2025. Verifact: Verifying facts in llm-generated clinical text with electronic health records. *arXiv preprint arXiv:2501.16672*.

Jacob Devlin, Ming-Wei Chang, Kenton Lee, and Kristina Toutanova. 2019. Bert: Pre-training of deep bidirectional transformers for language understanding. In *Proceedings of the 2019 conference of the North American chapter of the association for computational linguistics: human language technologies, volume 1 (long and short papers)*, pages 4171–4186.

Ran Elgedawy, Ioana Danciu, Maria Mahbub, and Sudarshan Srinivasan. 2024. Dynamic q&a of clinical documents with large language models. *arXiv preprint arXiv:2401.10733*.

Lei Huang, Weijiang Yu, Weitao Ma, Weihong Zhong, Zhangyin Feng, Haotian Wang, Qianglong Chen, Weihua Peng, Xiaocheng Feng, Bing Qin, and 1 others. 2025. A survey on hallucination in large language models: Principles, taxonomy, challenges, and open questions. *ACM Transactions on Information Systems*, 43(2):1–55.

Minbyul Jeong, Jiwoong Sohn, Mujeen Sung, and Jaewoo Kang. 2024. Improving medical reasoning through retrieval and self-reflection with retrieval-augmented large language models. *Bioinformatics*, 40(Supplement_1):i119–i129.

Albert Q. Jiang, Alexandre Sablayrolles, Arthur Mensch, Chris Bamford, Devendra Singh Chaplot, Diego de las Casas, Florian Bressand, Gianna Lengyel, Guillaume Lample, Lucile Saulnier, Lélio Renard Lavaud, Marie-Anne Lachaux, Pierre Stock, Teven Le Scao, Thibaut Lavril, Thomas Wang, Timothée Lacroix, and William El Sayed. 2023. Mistral 7b. *Preprint*, arXiv:2310.06825.

Emily Jiang, Alice Chen, Irene Tenison, and Lalana Kagal. 2024. Medirag: Secure question answering for healthcare data. In *2024 IEEE International Conference on Big Data (BigData)*, pages 6476–6485. IEEE.

Vladimir Karpukhin, Barlas Oguz, Sewon Min, Patrick SH Lewis, Ledell Wu, Sergey Edunov, Danqi Chen, and Wen-tau Yih. 2020. Dense passage retrieval for open-domain question answering. In *EMNLP (1)*, pages 6769–6781.

Jinhyuk Lee, Wonjin Yoon, Sungdong Kim, Donghyeon Kim, Sunkyu Kim, Chan Ho So, and Jaewoo Kang. 2020. Biobert: a pre-trained biomedical language representation model for biomedical text mining. *Bioinformatics*, 36(4):1234–1240.

Patrick Lewis, Ethan Perez, Aleksandra Piktus, Fabio Petroni, Vladimir Karpukhin, Naman Goyal, Heinrich Küttler, Mike Lewis, Wen-tau Yih, Tim Rocktäschel, and 1 others. 2020. Retrieval-augmented generation for knowledge-intensive nlp tasks. *Advances in neural information processing systems*, 33:9459–9474.

Chin-Yew Lin. 2004. Rouge: A package for automatic evaluation of summaries. In *Text summarization branches out*, pages 74–81.

Kathryn A Martinez, Rebecca Schulte, Michael B Rothberg, Maria Charmaine Tang, and Elizabeth R Pfoh. 2024. Patient portal message volume and time spent on the ehr: an observational study of primary care clinicians. *Journal of General Internal Medicine*, 39(4):566–572.

Joshua Maynez, Shashi Narayan, Bernd Bohnet, and Ryan McDonald. 2020. On faithfulness and factuality in abstractive summarization. *arXiv preprint arXiv:2005.00661*.

Sungrim Moon, Huan He, Heling Jia, Hongfang Liu, Jungwei Wilfred Fan, and 1 others. 2023. Extractive clinical question-answering with multianswer and multifocus questions: data set development and evaluation study. *JMIR AI*, 2(1):e41818.

Rodrigo Nogueira and Kyunghyun Cho. 2019. Passage re-ranking with bert. *arXiv preprint arXiv:1901.04085*.

Anusri Pampari, Preethi Raghavan, Jennifer Liang, and Jian Peng. 2018. emrQA: A large corpus for question answering on electronic medical records. In *Proceedings of the 2018 Conference on Empirical Methods in Natural Language Processing*, pages 2357–2368, Brussels, Belgium. Association for Computational Linguistics.

Kishore Papineni, Salim Roukos, Todd Ward, and Wei-Jing Zhu. 2002. Bleu: a method for automatic evaluation of machine translation. In *Proceedings of the 40th annual meeting of the Association for Computational Linguistics*, pages 311–318.

Rafael Rafailov, Archit Sharma, Eric Mitchell, Christopher D Manning, Stefano Ermon, and Chelsea Finn. 2023. Direct preference optimization: Your language model is secretly a reward model. *Advances in Neural Information Processing Systems*, 36:53728–53741.

Nils Reimers and Iryna Gurevych. 2019. Sentence-bert: Sentence embeddings using siamese bert-networks. *arXiv preprint arXiv:1908.10084*.

Tait D Shanafelt, Lotte N Dyrbye, and Colin P West. 2017. Addressing physician burnout: the way forward. *Jama*, 317(9):901–902.

Cynthia J Sieck, Jennifer L Hefner, Jeanette Schnierle, Hannah Florian, Aradhna Agarwal, Kristen Rundell, and Ann Scheck McAlearney. 2017. The rules of engagement: perspectives on secure messaging from experienced ambulatory patient portal users. *JMIR medical informatics*, 5(3):e7516.

Karan Singhal, Tao Tu, Juraj Gottweis, Rory Sayres, Ellery Wulczyn, Mohamed Amin, Le Hou, Kevin Clark, Stephen R Pfohl, Heather Cole-Lewis, and 1 others. 2025. Toward expert-level medical question answering with large language models. *Nature Medicine*, pages 1–8.

Sarvesh Soni and Dina Demner-Fushman. 2025a. A dataset for addressing patient's information needs related to clinical course of hospitalization. *arXiv preprint*.

Sarvesh Soni and Dina Demner-Fushman. 2025b. Overview of the archehr-qa 2025 shared task on grounded question answering from electronic health records. In *The 24th Workshop on Biomedical Natural Language Processing and BioNLP Shared Tasks*, Vienna, Austria. Association for Computational Linguistics.

Michael Stillman. 2023. Death by patient portal. *JAMA*, 330(3):223–224.

Yubo Wang, Xueguang Ma, Ge Zhang, Yuansheng Ni, Abhranil Chandra, Shiguang Guo, Weiming Ren, Aaran Arulraj, Xuan He, Ziyan Jiang, and 1 others. 2024. Mmlu-pro: A more robust and challenging multi-task language understanding benchmark. In *The Thirty-eight Conference on Neural Information Processing Systems Datasets and Benchmarks Track*.

Ran Xu, Wenqi Shi, Yue Yu, Yuchen Zhuang, Bowen Jin, May D Wang, Joyce C Ho, and Carl Yang. 2024. Ram-ehr: Retrieval augmentation meets clinical predictions on electronic health records. *arXiv preprint arXiv:2403.00815*.

Wei Xu, Courtney Napoles, Ellie Pavlick, Quanze Chen, and Chris Callison-Burch. 2016. Optimizing statistical machine translation for text simplification. *Transactions of the Association for Computational Linguistics*, 4:401–415.

Wen-wai Yim, Yujuan Fu, Asma Ben Abacha, Neal Snider, Thomas Lin, and Meliha Yetisgen. 2023. Aci-bench: a novel ambient clinical intelligence dataset for benchmarking automatic visit note generation. *Scientific data*, 10(1):586.

Yuheng Zha, Yichi Yang, Ruichen Li, and Zhiting Hu. 2023. Alignscore: Evaluating factual consistency with a unified alignment function. *arXiv preprint arXiv:2305.16739*.

Tianyi Zhang, Varsha Kishore, Felix Wu, Kilian Q Weinberger, and Yoav Artzi. 2019. Bertscore: Evaluating text generation with bert. *arXiv preprint arXiv:1904.09675*.

AEHRC at BioLaySumm 2025: Leveraging T5 for Lay Summarisation of Radiology Reports

Wenjun Zhang[⋆†] **Shekhar S. Chandra**[⋆] **Bevan Koopman**[†]
Jason Dowling[†] **Aaron Nicolson**[†]
[⋆]The University of Queensland, Brisbane, Australia
[†]Australian e-Health Research Centre, CSIRO Health and Biosecurity, Brisbane, Australia
wenjun.zhang@uq.edu.au

Abstract

Biomedical texts, such as research articles and clinical reports, are often written in highly technical language, making them difficult for patients and the general public to understand. The **BioLaySumm 2025 Shared Task** addresses this challenge by promoting the development of models that generate lay summarisations of biomedical content. This paper focuses on Subtask 2.1: Radiology Report Generation with Layman's Terms. In this work, we evaluate two large language model (LLM) architectures, T5-large (700M parameter encoder–decoder model) and LLaMA-3.2-3B (3B parameter decoder-only model). Both models are trained under fully-supervised conditions using the task's multi-source dataset. Our results show that T5-large consistently outperforms LLaMA-3.2-3B across nine out of ten metrics, including relevance, readability, and clinical accuracy, despite having only a quarter of the parameters. **Our T5-based model achieved the top rank in both the open-source and close-source tracks of the subtask 2.1.**

1 Introduction

Biomedical texts, ranging from research articles to clinical reports, are often written in highly technical language. This presents a major barrier for patients and the general public, limiting their ability to understand health information and make informed decisions. As a response, the field of biomedical lay summarisation has emerged to simplify expert-level content into language that non-experts can understand (Guo et al., 2021). A particular application of this is the translation of radiology reports. A recent analysis found that only about 4% (Martin-Carreras et al., 2019) of radiology reports are written below an 8th-grade reading level (the average adult reading level). Without clear explanations, the information can be confusing or even misleading, creating barriers to understanding one's own health.

To advance research in this area, the BioLaySumm 2025 Shared Task hosted at the BioNLP Workshop at ACL 2025, introduces two primary tasks aimed at improving the accessibility of biomedical information for non-expert audiences (Xiao et al., 2025). Task 1 focuses on the lay summarisation of biomedical research articles, and Task 2 centers on lay summarisation of radiology reports. Participants are provided with a dataset containing paired radiology reports and lay summarisations from sources such as Open-i (Demner-Fushman et al., 2016), PadChest (Bustos et al., 2020), BIMCV-COVID19 (de la Iglesia Vayá et al., 2020), and MIMIC-CXR (Johnson et al., 2019). The shared task includes two evaluation tracks. The open-source track uses test data from three public datasets (Open-i, PadChest, and BIMCV-COVID19), while the close-source track adds the additional MIMIC-CXR dataset. In this work, we trained our models exclusively on the open-source datasets and submitted predictions for both the open-source and close-source tracks.

While language models have become central to biomedical lay summarisation (Xie et al., 2023), there remains no clear consensus on whether encoder-decoder or decoder-only architectures are better suited to this task. Prior studies have shown both to be effective in different contexts, but few have directly compared them under controlled, fully-supervised conditions. Motivated by this, we conduct a comparison between an encoder–decoder model T5-large (Raffel et al., 2020) and a decoder-only LLaMA-3.2-3B model (Meta AI, 2024). Our final submission, based on T5-large, achieved the highest overall score in Subtask 2.1, ranking first among all participating teams. It consistently outperforms LLaMA-3.2-3B across nine of ten evaluation metrics, despite having only a quarter of the parameters.

BioNLP 2025 Shared Tasks, pages 171–178
August 1, 2025

2 Related Work

2.1 Biomedical Lay Summarisation

Biomedical lay summarisation is a sequence-to-sequence task that translates specialised medical language into accessible explanations for non-experts, such as patients (Xie et al., 2023). While biomedical summarisation has traditionally focused on preserving factual accuracy and completeness, much less attention has been given to simplifying language for broader public understanding. In response to this need, the BioLaySumm Shared Task series (2023–present) was introduced to encourage the development of models that generate lay summaries from biomedical content (Goldsack et al., 2023, 2024; Xiao et al., 2025).

Over time, the field of biomedical summarisation has evolved from rule-based and statistical methods to neural architectures and, more recently to language models (Xie et al., 2023). A similar trend is evident in the BioLaySumm Shared Tasks: in 2023, only 3 teams used LLMs, while in 2024, 18 teams used decoder-only LLMs (BioGPT (Luo et al., 2022), BioMistral (Labrak et al., 2024), GPT-4 (OpenAI, 2023), LLaMA (Touvron et al., 2023)) and 13 teams using encoder-decoder LLMs (T5 (Raffel et al., 2020), FLAN-T5 (Chung et al., 2024)), reflecting a growing interest in applying both architectures to the task of lay summarisation (Goldsack et al., 2024).

2.2 Encoder-Decoder and Decoder-only Language Models

Both encoder–decoder and decoder-only LLMs are based on the Transformer architecture (Vaswani et al., 2017), which was originally introduced for sequence-to-sequence tasks such as machine translation. The Transformer consists of two components: an encoder that processes the input into a latent space, and a decoder that generates the output based on that representation and the previously generated tokens. Encoder–decoder models, such as T5 (Raffel et al., 2020) and BART (Lewis et al., 2020), leverage both components to perform a wide range of text-to-text tasks, including translation, summarization, and question answering. In contrast, decoder-only models, such as the GPT series (OpenAI, 2022, 2023) and LLaMA (Touvron et al., 2023), comprise only the decoder component. These models are trained using an autoregressive objective, predicting the next token based solely on preceding tokens in a unidirectional manner. In the context of biomedical summarisation, both encoder–decoder and decoder-only models remain active areas of research, with no definitive answer as to which performs better (Goldsack et al., 2024). In this work, we will compare an encoder-decoder model T5 to a decoder-only model LLaMA on the task of lay summarisation of radiology reports.

2.3 Lay Summarisation of Radiology Reports

Due to the lack of alignment data, initial studies investigated the feasibility of direct prompting LLMs such as ChatGPT (OpenAI, 2022) and GPT-4 (OpenAI, 2023), to simplify radiology reports. The generated contents are reviewed by radiologists to measure accuracy, completeness, and simplicity. For instance, lyu (2023) showed that while ChatGPT could generate understandable plain-language summaries, it often over-simplified or introduced inaccuracies. Similarly, Kuckelman et al. (2024) found that ChatGPT's summaries of musculoskeletal radiology reports were generally accurate but sometimes lacked completeness. In 2024, Zhao et al. (2025) introduced the LaymanRRG framework, utilised GPT-4o to translate professional radiology reports from MIMIC-CXR into layman's terms. BioLaySumm Shared Task 2 (Xiao et al., 2025) built on this dataset to establish a standardised benchmark for evaluating lay summarisation systems across four datasets Open-i, PadChest, and BIMCV-COVID19 and MIMIC-CXR (Demner-Fushman et al., 2016; Bustos et al., 2020; de la Iglesia Vayá et al., 2020; Johnson et al., 2019).

3 Methodology

Our primary goal was to develop a model for Subtask 2.1 of the BioLaySumm 2025 Shared Task, which involved generating lay summarisations from radiology reports. As part of this effort, we fine-tuned T5-large (Raffel et al., 2020) and LLaMA-3.2-3B (Meta AI, 2024) on the shared task dataset. The model with the highest validation BERTScore-F1 (Zhang et al., 2020) was submitted for testing on both the open-source track and the close-source track, described in Section 3.2.

3.1 Model Selection

We selected two prominent open-source LLM architectures:

- **T5-large (Raffel et al., 2020)**: An encoder–decoder model pretrained on Colossal Clean Crawled Corpus (C4) dataset (comprises 156 billion tokens) (Dodge et al., 2021).

- **LLaMA-3.2-3B (Meta AI, 2024)**: A decoder-only model pretrained on up to 9 trillion tokens from publicly available sources, including Common Crawl, Wikipedia, and other web-based corpora.

3.2 Datasets

Task 2.1 includes two evaluation tracks. The open-source track evaluates models on test sets from three publicly available datasets: Open-i, PadChest, and BIMCV-COVID19 (Demner-Fushman et al., 2016; Bustos et al., 2020; de la Iglesia Vayá et al., 2020). The close-source track adds an additional dataset, namely MIMIC-CXR (Johnson et al., 2019). We trained our models exclusively on the three open datasets, and submitted the same model to both the open-source track and close-source track. This setup enables us to evaluate in-domain performance (on seen datasets) as well as generalisation to unseen data (MIMIC-CXR). Table 1 summarises the dataset composition across tracks.

Table 1: Composition of datasets for Subtask 2.1. We train on the three open-source track datasets (top). The MIMIC-CXR dataset (bottom) is used exclusively for the close-source track and is not included in our training or validation.

Dataset	Train	Val	Test
Open-source track datasets (training)			
Open-i	2,243	134	186
PadChest	116,847	7,824	7,130
BIMCV-COVID19	31,364	2,042	3,221
Open-source total	**150,454**	**10,000**	**10,537**
Close-source track addition (evaluation only)			
MIMIC-CXR	45,000*	5,000*	500
Close-source test total	–	–	**11,037**

* Provided for reference only; not used in training or validation.

3.3 Data Processing and Tokenisation

For training T5, input and output sequences were tokenised separately using a SentencePiece tokeniser. The input (e.g., a radiology report) was tokenised without an end-of-sequence token, while the output (e.g., a lay summarisation) includes the `<eos>` token at the end. For example, the input "The chest X-ray shows pneumonia." was tokenised as `['_The', '_chest', ..., '_pneumonia.']`, and the output "There is an infection in the lungs." was tokenised as `['_There', '_is', ..., '_lungs.', '<eos>']`. In contrast, LLaMA-3.2-3B used a byte-level BPE tokeniser and expected the input and output to be concatenated into a single sequence for decoder-only training. We used a special token <s> to separate the lay summarisation from the radiology report. Specifically, the input before tokenisation was "The chest X-ray shows pneumonia.<s>There is an infection in the lungs.</s>", where </s> is the end-of-sentence token.

3.4 Training

We trained both models for six epochs using PyTorch Distributed Data Parallel (DDP) on a single H100 compute node with four GPUs. Each GPU processed a local batch size of 2, resulting in a global batch size of 8. We used the AdamW optimiser (Loshchilov and Hutter, 2019) with a learning rate of 2e-5 and a constant learning rate schedule with 500 warmup steps. Input sequences were truncated or padded to a maximum length of 2048 tokens, and during inference, the models were allowed to generate up to 300 tokens. BERTScore-F1 (Zhang et al., 2020) was used as the primary metric for model selection.

3.5 Metrics

The evaluation is based on three key categories defined by the BioLaySumm 2025 Shared Task: relevance, readability, and clinical accuracy.

Relevance measures how well the generated summaries align with the reference texts. ROUGE-1/2/L (Lin, 2004) are computed for lexical overlap; BLEU (Papineni et al., 2002) for 1-to-4-gram precision with brevity penalty; METEOR (Banerjee and Lavie, 2005) for word-level matches and ordering; BERTScore (Zhang et al., 2020) for embedding-based semantic similarity. Semantic similarity between radiology report and lay summarisations is also measure.

Readability evaluates how easy the generated text is to understand for a general audience. The Flesch–Kincaid Grade Level (FKGL) (Kincaid et al., 1975), Dale–Chall Readability Score (DCRS) (Dale and Chall, 1948), and the Coleman–Liau Index (CLI) (Coleman and Liau, 1975) are reported; lower values indicate simpler language.

Clinical accuracy assesses the factual correctness of medical content. This is captured by

Table 2: Performance of LLaMA-3.2-3B and T5-large across relevance, readability, and clinical-accuracy metrics (open-source track). Boldface indicates best score.

Model	ROUGE	BLEU	MET.	BERT.	FKGL†	DCRS†	CLI†	Sim.	RadF1	ChexF1
LLaMA-3.2-3B	64.20	42.90	67.70	94.90	7.46	9.38	**8.04**	87.90	37.20	85.00
T5-large-770M	**66.9**	**45.6**	**70.2**	**95.3**	**7.43**	**9.34**	8.11	**88.9**	**40.0**	**85.6**

† Lower values indicate better readability.

CheXbert-F1 (Smit et al., 2020), which gauges agreement on clinical findings, and RadGraph-F1 (Jain et al., 2021), which measures overlap of medical entities and their relations.

4 Results and Discussion

4.1 Comparing Encoder–Decoder Models and Decoder-only Models Results

We present our submission results for the open-source track in Table 2, comparing our fine-tuned models: T5-large and LLaMA-3.2-3B. T5-large outperforms LLaMA-3.2-3B across 9 out of 10 metrics, including all relevance (ROUGE, BLEU, METEOR, BERTScore, and semantic similarity) and clinical accuracy metrics (RadGraph-F1 and CheXbert-F1). Readability scores are similar across models, but T5 achieves lower Dale–Chall and FKGL scores, indicating slightly simpler vocabulary.

Notably, T5-large achieves these results with only one-quarter the parameters of LLaMA-3.2-3B, suggesting that encoder–decoder architectures may be more efficient for the supervised lay summarization task. Several factors may explain this performance difference. First, the encoder–decoder structure of T5 is inductively biased toward input–output transformation tasks such as summarisation and paraphrasing, potentially making it more effective in low-resource, domain-specific settings. Second, the larger capacity of LLaMA-3.2-3B may require more training data to fully optimise, and in limited-data scenarios, its performance could suffer due to underfitting or instability. Indeed, we observed more fluctuation in validation scores during LLaMA training, whereas T5's training curve was smoother and more consistent.

These quantitative results are further supported by qualitative case studies in Appendix A and B, where T5's outputs consistently demonstrate higher fidelity to the original clinical findings while offering clearer and more accessible lay-language paraphrasing. For example, in Case 1, T5 translates technical terms like "atelectasis and consolidation" into "lung collapse and solid areas," while LLaMA retains the original jargon, making the output less readable for general audiences. In Case 2, T5 maintains more specific references such as "heart problems related to the coronary arteries" and "chronic obstructive pulmonary disease (COPD)," whereas LLaMA tends to generalise or simplify more loosely.

Table 3: Performance of T5-large on the open-source track evaluation and close-source track evaluation. Boldface indicates best score.

Metric	Open	Close
ROUGE-L	**66.90**	58.66
BLEU	**45.61**	32.08
METEOR	**70.17**	62.68
BERTScore	**95.30**	94.33
FKGL†	**7.43**	7.65
DCRS†	9.34	**8.60**
CLI†	8.11	**7.88**
Semantic Score	88.88	**89.52**
RadGraph-F1	**39.96**	34.81
CheXbert-F1	**85.64**	68.20

4.2 Comparing Open-source Track and Close-source Track Submission Results

We submitted our model to both the open-source track and close-source track for evaluation, with results shown in Table 3. The close-source track evaluates model performance on a broader test set than the open-source track. Specifically, the open-source track includes test samples from three public datasets—Open-i, PadChest, and BIMCV-COVID19—while the close-source track adds MIMIC-CXR.

Our model was trained exclusively on the open-source track training set, which means that its performance on the close-source track reflects both in-domain evaluation (on seen sources) and out-of-domain generalisation (on unseen MIMIC-CXR examples). We can observe overall trends: compared to the open-source track results, the close-source track metrics drop across nearly all categories, especially BLEU, ROUGE-L, and CheXbert-F1. However, because the closed test set is a mixture of all four datasets and aggregate scores are reported, we cannot isolate exact performance on MIMIC-CXR. Also, due to the absence of comparison with T5-large trained on the closed-source track, the relative performance drop remains unknown. Therefore, we cannot fully conclude on the model's generalisation to unseen data.

Nevertheless, our model achieved the top rank in both the open-source and close-source tracks, according to the competition leaderboard, outperforming all other submissions, including those trained on the full close-source dataset.

4.3 Conclusion

This study investigated the task of translating radiology reports into lay summarisation as part of Subtask 2.1 of the BioLaySumm 2025 Shared Task. We compare the performance of two large language model (LLM) architectures: the decoder-only LLaMA-3.2-3B and the encoder–decoder T5-large. Our results demonstrate that the T5-large model consistently outperforms LLaMA-3.2-3B across a broad range of metrics, including relevance, readability, and clinical accuracy, despite having significantly fewer parameters. This finding suggests that encoder–decoder models may be more effective and parameter-efficient for fully-supervised summarisation tasks, such as lay summarisation of radiology reports.

References

2023. Translating radiology reports into plain language using chatgpt and gpt-4 with prompt learning: results, limitations, and potential. *Visual Computing for Industry, Biomedicine, and Art*, 6(1).

Satanjeev Banerjee and Alon Lavie. 2005. METEOR: An automatic metric for MT evaluation with improved correlation with human judgments. In *ACL*, pages 65–72.

Aurelia Bustos, Antonio Pertusa, Jose-Maria Salinas, and María de la Iglesia-Vayá. 2020. Padchest: A large chest x-ray image dataset with multi-label annotated reports. *Medical Image Analysis*, 66:101797.

Hyung Won Chung, Le Hou, Shayne Longpre, Jason Wei, and 1 others. 2024. Scaling instruction-finetuned language models. *Journal of Machine Learning Research*, 25:1–53.

Meri Coleman and T. L. Liau. 1975. A computer readability formula designed for machine scoring. *Journal of Applied Psychology*, 60(2):283–284.

Edgar Dale and Jeanne S. Chall. 1948. A formula for predicting readability. *Educational Research Bulletin*, 27(1):11–20, 28.

María de la Iglesia Vayá, Jose Manuel Saborit, Joaquim Angel Montell, Antonio Pertusa, Aurelia Bustos, Miguel Cazorla, Joaquin Galant, Xavier Barber, Domingo Orozco-Beltrán, Francisco García-García, Marisa Caparrós, Germán González, and Jose María Salinas. 2020. BIMCV COVID-19+: A large annotated dataset of rx and ct images from covid-19 patients. *arXiv preprint arXiv:2006.01174*.

Dina Demner-Fushman, Marc D Kohli, Marc B Rosenman, Sonya E Shooshan, Laritza Rodriguez, Sameer Antani, George R Thoma, and Clement J McDonald. 2016. Preparing a collection of radiology examinations for distribution and retrieval. *Journal of the American Medical Informatics Association*, 23(2).

Jesse Dodge, Maarten Sap, Ana Marasović, William Agnew, Gabriel Ilharco, Dirk Groeneveld, Margaret Mitchell, and Matt Gardner. 2021. Documenting large webtext corpora: A case study on the colossal clean crawled corpus. In *In EMNLP*, pages 1286–1305.

Tomas Goldsack, Zheheng Luo, Qianqian Xie, Carolina Scarton, Matthew Shardlow, Sophia Ananiadou, and Chenghua Lin. 2023. Overview of the biolaysumm 2023 shared task on lay summarization of biomedical research articles. In *The 22nd Workshop on Biomedical Natural Language Processing and BioNLP Shared Tasks*, pages 468–477, Toronto, Canada. Association for Computational Linguistics.

Tomas Goldsack, Carolina Scarton, Matthew Shardlow, and Chenghua Lin. 2024. Overview of the BioLaySumm 2024 shared task on the lay summarization of biomedical research articles. In *Proceedings of the 23rd Workshop on Biomedical Natural Language Processing*, pages 122–131, Bangkok, Thailand. Association for Computational Linguistics.

Yue Guo, Wei Qiu, Yizhong Wang, and Trevor Cohen. 2021. Automated lay language summarization of biomedical scientific reviews. In *AAAI*, pages 160–168.

Saahil Jain, Ashwin Agrawal, Adriel Saporta, Steven Q. H. Truong, Du Nguyen Duong, Tan Bui, Pierre J. Chambon, Yuhao Zhang, Matthew P. Lungren, Andrew Y. Ng, Curtis P. Langlotz, and Pranav Rajpurkar. 2021. Radgraph: Extracting clinical entities and relations from radiology reports. In *NeurIPS*.

Alistair E. W. Johnson, Tom J. Pollard, Seth J. Berkowitz, Nathaniel R. Greenbaum, Matthew P. Lungren, Chih-ying Deng, Roger G. Mark, and Steven Horng. 2019. MIMIC-CXR: A de-identified publicly available database of chest radiographs with free-text reports. *Scientific Data*, 6:317.

J. Peter Kincaid, Robert P. Fishburne, Richard L. Rogers, and Brad S. Chissom. 1975. Derivation of new readability formulas (automated readability index, fog count and flesch reading ease formula) for navy enlisted personnel. Research Branch Report 8-75.

Ian J. Kuckelman, Karla Wetley, Paul Hyunsoo Yi, and Andrew Bailey Ross. 2024. Translating musculoskeletal radiology reports into patient-friendly summaries using ChatGPT-4. *Skeletal Radiology*, 53:1621–1624.

Yanis Labrak, Adrien Bazoge, Emmanuel Morin, Pierre-Antoine Gourraud, Mickael Rouvier, and Richard Dufour. 2024. Biomistral: A collection of open-source pre-trained large language models for medical domains. In *ACL*.

Mike Lewis, Yinhan Liu, Naman Goyal, Marjan Ghazvininejad, Abdelrahman Mohamed, Omer Levy, Veselin Stoyanov, and Luke Zettlemoyer. 2020. BART: Denoising sequence-to-sequence pre-training for natural language generation, translation, and comprehension. In *ACL*, pages 7871–7880.

Chin-Yew Lin. 2004. ROUGE: A package for automatic evaluation of summaries. In *Text Summarization Branches Out*, pages 74–81, Barcelona, Spain. Association for Computational Linguistics.

Ilya Loshchilov and Frank Hutter. 2019. Decoupled weight decay regularization. In *ICLR*.

Renqian Luo, Liai Sun, Yingce Xia, Tao Qin, Sheng Zhang, Hoifung Poon, and Tie-Yan Liu. 2022. Biogpt: Generative pre-trained transformer for biomedical text generation and mining. *Briefings in Bioinformatics*, 23(6):bbac409.

Teresa Martin-Carreras, Tessa S. Cook, and Charles E. Kahn. 2019. Readability of radiology reports: implications for patient-centered care. *Clinical Imaging*, 54:116–120.

Meta AI. 2024. Llama 3.2: Multilingual and multimodal foundation models. `https://github.com/meta-llama/llama-models/blob/main/models/llama3_2/MODEL_CARD.md`. Model card, version 3.2 (release date: 2024-09-25).

OpenAI. 2022. Chatgpt: Optimizing language models for dialogue. `https://openai.com/blog/chatgpt`.

OpenAI. 2023. Gpt-4 technical report. *arXiv preprint arXiv:2303.08774*.

Kishore Papineni, Salim Roukos, Todd Ward, and Wei-Jing Zhu. 2002. BLEU: a method for automatic evaluation of machine translation. In *ACL*, pages 311–318, Philadelphia, PA.

Colin Raffel, Noam Shazeer, Adam Roberts, Katherine Lee, Sharan Narang, Michael Matena, Yanqi Zhou, Wei Li, and Peter J. Liu. 2020. Exploring the limits of transfer learning with a unified text-to-text transformer. *Journal of Machine Learning Research*, 21(140):1–67.

Akshay Smit, Saahil Jain, Pranav Rajpurkar, Anuj Pareek, Andrew Y. Ng, and Matthew P. Lungren. 2020. Chexbert: Combining automatic labelers and expert annotations for accurate radiology report labeling using BERT. In *EMNLP*, pages 1500 – 1519.

Hugo Touvron, Thibaut Lavril, Gautier Izacard, and 1 others. 2023. Llama: Open and efficient foundation language models. *arXiv preprint arXiv:2302.13971*.

Ashish Vaswani, Noam Shazeer, Niki Parmar, Jakob Uszkoreit, Llion Jones, Aidan N. Gomez, Łukasz Kaiser, and Illia Polosukhin. 2017. Attention is all you need. In *NeurIPS*, pages 5998–6008.

Chenghao Xiao, Kun Zhao, Xiao Wang, Siwei Wu, Sixing Yan, Tomas Goldsack, Sophia Ananiadou, Noura Al Moubayed, Liang Zhan, William Cheung, and Chenghua Lin. 2025. Overview of the biolaysumm 2025 shared task on lay summarization of biomedical research articles and radiology reports. In *The 24th Workshop on Biomedical Natural Language Processing and BioNLP Shared Tasks*, Vienna, Austria. Association for Computational Linguistics.

Qianqian Xie, Zheheng Luo, Benyou Wang, and Sophia Ananiadou. 2023. A survey for biomedical text summarization: From pre-trained to large language models. *Preprint*, arXiv:2304.08763.

Tianyi Zhang, Varsha Kishore, Felix Wu, Kilian Q. Weinberger, and Yoav Artzi. 2020. Bertscore: Evaluating text generation with bert. In *ICLR*.

Kun Zhao, Chenghao Xiao, Sixing Yan, William K. Cheung, Kai Ye, Noura Al Moubayed, Liang Zhan, and Chenghua Lin. 2025. X-ray made simple: Lay radiology report generation and robust evaluation. *Preprint*, arXiv:2406.17911.

A Case Study 1

Table 4: Comparison of T5 and LLaMA outputs for a radiology report indicating COVID-19-type pneumonic infiltrates. T5 better paraphrases technical terms like "atelectasis" into lay-friendly expressions like "lung collapse," improving accessibility. LLaMA preserves more clinical terminology, which may limit interpretability. T5 also mirrors the structure of the ground truth better, enhancing relevance and readability.

Radiology Report	**Ground Truth Lay Summary**
Bands of atelectasis and consolidation in the mid-lung fields, along with confluent opacities in the lower mid-lung fields, also showing bilateral involvement. Findings, considering the reported history, could correspond to **COVID-19-type pneumonic infiltrates**. Evaluate progression and correlate with clinical and laboratory data.	There are areas of lung collapse and thickening in the middle parts of the lungs, along with large blurry spots in the lower middle areas, affecting both lungs. These findings, considering the patient's history, could be related to **the type of lung inflammation seen in COVID-19**. Further evaluation of the progression and correlation with clinical and lab data is needed.
T5-large Output	**LLaMA-3.2-3B Output**
There are bands of lung collapse and solid areas in the middle parts of the lungs, along with merged cloudiness in the lower middle parts of the lungs, affecting both sides. These findings, considering the reported history, could be signs of **COVID-19-type lung inflammation**. It is important to evaluate the progression and compare with clinical and lab data.	The middle areas of the lungs show signs of atelectasis and consolidation, and the lower middle areas of the lungs have merged opacities on both sides. These findings, considering the patient's history, could be related to **COVID-19-like lung infiltrates**. It is recommended to evaluate the progression and correlate this with clinical and laboratory data.

B Case Study 2

Table 5: Comparison of T5 and LLaMA outputs for a radiology report covering dyspnea, coronary artery disease, scoliosis, and COPD. T5 demonstrates better consistency with the ground truth by retaining specific phrases like "coronary artery disease" and clearly expressing medical conditions. LLaMA tends to generalise phrases (e.g., "heart problems"), which may lose important details for patient understanding.

Radiology Report	**Ground Truth Lay Summary**
Reason for consultation: **dyspnea. Old coronary artery disease.** Significant dorsal scoliosis. **Aortic elongation with descending aortic dilatation. Right pleural thickening.** Radiological signs of **chronic obstructive pulmonary disease (COPD).**	The person came in because they were having **trouble breathing.** They have a history of **heart problems related to the coronary arteries.** They also have a significant curve in their spine. Their **main blood vessel, the aorta, is longer than usual and the part of it that goes down from the heart is wider.** The **outer layer of the lung on the right side is thicker than it should be.** The x-ray shows signs of a **lung disease called COPD, which makes it hard to breathe.**
T5-large Output	**LLaMA-3.2-3B Output**
The patient came in because they were having **trouble breathing.** They have a history of **heart problems related to the coronary arteries.** They also have a significant curve in their spine. The **main blood vessel coming out of the heart is stretched out and wider than normal.** The **outer lining of the lungs on the right side is thicker than it should be.** The x-ray shows signs of a **long-term lung condition called chronic obstructive pulmonary disease (COPD).**	The patient came in because they were having **trouble breathing.** They have a history of **heart problems** and a curved spine. The **main blood vessel leaving the heart, called the aorta, is stretched out and the part that goes down from the heart is widened.** The **outer lining of the lungs on the right side is thicker than normal.** The x-ray shows signs of a **long-term lung condition called COPD.**

MetninOzU at BioLaySumm2025: Text Summarization with Reverse Data Augmentation and Injecting Salient Sentences

Egecan Çelik Evgin[1], **İlknur Karadeniz**[1,2], **Olcay Taner Yıldız**[1,2]
[1]Department of Artificial Intelligence and Data Engineering, Özyeğin University, Türkiye
[2]Department of Computer Science, Özyeğin University, Türkiye
egecan.evgin@ozu.edu.tr,
{ilknur.karadeniz, olcay.yildiz}@ozyegin.edu.tr

Abstract

In this paper, we present our approach to the BioLaySumm 2025 Shared Task on lay summarization of biomedical research articles, which was conducted as part of the BioNLP Workshop 2025. This marks the third edition of the BioLaySumm Shared Task (Goldsack et al., 2023, 2024; Xiao et al., 2025). The aim of the task is to create lay summaries from scientific articles to improve accessibility for a non-expert audience. To this end, we applied preprocessing techniques to clean and standardize the input texts, and fine-tuned Qwen2.5 (Team, 2024; Team) and Qwen3-based language models (Yang et al., 2025; Team, 2025) for the summarization task. For abstract-based fine-tuning, we investigated whether we can insert salient sentences from the main article into the summary to enrich the input. We also curated a dataset of child-friendly articles with corresponding gold-standard summaries and used large language models to rewrite them into more complex scientific variants to augment our training data with more examples.

1 Introduction

Interdisciplinary collaboration is a major challenge, especially in the biomedical field, where the number of scientific publications is increasing rapidly and the language used is often highly technical. This complexity poses significant obstacles not only for researchers from other disciplines, but also for the general public, making it difficult to access and understand new scientific findings. One promising solution to this problem is the inclusion of lay summaries in biomedical research articles. These summaries serve as a bridge between specialized content and a broader audience, allowing students, interdisciplinary researchers, and laypeople to better understand and engage with biomedical advances. The BioLaySumm 2025 Shared Task aims to improve automated systems for generating summaries of biomedical research articles. The focus is on producing summaries that are factually accurate, accessible to non-specialists and faithful to the original scientific content, thus supporting the wider dissemination and understanding of biomedical knowledge.

Previously, Bao et al. (2024) investigated simple preprocessing techniques such as hard truncation and text fragmentation and showed that large language models can produce effective lay summaries of biomedical texts even without complex pipelines. Stefanou et al. (2024) developed a child-friendly summarization method by fine-tuning biomedical models to simplified summaries. They used specialized tokens and data augmentation to improve accessibility for younger readers, using training data from the Science Journal for Kids (Science Journal for Kids, 2024). Modi and Karthikeyan (2024) showed that minimal preprocessing of summaries such as removing parenthetical content can significantly improve LLM performance in lay biomedical summarization. You et al. (2024) applied an extract-then-summarize strategy and tuned GPT-3.5 (OpenAI, 2023) on salient sentences to achieve strong relevance and overall performance. These studies show how different approaches, from basic cleanup to structured extraction, aim to make biomedical lay summaries clearer and more accessible.

2 Datasets

The task included two datasets, PLOS and eLife (Goldsack et al. (2024) (Goldsack et al., 2022) (Luo et al., 2022)). *PLOS* is the largest dataset derived from the Public Library of Science, comprising 24,773 training instances and 1,376 for validation, while the *eLife* dataset was derived from the peer-reviewed *eLife* journal and contains 4,346 instances for training and 241 for validation. The test data used for evaluation consisted of examples from both sources and was kept hidden by the organizers.

BioNLP 2025 Shared Tasks, pages 179–184
August 1, 2025 ©2025 Association for Computational Linguistics

3 Methodology

We investigated the pre-processing of full texts, the use of summaries and full articles for lay summarization, the generation of synthetic data from child-friendly texts with LLMs, and the extraction of key phrases by clustering.

3.1 Preprocessing

Before fine-tuning, we evaluated the performance of zero-shot and few-shot models using raw text input. The models tested include DeepSeek-Qwen (Lyu et al., 2025) and Qwen2.5 (Yang et al., 2024) with either full articles or abstracts provided as input. Building on the principles of PoA (Preprocessing over Abstract) from Modi and Karthikeyan (2024), we introduce a preprocessing step PoWA (Preprocessing over Whole Article) that improves the performance in both zero-shot and few-shot scenarios. PoWA involves removing all content enclosed in square, round or curly brackets from the input text including those in the training and test sets.

As with many systems submitted in previous years, our initial strategy focused on using only the abstract as input text for the summary. However, due to the varying lengths of the abstracts, we adopted a consistent approach by selecting the first 10 sentences from each abstract. The sentence boundaries were determined by splitting on periods, and applied uniformly to both the training and validation sentences. Unless otherwise specified (e.g. the condition "Full test" in Table 1), we only used the first 10 sentences of each test instance during the tests. This ensured comparability between different models and configurations.

3.2 Reverse Data Augmentation

Following the approach of Stefanou et al. (2024), we adopted a fine-tuning enhancement strategy that incorporates external data. Specifically, we used Frontiers for Young Minds(Frontiers for Young Minds, 2024), a child-friendly branch of the Frontiers journal series(Frontiers, 2024), which features simplified scientific articles written for young audiences. We collected 373 articles from the 'Human Health' section using a web scraping script built with the Selenium library (Selenium Project, 2025). Each article includes an abstract and spans approximately 500–1000 words. Designed for readers aged 8 to 12, these texts employ low FKGL (Flesch-Kincaid Grade Level) language (Flesch, 1975), with accompanying abstracts that provide even more simplified summaries. Each abstract was treated as a golden summary, resulting in a data set with two columns: Article and Summary. However, since both the article texts and their summaries were already simplified, the resulting pairs did not reflect the input-output complexity of the task. To address this gap, we used the DeepSeek-R1-Distill-Qwen-32B model (DeepSeek-AI, 2024) to rewrite the simplified articles in a more scientific tone, following the method described by DeepSeek-AI (2024). We used the following prompt: "Rewrite the given text so that it is more scientific and suitable for publication." The generation was limited to 1024 tokens with a temperature of 0.01 and a repetition penalty of 1.2. As mentioned in DeepSeek-AI (2024), DeepSeek models often produce internal thoughts before generating the final output. To address this, we extract the content following the </think> tag, along with minimal pre- and post-processing to format the results.

The gold summaries from Frontiers for Young Minds typically had FKGL scores between 8 and 10 (Flesch, 1975), and were notably shorter than the summaries found in the eLife and PLOS training sets (Task, 2025a,b). To address this length and complexity mismatch, we incorporated a curriculum learning strategy (Bengio et al., 2009), which is discussed further in Section 4.4 on model fine-tuning.

3.3 Injecting Salient Sentences

Using only the abstract to summarize an entire article was found to be insufficient. To improve this and build on strategies observed in our earlier literature review, we appended key sentences from the full text to the end of each abstract. To process sentences beyond the initial 10 in each article, we developed a function that encodes these sentences using the all-MiniLM-L6-v2 model (Wang et al., 2020), which is accessible via the Hugging Face repository (Reimers and Gurevych, 2021).

We trained a K-Means clustering model with $k = 3$ on encoded sentence representations to identify the salient content (Lloyd, 1982). A sentence closest to each centroid was selected, resulting in three sentences in total, which were then appended to the end of the article's abstract. Transformers and Scikit-learn libraries were used for this phase (Wolf et al., 2020; Pedregosa et al., 2011).

3.4 Model Fine-tuning

First, fine-tuning was performed only on the abstract and lay summary pairs using the Qwen2.5: 1.5B and Qwen2.5: 3B3B models (Team, 2024; Team), prompted with a very short instruction: *"Summarize the following:"* The Qwen2.5 models were fine-tuned using low-rank adaptation (LoRA) (Hu et al., 2021).

For Qwen3 models (Yang et al., 2025; Team, 2025), we applied LoRA for parameter-efficient fine-tuning, using a rank of 8, a scaling factor of 16, and a dropout rate of 0.05. Adaptation was limited to the `q_proj` and `v_proj` attention layers, without any bias terms, under a causal language modeling setup (Hu et al., 2021; Dettmers et al., 2023).

After preprocessing steps such as trimming, salient sentence injection, curriculum learning, and adding Frontiers for Young Minds articles, the data was converted into ChatML format (OpenAI, 2023) and used for fine-tuning.

Training hyperparameters were slightly adjusted based on the dataset. For eLife, we fine-tuned the model for 3 epochs with a learning rate of 1×10^{-4} and 6 gradient accumulation steps. For PLOS, we used 2 epochs, a higher learning rate of 1.5×10^{-4}, and 8 accumulation steps. For other datasets, we set the learning rate to 1.25×10^{-4}, trained for 2 epochs, and used 7 accumulation steps. These values were chosen after a few initial trials to balance training time and performance. All models were trained with a per-device batch size of 2 and FP16 precision using Hugging Face Transformers and PEFT libraries (Wolf et al., 2020; Dettmers et al., 2023).

We applied curriculum learning (Bengio et al., 2009), which is presented in Table 1 with "Aug" label, in which 373 articles from *Frontiers for Young Minds* were placed at the beginning of the training dataset (Frontiers for Young Minds, 2024),as explained in Section 3.2. The remaining articles were then sorted by word count in ascending order, resulting in a training sequence that gradually progressed from simpler to more complex texts. In the Salient Sentence Injection strategy (see Section 3.3), the three most important sentences following the abstract were added to it, and fine-tuning was done on this updated version of the dataset. The part marked as *Full Text* in Table 1 refers to the evaluation of the two 142-entry test sets *without any trimming*, prior to fine-tuning. The experiment labeled as "Post Processing" in the same table refers to the action taken after fine-tuning, as described in Section 3.5.

3.5 Post-processing for Readability

To slightly reduce the FKGL (Flesch, 1975) score of the summaries generated by the fine-tuned LLMs, a post-processing step was applied. Using the DeepSeek-R1-Distill-Qwen-32B model (DeepSeek-AI, 2024) in a zero-shot setting, we prompted it with: "Reduce the FKGL score of the text. Simplify while preserving the scientific content" DeepSeek-AI (2024). As in Section 3.2, post-processing was also applied to the outputs of the DeepSeek model (DeepSeek-AI, 2024). In most experiments, additional steps and alternative prompts were needed due to the model frequently disrupting the structure of the article.

4 Experimental Setup

The training was performed on an NVIDIA A100 GPU (Corporation, 2020) provided by Google Colaboratory (Bisong, 2019). Several automatic metrics to measure relevance were used for evaluation, with a focus on comparing system output with human-written references. ROUGE (Lin, 2004) evaluates recall by measuring the overlap of n-grams between the generated text and the reference text. BLEU (Papineni et al., 2002) focuses on the precision of the n-grams and applies a penalty for brevity to prevent overly short outputs. METEOR (Banerjee and Lavie, 2005) considers synonym matching, stemming and word order, balances precision and recall, and penalizes disjointed output. BERTScore (Zhang et al., 2020) captures semantic similarity by calculating cosine similarity between contextualized token embeddings from models such as BERT (Devlin et al., 2019), enabling a deeper evaluation of meaning beyond surface-level overlaps.

The Flesch-Kincaid Grade Level (FKGL)(Flesch, 1975) assesses the reading difficulty of a text based on sentence length and word syllables and provides a score that corresponds to US school levels. The Coleman-Liau Index (CLI)(Coleman and Liau, 1975) provides a similar assessment of readability, but is based on the number of characters rather than the number of syllables, making it more suitable for automatic processing of digital texts. The D-Level Sentence Complexity Rating Scheme (DCRS)(Rambow

Model	ROUGE	BLEU	METEOR	BERTScore	FKGL	DCRS	CLI	LENS	AlignScore	SummaC
Qwen3:4B Trim + Aug + SSI + PostP	0.3061	5.3966	0.2555	0.8537	16.7644	11.2446	16.0117	60.0364	0.7837	0.6858
Qwen3:4B Trim + Aug + SSI + Full Test	0.2576	4.2385	0.3296	0.8493	15.0595	10.0385	15.5170	22.2383	0.9025	0.9369
Qwen3:4B Trim + Aug + SSI	0.3261	6.6388	0.2910	0.8560	16.3742	11.0955	16.9846	34.2622	0.8748	0.9195
Qwen3:4B Trim + Aug	0.3279	6.7490	0.2928	0.8560	16.3242	11.0915	16.9893	34.0978	0.8679	0.9203
Qwen3:4B Trim	0.3300	6.9466	0.2903	0.8567	16.4528	11.2157	17.0054	34.8577	0.8807	0.9203
Qwen2.5:3B Trim	0.3127	6.2905	0.3036	0.8486	14.7591	9.8484	15.4835	23.1406	0.7937	0.9172
Qwen2.5:1.5B Trim	0.3108	6.2470	0.3014	0.8484	14.8767	9.7678	15.6298	23.1043	0.8047	0.9170

Table 1: Evaluation metrics of Qwen models on various configurations. Trim: Trimming top 10 sentences of the article, Aug: Reverse data augmentation using Frontiers for Young Minds, SSI: Salient Sentence Injection, PostP: Postprocessing for lower FKGL using DeepSeek, Full Test: Full test set in the inference without trimming

et al., 2004) assesses grammatical complexity by analyzing syntactic features such as sentence structure and part-of-speech patterns. More recently, LENS(Tan et al., 2023) uses a comprehensive language model to estimate how difficult a passage is to understand, providing a neural-based alternative to traditional readability metrics.

To assess factuality, AlignScore (Jia et al., 2022) was used to determine whether the generated summary remains faithful to the content of the source. It applies a Natural Language Inference (NLI) model (Bowman et al., 2015) to assess whether each sentence in the summary is implied by the source text. Similarly, SummaC (Laban et al., 2022) checks the factual consistency between the summary and the source by applying sentence-level entailment models to ensure logical consistency.

5 Results

The Qwen2.5-1.5B and 3B (Team, 2024; Team) models were fine-tuned with LoRA, reducing the training and validation sentences to their first 10 sentences. They were then tested with zero shot on similarly trimmed test sets, and the results were surprising. After the experiments, the lowest FKGL values were observed for the two Qwen2.5 models.

The Qwen3-4B model (Yang et al., 2025; Team, 2025) was fine-tuned with LoRA, reducing the training and validation sentences to their first 10 sentences. The highest ROUGE score was observed in the scenario where only the test set was trimmed, with no data augmentation, injection of salient sentences, post-processing, or use of the full test data (labeled 'Qwen3: 4B Trim' in Table 1). With augmentation, the FKGL score decreased slightly and the METEOR score increased slightly, but ROUGE, BLEU, BERTScore and AlignScore all decreased in the Qwen3:4B Trim + Aug setting. With the addition of Salient Sentence Injection (SSI), most relevance scores decreased and AlignScore increased slightly, which is shown in Table 1 as Qwen3:4B Trim + Aug + SSI.

In the Qwen3:4B Trim + Aug + SSI + Full Test experiment, the test set without trimming was used. As a result, ROUGE and BLEU scores decreased significantly, while METEOR, AlignScore and SummaC were higher than in all other experiments. The FKGL, CLI and LENS scores also decreased, suggesting that higher factuality could be achieved in this setting.

In our comparative analysis of the different techniques, we found that data augmentation consistently improves readability, but leads to a decrease in relevance and factuality. Salient Sentence Injection led to a decrease in all three evaluation criteria. Full fine-tuning also decreased performance in relevance and readability, but scored highest in factuality. Post-processing with external LLMs performed worst overall, scoring lowest in all experiments.

6 Conclusion

In this paper, we present our participation in the BioLaySumm 2025. Our results show that the performance of the Qwen 1.5B model with low parameters was particularly promising and shows that even smaller models can be competitive if they have sufficient input data and the hyperparameters are set appropriately. With additional input data and further optimization, this model has the potential to outperform larger counterparts, especially in terms of readability. In particular, the use of untrimmed test data significantly improved factuality, on the other hand it led to a decrease in core relevance scores. This suggests that an intermediate strategy (e.g. using a higher value for the first sentences instead of first 10 sentences) might provide a better balance between factuality and relevance. Although techniques such as salient sentence injection, reverse data augmentation, and postprocessing with auxiliary LLMs did not yield the expected gains, they remain promising for future exploration.

References

Satanjeev Banerjee and Alon Lavie. 2005. Meteor: An automatic metric for mt evaluation with improved correlation with human judgments. In *Proceedings of the ACL workshop on intrinsic and extrinsic evaluation measures for machine translation and/or summarization*, pages 65–72.

Siyu Bao, Ruijing Zhao, Siqin Zhang, Jinghui Zhang, Weiyin Wang, and Yunian Ru. 2024. Ctyun ai at biolaysumm: Enhancing lay summaries of biomedical articles through large language models and data augmentation. In *Proceedings of the 23rd Workshop on Biomedical Natural Language Processing*, pages 837–844, Bangkok, Thailand. Association for Computational Linguistics.

Yoshua Bengio, Jérôme Louradour, Ronan Collobert, and Jason Weston. 2009. Curriculum learning. In *Proceedings of the 26th Annual International Conference on Machine Learning*, pages 41–48. ACM.

Ekaba Bisong. 2019. Google colaboratory. In *Building Machine Learning and Deep Learning Models on Google Cloud Platform*, pages 59–64. Apress, Berkeley, CA.

Samuel R Bowman, Gabor Angeli, Christopher Potts, and Christopher D Manning. 2015. A large annotated corpus for learning natural language inference. In *Proceedings of the 2015 Conference on Empirical Methods in Natural Language Processing (EMNLP)*, pages 632–642.

Meri Coleman and T L Liau. 1975. A computer readability formula designed for machine scoring. *Journal of Applied Psychology*, 60(2):283.

NVIDIA Corporation. 2020. Nvidia a100 tensor core gpu. Accessed: 2025-05-19.

DeepSeek-AI. 2024. Deepseek-r1-distill-qwen-32b. `https://huggingface.co/deepseek-ai/DeepSeek-R1-Distill-Qwen-32B`. Accessed: 2025-04-18.

DeepSeek-AI. 2024. Deepseek-r1: Incentivizing reasoning capability in llms via reinforcement learning. *arXiv preprint arXiv:2501.12948*. Accessed: 2025-04-18.

Tim Dettmers, Artidoro Pagnoni, Arjun Guha, and Luke Zettlemoyer. 2023. Peft: State-of-the-art parameter-efficient fine-tuning methods. `https://github.com/huggingface/peft`. Accessed: 2024-05-18.

Jacob Devlin, Ming-Wei Chang, Kenton Lee, and Kristina Toutanova. 2019. Bert: Pre-training of deep bidirectional transformers for language understanding. In *Proceedings of NAACL-HLT*, pages 4171–4186.

Rudolf Flesch. 1975. Flesch–kincaid grade level. `https://en.wikipedia.org/wiki/FleschâĂŞKincaid_readability_tests`. Accessed: 2025-04-18.

Frontiers. 2024. Frontiers. `https://www.frontiersin.org`. Accessed: 2024-12-01.

Frontiers for Young Minds. 2024. Frontiers for young minds. `https://kids.frontiersin.org`. Accessed: 2024-12-01.

Tomas Goldsack, Zheheng Luo, Qianqian Xie, Carolina Scarton, Matthew Shardlow, Sophia Ananiadou, and Chenghua Lin. 2023. Overview of the biolaysumm 2023 shared task on lay summarization of biomedical research articles. In *The 22nd Workshop on Biomedical Natural Language Processing and BioNLP Shared Tasks*, pages 468–477, Toronto, Canada. Association for Computational Linguistics.

Tomas Goldsack, Carolina Scarton, Matthew Shardlow, and Chenghua Lin. 2024. Overview of the BioLaySumm 2024 shared task on the lay summarization of biomedical research articles. In *Proceedings of the 23rd Workshop on Biomedical Natural Language Processing*, Bangkok, Thailand. Association for Computational Linguistics.

Tomas Goldsack, Zhihao Zhang, Chenghua Lin, and Carolina Scarton. 2022. Making science simple: Corpora for the lay summarisation of scientific literature. In *Proceedings of the 2022 Conference on Empirical Methods in Natural Language Processing*, pages 10589–10604, Abu Dhabi, United Arab Emirates. Association for Computational Linguistics.

Edward J. Hu, Yelong Shen, Phillip Wallis, Zeyuan Allen-Zhu, Yuanzhi Li, Shean Wang, and Weizhu Chen. 2021. Lora: Low-rank adaptation of large language models. *Preprint*, arXiv:2106.09685.

Qingxiu Jia, Qipeng Xu, Weiting Yu, Yitong Duan, Jian-Yun Nie, and Zhiyuan Liu. 2022. Alignscore: Evaluating factual consistency with contextual alignment. In *Proceedings of the 2022 Conference on Empirical Methods in Natural Language Processing (EMNLP)*.

Philippe Laban, Florian Trummer, and Marti A Hearst. 2022. Summac: Re-visiting nli-based models for consistency evaluation. In *Transactions of the Association for Computational Linguistics*, volume 10, pages 163–177.

Chin-Yew Lin. 2004. Rouge: A package for automatic evaluation of summaries. In *Text summarization branches out*, pages 74–81.

Stuart Lloyd. 1982. Least squares quantization in pcm. *IEEE Transactions on Information Theory*, 28(2):129–137.

Zheheng Luo, Qianqian Xie, and Sophia Ananiadou. 2022. Readability controllable biomedical document summarization. In *Findings of the Association for Computational Linguistics: EMNLP 2022*, pages 4667–4680, Abu Dhabi, United Arab Emirates. Association for Computational Linguistics.

Yuxuan Lyu and 1 others. 2025. Deepseek-r1: Incentivizing reasoning capability in llms via reinforcement learning. *arXiv preprint arXiv:2501.12948*.

Satyam Modi and T Karthikeyan. 2024. Eulerian at BioLaySumm: Preprocessing over abstract is all you need. In *Proceedings of the 23rd Workshop on Biomedical Natural Language Processing*, pages 826–830, Bangkok, Thailand. Association for Computational Linguistics.

OpenAI. 2023. Chatml prompting format. https://platform.openai.com/docs/guides/gpt/chat-completions-api. Accessed: 2025-05-19.

OpenAI. 2023. Gpt-3.5. https://platform.openai.com/docs/models/gpt-3-5. Accessed: 2024-12-01.

Kishore Papineni, Salim Roukos, Todd Ward, and Wei-Jing Zhu. 2002. Bleu: a method for automatic evaluation of machine translation. In *Proceedings of the 40th annual meeting of the Association for Computational Linguistics*, pages 311–318.

Fabian Pedregosa, Gaël Varoquaux, Alexandre Gramfort, Vincent Michel, Bertrand Thirion, Olivier Grisel, Mathieu Blondel, Peter Prettenhofer, Ron Weiss, Vincent Dubourg, Jake Vanderplas, Alexandre Passos, David Cournapeau, Matthieu Brucher, Matthieu Perrot, and Édouard Duchesnay. 2011. Scikit-learn: Machine learning in Python. *Journal of Machine Learning Research*, 12:2825–2830.

Owen Rambow, Lokesh Liu, Lance Johnson, Nathaniel Fillmore, and Benoit Lavoie. 2004. Summarizing multiple news articles using readability-based evaluation. In *Proceedings of the Human Language Technology Conference of the NAACL*.

Nils Reimers and Iryna Gurevych. 2021. all-minilm-l6-v2. https://huggingface.co/sentence-transformers/all-MiniLM-L6-v2. Accessed: 2025-05-19.

Science Journal for Kids. 2024. Science journal for kids. https://www.sciencejournalforkids.org. Accessed: 2024-12-01.

Selenium Project. 2025. Selenium webdriver. https://www.selenium.dev. Accessed: 2025-04-18.

Loukritia Stefanou, Tatiana Passali, and Grigorios Tsoumakas. 2024. AUTH at BioLaySumm 2024: Bringing scientific content to kids. In *Proceedings of the 23rd Workshop on Biomedical Natural Language Processing*, pages 793–803, Bangkok, Thailand. Association for Computational Linguistics.

Zhengyuan Tan, Mounica Maddela, Wei Xu, Xiaojun Wan, and Fei Wu. 2023. Lens: A learned evaluation metric for text simplification. In *Proceedings of the 2023 Conference of the North American Chapter of the Association for Computational Linguistics (NAACL)*.

BioLaySumm Shared Task. 2025a. Biolaysumm2025-elife. https://huggingface.co/datasets/BioLaySumm/BioLaySumm2025-eLife. Accessed: 2025-05-19.

BioLaySumm Shared Task. 2025b. Biolaysumm2025-plos. https://huggingface.co/datasets/BioLaySumm/BioLaySumm2025-PLOS. Accessed: 2025-05-19.

Qwen Team. Qwen2.5 models on hugging face. https://huggingface.co/Qwen. Accessed: 2024-05-18.

Qwen Team. 2024. Qwen2 technical report. https://qwen.aliyun.com/. Accessed: 2024-05-18.

Qwen Team. 2025. Qwen3 models on hugging face. https://huggingface.co/Qwen. Accessed: 2025-05-19.

Wenhui Wang, Furu Wei, Li Dong, Hangbo Bao, Nan Yang, and Ming Zhou. 2020. Minilm: Deep self-attention distillation for task-agnostic compression of pre-trained transformers. *arXiv preprint arXiv:2002.10957*.

Thomas Wolf, Lysandre Debut, Victor Sanh, Julien Chaumond, Clement Delangue, Anthony Moi, Pierric Cistac, Tim Rault, Remi Louf, Morgan Funtowicz, and Jamie Brew. 2020. Transformers: State-of-the-art natural language processing. In *Proceedings of the 2020 Conference on Empirical Methods in Natural Language Processing: System Demonstrations*, pages 38–45, Online. Association for Computational Linguistics.

Chenghao Xiao, Kun Zhao, Xiao Wang, Siwei Wu, Sixing Yan, Sophia Ananiadou, Noura Al Moubayed, Liang Zhan, William Cheung, and Chenghua Lin. 2025. Overview of the biolaysumm 2025 shared task on lay summarization of biomedical research articles and radiology reports. In *The 24nd Workshop on Biomedical Natural Language Processing and BioNLP Shared Tasks*, Vienna, Austria. Association for Computational Linguistics.

An Yang, Anfeng Li, Baosong Yang, Beichen Zhang, Binyuan Hui, Bo Zheng, Bowen Yu, Chang Gao, and 1 others. 2025. Qwen3 technical report. *Preprint*, arXiv:2505.09388. Available at: https://arxiv.org/abs/2505.09388.

An Yang and 1 others. 2024. Qwen2.5 technical report. *arXiv preprint arXiv:2412.15115*.

Zhiwen You, Shruthan Radhakrishna, Shufan Ming, and Halil Kilicoglu. 2024. UIUC_BioNLP at BioLaySumm: An extract-then-summarize approach augmented with Wikipedia knowledge for biomedical lay summarization. In *Proceedings of the 23rd Workshop on Biomedical Natural Language Processing*, pages 132–143, Bangkok, Thailand. Association for Computational Linguistics.

Tianyi Zhang, Varsha Kishore, Felix Wu, Kilian Q Weinberger, and Yoav Artzi. 2020. Bertscore: Evaluating text generation with bert. In *International Conference on Learning Representations (ICLR)*.

Shared Task at Biolaysumm2025 : Extract then summarize approach Augmented with UMLS based Definition Retrieval for Lay Summary generation.

Aaradhya Gupta and Dr Parameswari Krishnamurthy
LTRC, International Institute of Information Technology, Hyderabad
aaradhya.gupta@research.iiit.ac.in and param.krishna.iiit.ac.in

Abstract

We present LayForge, a two-track lay summary generation system developed for the BioLaySumm2025 shared task(Xiao et al., 2025) . Task 1.1 addresses the lay summarization using only the internal content of the article, while Task 1.2 augments this process with domain knowledge such as biomedical definitions and concept explanations. BioLaySumm employs a modular architecture that leverages large language models (LLMs), a BioBERT-based named entity recognizer (NER), and the UMLS(Bodenreider, 2004) knowledge base to create readable, informative, and faithful lay summaries. Our system shows strong performance on both tasks when evaluated on the PLOS and elife subset(Goldsack et al., 2022), particularly in readability and factuality metrics. The architecture illustrates how modularity and domain adaptation can be effectively combined for accessible biomedical communication.

1 Introduction

Lay summaries are a critical bridge between dense biomedical literature and non-specialist audiences, including patients, caregivers, and policy makers. These summaries must balance clarity, completeness, and technical accuracy. The BioLaySumm2025 shared task(Xiao et al., 2025) presents two summarization challenges:

- **Task 1 (Internal-only):** Generate a lay summary using only the content of the original article.
- **Task 2 (Augmented):** Improve the lay summary by incorporating external biomedical knowledge such as terminology definitions.

We introduce **LayForge**, a flexible and extensible system designed to address both tracks. Our design is rooted in modular NLP techniques - chunk extraction, LLM-based draft generation, and iterative rewriting—with additional augmentation for Task 2 using BioBERT-based NER (Lee et al., 2019) and UMLS-based concept simplification.

Our contributions include:

- A two-tiered summarization pipeline that integrates pretrained LLMs with biomedical NER and knowledge retrieval.
- A task-specific rewriting mechanism for increasing the readability and accessibility of summaries.
- A detailed performance comparison across readability, fidelity, and factuality metrics.

2 Related Work

The BioLaySumm shared task series began in 2023 (Goldsack et al., 2023), with a follow-up edition in 2024 (Goldsack et al., 2024), laying the groundwork for consistent evaluation and dataset development in biomedical lay summarization. Our work builds on the methodologies and evaluation frameworks introduced in these earlier editions, Biomedical summarization has traditionally leveraged sequence-to-sequence architectures and domain-specific pretrained models such as BioBERT and PubMedBERT(Beltagy et al., 2020). Recent trends in summarization, including the use of large language models and retrieval-augmented generation (RAG)(Lewis et al., 2020), show promise in improving factuality and reducing hallucination. Entity-level simplification is another important strand, where domain terms are replaced or explained using biomedical ontologies. However, most prior work stops at simple substitutions, while our system integrates retrieved definitions into fluent rewrites. Instruction tuning for LLMs is also a promising avenue of research. (Tran et al., 2024) introduced a corpus of 25,005 human-crafted prompts to instruction-tune LLaMA models on biomedical tasks, yielding QA gains and generation improvements There have also been

BioNLP 2025 Shared Tasks, pages 185–189
August 1, 2025 ©2025 Association for Computational Linguistics

3 System Architecture

We ensure that the system architecture is modular and easy to understand. The 2 tasks share a common pipeline in the beginning. The augmentation using the UMLS backed definitions is performed for task 2 at the end.

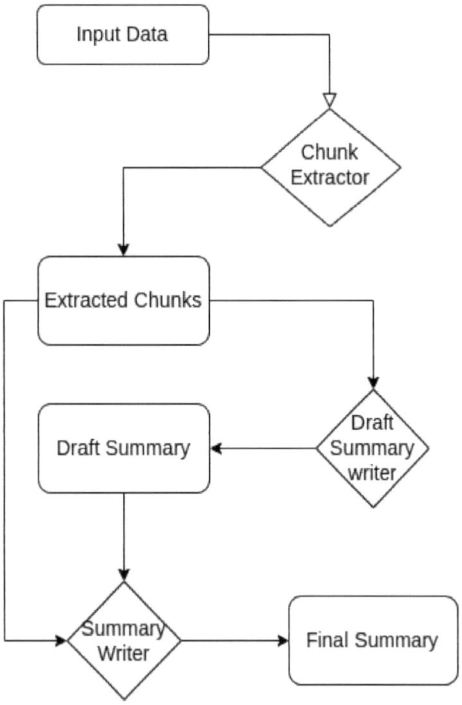

Figure 1: Task 1.1 : Framework for lay summary generation with no external information

3.1 Shared Architecture (Tasks 1 & 2)

Preprocessing and Chunking Articles are segmented into overlapping text chunks (3,000 tokens with 200-token overlap) to accommodate LLM context windows and ensure semantic continuity.

Top-k Sentence Extraction For each chunk, salient sentences are extracted using an LLM (LLaMA 3-70B) prompted to select informative statements. The resulting sentence pool contains all the key findings and methods.

Draft Generation We conditioned the LLM with article metadata, keywords, and extracted sentences to generate a draft lay summary. Prompts guide the model to assume a "science teacher" persona to ensure accessibility.

Iterative Rewriting Two rewriting passes are applied:

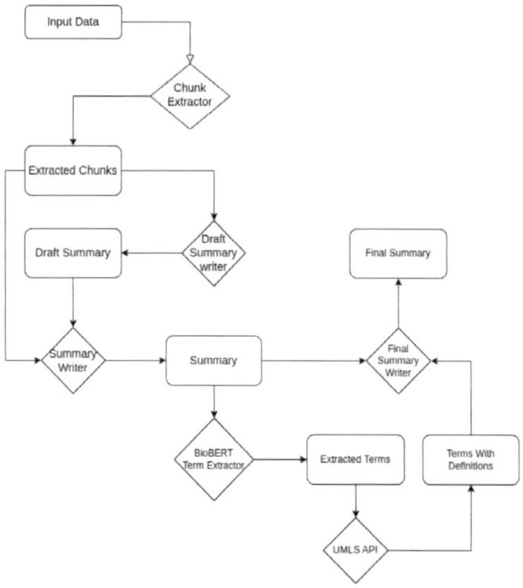

Figure 2: Task 1.2 : Framework for lay summary generation with external information

- **Reader Rewrite:** A persona-based prompt enhances flow and readability.
- **Jargon Softening:** Phrases are simplified and clarified, guided by syntactic and lexical heuristics.

3.2 Knowledge Augmentation (Task 2 Only)

Domain NER A BioBERT model fine-tuned for NER identifies and extracts biomedical terms that are not layperson friendly in the summary.

Definition Retrieval Each detected term is passed to a UMLS-backed lookup API, which retrieves lay definitions.

Definition-Guided Rewrite These definitions are incorporated into the summary through guided LLM prompts, either by appending explanations or paraphrasing terms inline.

3.3 Model Selection

We chose LLaMA 3–70B as our backbone because it offers a strong balance between model capacity and computational cost, while remaining fully open-source under a permissive license. In preliminary experiments (not shown), LLaMA 3–70B outperformed smaller variants (e.g., 13B) on zero-shot biomedical QA benchmarks. Additionally, its 8K-token context window accommodates long article chunks without resorting to expensive retrieval passes, which was critical for processing 3,000-token windows in our pipeline.

4 Implementation Details

We implemented LayForge in Python, using Langchain for orchestrating LLM calls and LangGraph for managing pipeline state. Sentence extraction and summarization use the Groq-hosted LLaMA 3-70B model. BioBERT NER is handled using the SimpleTransformers library.

UMLS queries are made via a RESTful endpoint returning short, simplified definitions. All components are containerized and run using Google Colab with GPU acceleration for efficiency.

4.1 Handling UMLS Definition Ambiguity

When a detected term has multiple definitions in UMLS, our lookup strategy resolves ambiguity by:

- **Source prioritization:** We only accept definitions whose rootSource is in (MSH, PDQ, NCI, MEDLINEPLUS), in that order.
- **Conciseness heuristic:** If multiple definitions remain, we choose the one with the fewest tokens, assuming brevity aids lay understanding.
- **Fallback:** If no preferred definition is found, we leave the term unchanged and rely on the LLM's paraphrasing step to "soften" it.

5 Experimental Setup

We evaluate both tasks using the BioLaySumm2025 PLOS and elife datasets.

Evaluation Metrics Various evaluation metrics were used to evaluate performance of the system in different fields.(Luo et al., 2022)

- **Readability:** FKGL, DCRS, CLI and LENS.
- **Content Fidelity:** ROUGE-L, BLEU-4, METEOR, BERTScore.
- **Factuality:** SummaC and AlignScore.

6 Results and Discussion

Our results show that augmentation with external definitions significantly improves readability metrics, with FKGL decreasing by over 3 points and DCRS/CLI also showing similar gains. The LENS metric confirms slightly longer outputs, likely due to inserted definitions and the model being more verbose to avoid using technical terms

Metric	Task 1	Task 2
ROUGE	0.32	0.29
BLEU	5.45	4.32
METEOR	0.29	0.26
BERTScore	0.85	0.85
FKGL	14.56	11.15
DCRS	10.01	8.36
CLI	15.36	11.93
LENS	71.51	81.50
AlignScore	0.69	0.61
SummaC	0.50	0.53

Table 1: Performance across BioLaySumm track as per the Leaderboard

Interestingly, although Task-2 reduces ROUGE and BLEU slightly, this can be attributed to paraphrasing and definition insertion changes that promote lay understanding at the cost of n-gram overlap. Semantic paraphrase or added explanatory phrase tend to reduce these metrics despite improving readability and fidelity(Xu et al., 2016; Ganesan, 2018; Wang et al., 2022; Elliott et al., 2019).

SummaC(Tian et al., 2022) improves in Task2, suggesting factual information is more clearly and accurately preserved with domain support. AlignScore, however, drops slightly, indicating a potential loss in extractive alignment.

7 Conclusion and Future Work

We presented LayForge, a modular, domain-aware pipeline for lay summary generation in biomedical research. Our two-track approach enables detailed comparison between internal-only and knowledge-augmented strategies.

In future work, we aim to incorporate dynamic chunk prioritization, retrieval-based augmentation, and human-in-the-loop feedback. We also plan to evaluate generalizability across different journals and medical domains. We also see great potential in Reinforcement Learning from Human Feedback (RLHF)(Natarajan et al., 2024) to directly optimize summary-quality objectives (e.g. readability, factuality) by defining appropriate reward functions.

Acknowledgments

We thank the BioLaySumm organizers and annotators for their support. This work is supported by the LTRC Lab at IIIT Hyderabad

Step	Text
Original Paragraph	"DPAGT1-CDG is a Congenital Disorder of Glycosylation (CDG) that ..."
Top-k Sentences	1. DPAGT1-CDG is a Congenital Disorder of Glycosylation (CDG) that lacks effective therapies. 2. DPAGT1 synthesizes dolichol-PP-GlcNAc, which is the first step in N-glycosylation. 3. We used a Drosophila rough eye model of DPAGT1-CDG with an improperly developed, small eye phenotype.
Draft Simple Passage	"DPAGT1-CDG is a rare genetic disorder that affects the synthesis of dolichol-PP-GlcNAc, a crucial step in N-glycosylation. This disorder leads to developmental delay, muscle weakness, and seizures, among other symptoms. To better understand the disease and identify potential therapeutic options..."
Retrieved Definitions	developmental delay: 'Failure to meet, or late achievement of developmental milestones.' dolichol-PP-GlcNAc: 'A compound involved in the first step of N-glycosylation.' muscle weakness: 'A reduction in the strength of one or more muscles.'
Final Rewrite	"DPAGT1-CDG is a rare genetic disorder that affects the production of a crucial molecule, essential for a process called N-glycosylation. This disorder causes symptoms including delayed developmental milestones, muscle weakness, and seizures. To study the disease and identify treatments, researchers used a fruit fly model with small, improperly developed eyes and tested 1,520 existing drugs from a library of FDA- or EMA-approved compounds..."

Table 2: Example of the Extract–Draft–Rewrite Flow

References

Iz Beltagy, Kyle Lo, and Arman Cohan. 2020. Pubmedbert: Domain-specific language model pretraining for biomedical natural language processing. In *Findings of EMNLP*, pages 3615–3620.

Olivier Bodenreider. 2004. The unified medical language system (umls): integrating biomedical terminology. *Nucleic Acids Research*, 32(Database issue):D267–D270.

Dan Elliott and 1 others. 2019. Bleu is not suitable for the evaluation of text simplification. Technical blog post, ResearchGate.

Kavita Ganesan. 2018. Rouge 2.0: Updated and improved measures for evaluation of summarization tasks. *arXiv preprint arXiv:1803.01937*.

Tomas Goldsack, Zheheng Luo, Qianqian Xie, Carolina Scarton, Matthew Shardlow, Sophia Ananiadou, and Chenghua Lin. 2023. Overview of the biolaysumm 2023 shared task on lay summarization of biomedical research articles. In *The 22nd Workshop on Biomedical Natural Language Processing and BioNLP Shared Tasks*, pages 468–477, Toronto, Canada. Association for Computational Linguistics.

Tomas Goldsack, Carolina Scarton, Matthew Shardlow, and Chenghua Lin. 2024. Overview of the biolaysumm 2024 shared task on the lay summarization of biomedical research articles. *Preprint*, arXiv:2408.08566.

Tomas Goldsack, Zhihao Zhang, Chenghua Lin, and Carolina Scarton. 2022. Making science simple: Corpora for the lay summarisation of scientific literature. In *Proceedings of the 2022 Conference on Empirical Methods in Natural Language Processing*, pages 10589–10604, Abu Dhabi, United Arab Emirates. Association for Computational Linguistics.

Jinhyuk Lee, Wonjin Yoon, Sungdong Kim, Donghyeon Kim, Sunkyu Kim, Chanho So, and Jaewoo Kang. 2019. Biobert: a pre-trained biomedical language representation model for biomedical text mining. *Bioinformatics*, 36(4):1234–1240.

Patrick Lewis, Ethan Perez, Aleksandra Piktus, Fabio Petroni, Vladimir Karpukhin, Naman Goyal, Heinrich Küttler, Mike Lewis, Wen-tau Yih, Tim Rocktäschel, Sebastian Riedel, and Douwe Kiela. 2020. Retrieval-augmented generation for knowledge-intensive nlp tasks. In *Advances in Neural Information Processing Systems (NeurIPS)*, volume 33, pages 9459–9474.

Zheheng Luo, Qianqian Xie, and Sophia Ananiadou. 2022. Readability controllable biomedical document summarization. In *Findings of the Association for Computational Linguistics: EMNLP 2022*, pages 4667–4680, Abu Dhabi, United Arab Emirates. Association for Computational Linguistics.

Sriraam Natarajan, Saurabh Mathur, Sahil Sidheekh, Wolfgang Stammer, and Kristian Kersting. 2024. Human-in-the-loop or ai-in-the-loop? automate or collaborate? *Preprint*, arXiv:2412.14232.

Tian Tian, Emma Reiter, and 1 others. 2022. Summac: Revisiting nli-based models for inconsistency detection in summaries. *Transactions of the Association for Computational Linguistics (TACL)*, 10:350–365.

Hieu Tran, Zhichao Yang, Zonghai Yao, and Hong Yu. 2024. Bioinstruct: Instruction tuning of large language models for biomedical natural language processing. *Preprint*, arXiv:2310.19975.

Zhuohan Wang and 1 others. 2022. On the evaluation metrics for paraphrase generation. In *Proceedings of EMNLP*, pages 208–221.

Chenghao Xiao, Kun Zhao, Xiao Wang, Siwei Wu, Sixing Yan, Sophia Ananiadou, Noura Al Moubayed, Liang Zhan, William Cheung, and Chenghua Lin. 2025. Overview of the biolaysumm 2025 shared task on lay summarization of biomedical research articles and radiology reports. In *The 24nd Workshop on Biomedical Natural Language Processing and BioNLP Shared Tasks*, Vienna, Austria. Association for Computational Linguistics.

Wei Xu, Courtney Napoles, Claire O'Connor, Chris Callison-Burch, and Ellie Loper. 2016. Optimizing statistical machine translation for text simplification. In *Proceedings of NAACL-HLT*, pages 85–95.

RainCityNLP at BioLaySumm2025: Extract then Summarize at Home

Jen Wilson **Avery Bellamy** **Rachel Edwards** **Michael Pollack** **Helen Salgi**
Department of Linguistics, University of Washington
(jenwils, elshire, rke4, uwmpp, hvs278)@uw.edu

Abstract

As part of the BioLaySumm shared task at ACL 2025, we developed a summarization tool designed to translate complex biomedical texts into layperson-friendly summaries. Our goal was to enhance accessibility and comprehension for patients and others without specialized medical knowledge. The system employed an extractive-then-abstractive summarization pipeline. For the abstractive component, we experimented with two models: Pegasus-XSum and a Falcons.ai model pre-trained on medical data. Final outputs were evaluated using the official BioLaySumm 2025 metrics. To promote practical accessibility, we completed all experimentation on consumer-grade hardware, demonstrating the feasibility of our approach in low-resource settings.

1 Introduction

The BioLaySumm shared task *Lay Summarization of Biomedical Research Articles and Radiology Reports @ BioNLP Workshop, ACL 2025* (Xiao et al., 2025) is conducting its third iteration this year. The goal of the shared task is to improve techniques for summarizing biomedical texts in non-scientific lay-terms, in order to increase the accessibility and understanding of medical texts for patients and others who are not in the medical field. We used the data from the shared task as well as their evaluation methods to create and evaluate our models and referenced previous participants' work for inspiration. We used an extractive-then-abstractive summarization technique. Beginning with extractive summarization and followed by training both the Pegasus-XSum model and the Falconsai/medical_summarization model to produce abstractive summaries. As a step towards future iterations of summarization, we have also created a dictionary of medical terms translated to lay-terms for injection.[1]

[1]https://github.com/michael-pollack/573Project.git

2 Related Work

Our pipeline of extractive-to-abstractive summarization was inspired by previous iterations of this workshop (Goldsack et al., 2023), (Goldsack et al., 2024) and the winning paper from 2024 (You et al., 2024). Our work is also influenced by the datasets used in this task (eLife and PLOS), which were developed by (Goldsack et al., 2022) and (Luo et al., 2022).

3 Description of Data

The dataset 'BioLaySumm2025-PLOS' consists of 26,291 rows and the dataset 'BioLaySumm2025-eLife' consists of 4,729 rows. Each row consists of the following information: the original text of a biomedical article, a gold-standard lay-terms summary, a list of section headings, a list of keywords, the year of publication, and the article title. Both datasets are already split into training, validation, and test.

We created a lay-term dictionary to add lay-term injection to our pipeline in the future. The dictionary consists of medical terms and their corresponding lay-term alternative based on a Stanford Glossary of medical terms (Stanford Research Compliance Office, n.d.). We were careful to start definitions with a consonant if the original word began with a consonant, and extended this to vowels. This premeditated measure was taken to facilitate smoother substitutions in the future with lay-term injections in the abstractive summaries.

3.1 Pre-Processing Data

Analysis showed that there are a large number of citations in academic text, which tend not to contribute significantly to the actual meaning of the document and are laden with complicated punctuation that affected our sentence tokenizer. We removed all information enclosed in parentheses

BioNLP 2025 Shared Tasks, pages 190–195
August 1, 2025 ©2025 Association for Computational Linguistics

using regex and acknowledge that it removes more than just citations.

Since TF-IDF relies on vocabulary counts to calculate the importance of words, it is beneficial to remove stopwords and lemmatize the data first to reduce vocabulary size and establish obvious connections between different morphological variations of the same word. We used the built-in NLTK list of English stopwords, as well as our own short list of custom stopwords to target and remove stopwords from the data. NLTK's WordNet Lemmatizer was used to lemmatize remaining words in the document. The data we used to create the extractive summaries consists of both the clean lemmatized data resulting from these preprocessing techniques as well as the un-lemmatized version.

4 Model

4.1 Total Pipeline Overview

We use data cleaning and TF-IDF for preprocessing and the creation of extractive summaries. The extractive summaries are fed into an abstractive summary model.

4.2 TF-IDF

Term Frequency – Inverse Document Frequency (TF-IDF) (Sparck Jones, 1988) gives each word in a document a score based on importance to the document's meaning relative to the collection of documents. We chose TF-IDF because it allows us to numerically calculate the importance of words and sentences in a systematic way, thereby allowing us to rank which sentences should appear in the final extractive summary.

We used Scikit-Learn's prebuilt TF-IDF vectorizer model with the cleaned and lemmatized Elife training data as input to calculate the numerical importance of every word in each document in the data set. This produces a set of (word, vector) pairs for each document, where the larger the vector number, the higher the importance of the word. We calculated the importance of each sentence within a document by summing the TF-IDF scores of each word in the current sentence and dividing by the sentence's total word count. A higher score means that the sentence has a greater relevance to the meaning of the document.

We then return the 40% top-scoring sentences as an extractive summary.

4.3 Pegasus-XSum

Pegasus is an abstractive text summarization model developed by Google Research (Zhang et al., 2020). It is based on the Transformer architecture and was specifically pre-trained for summarization tasks using a "gap-sentence" technique, where whole sentences are masked and the model learns to predict them from the remaining text. This model was chosen because it could be fine-trained on our hardware.

4.4 T5 for Medical Text Summarization

Parallel to Pegasus-XSum, we also used the Falconsai/medical_summarization model (Wolf et al., 2020). This T5 Large for Medical Text Summarization model is fine-tuned specifically for medical domain summarization tasks. This model was selected for its strong performance on domain-specific texts and its ability to run efficiently on consumer-grade hardware, making it suitable for reproducible and accessible NLP research.

4.5 Computing Limitations

Limited access to high-end computing made it unrealistic to fine-tune hyper-parameters during the data validation. This is discussed in Section 7.1.

5 Evaluation

Relevance is measured using ROUGE (1, 2, and L), BLEU, METEOR, and BERTScore. ROUGE (Recall-Oriented Understudy for Gisting Evaluation) includes measures to automatically determine the quality of a summary by comparing it to other (ideal) summaries created by humans. BLEU (Bilingual Evaluation Understudy) is a measurement of an automatic translation and a human written translation of the same material. METEOR (Metric for Evaluation of Translation with Explicit ORdering) is based on the harmonic mean of unigram precision and recall, with recall weighted higher than precision. For BERTScore, a neural evaluation metric uses contextual embeddings from pre-trained language models (like BERT) to calculate similarity scores between candidate and reference texts.

Readability is measured using Flesch-Kincaid Grade Level (FKGL) and Dale-Chall Readability Score (DCRS), Coleman-Liau Index (CLI), and LENS. Factuality is measured using AlignScore, and SummaC.

6 Results

For the medical summarization model, we used 'Summarize the following scientific article" as the prompt followed by the summary of the TF-IDF. For the pegasus models, we pass the TF-IDF summary through the model.

The results table of our summarizations are shown in 1.

6.1 Relevancy

As shown our evaluation data, medical_summarization scored highest for relevance. This model was developed specifically for summarization of medical text and it follows reasonably that it would score the highest. The extractive summaries coming in second in relevancy, beating out the Pegasus-XSum model (another specifically trained model for summarization) The extractive summaries were based mainly on frequency of word occurrence, using TF-IDF, which may make sense in context as a word that occurs frequently in the documents but is not an extremely general word of English, has a high likelihood of being relevant.

6.2 Readability

The Pegasus-XSum model dominated in the results for Readability in the FKGL evaluation metric, with the fine-tuned version of the model performing extremely well under the LENS evaluation metric. For DCRS the Pegasus-XSum model seems to perform slightly better than the rest, with the fine-tuned version actually performing the worst, and for CLI all three models other than the fine-tuned pegasus model perform at extremely similar levels, with the medical_sumarization model just barely performing a bit better than the rest. Readability tends to focus on word complexity and overall clarity of a summary. The Pegasus model is trained on medical texts, along with a wider variety of text to produce summarizations. This wider expanse of data could contribute to its readability scores as opposed to relevance because it is trained to create well-made abstractive summaries.

6.3 Factuality

Our original extractive summaries performed better than other models using both evaluation metrics. This is a relatively unsurprising result as the extractive summaries utilize the original sentences from the documents. Consequently, the summaries will be more factual than for the pegasus or medical_summarization models because the text is coming straight from the source.

7 Discussion

7.1 Accessible AI

In this section, we discuss how medical summarization systems can be made more accessible to a broader range of users. While recent advancements in medical NLP have demonstrated impressive capabilities, they often come with steep computational requirements, limiting their practical use outside well-resourced research institutions. Bednarczyk et al.(2025) report that the success of using an LLM for summarization relies on the computational resources available and future research should "evaluate the economic impact of deployment to ensure that LLM adoption is both technically and financially sustainable in clinical settings." We argue that accessibility - in both economic and practical terms - is essential if these technologies are to benefit clinicians, medical researchers, and public health professionals operating in low-resource environments or institutions without dedicated computing clusters.

7.2 Economic Accessibility

The growing trend toward large-scale models has created a barrier to entry for many who wish to apply or replicate state-of-the-art NLP techniques. We quickly ran into computational resource barriers because our plans of replicating previous work required computing resources that we did not have. As a result, our final choice of models and data processing were simpler to run, and can be used by people who do not have access to high-end computing resources. Our approach eliminates the need for expensive GPU clusters that are often used in academic settings.

All experiments in this study were conducted on one of two laptops. We used either a laptop with an Intel 13th Gen Core i9-13900H CPU, 32GB of RAM and a NVIDIA GeForce RTX 4060 Laptop GPU with 8GB RAM. This configuration, while not trivial, remains within reach of many individuals with limited budgets and does not rely on a distributed GPU cluster or a cloud-based API that incurs costs.

By using moderately sized, open-access models, such as Pegasus-XSum, and optimizing evaluation tools, we demonstrate that it is feasible to run sum-

	Relevance				Readability				Factuality	
model	**ROUGE**	**BLEU**	**METEOR**	**BERTScore**	**FKGL**	**DCRS**	**CLI**	**LENS**	**AlignScore**	**SummaC**
fine tuned pegasus-x sum	0.2119	1.8929	0.1453	0.8395	11.3870	8.6415	11.6500	45.9017	0.7654	0.6347
extractive summaries	0.2123	2.6039	0.2823	0.8276	19.7827	9.6537	15.4580	9.5067	0.9494	0.7874
pegasus-x sum hugging face	0.1816	2.0111	0.2122	0.8039	62.7306	13.4768	15.5227	9.1487	0.9029	0.6957
falconsai/ medical_s ummarizati on	0.2845	4.8686	0.2405	0.8396	16.7401	11.6596	16.2369	9.4109	0.6118	0.6531

Figure 1: Evaluation Results

marization pipelines and evaluate results locally. This makes the system viable for clinicians, medical students, or NGOs who may wish to generate or verify lay summaries in real-world medical communication contexts.

7.3 Practical Reproducibility and Implementation

An important tenet of scientific research is reproducibility. Methods that can be implemented on accessible hardware can more easily be reproduced by scholars and others who want to learn. An analysis by Belz et al. (2021) demonstrates that replication and reproducibility are critical to scientific research, yet "it is surprisingly hard to achieve, 70% of scientist reporting failure to reproduce someone else's results, and more than half reporting failure to reproduce their own."

In this study, reproducibility was hindered by the complexity of previous configurations, conflicting dependencies, and reliance on costly computing environments. By deliberately choosing lightweight tools and open-source resources, we designed a summarization and evaluation pipeline that can be easily shared, executed, and adapted.

7.4 The Downside of Small Computing

While our emphasis on accessibility enables broader participation in this task, it also introduces notable limitations. Fine-tuning Pegasus-XSum on our consumer-grade hardware required approximately 40 hours, significantly slowing experimentation cycles. These experimentation cycles were slow and had to be run sequentially, instead of in parallel as could be done on a distributed GPU cluster. Due to hardware constraints, we were unable to explore larger or more recent models which would likely to produce results that score higher on the leaderboard.

Time and resource constraints prevented us from fine-tuning with separate validation data, limiting our ability to tune hyper-parameters effectively. These trade-offs demonstrate the challenges faced by low-resource researchers and scientists while highlighting the need to develop lightweight, efficient models that perform well without requiring extensive investments in hardware.

8 Conclusion

Our work demonstrates that medical summarization is achievable even with limited computational resources. By leveraging models like Pegasus-XSum, we were able to develop and evaluate effective summarization systems on a standard laptop, highlighting the potential for accessible and reproducible research in this space. Our findings support the idea that meaningful contributions to biomedical NLP can be made without relying on large-scale infrastructure, paving the way for more inclusive and resource-efficient approaches to language technology.

Limitations

While our system demonstrates results in generating readable and relevant lay summaries of biomedical texts using consumer-grade hardware, several limitations should be acknowledged.

Dataset Diversity

This work relies exclusively on two open-acess datasets: BioLaySumm2025-eLife and BioLaySumm2025-PLOS, both of which consist entirely of English language documents from a set of biomedical articles. As such, our model's generalizability to other medical domains or languages is untested.

Lay-Term Dictionary Coverage

The lay-term lexicon that we created, while a valuable resource for term injection, is limited in scope. It is derived from a single source (Stanford Research Compliance Office) and does not cover all relevant terminology. It requires an additional step and is not part of the summarization pipeline.

Pre-Processing

Our pre-processing decisions, particularly the removal of all parenthetical content using regular expressions, may have inadvertently discarded meaningful information. Although our rationale was that parenthetical content typically contains citations or supplementary material, this approach may have led to the loss of scientific details.

Experimentation Bottlenecks

Because all experiments were conducted on consumer-grade hardware without parallel GPU resources, experimentation had to proceed sequentially and in a time-consuming manner. This significantly limited our ability to iterate on model design or integrate new features (such as lay-term injection).

Validation and Fine-Tuning

Time and hardware constraints prevented us from fine-tuning using dedicated validation data. This limited our capacity to adapt the models. As a result, our models may not be optimally calibrated for the data distributions.

Ethical Considerations

The goal of this project and of the shared task as a whole is to expand the reach of biomedical text and make this information more approachable to people outside of the medical field. However, it is important to acknowledge that this task is not without its risks. For example, a flawed summarization system has the potential to give false information or omit important details from the original text, which is fundamentally opposed to the goal of the project. Additionally, it is important to include a diverse selection of texts when training a model of this kind, in order to reduce biases and create a model that can adapt and be used for a variety of new documents.

For this type of project, it is important to know that private information is not included in training documents, as that would be a violation of the privacy of individuals. The data included in our project was provided by the creators of the BioLaySumm shared task and comes from an open-access publisher (PLOS) and journal (eLife) so this is not a major concern for us.

References

Lydie Bednarczyk, Daniel Reichenpfader, Christophe Gaudet-Blavignac, Amon Kenna Ette, Jamil Zaghir, Yuanyuan Zheng, Adel Bensahla, Mina Bjelogrlic, and Christian Lovis. 2025. Scientific evidence for clinical text summarization using large language models: Scoping review. *Journal of Medical Internet Research*, 27:e68998.

Anya Belz, Shubham Agarwal, Anastasia Shimorina, and Ehud Reiter. 2021. A systematic review of reproducibility research in natural language processing. In *Proceedings of the 16th Conference of the European Chapter of the Association for Computational Linguistics: Main Volume*, pages 381–393, Online. Association for Computational Linguistics.

Tomas Goldsack, Zheheng Luo, Qianqian Xie, Carolina Scarton, Matthew Shardlow, Sophia Ananiadou, and Chenghua Lin. 2023. Overview of the biolaysumm 2023 shared task on lay summarization of biomedical research articles. In *The 22nd Workshop on Biomedical Natural Language Processing and BioNLP Shared Tasks*, pages 468–477, Toronto, Canada. Association for Computational Linguistics.

Tomas Goldsack, Carolina Scarton, Matthew Shardlow, and Chenghua Lin. 2024. Overview of the BioLaySumm 2024 shared task on the lay summarization of biomedical research articles. In *Proceedings of the 23rd Workshop on Biomedical Natural Language Processing*, pages 122–131, Bangkok, Thailand. Association for Computational Linguistics.

Tomas Goldsack, Zhihao Zhang, Chenghua Lin, and Carolina Scarton. 2022. Making science simple: Corpora for the lay summarisation of scientific literature. In *Proceedings of the 2022 Conference on Empirical Methods in Natural Language Processing*, pages 10589–10604, Abu Dhabi, United Arab Emirates. Association for Computational Linguistics.

Zheheng Luo, Qianqian Xie, and Sophia Ananiadou. 2022. Readability controllable biomedical document summarization. In *Findings of the Association for Computational Linguistics: EMNLP 2022*, pages 4667–4680, Abu Dhabi, United Arab Emirates. Association for Computational Linguistics.

Karen Sparck Jones. 1988. *A statistical interpretation of term specificity and its application in retrieval*, page 132–142. Taylor Graham Publishing, GBR.

Stanford Research Compliance Office. n.d. Definitions & lay glossary of medical terms. `https://researchcompliance.stanford.edu/panels/hs/for-all-researchers/definitions`. Accessed: 2025-05-23.

Thomas Wolf, Lysandre Debut, Victor Sanh, Julien Chaumond, Clement Delangue, Anthony Moi, Pierric Cistac, Tim Rault, Rémi Louf, Morgan Funtowicz, Joe Davison, Sam Shleifer, Patrick von Platen, Clara Ma, Yacine Jernite, Julien Plu, Canwen Xu, Teven Le Scao, Sylvain Gugger, and 3 others. 2020. Huggingface's transformers: State-of-the-art natural language processing. *Preprint*, arXiv:1910.03771.

Chenghao Xiao, Kun Zhao, Xiao Wang, Siwei Wu, Sixing Yan, Tomas Goldsack, Sophia Ananiadou, Noura Al Moubayed, Liang Zhan, William Cheung, and Chenghua Lin. 2025. Overview of the biolaysumm 2025 shared task on lay summarization of biomedical research articles and radiology reports. In *The 24th Workshop on Biomedical Natural Language Processing and BioNLP Shared Tasks*, Vienna, Austria. Association for Computational Linguistics.

Zhiwen You, Shruthan Radhakrishna, Shufan Ming, and Halil Kilicoglu. 2024. UIUC_BioNLP at BioLaySumm: An extract-then-summarize approach augmented with Wikipedia knowledge for biomedical lay summarization. In *Proceedings of the 23rd Workshop on Biomedical Natural Language Processing*, pages 132–143, Bangkok, Thailand. Association for Computational Linguistics.

Jingqing Zhang, Yao Zhao, Mohammad Saleh, and Peter J. Liu. 2020. Pegasus: pre-training with extracted gap-sentences for abstractive summarization. In *Proceedings of the 37th International Conference on Machine Learning*, ICML'20. JMLR.org.

TLPIQ at BioLaySumm: Hide and Seq, a FLAN-T5 Model for Biomedical Summarization

Melody Bechler[1], Carly Crowther[1], Emily Luedke[1], Natasha Schimka[1], Ibrahim Sharaf[1]
[1]Department of Linguistics, University of Washington, Seattle, WA, USA
{mbechler, carlyc88, eluedke, nschim2, ibshar}@uw.edu

Abstract

BioLaySumm 2025 is a shared task that aims to automatically generate lay summaries of scientific papers for a wider audience of readers without domain-specific knowledge, making scientific discoveries in the domain of biology and medicine more accessible to the general public. Our submission to the task is a FLAN-T5 base model fine-tuned on the abstract and conclusion of articles and expert-written lay summaries from the shared task's provided datasets. We find that our system performs competitively in terms of relevance, exceeds the baseline on factuality, but falls short on readability.[1]

1 Introduction

Lay summarization is the task of summarizing domain specific texts into simplified summaries non-experts can understand. In these types of summaries, complex jargon is eliminated and information is summarized in a clear and concise manner for easy readability. Biomedical literature is an example of highly technical, jargon-rich texts that are difficult to understand by those outside of the field, but are invaluable resources for interested researchers, professionals, and the general public. Unfortunately, this wealth of knowledge has limited accessibility and comprehension due to length and complexity. Lay summaries can improve science literacy, help limit the spread of misinformation, and invite interdisciplinary work (King et al., 2017).

To address these persistent issues, the Biomedical Lay Summarization task (BioLaySumm) 2025 (Xiao et al., 2025) shared task at the BioNLP Workshop at ACL 2025[2] focuses on various biomedical lay summarization tasks, from plain lay summarization to multimodal lay summarization. The Lay People in Question (TLPIQ) team focuses on plain lay summarization as a baseline model to summarize biomedical texts. This model aims to improve accessibility and understanding of these complex texts, while maintaining factuality and domain relevance.

2 Related Work

Previous work has evaluated two types of summarization: extractive and abstractive. Extractive summarization aims to select verbatim components of a document to create a summary, whereas abstractive summarization generates novel summaries. Overviews of the past two years of the task can be found in Goldsack et al. (2023) and Goldsack et al. (2024). Particularly, using an extract-then-summarize approach with TextRank (Mihalcea and Tarau, 2004) and BERT (Devlin et al., 2019), You et al. (2024) extracted text to reduce input length to separately fine-tune GPT-3.5 and a Longformer Encoder Decoder model to achieve the best performance in the task last year.

Preprocessing techniques showcased positive summarization results. Zhao et al. (2024) indicated that hard truncation and text-chunking resulted in better quality and efficiency compared to data augmentation and prompt engineering techniques. Modi and Karthikeyan (2024) utilized a preprocessing over the abstract technique to extract initial sentences from a document and remove punctuation and enclosed text to successfully increase summary readability.

Previous work has utilized smaller parameter sequence-to-sequence models with varying results. Malik et al. (2024) utilized a FLAN-T5 model with a basic prompt structure, but the lack of constraints, limited training, and context length of the model resulted in poor lay summarization output. Modi and Karthikeyan (2024) also fine-tuned a FLAN-T5-base, but focused on preprocessing over the

[1]Our code is made available in a public repository: https://github.com/nschimka/TLPIQ—BioLaySumm-2025
[2]https://aclweb.org/aclwiki/BioNLP_Workshop

BioNLP 2025 Shared Tasks, pages 196–201
August 1, 2025 ©2025 Association for Computational Linguistics

abstract and a cosine scheduler to generate lay summaries.

In this task, we train a sequence-to-sequence FLAN-T5-base [3] model with abstract extraction, instruction tuning with dataset tags, and a specialized prompt template to improve upon previous T5 lay summarization methods. Sequence-to-sequence models handle input and output sequences better than other larger models while being computationally efficient, making the T5 model a strong choice for this summarization task.

3 Data

The dataset for the shared task is from Goldsack et al. (2022) which includes articles from two different biomedical resources. The Public Library of Science[4] (PLOS) is an open-access non-profit publisher of articles from various peer-reviewed journals in a wide variety of scientific fields. eLife[5] is an open-access peer-reviewed journal of biomedical and life sciences. Of the two, PLOS is longer, with 24,773 instances for training and 1,376 for validation. eLife contains 4,346 instances for training and 241 for validation.

We performed exploratory data analysis (EDA) on the two data sets to better understand the quantitative and qualitative features of both the articles and the summaries. See Appendix A for the results of the EDA.

4 Methods

4.1 Preprocessing

Because the FLAN-T5-base model has a maximum input length of 1,024 tokens, the original articles needed to be shortened significantly from the average token lengths of 6,981 tokens for PLOS and 10,428 tokens for eLife (see Figure 1 in Appendix A.1).

We segmented each article into sections using newline characters, appended the dataset-provided keywords to each input to enrich contextual information, and removed in-text citations with a regular expression.

We implemented a TF-IDF scoring function with scikit-learn's `TfidfVectorizer` class (Pedregosa et al., 2012) to find the most important sentences in each section, rank the sentences by importance, then take the first n allotted tokens starting at the top of the list of sentences. We found the most success allotting 50% of the input tokens to the abstract and 50% to the conclusion/discussion section (depending on each article's naming conventions).

4.2 Model

We fine-tuned FLAN-T5-base (248M parameters), an instruction-tuned variant of the T5 architecture (Raffel et al., 2020; Chung et al., 2022) to balance the compute cost and performance in our combined 30K sample biomedical corpus. Its Transformer backbone with multi-headed attention (Vaswani et al., 2017) captures long-range dependencies in scientific text, enabling accurate and accessible lay summaries.

At inference time, we steer our fine-tuned FLAN-T5-base model with a diverse controlled beam search setup to balance faithfulness, readability, and coverage. We generate up to 400 new tokens (minimum 120) beyond the input prompt to ensure complete summaries without truncation, using 8 beams divided into 4 diversity groups (diversity penalty = 0.8) to explore varied phrasings. To avoid repetition of three-gram patterns, we enforce `no_repeat_ngram_size=3` to avoid repeating n-grams and apply a mild repetition penalty of 1.2. A length penalty of 0.9 encourages more comprehensive output.

Details of our model approach can be found in Appendix B.

5 Evaluation

Evaluation for this task cover three areas: relevance of the summary to the original article, readability, and factuality.

Relevance is measured with ROUGE (1, 2, and L), BLEU, METEOR, and BERTScore; readability is measured with Flesch-Kincaid Grade Level (FKGL) and Dale-Chall Readability Score (DCRS), Coleman-Liau Index (CLI), and LENS; factuality is measured with AlignScore and SummaC.

BioLaySumm 2025 utilizes the Codabench (Xu et al., 2022) platform for participants to submit their predicted summary results. Our final model output consisted of the predicted summaries, quality scores, token counts, and input text identifiers. We created a script to retain only the predicted summaries in the appropriate submission format to then evaluate model performance.

[3]https://huggingface.co/docs/transformers/model_doc/flan-t5

[4]https://huggingface.co/datasets/BioLaySumm/BioLaySumm2025-PLOS

[5]https://huggingface.co/datasets/BioLaySumm/BioLaySumm2025-eLife

Source	ROUGE	BLEU	METEOR	BERTScore
FLAN-T5 base	0.34	7.16	0.27	0.86
llama3-8B-sft	**0.37**	**9.86**	**0.31**	0.86
qwen2.5-7B-sft	0.35	8.74	0.3	**0.87**

Table 1: Comparison of our relevance scores across evaluation metrics compared to the baselines. Best score for each metric is in bold.

Source	FKGL	DCRS	CLI	LENS
FLAN-T5 base	13.44	10.59	13.43	43.68
llama3-8B-sft	**12.21**	**9.23**	**12.98**	72.86
qwen2.5-7B-sft	12.71	9.65	13.7	**60.22**

Table 2: Readability scores across metrics

6 Results

Tables 1, 2, and 3 present our FLAN-T5 Base model's performance alongside the shared-task baselines—Llama3 (8B params) (Grattafiori et al., 2024) and Qwen2.5 (7B params) (Qwen Team, 2024)—as reported on Codabench[6].

Our combined-dataset FLAN-T5 system (248M params) achieves a ROUGE of 0.34 and a BERTScore of 0.86, compared to Llama3's ROUGE of 0.37 (BERTScore = 0.86) and Qwen2.5's ROUGE of 0.35 (BERTScore = 0.87) (Table 1). In factuality metrics (Table 3), we match or exceed these baselines, with an AlignScore of 0.76 (vs. 0.72/0.75) and SummaC of 0.64.

However, our readability scores (Table 2) reveal a larger gap: our FKGL of 13.44 and LENS of 43.68 lag behind Llama3 (12.21/72.86) and Qwen2.5 (12.71/60.22).

These results demonstrate that a lightweight 248M-parameter FLAN-T5 model can achieve relevance and factuality on par with much larger 7–8 B–parameter systems, but still requires further refinement to match their readability.

Source	AlignScore	SummaC
FLAN-T5 base	**0.76**	**0.64**
llama3-8B-sft	0.72	0.64
qwen2.5-7B-sft	0.75	0.64

Table 3: Factuality scores across metrics

[6]Accessed May 22, 2025

7 Discussion

In our error analysis, we identified two main shortcomings of the combined-dataset model. First, eLife summaries were sometimes truncated mid-sentence; key findings would abruptly end because the model had internalized a compression ratio dominated by the shorter PLOS summaries (see Figure 3 in Appendix A.1). Second, despite our diverse beam search and generation strategies, occasional technical terms still slipped through, subtly raising both Flesch–Kincaid and LENS scores and detracting from true lay readability.

Looking ahead, we see three promising directions. First, training separate, dataset-specific models would let each learn its own optimal compression ratio and vocabulary constraints, eliminating length-bias effects. Second, a two-stage pipeline, initially generating a faithful summary and then passing it through a lightweight simplification model or rule-based rewriter, could ensure factual accuracy while improving clarity. Finally, integrating a post hoc lexical simplification step, via curated synonym lists or a small neural simplifier, would remove residual jargon and bring reading levels down to our grade 8-9 target. Together, these refinements promise to restore full-sentence integrity and markedly boost readability without sacrificing domain fidelity.

Limitations

While our system achieves strong relevance and factuality scores, it exhibits several limitations that affect its overall performance—particularly in terms of readability. First, the use of a single model trained on both PLOS and eLife datasets introduced a compression mismatch: summaries generated from longer eLife articles were occasionally truncated mid-sentence, likely due to the model internalizing an average summary length skewed by the shorter PLOS samples. This resulted in incomplete outputs and diminished coherence for eLife inputs.

Second, despite instruction-tuning and controlled decoding strategies, technical vocabulary and complex syntax persisted in some outputs. This limited the model's ability to consistently produce content aligned with the target 8th–9th grade reading level, as evidenced by elevated FKGL and LENS scores.

Moreover, due to time constraints, we did not explore more advanced strategies such as multi-stage summarization, dataset-specific modeling, or post-hoc simplification pipelines. These approaches may have mitigated the readability issues while preserving factual accuracy.

Finally, all evaluations rely on automatic metrics. While useful for benchmarking, they may not fully capture nuance in accessibility, clarity, or human comprehension—factors that are especially critical in the biomedical lay summarization context.

Acknowledgments

This work was produced as part of a class project for the Professional MS in Computational Linguistics during Spring 2025 at the University of Washington.

References

Hyung Won Chung, Le Hou, Shayne Longpre, Barret Zoph, Yi Tay, William Fedus, Eric Li, Xuezhi Wang, Mostafa Dehghani, Siddhartha Brahma, and 1 others. 2022. Scaling Instruction-Finetuned Language Models. *arXiv preprint arXiv:2210.11416*.

Jacob Devlin, Ming-Wei Chang, Kenton Lee, and Kristina Toutanova. 2019. BERT: Pre-training of Deep Bidirectional Transformers for Language Understanding. In *Proceedings of the 2019 Conference of the North American Chapter of the Association for Computational Linguistics: Human Language Technologies, Volume 1 (Long and Short Papers)*, pages 4171–4186, Minneapolis, Minnesota. Association for Computational Linguistics.

Tomas Goldsack, Zheheng Luo, Qianqian Xie, Carolina Scarton, Matthew Shardlow, Sophia Ananiadou, and Chenghua Lin. 2023. Overview of the BioLaySumm 2023 Shared Task on Lay Summarization of Biomedical Research Articles. In *The 22nd Workshop on Biomedical Natural Language Processing and BioNLP Shared Tasks*, pages 468–477, Toronto, Canada. Association for Computational Linguistics.

Tomas Goldsack, Carolina Scarton, Matthew Shardlow, and Chenghua Lin. 2024. Overview of the BioLaySumm 2024 Shared Task on the Lay Summarization of Biomedical Research Articles. In *Proceedings of the 23rd Workshop on Biomedical Natural Language Processing*, pages 122–131, Bangkok, Thailand. Association for Computational Linguistics.

Tomas Goldsack, Zhihao Zhang, Chenghua Lin, and Carolina Scarton. 2022. Making Science Simple: Corpora for the Lay Summarisation of Scientific Literature. In *Proceedings of the 2022 Conference on Empirical Methods in Natural Language Processing*, pages 10589–10604, Abu Dhabi, United Arab Emirates. Association for Computational Linguistics.

Aaron Grattafiori, Abhimanyu Dubey, Abhinav Jauhri, Abhinav Pandey, Abhishek Kadian, Ahmad Al-Dahle, Aiesha Letman, Akhil Mathur, Alan Schelten, Alex Vaughan, Amy Yang, Angela Fan, Anirudh Goyal, Anthony Hartshorn, Aobo Yang, Archi Mitra, Archie Sravankumar, Artem Korenev, Arthur Hinsvark, and 542 others. 2024. The Llama 3 Herd of Models. *Preprint*, arXiv:2407.21783.

Stuart RF King, Emma Pewsey, and Sarah Shailes. 2017. Plain-language Summaries of Research: An inside guide to eLife digests. *eLife*, 6(e25410).

Hemang Malik, Gaurav Pradeep, and Pratinav Seth. 2024. HGP-NLP at BioLaySumm: Leveraging LoRA for Lay Summarization of Biomedical Research Articles using Seq2Seq Transformers. In *Proceedings of the 23rd Workshop on Biomedical Natural Language Processing*, pages 831–836, Bangkok, Thailand. Association for Computational Linguistics.

Rada Mihalcea and Paul Tarau. 2004. TextRank: Bringing Order into Text. In *Proceedings of the 2004 Conference on Empirical Methods in Natural Language Processing*, pages 404–411, Barcelona, Spain. Association for Computational Linguistics.

Satyam Modi and T Karthikeyan. 2024. Eulerian at BioLaySumm: Preprocessing Over Abstract is All You Need. In *Proceedings of the 23rd Workshop on Biomedical Natural Language Processing*, pages 826–830, Bangkok, Thailand. Association for Computational Linguistics.

Fabian Pedregosa, Gaël Varoquaux, Alexandre Gramfort, Vincent Michel, Bertrand Thirion, Olivier Grisel, Mathieu Blondel, Peter Prettenhofer, Ron Weiss, Vincent Dubourg, Jake VanderPlas, Alexandre Passos, David Cournapeau, Matthieu Brucher, Matthieu Perrot, and Edouard Duchesnay. 2012. Scikit-learn: Machine Learning in Python. *CoRR*, abs/1201.0490.

Qwen Team. 2024. Qwen2.5 Technical Report. *arXiv preprint arXiv:2412.15115*.

Colin Raffel, Noam Shazeer, Adam Roberts, Katherine Lee, Sharan Narang, Michael Matena, Yanqi Zhou, Wei Li, and Peter J. Liu. 2020. Exploring the Limits of Transfer Learning with a Unified Text-to-Text Transformer. *Journal of Machine Learning Research*, 21(140):1–67.

Hai Tran, Zhengyun Yang, Zeya Yao, and Hong Yu. 2024. BioInstruct: Instruction Tuning of Large Language Models for Biomedical Natural Language Processing. *Journal of the American Medical Informatics Association*, 31(9):1821–1832.

Ashish Vaswani, Noam Shazeer, Niki Parmar, Jakob Uszkoreit, Llion Jones, Aidan N. Gomez, Łukasz Kaiser, and Illia Polosukhin. 2017. Attention Is All You Need. In *Advances in Neural Information Processing Systems*, pages 5998–6008.

Daniel Wan, Mengwen Liu, Kathleen McKeown, Markus Dreyer, and Mohit Bansal. 2023. Faithfulness-Aware Decoding Strategies for Abstractive Summarization. In *Proceedings of the 17th Conference of the European Chapter of the Association for Computational Linguistics*, pages 2864–2880.

Chenghao Xiao, Kun Zhao, Xiao Wang, Siwei Wu, Sixing Yan, Tomas Goldsack, Sophia Ananiadou, Noura Al Moubayed, Liang Zhan, William Cheung, and Chenghua Lin. 2025. Overview of the BioLaySumm 2025 Shared Task on Lay Summarization of Biomedical Research Articles and Radiology Reports. In *The 24nd Workshop on Biomedical Natural Language Processing and BioNLP Shared Tasks*, Vienna, Austria. Association for Computational Linguistics.

Zhen Xu, Sergio Escalera, Adrien Pavão, Magali Richard, Wei-Wei Tu, Quanming Yao, Huan Zhao, and Isabelle Guyon. 2022. Codabench: Flexible, easy-to-use, and reproducible meta-benchmark platform. *Patterns*, 3(7):100543.

Zhiwen You, Shruthan Radhakrishna, Shufan Ming, and Halil Kilicoglu. 2024. UIUC_BioNLP at BioLaySumm: An Extract-then-summarize Approach Augmented with Wikipedia Knowledge for Biomedical Lay Summarization. In *Proceedings of the 23rd Workshop on Biomedical Natural Language Processing*, pages 132–143, Bangkok, Thailand. Association for Computational Linguistics.

Ruijing Zhao, Siyu Bao, Siqin Zhang, Jinghui Zhang, Weiyin Wang, and Yunian Ru. 2024. Ctyun AI at BioLaySumm: Enhancing Lay Summaries of Biomedical Articles Through Large Language Models and Data Augmentation. In *Proceedings of the 23rd Workshop on Biomedical Natural Language Processing*, pages 837–844, Bangkok, Thailand. Association for Computational Linguistics.

A Data

A.1 Text Length

Figure 1 shows the distribution of tokens per article in both datasets. We found the PLOS articles to be shorter on average with a mean of 6981 tokens per article. The eLife articles were longer with an mean of 10,428 tokens with a greater variability in length.

Figure 2 compares the number of tokens across the gold standard summaries for the two datasets. A similar trend appears, with the PLOS lay summaries containing fewer (mean of 195) tokens than the eLife lay summaries (mean of 386), and the eLife distribution is again wider.

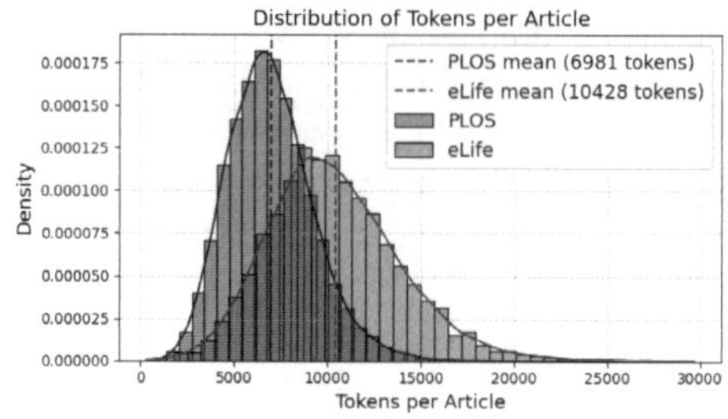

Figure 1: Distribution of tokens per article in the PLOS and eLife datasets.

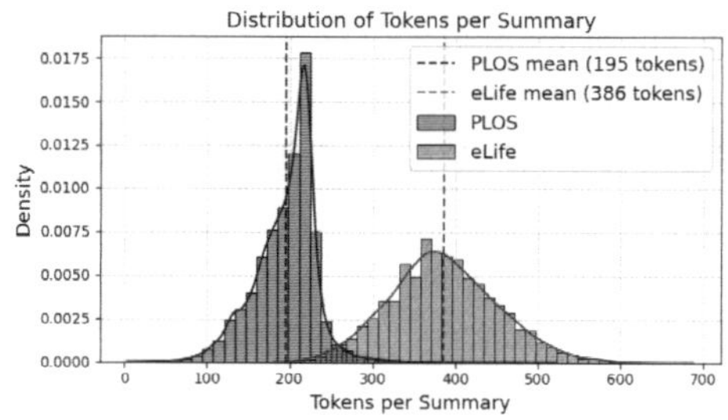

Figure 2: Distribution of tokens per summary in the PLOS and eLife datasets.

A.2 Section Relevance

You et al. (2024) compared each section of an article's relevance to the summary via cosine similarity. Across both datasets, they found the abstract, background, and conclusion to be the most relevant to the summary, in that order.

The existing dataset does not retain the section headings in place in the article text. The are extracted into a section headings list for each instance. We found that the article could be split on '\n' into a list of the different sections. We compared the listed sections across all instances and found that across PLOS instances, 100% contained an abstract, 99.85% contained an introduction, and 95.83% contained a discussion section (with another 3.53% containing a combined results/discussion section). Across eLife instances, 100% contained an abstract, 99.33% contained an introduction, and 98.62% contained a discussion. The compression ratio refers to the difference in length of an article and its lay summary. Figure 3 demonstrates that on average, the PLOS articles are less compressed than the eLife articles. While the eLife summaries are still longer on average than PLOS summaries, their articles are much longer, necessitating more compression of their information into a summary.

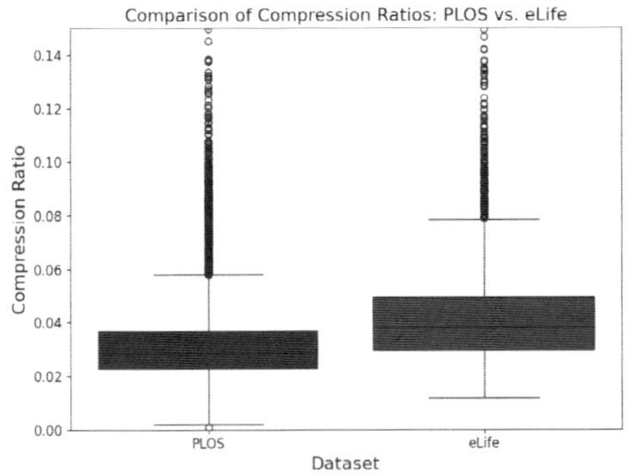

Figure 3: Distribution of compression ratios (article token length divided by summary token length).

B Model Settings

We framed the task using the prompt: *"Create a lay summary of this scientific research for a general audience who has no background in biology,"* leveraging Flan-T5's instruction-tuning capabilities. This approach aligns with recent work showing prompt-based task framing enhances performance in biomedical applications (Tran et al., 2024). We structured inputs with source-specific tags (e.g., <plos> [TITLE]... [ABSTRACT]...) as lightweight semantic cues. Input documents were truncated to 1024 tokens, with output summaries capped at 400 tokens.

Training used AdamW with a learning rate of 3e-5, weight decay of 0.01, and warmup ratio of 0.1. We employed a batch size of 12 without gradient accumulation, using PyTorch with expandable segment configuration for memory efficiency. Early stopping was applied with a patience of 2 evaluation steps. For generation, we utilized beam search with 4 beams, shown to produce more faithful summaries than sampling-based approaches (Wan et al., 2023).

The gradual decrease of both training and validation loss indicate that our model was able to learn and generalize effectively, as shown in Figure 4.

Figure 4: Comparison between the training and validation loss values across all model runs.

LaySummX at BioLaySumm: Retrieval-Augmented Fine-Tuning for Biomedical Lay Summarization Using Abstracts and Retrieved Full-Text Context

Fan Lin* **and** **Dezhi Yu***
School of Information, University of California, Berkeley, USA
{fan.lin, dezhi.yu}@berkeley.edu

Abstract

Generating lay summaries of biomedical research remains a time-intensive task, despite their importance in bridging the gap between scientific findings and non-expert audiences. This study introduces a retrieval-augmented fine-tuning framework for biomedical lay summarization, integrating abstract-driven semantic retrieval with LoRA-tuned LLaMA 3.1 models. Abstracts are used as queries to retrieve relevant text segments from full-text articles, which are then incorporated into prompts for supervised fine-tuning. Evaluations on the PLOS and eLife datasets show that this hybrid approach significantly improves relevance and factuality metrics compared to both base models and those tuned individually, while maintaining competitive readability. Prompt design experiments highlight a trade-off between readability and factual accuracy. Our fine-tuned model demonstrates strong performance in relevance and factuality among open-source systems and rivals closed-source models such as GPT, providing an efficient and effective solution for domain-specific lay summarization.

1 Introduction

Biomedical research is essential to advancing human health and societal well-being. However, with over 1.5 million articles published annually (González-Márquez et al., 2024), it is increasingly difficult for readers to absorb new findings efficiently. Although abstracts are designed to summarize key results, their technical language often limits accessibility for non-experts. Lay summaries help bridge this gap by presenting core contributions in clear, non-technical language, yet they remain uncommon due to the manual effort required. The BioLaySumm shared task addresses this challenge by promoting the automatic generation of high-quality lay summaries to support broader understanding of biomedical research (Xiao et al., 2025).

Recent advances in large language models (LLMs) have enabled zero- and few-shot summarization, reshaping the field through strong language understanding and instruction-following capabilities (Zhang et al., 2024). Results from the BioLaySumm shared task further demonstrate that LLM-based methods perform well in generating lay summaries of biomedical texts (Goldsack et al., 2024, 2023).

One of the key challenges in the BioLaySumm shared task is the computational cost of feeding an entire research article into a large language model (LLM), even though many recent LLMs support extended context windows (e.g., up to 128k tokens in LLaMA 3.1). Prior research has investigated several strategies to address this issue, including text chunking, which segments lengthy documents into smaller, more manageable units for summarization by models such as Mixtral 8x7B (Bao et al., 2024), or extractive summarization techniques that identify and select salient sentences from the full text (You et al., 2024).

In this study, we developed a workflow that integrates retrieval-augmented generation (RAG) with LoRA-based fine-tuning to improve the performance of LLaMA 3.1 on the biomedical lay summarization task (Figure 1) [1]. To address input length constraints imposed by limited GPU memory, we used the abstract of each article as a query to retrieve relevant but complementary content from the full text. Both the abstract and the retrieved information were used to fine-tune the model, enabling it to generate lay summaries that match the editorial style of the target journals, PLOS and eLife.

*These authors contributed equally.

[1] https://github.com/ACL-LLM-Research/BioLaySummarization

BioNLP 2025 Shared Tasks, pages 202–214
August 1, 2025 ©2025 Association for Computational Linguistics

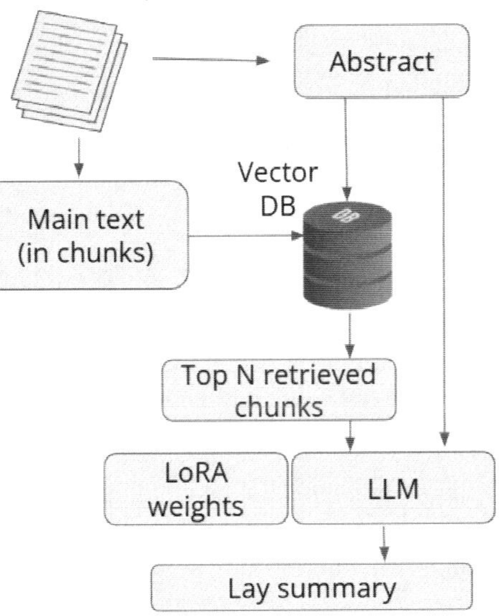

Figure 1: Overview of the proposed workflow for biomedical lay summarization. The abstract is used to query a vector database constructed from the segmented main text of the article. The retrieved content is then combined with the abstract and processed by a fine-tuned language model to generate a lay summary.

2 Methods

2.1 Datasets

In this study, we used a publicly available PLOS and eLife dataset (Goldsack et al., 2022), which includes both full research articles and their corresponding lay summaries written by the original authors or editors. Summary statistics of the data set can be found in the Appendix A.

2.2 Supervised Fine-Tuning

The LLaMA 3.1 8B model was used as the base model for supervised fine-tuning (Grattafiori et al., 2024). Given the size of the training set, we adopted Low-Rank Adaptation (LoRA), a parameter-efficient fine-tuning approach. A brief hyperparameter search was conducted based on the autoregressive loss. Further details are provided in the Appendix B.

2.3 Retrieval-Augmented Generation (RAG)

The vector database was constructed using the main text of each article. The text was segmented into 500-character chunks with a 50-character overlap. Each chunk was embedded using the all-MiniLM-L6-v2 model from the Sentence Transformers library, which encodes sentences and short paragraphs into dense vectors optimized for semantic similarity and retrieval (Reimers and Gurevych, 2019). The resulting embeddings were indexed using FAISS (Douze et al., 2025).

During the retrieval phase, each article's abstract was used as a query to retrieve semantically similar and contextually relevant content from the corresponding document in the vector database. The top five most relevant chunks, ranked by embedding similarity, were incorporated into the prompt alongside the original abstract.

2.4 Evaluation Metrics

We evaluated summary quality using three metric categories: relevance, readability, and factuality.

Relevance was evaluated using ROUGE (Lin, 2004), BLEU(Papineni et al., 2002), METEOR (Banerjee and Lavie, 2005) and, BERTScore (Zhang et al., 2020), which quantify lexical and semantic overlap between the generated and reference summaries.

Readability was evaluated using the Flesch-Kincaid Grade Level (FKGL) (Kincaid et al., 1975), Dale-Chall Readability Score (DCRS) (Dale and Chall, 1948), Coleman-Liau Index (CLI) (Coleman and Liau, 1975) and LENS (Maddela et al., 2023).

Factuality metrics include AlignScore (Zha et al., 2023) and SummaC (Laban et al., 2022), which estimate the consistency of generated summaries with the source content.

Additionally, we explored the use of G-Eval, an LLM-based evaluator that provides a holistic assessment by jointly considering relevance, readability, and factuality. However, it requires further development and was not included in this study. Further details are provided in Appendix C.

3 Result

We first explored several strategies to improve model performance, including retrieval-augmented generation (RAG), LoRA-based fine-tuning, and prompt engineering using the validation set. Based on these evaluations, we selected the best-performing approach and compared its performance on the test set against that of general-purpose large language models.

3.1 Retrieval-Augmented Fine-Tuning

Our study utilizes LLaMA3.1-8B-Instruct as the primary baseline model. To assess the impact of scaling model size by an order of magnitude, we also included the LLaMA3-70B-Instruct base model. However, the performance of the 70B model was only marginally better than that of the 8B model (Table 1).

We then incorporated retrieval-augmented generation (RAG) to evaluate whether retrieved text chunks from the full text could enhance summary quality. The underlying hypothesis is that retrieved content may contain contextual information relevant to key points mentioned in the abstract, thereby providing additional background for generating more comprehensive and informative summaries. Compared to the base model, the RAG approach achieved higher scores in relevance metrics such as ROUGE and METEOR, as well as in most readability metrics across both datasets, although it underperformed in factuality metrics.

The main rationale for using retrieved text in place of the full article is to minimize computational overhead. We compared the performance of the RAG approach with models using full-text input. Interestingly, the results were dataset-dependent. On the PLOS dataset, the RAG-based model outperformed the full-text model across all relevance and factuality metrics. In contrast, on the eLife dataset, the RAG-based summaries underperformed relative to the full-text model in both relevance and readability metrics.(Table 1).

Next, we evaluated whether supervised fine-tuning using LoRA could enhance summary quality. Compared to the base model, the LoRA fine-tuned model demonstrated improvements across all evaluation metrics on both the PLOS and eLife datasets, with the exception of the readability metric LENS. (Table 1).

Finally, we assessed a combined approach using both LoRA and RAG to determine whether the two strategies are complementary. On the PLOS dataset, this combined model outperformed the base model as well as models using LoRA or RAG alone in both relevance and factuality metrics, though not in all readability metrics. On the eLife dataset, the combined model outperformed others in most relevance metrics and achieved higher scores in one factuality metric, AlignScore (Table 1).

3.2 Prompt-Based Trade-off between factuality and readability

Given the low scores observed in readability metrics such as FKGL, CLI, and DCRS, we modified the prompt instructions to enhance readability. Prompts 1 through 4 (B.2 to B.5) were designed to incrementally increase emphasis on readability, while progressively reducing focus on factual accuracy

The results from Prompt 1 to Prompt 4 without LoRA exhibit a consistent upward trend across all readability metrics (Table 2). However, this improvement in readability is accompanied by a decline in factuality, as evidenced by decreasing scores in AlignScore and SummaC. In contrast, for models fine-tuned using LoRA, the increase in readability is less consistent compared to models relying solely on RAG.

We also estimated the average readability metrics of the reference summaries using 100 examples from the eLife training set. These editor-written summaries achieved average scores of FKGL = 11.1694, CLI = 12.2691, DCRS = 11.1068, and LENS = 58.9321. In comparison, our generated summaries using the RAG approach with Prompt 4 produced slightly lower, but comparable results: FKGL = 11.6004, CLI = 12.5959, DCRS = 14.3942, and LENS = 52.1310.

3.3 Comparison against other pretrained LLMs

Given the strong overall performance of LLaMA 3.1–8B with retrieval-augmented fine-tuning using prompts that emphasize factual accuracy, we submitted its results for test set evaluation and compared them to summaries generated by various types of LLMs, including GPT, Qwen (a hybrid reasoning model), and Seed/Doubao (a mixture-of-experts model). The aggregated results across both datasets are presented in Table 3 .

Our model achieves the highest average scores in ROUGE, BLEU, METEOR, AlignScore, and SummaC, demonstrating superior performance compared to other LLMs, including GPT-4. Among the general-purpose systems, GPT-4 performs second-best on relevance metrics, but still lags behind our approach by over 0.02 in ROUGE and 0.04 in METEOR. In contrast, the Doubao and Qwen-3-32B models perform significantly worse, highlighting the effectiveness of retrieval-augmented LoRA fine-tuning for domain-specific summarization. In readability metrics, our system achieves stronger performance than GPT-3.5 on CLI, FKGL, and DCRS, although it underperforms relative to GPT-4.

Approach	Dataset	ROUGE	BLEU	METEOR	BERTScore	FKGL ↓	CLI ↓	DCRS ↓	LENS	AlignScore	SummaC
LLaMA3.1-8B base	PLOS	0.3015	6.5620	0.2523	0.8472	15.9551	14.3183	17.6025	43.7496	0.7888	0.6043
LLaMA3.1-70B base	PLOS	0.3177	5.9399	0.2732	0.8482	16.6955	14.6441	18.4586	56.0141	0.7838	0.6110
LLaMA3.1-8B +RAG	PLOS	0.3111	6.4173	0.2757	0.8448	15.6542	14.0322	16.3438	37.9543	0.7716	0.5969
LLaMA3.1-8B, full text	PLOS	0.2868	5.5198	0.2696	0.8386	**14.0785**	13.7034	**15.3931**	39.2586	0.7406	0.4899
LLaMA3.1-8B +LoRA	PLOS	0.3125	8.0553	0.2684	0.8483	14.2926	13.6650	15.7595	43.5402	0.7961	0.6606
LLaMA3.1-8B +RAG+LoRA	PLOS	**0.3682**	**13.1528**	**0.3294**	**0.8589**	16.0238	**13.6458**	16.2309	**59.2123**	**0.8905**	**0.8325**
LLaMA3.1-8B base	eLife	0.1938	0.9549	0.1247	0.8250	15.0991	14.1559	17.8718	**50.6078**	0.8171	0.5587
LLaMA3.1-70B base	eLife	0.2583	2.6717	0.2026	0.8237	16.2511	14.2092	17.7906	41.9114	0.8145	0.5141
LLaMA3.1-8B +RAG	eLife	0.2357	1.6102	0.1654	0.8208	14.7901	13.8905	16.3313	35.0491	0.7680	0.4821
LLaMA3.1-8B, full text	eLife	0.2475	2.7377	0.2267	0.8124	**12.8411**	13.7393	**14.9535**	16.7232	0.7919	0.4650
LLaMA3.1-8B +LoRA	eLife	0.2276	1.2622	0.1467	**0.8283**	14.2445	13.5854	16.2066	46.0314	0.8103	**0.5900**
LLaMA3.1-8B +RAG+LoRA	eLife	**0.3093**	**4.8882**	**0.2404**	0.8277	16.0863	**13.5323**	17.1463	49.9853	**0.8187**	0.5412

Table 1: Performance of models with RAG and LoRA on the validation set. ↓ Indicates that lower values correspond to better performance. Bold indicates the best score in each dataset. All metrics were computed on the full validation set (PLOS, $n = 1376$. eLife, $n = 271$).

Approach	Dataset	ROUGE	BLEU	METEOR	BERTScore	FKGL ↓	CLI ↓	DCRS ↓	LENS	AlignScore	SummaC
RAG, prompt 1	plos	0.3111	6.4173	0.2757	0.8448	15.6542	14.0322	16.3438	37.9543	0.7716	0.5969
RAG, prompt 2	plos	0.3139	6.3055	0.2856	0.8456	14.3461	13.6433	15.6403	46.0194	0.7479	0.5559
RAG, prompt 3	plos	0.3088	6.3005	0.2632	0.8492	13.0738	12.9733	14.2286	62.0774	0.7013	0.5466
RAG, prompt 4	plos	0.2966	4.5010	0.2493	0.8467	**11.7158**	**12.1700**	**12.5419**	66.4741	0.5951	0.5133
RAG+LoRA, prompt 1	plos	**0.3682**	**13.1528**	0.3294	**0.8589**	16.0238	13.6458	16.2309	59.2123	**0.8905**	**0.8325**
RAG+LoRA, prompt 2	plos	0.3485	9.8177	0.3227	0.8550	16.5375	13.5550	16.3920	66.1999	0.7753	0.6310
RAG+LoRA, prompt 3	plos	0.3601	10.1433	**0.3315**	0.8561	15.1839	13.3211	15.7658	69.0966	0.7598	0.5718
RAG+LoRA, prompt 4	plos	0.3434	8.2041	0.3230	0.8560	15.1830	12.8227	14.9630	**73.0403**	0.6322	0.5242
RAG, prompt 1	elife	0.2357	1.6102	0.1654	0.8208	14.7901	13.8905	16.3313	35.0491	0.7680	0.4821
RAG, prompt 2	elife	0.2638	2.2989	0.1846	0.8271	13.5047	13.3958	15.5217	46.3963	0.7802	0.5234
RAG, prompt 3	elife	0.2739	3.1185	0.2050	0.8283	11.8284	12.7894	14.3307	47.2025	0.7419	0.5303
RAG, prompt 4	elife	0.2771	3.2216	0.2031	0.8296	**11.6004**	**12.5959**	14.3942	52.1310	0.7366	0.5375
RAG+LoRA, prompt 1	elife	**0.3093**	4.8882	**0.2404**	0.8277	16.0863	13.5323	17.1463	49.9853	**0.8187**	**0.5412**
RAG+LoRA, prompt 2	elife	0.2886	4.4900	0.2186	0.8241	15.8230	15.0335	14.0296	52.6318	0.7171	0.5104
RAG+LoRA, prompt 3	elife	0.2957	4.8307	0.2215	0.8252	15.6024	14.8157	13.7995	52.2966	0.7302	0.5364
RAG+LoRA, prompt 4	elife	0.3061	**5.0620**	0.2317	**0.8303**	15.3697	13.9646	**13.3087**	**60.8793**	0.6404	0.4672

Table 2: The impact of prompt design on generated summaries using augmented LLaMA 3.1 models. Prompts 1 through 4 progressively increase emphasis on readability while reducing emphasis on factuality. ↓ indicates that lower values correspond to better performance. Bold values indicate the best score within each dataset. All metrics were computed on the full validation set (PLOS, $n = 1376$; eLife, $n = 271$).

Model	ROUGE	BLEU	METEOR	BERTScore	FKGL ↓	CLI ↓	DCRS ↓	LENS	AlignScore	SummaC
LLaMA3.1-8B +RAG+LoRA, prompt 1	**0.3469**	**8.6382**	**0.2978**	0.8534	16.9472	10.9176	17.2120	57.6922	**0.8801**	**0.7471**
LLaMA3.1-8B +RAG, prompt 4	0.2985	4.6963	0.2499	0.8457	12.9965	10.3171	14.5694	53.3393	0.7646	0.5704
Seed/Doubao-1.5-pro, RAG, prompt 1	0.1371	0.4055	0.1202	0.8052	12.3599	11.0582	15.6888	71.4021	0.3423	0.4382
Qwen3-32B, RAG, prompt 1	0.1926	1.4937	0.1396	0.8338	16.4236	14.2241	19.7064	40.6607	0.6860	0.5315
GPT3.5, RAG, prompt 1	0.2918	3.9624	0.2076	0.8536	17.5771	12.1847	18.9784	66.3074	0.8047	0.5118
GPT3.5, RAG, prompt 4	0.2543	2.2707	0.1709	0.8544	14.7574	11.7962	16.9538	74.9194	0.7850	0.5180
GPT4, RAG, prompt 4	0.3207	5.4428	0.2532	**0.8554**	**12.2789**	**9.5065**	**13.3833**	**80.4591**	0.6754	0.5210

Table 3: Final submission and test set performance compared to other general-purpose LLMs. The table reports average results across the PLOS and eLife datasets. ↓ indicates that lower values correspond to better performance. Bold values indicate the best scores. All metrics were computed on the test set (PLOS, $n = 142$. eLife, $n = 142$).

4 Discussion and Conclusion

Applying both LoRA and RAG to fine-tune LLaMA3.1 resulted in superior overall performance on the biomedical lay summarization task compared to using the base model or applying LoRA or RAG individually. This combined approach substantially improved relevance and factuality metrics, though it slightly reduced performance on most readability metrics. The gains in relevance and factuality are likely attributable to the additional contextual information retrieved from the full text, which often contains factual content present in the reference summaries but absent from the abstract. The slight decline in readability metrics, such as FKGL, CLI, and DCRS, may result from the introduction of new concepts via the retrieved content or from the integration of additional information using more complex sentence structures, such as subordinate clauses.

Our prompt design experiments revealed a trade-off between factuality and readability in LLM-generated summaries, suggesting that efforts to simplify language or meet brevity constraints may compromise the accurate representation of complex scientific content. It may be challenging for a single model to simultaneously enforce simplified vocabulary and sentence structures, comply with word count constraints, and extract essential information while preserving technical precision. A potential solution is to adopt a two-stage summarization framework (Goldsack et al., 2025), where an "author" model first extracts key factual content, followed by a "writer" model that generates a more readable summary while preserving that information.

Our RAG fine-tuned LLaMA3.1 model demonstrated superior performance in relevance and factuality metrics compared to pretrained general-purpose LLMs in this summarization task. However, we also observed that the pretrained GPT-4 model excels in readability metrics while maintaining competitive performance in relevance and factuality. This suggests that GPT-4 may serve as a strong base model for fine-tuning, potentially achieving well-balanced performance across all evaluation criteria, as demonstrated in previous work (You et al., 2024). Nevertheless, leveraging GPT-4 for fine-tuning and inference entails significantly higher computational and financial costs.

Limitations

This study has several limitations. First, the RAG component relied exclusively on the main text of each article. As a result, it may have omitted essential background information, such as fundamental biological concepts, which are critical for generating accurate and accessible lay summaries. Future work could enhance summary quality by incorporating external domain-specific resources—such as biomedical ontologies or reference texts—into the RAG pipeline. Additionally, the embedding model used in our RAG implementation was a small, cost-efficient variant. Employing larger and more powerful models, such as all-mpnet-base-v2, may further improve retrieval quality and overall summarization performance. Third, we used prompt templates optimized for LLaMA 3.1 to evaluate other LLMs, which may disadvantage models whose optimal prompts differ in structure or emphasis.

Acknowledgments

We are grateful for Peter Grabowski's insightful feedback regarding the LLM based evaluation and experimental design.

References

Satanjeev Banerjee and Alon Lavie. 2005. METEOR: An automatic metric for MT evaluation with improved correlation with human judgments. In *Proceedings of the ACL Workshop on Intrinsic and Extrinsic Evaluation Measures for Machine Translation and/or Summarization*, pages 65–72, Ann Arbor, Michigan. Association for Computational Linguistics.

Siyu Bao, Ruijing Zhao, Siqin Zhang, Jinghui Zhang, Weiyin Wang, and Yunian Ru. 2024. Ctyun AI at BioLaySumm: Enhancing lay summaries of biomedical articles through large language models and data augmentation. In *Proceedings of the 23rd Workshop on Biomedical Natural Language Processing*, pages 837–844, Bangkok, Thailand. Association for Computational Linguistics.

Meri Coleman and T. L. Liau. 1975. A computer readability formula designed for machine scoring. *Journal of Applied Psychology*, 60(2):283–284.

Edgar Dale and Jeanne S. Chall. 1948. A formula for predicting readability: Instructions. *Educational Research Bulletin*, 27(1):37–54.

Matthijs Douze, Alexandr Guzhva, Chengqi Deng, Jeff Johnson, Gergely Szilvasy, Pierre-Emmanuel Mazaré, Maria Lomeli, Lucas Hosseini, and Hervé Jégou. 2025. The faiss library. *Preprint*, arXiv:2401.08281.

Tomas Goldsack, Zheheng Luo, Qianqian Xie, Carolina Scarton, Matthew Shardlow, Sophia Ananiadou, and Chenghua Lin. 2023. Overview of the biolaysumm 2023 shared task on lay summarization of biomedical research articles. In *The 22nd Workshop on Biomedical Natural Language Processing and BioNLP Shared Tasks*, pages 468–477, Toronto, Canada. Association for Computational Linguistics.

Tomas Goldsack, Carolina Scarton, and Chenghua Lin. 2025. Leveraging large language models for zero-shot lay summarisation in biomedicine and beyond. *Preprint*, arXiv:2501.05224.

Tomas Goldsack, Carolina Scarton, Matthew Shardlow, and Chenghua Lin. 2024. Overview of the BioLaySumm 2024 shared task on the lay summarization of biomedical research articles. In *Proceedings of the 23rd Workshop on Biomedical Natural Language Processing*, pages 122–131, Bangkok, Thailand. Association for Computational Linguistics.

Tomas Goldsack, Zhihao Zhang, Chenghua Lin, and Carolina Scarton. 2022. Making science simple: Corpora for the lay summarisation of scientific literature. In *Proceedings of the 2022 Conference on Empirical Methods in Natural Language Processing*, pages 10589–10604, Abu Dhabi, United Arab Emirates. Association for Computational Linguistics.

Rita González-Márquez, Luca Schmidt, Benjamin M. Schmidt, Philipp Berens, and Dmitry Kobak. 2024. The landscape of biomedical research. *Patterns*, 5(6):100968.

Aaron Grattafiori, Abhimanyu Dubey, Abhinav Jauhri, Abhinav Pandey, Abhishek Kadian, Ahmad Al-Dahle, Aiesha Letman, Akhil Mathur, Alan Schelten, Alex Vaughan, Amy Yang, Angela Fan, Anirudh Goyal, Anthony Hartshorn, Aobo Yang, Archi Mitra, Archie Sravankumar, Artem Korenev, Arthur Hinsvark, and 542 others. 2024. The llama 3 herd of models. *Preprint*, arXiv:2407.21783.

Jiawei Gu, Xuhui Jiang, Zhichao Shi, Hexiang Tan, Xuehao Zhai, Chengjin Xu, Wei Li, Yinghan Shen, Shengjie Ma, Honghao Liu, Saizhuo Wang, Kun Zhang, Yuanzhuo Wang, Wen Gao, Lionel Ni, and Jian Guo. 2025. A survey on llm-as-a-judge. *Preprint*, arXiv:2411.15594.

J. Peter Kincaid, Robert P. Fishburne, Richard L. Rogers, and Brad S. Chissom. 1975. Derivation of new readability formulas (automated readability index, fog count and flesch reading ease formula) for navy enlisted personnel. Technical Report Research Branch Report 8-75, Naval Technical Training Command, Millington TN Research Branch.

Philippe Laban, Tobias Schnabel, Paul N. Bennett, and Marti A. Hearst. 2022. SummaC: Re-visiting NLI-based models for inconsistency detection in summarization. *Transactions of the Association for Computational Linguistics*, 10:163–177.

Chin-Yew Lin. 2004. ROUGE: A package for automatic evaluation of summaries. In *Text Summarization Branches Out*, pages 74–81, Barcelona, Spain. Association for Computational Linguistics.

Yang Liu, Dan Iter, Yichong Xu, Shuohang Wang, Ruochen Xu, and Chenguang Zhu. 2023. G-eval: NLG evaluation using gpt-4 with better human alignment. In *Proceedings of the 2023 Conference on Empirical Methods in Natural Language Processing*, pages 2511–2522, Singapore. Association for Computational Linguistics.

Mounica Maddela, Yao Dou, David Heineman, and Wei Xu. 2023. LENS: A learnable evaluation metric for text simplification. In *Proceedings of the 61st Annual Meeting of the Association for Computational Linguistics (Volume 1: Long Papers)*, pages 16383–16408, Toronto, Canada. Association for Computational Linguistics.

Kishore Papineni, Salim Roukos, Todd Ward, and Wei jing Zhu. 2002. Bleu: a method for automatic evaluation of machine translation. In *Proceedings of the 40th Annual Meeting of the Association for Computational Linguistics*, pages 311–318.

Nils Reimers and Iryna Gurevych. 2019. Sentence-bert: Sentence embeddings using siamese bert-networks. In *Proceedings of the 2019 Conference on Empirical Methods in Natural Language Processing*. Association for Computational Linguistics.

Chenghao Xiao, Kun Zhao, Xiao Wang, Siwei Wu, Sixing Yan, Tomas Goldsack, Sophia Ananiadou, Noura Al Moubayed, Liang Zhan, William Cheung, and Chenghua Lin. 2025. Overview of the biolaysumm 2025 shared task on lay summarization of biomedical research articles and radiology reports. In *The 24th Workshop on Biomedical Natural Language Processing and BioNLP Shared Tasks*, Vienna, Austria. Association for Computational Linguistics.

Zhiwen You, Shruthan Radhakrishna, Shufan Ming, and Halil Kilicoglu. 2024. UIUC_BioNLP at BioLaySumm: An extract-then-summarize approach augmented with Wikipedia knowledge for biomedical lay summarization. In *Proceedings of the 23rd Workshop on Biomedical Natural Language Processing*, pages 132–143, Bangkok, Thailand. Association for Computational Linguistics.

Yuheng Zha, Yichi Yang, Ruichen Li, and Zhiting Hu. 2023. AlignScore: Evaluating factual consistency with a unified alignment function. In *Proceedings of the 61st Annual Meeting of the Association for Computational Linguistics (Volume 1: Long Papers)*, pages 11328–11348, Toronto, Canada. Association for Computational Linguistics.

Haopeng Zhang, Philip S. Yu, and Jiawei Zhang. 2024. A systematic survey of text summarization: From statistical methods to large language models. *Preprint*, arXiv:2406.11289.

Dataset	Train	Validation	Test
PLOS	24,773	1,376	142
eLife	4,346	242	142

Table 4: Number of examples in the training, validation, and test sets of the PLOS and eLife lay summary dataset.

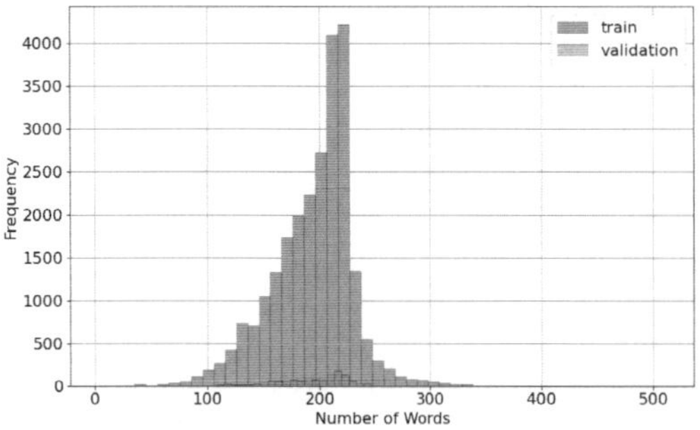

Figure 2: Word counts of PLOS reference summaries in the training and validation sets.

Tianyi Zhang, Varsha Kishore, Felix Wu, Kilian Q. Weinberger, and Yoav Artzi. 2020. Bertscore: Evaluating text generation with bert. *Preprint*, arXiv:1904.09675.

A Dataset Summary Statistics

The dataset was divided into training, validation, and test sets, as shown in Table 4.

Summary statistics of word counts are presented in Tables 5 and 6 to confirm that the validation and test splits are representative of the dataset. The training, validation, and test sets display comparable mean and median word counts. However, 13 instances in the PLOS training set contain incorrectly phrased abstracts, each comprising fewer than 500 tokens according to the LLaMA 3 tokenizer. These instances were identified as having improperly parsed abstracts and were removed prior to training.

The reference summaries typically range from 100–300 words for PLOS and 200–600 words for eLife (Figure 2 and Figure 3). These ranges informed the prompt design, enabling the model to generate summaries of comparable lengths .

B Fine-Tuning

The prompts used for LoRA fine-tuning—with and without RAG—are provided in B.1 and B.2, respectively. Similar prompts were used during inference, except that the reference summary part was omitted.

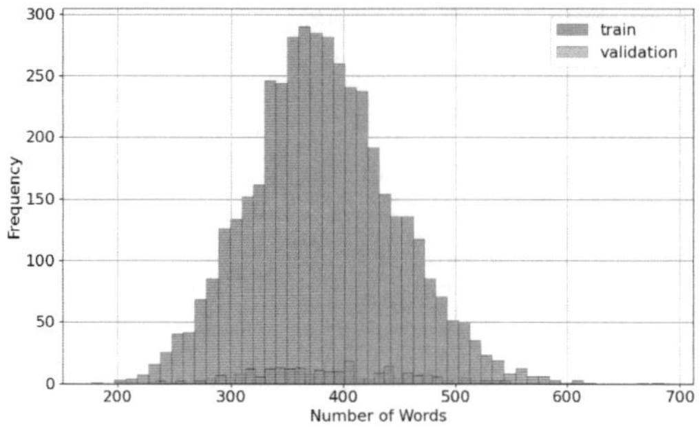

Figure 3: Word counts of eLife reference summaries in the training and validation sets.

The final LoRA configuration employed a LoRA rank of 8, a LoRA alpha of 16, a LoRA dropout of 0.1, and a learning rate of 1×10^{-5}. Hyperparameters such as the LoRA rank (8, 16) were explored using the PLOS dataset, but performance differences were minimal. The number of epochs (1, 2, 4) was also explored. The optimal number of training epochs was found to be 1 for the PLOS dataset and 2 for the eLife dataset. The training and validation loss curves are available in the GitHub repository.

A series of prompts with progressively increased emphasis on readability were explored: Prompt 1 (B.2), Prompt 2 (B.3), Prompt 3 (B.4), and Prompt 4 (B.5). Blue color coding indicates instructions related to accuracy, while orange color coding highlights instructions aimed at improving readability.

C Evaluation of G-Eval

Multiple classical metrics were employed in this study, some of which exhibited contradictory behavior during prompt optimization. This highlights the challenge of determining appropriate weights for each metric in order to construct a meaningful overall evaluation score. Previously, equal weights were assigned to each metric to calculate average performance within each evaluation aspects (Goldsack et al., 2024). Recent advancements have introduced the "LLM-as-a-Judge" paradigm, wherein large language models are employed as evaluators for complex tasks, offering scalable, cost-effective, and consistent assessments across diverse domains (Gu et al., 2025).

In this preliminary study, we employed G-Eval, an LLM-based evaluation framework that prompts a language model to assign scores and provide justifications based on criteria such as relevance, readability, and factuality (Liu et al., 2023). G-Eval

	Text	Min	Max	Mean	Median
Train	Abstract	71	509	166	165
	Main text	324	28,696	10,200	9.890
Validation	Abstract	76	306	166	165
	Main text	3,408	23,048	10,031	9,707
Test	Abstract	83	464	267	220
	Main text	2,666	16,954	8,157	8,032

Table 5: Words counts of abstracts and articles from the eLife dataset.

	Text	Min	Max	Mean	Median
Train	Abstract	2*	701	268	269
	Main text	748	26,643	6,754	6.581
Validation	Abstract	93	561	271	273
	Main text	933	24,751	8,869	8,649
Test	Abstract	97	377	245	245
	Main text	3,316	17,330	7,735	7,521

Table 6: Word counts of abstracts and articles from the PLOS dataset. * indicate instances with unusually low word counts due to incorrectly parsed abstracts, which were removed prior to training.

was implemented using the GPT-3.5-turbo model, and the evaluation criteria are detailed in Box C.1.

To evaluate the effectiveness of G-Eval for this task, we conducted a controlled experiment using synthesized data. Specifically, we examined whether G-Eval scores could differentiate among positive controls, negative controls, and standard summaries generated with the LLaMA3.1 model using various prompts. The positive controls (Paraphrased Gold Summaries) were created by paraphrasing the reference summaries to preserve their factual content while altering surface form. The negative controls (Intentionally Degraded Summaries) were generated by prompting the model to produce outputs characterized by vague language, poor structure, and incorrect terminology. The standard summaries were generated directly from abstracts using a conventional prompt. All prompts used to generate this synthetic data are listed in Table 7.

The results showed that paraphrased reference summaries achieved the highest median G-Eval scores, while intentionally degraded summaries received the lowest scores. Summaries generated using the standard prompt fell between these two extremes (Figure 4). These findings suggest that G-Eval is effective in distinguishing between summaries of varying quality.

In addition, summaries that received low G-Eval scores were manually reviewed to assess the justifications provided by the G-Eval framework for their evaluation.

We applied G-Eval scoring to 20 test set examples across four model configurations: the LLaMA 3.1 baseline, LLaMA 3.1 + LoRA, LLaMA 3.1 + RAG, and LLaMA 3.1 + RAG + LoRA (Figure 5). Consistent with the results in Table 1, the model fine-tuned with both LoRA and RAG achieved the highest median G-Eval score, suggesting that G-Eval is capable of distinguishing higher-performing models from lower-performing ones. However, the boxplot reveals substantial variance across the 20 evaluated samples, indicating that a larger sample size would be necessary to establish statistical significance.

Upon reviewing examples with low G-Eval scores, we found that the most common reason for low performance was the omission of key details present in the reference summary. This issue likely stems from limitations in the abstract, which may lack sufficient context, and from retrieved chunks that failed to supplement the missing information.

Box B.1: LoRA without RAG, prompt 1

```
#system:You are an expert science communicator. Your task is to generate a clear,
accurate, and formal summary of biomedical research articles.
The summary should be accessible to a general audience while maintaining
scientific rigor.
#user:
Title: (...)
Abstract: (...)

Provide a formal summary of the article in {summary_word_len} words.
Do not include explanations, self-reflections, or additional notes.
Keep the response strictly to the summary. The output should begin directly with
the summary text itself.

#assistant:
(ref summary...)
```

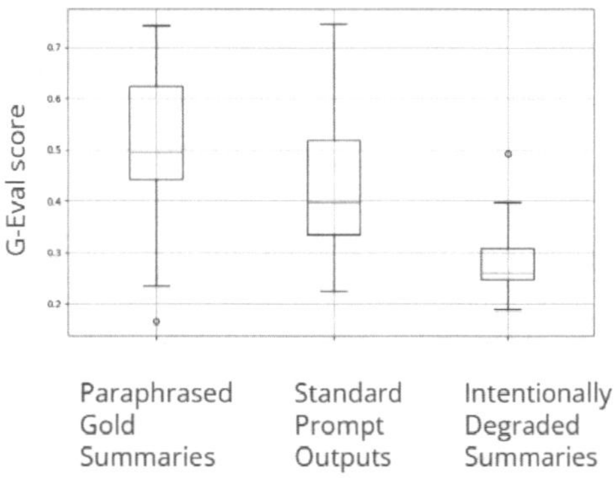

Figure 4: G-Eval scores for summaries generated by the LLaMA3.1-Instruct model and control conditions. Paraphrased Gold Summaries were created by rephrasing the original lay summaries while preserving their meaning. Intentionally Degraded Summaries were generated by explicitly prompting LLaMA 3.1 to produce outputs with vague language, poor structure, and incorrect terminology. n=20

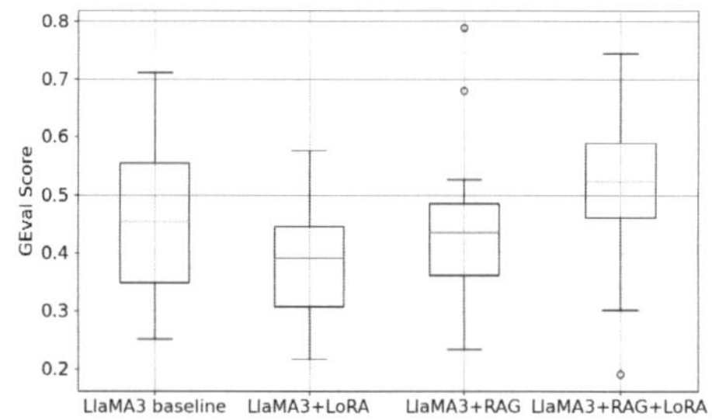

Figure 5: G-Eval scores for summaries generated by the LLaMA 3.1 model and fine-tuned models, with or without RAG. n=20

Box B.2: LoRA with RAG, prompt 1

```
#system: You are an expert science communicator. Your task is to generate a
clear, accurate, and formal summary of biomedical research articles.
The summary should be accessible to a general audience while maintaining
scientific rigor.
#user:
Title: (...)
Abstract: (...)

Supporting Text:
(retrieved text chunks...)

Provide a formal summary of the article in {summary_word_len} words.
Do not include explanations, self-reflections, or additional notes.
Keep the response strictly to the summary. The output should begin directly with
the summary text itself.

#assistant:
(ref summary...)
```

Box B.3: LoRA with RAG, prompt 2

```
#system: You are an expert science communicator. Your task is to generate a
clear, accurate, and formal summary of biomedical research articles.
The summary should be accessible to a general audience using plain language,
short sentences, and avoiding technical jargon where possible, while maintaining
scientific accuracy.
#user:
Title: (...)
Abstract: (...)

Supporting Text:
(retrieved text chunks...)

Provide a formal summary of the article in {summary_word_len} words.
Do not include explanations, self-reflections, or additional notes.
Keep the response strictly to the summary. The output should begin directly with
the summary text itself.

#assistant:
(ref summary...)
```

Box B.4: LoRA with RAG, prompt 3

```
#system: You are an expert science communicator. Your task is to generate a
clear, accurate, and formal summary of biomedical research articles.
The summary should be accessible to a general audience. Use simple sentence
structures, common words, and avoid long or complex clauses. Aim for a tone
similar to science communication articles in outlets like Scientific American or
NIH press releases.
#user:
Title: (...)
Abstract: (...)

Supporting Text:
(retrieved text chunks...)

Provide a formal summary of the article in {summary_word_len} words.
Do not include explanations, self-reflections, or additional notes.
Keep the response strictly to the summary. The output should begin directly with
the summary text itself.

#assistant:
(ref summary...)
```

Box B.5: LoRA with RAG, prompt 4

```
#system: You are an expert science communicator. Your task is to generate a
summary of biomedical research articles.
The summary should be accessible to a general audience. Use simple sentence
structures, common words, and avoid long or complex clauses. Aim for a tone
similar to science communication articles in outlets like Scientific American or
NIH press releases.
#user:
Title: (...)
Abstract: (...)

Supporting Text:
(retrieved text chunks...)

Provide a formal summary of the article in {summary_word_len} words.
Do not include explanations, self-reflections, or additional notes.
Keep the response strictly to the summary. The output should begin directly with
the summary text itself.

#assistant:
(ref summary...)
```

Box C.1: G-Eval Evaluation Criteria

Evaluate the generated lay summary on the following three criteria: 1. **Relevance (1-5)**: Does the summary retain all major findings and themes of the source abstract? Score higher if it covers key points, even if phrased differently. Penalize only if essential information is missing or incorrect topics are introduced. 2. **Readability (1-5)**: Is the summary easy to understand for a non-expert audience? Consider fluency, sentence structure, and clarity. Avoid penalizing for simplified language unless it introduces confusion. 3. **Factuality (1-5)**: Does the summary accurately reflect the scientific claims in the source abstract? Check for hallucinations or misinterpretations, not just omissions. Each criterion should be scored from 1 (poor) to 5 (excellent). Then provide a final **Overall Score**.

Synthetic Data Type	Prompt
Paraphrased Gold Summaries	#system You are a professional science communicator. Your role is to paraphrase lay summaries with precision, maintaining the original meaning and content without introducing interpretation or additional information. #user Title: (...) Summary: (...) Rephrase the summary in 100–300 words. Do not include explanations, commentary, or additional remarks. Keep the response strictly to the summary. #assistant
Intentionally Degraded Summaries	#system You are a deliberately ineffective science communicator. Your task is to generate an example of a poorly written summary of biomedical research. This summary should reflect common mistakes in science communication, such as vague language, poor structure, and misuse of terminology. The summary may also include minor factual inaccuracies or exaggerated claims to illustrate how misleading summaries might appear. This output will be used strictly for educational comparison with well-written summaries. #user Title: (...) Abstract: (...) Provide a poor-quality summary of the article in 100–300 words, reflecting issues like lack of clarity, overgeneralization, or scientific inaccuracy (intended for contrastive purposes only). At least some summary needs to be generated. Do not include explanations, self-reflections, or additional notes. Keep the response strictly to the summary. #assistant
Standard Prompt Outputs	#system You are an expert science communicator. Your task is to generate a clear, accurate, and formal summary of biomedical research articles. The summary should be accessible to a general audience while maintaining scientific rigor. #user Title: (...) Abstract: (...) Provide a formal summary of the article in 100–300 words. Do not include explanations, self-reflections, or additional notes. Keep the response strictly to the summary. #assistant

Table 7: Prompts used to generate paraphrased, degraded, and standard summaries for evaluating G-Eval.

5cNLP at BioLaySumm2025: Prompts, Retrieval, and Multimodal Fusion

Juan Antonio Lossio-Ventura[1], **Callum Chan**[2], **Arshitha Basavaraj**[3], **Hugo Alatrista-Salas**[4], **Francisco Pereira**[1], **Diana Inkpen**[2]

[1]Machine Learning Core, National Institute of Mental Health, National Institutes of Health, USA
[2]School of Electrical Engineering and Computer Science, University of Ottawa, Canada
[3]International Institute of Information Technology, Bangalore, India
[4]De Vinci Research Center, Paris, France,

juan.lossio@nih.gov, cchan073@uottawa.ca, arshitha.basavaraj@iiitb.ac.in
hugo.alatrista_salas@devinci.fr, francisco.pereira@nih.gov, dinkpen@uottawa.ca

Abstract

In this work, we present our approach to addressing all subtasks of the BioLaySumm 2025 shared task by leveraging prompting and retrieval strategies, as well as multimodal input fusion. Our method integrates: (1) zero-shot and few-shot prompting with large language models (LLMs); (2) semantic similarity-based dynamic few-shot prompting; (3) retrieval-augmented generation (RAG) incorporating biomedical knowledge from the Unified Medical Language System (UMLS); and (4) a multimodal fusion pipeline that combines images and captions using image-text-to-text generation for enriched lay summarization. Our framework enables lightweight adaptation of pretrained LLMs for generating lay summaries from scientific articles and radiology reports. Using modern LLMs, including Llama-3.3-70B-Instruct and GPT-4.1, our 5cNLP team achieved third place in Subtask 1.2 and second place in Subtask 2.1, among all submissions.

1 Introduction

BioLaySumm's third edition (Xiao et al., 2025b) introduces a new task focused on translating radiology reports into layperson-friendly language, while continuing its existing biomedical article summarization task from previous editions (Goldsack et al., 2024, 2023). Summaries are expected to include more background information and reduce technical jargon to improve accessibility.

Thus, BioLaySumm 2025 comprises two main tasks, each with two subtasks, aimed at improving biomedical communication for lay audiences. Task 1 focuses on generating accessible summaries of biomedical research articles from PLOS and eLife, either directly (Subtask 1.1) or with the integration of external knowledge sources (Subtask 1.2). Task 2 targets the translation of radiology reports into layperson-friendly language, using text alone (Subtask 2.1) or in combination with chest x-ray images (Subtask 2.2). This task was offered in both open and closed tracks, with the closed track additionally incorporating the MIMIC-CXR dataset. We opted for the closed track in our submission.

To address these tasks, we developed a unified and flexible framework that combines prompting, retrieval, and multimodal fusion techniques. It supports zero- and few-shot prompting with LLMs, dynamic few-shot selection via embedding-based nearest neighbors, retrieval-augmented generation using UMLS (Bodenreider, 2004), and multimodal processing through image-text-to-text generation for enriched lay summarization. Based on our previous experience, we adopted structured (compositional) prompting including task goals, instructions, formatting guidelines, and output specifications (Chan et al., 2025). Also, previous work shows that LLMs perform better with well-chosen in-context examples (Brown et al., 2020; Liu et al., 2021). Following (Liu et al., 2021), we chose most similar samples based on cosine similarity for few-shot prompting. We also explored varying the number and selection strategy of these examples. Moreover, we explored several LLMs of varying sizes, including Llama-3.1-8B, Llama-3.1-8B-Instruct (standard and 8-bit quantized), Llama-3.3-70B-Instruct, and GPT-4.1. A single approach was applied across all task datasets, without building data-specific models, to improve generalizability.

2 Shared Task Overview

Task 1: Lay Summarization: Participants were required to generate layperson-accessible summaries of biomedical articles from two datasets, PLOS and eLife, using two different approaches.

- **Subtask 1.1: Plain Lay Summarization**: Given an article's abstract and main text, systems had to produce a non-technical summary suitable for a general audience.

BioNLP 2025 Shared Tasks, pages 215–231
August 1, 2025

- **Subtask 1.2: Lay Summarization with External Knowledge**: This task extended Subtask 1.1 by permitting the use of additional knowledge sources (e.g., databases, medical ontologies) to enrich contextual understanding for lay readers.

Task 2: Radiology Report Translation to Layperson's Terms: This task was offered in open and closed tracks. The open track used PadChest (Bustos et al., 2020), Open-i, and BIMCV-COVID19 (de la Iglesia Vayá et al., 2020), while the closed track additionally included MIMIC-CXR (Johnson et al., 2019, 2024).

- **Subtask 2.1: Radiology Report Translation:** The goal was to build models to translate professional radiology reports to layperson's terms.
- **Subtask 2.2: Multimodal Radiology Report Translation:** This was a multi-modal task with the goal of achieving a lay translation of radiology reports. The input was chest x-ray images and radiology reports and the output should be a report in layperson's terms.

Datasets: All datasets were made available by the organizers on HuggingFace (Xiao et al., 2025a; Zhao et al., 2024) - except the imaging data from MIMIC-CXR used in Subtask 2.1. For Task 1, two datasets from biomedical journals, PLOS and eLife, were provided (Goldsack et al., 2022; Luo et al., 2022). For Task 2, four datasets were used: Open-i, PadChest, BIMCV-COVID19, and MIMIC-CXR (Zhao et al., 2025). Participants could choose between using only the first three (open track) or all four (closed track). The training, validation, and test splits are detailed in Appendix Tables 4 and 5.

Evaluation Metrics: Submissions were evaluated using task-specific metrics. For Task 1, summaries were assessed on relevance (ROUGE-1/2/L, BLEU, METEOR, BERTScore), readability (FKGL, DCRS, CLI, LENS), and factuality (AlignScore, SummaC). Task 2 used the same relevance metrics, similar readability measures (excluding LENS), and clinical-specific factuality metrics (CheXbert-F1, RadGraph-F1). All metrics were determined by the shared task organizers.

3 Methods

We used prompting, retrieval, and multimodal fusion with Llama and GPT models, outlined below.

TASK 1

We focused on text-to-text generation tasks, mainly using zero-shot, one-shot, and few-shot prompting. Building on our experience from previous shared tasks, we used structured (compositional) prompting, which included task goals, instructions, guidelines, and output formats (Chan et al., 2025). In this work, we extended our structured prompts by incorporating role-based instructions, directing the model to adopt specific personas, such as a teacher explaining complex concepts to students of varying ages (role prompting). We tried small models as baselines and larger models to increase performance. For instance, Llama-3.1-8B and its quantized variant support a combined input/output token limit of 8,192 tokens. Accordingly, we constrained model responses to 500 tokens and truncated input articles when necessary. Most experiments involving small models were conducted using zero-shot prompting.

Subtask 1.1

- **Zero-Shot Prompting on Initial and Final Article Segments**: To maximize the use of available tokens, we used only the beginning and end of each article. This approach was applied with small models only.
- **Zero-Shot Prompting on Summaries**: We divided long texts into chunks, summarized each chunk individually, and then combined them into a final summary.
- **Zero/One-shot Prompting on Section-Based Inputs**: Articles often contained diverse section structures. We extracted combinations such as: abstract only, abstract + introduction, abstract + discussion + conclusion (when available), or all four sections.
- **One-Shot Prompting with Random Sample**: Due to token constraints, we used a random example per prompt.
- **One-Shot Prompting with Most Similar Example**: We used cosine similarity (via Llama-3.1-8B embeddings) to find the most similar article-summary pair from the validation dataset. For long articles, we split them into chunks, computed embeddings, and averaged them. The most similar validation example was then included in the prompt.
- **Few-Shot Prompting with Five Examples (Lay Summaries Only)**: We selected five lay sum-

maries based on the most similar examples from the validation set in this few-shot prompting.

Subtask 1.2

This subtask aimed to improve upon Task 1.1 by incorporating external knowledge. It was based on the Retrieval-Augmented Generation (RAG) framework and focused on handling technical terms. Our process for this task included the following steps.

- **Extraction of Clinical Terms**: We used structured zero-shot prompting to extract technical terms from test articles.
- **Definition of Clinical Terms**: Each extracted term was queried using the Unified Medical Language System (UMLS) API. When available, the most suitable definition was selected.
- **Prompt Augmentation**: The resulting term-definition pairs were formatted and incorporated into the zero-shot, one-shot, and few-shot prompts used in Subtask 1.1. These refined prompts were then applied with larger models.

TASK 2

As in Task 1, we used prompt engineering to convert radiology reports (intended for healthcare professionals) into layperson-accessible summaries.

Subtask 2.1: Closed Track

We used the MIMIC-CXR dataset along with three public datasets: PadChest, Open-i, and BIMCV-COVID19. We used structured zero-shot and few-shot prompting approaches, incorporating examples selected either at random or based on cosine similarity of embeddings. Prompts explicitly defined the terms "radiology report" and "layman report" and included clear guidelines, as described as follows.

- **Zero-Shot Prompting on Radiology Report**: Our baseline used a structured prompt without examples.
- **Few-Shot Prompting with Five Random Examples**: We added five example pairs of radiology reports and lay summaries, one from each dataset, plus a fifth example illustrating variations of reports containing the phrase "No significant findings". This improved factuality and relevance, but caused a slight drop in readability.
- **Few-Shot Prompting with Ten Most Similar Examples**: For each test case, we used cosine similarity on BERT-large uncased embeddings to select the ten most similar examples from the validation set (approximately 20k samples).
- **Few-Shot Prompting with Twenty Most Similar Examples**: We extended the above method to include the top 20 most similar examples. Like the ten-example approach, this relied on the validation dataset to reduce computational costs while maintaining strong performance.

Subtask 2.2

We adopted an image-text-to-text model, BLIP (Bootstrapping Language-Image Pretraining) (Li et al., 2022), that combines a Vision Transformer with a Transformer-based text decoder to generate text from images and optional textual prompts. While less advanced than newer models like BLIP-2 (Li et al., 2023) and LLaVA (Zhang et al., 2025), it offers an efficient solution for descriptive image captioning. For the experiments, we used images and corresponding radiology reports, lay summaries, and metadata from OpenI, PadChest, and BIMCV-COVID19. The Hugging Face test set included 10,537 records, though actual image counts varied (e.g., OpenI often includes two images per record), and some images were missing. After aligning the metadata with the available images, the final dataset comprised 9,865 entries. Therefore, we were unable to submit official results due to mismatches between the number of processed records and the expected count.

4 Results

We report the results of our official submissions on the test data, as evaluated by the official evaluation server. The results for Subtask 1.1, Subtask 1.2, and Subtask 2.1 are presented in Tables 1, 2, and 3, respectively.

5 Discussion

We officially submitted approaches for three Subtasks 1.1, 1.2, and 2.1. Our approaches focused on the generalization of a single method (using the same model) across different datasets. For Task 1, we used a single approach for both eLife and PLOS. Similarly, for Subtask 2.1, we adopted a unified model for MIMIC, COVID, PadChest, and OpenI. We also conducted experiments for Subtask 2.2; however, due to issues related to dataset download and size, we were unable to submit our results for evaluation. Our experiments provided several key insights regarding the performance of

		Llama-3.1 (8-bit quantized)			Llama-3.3-70B-Inst.		GPT-4.1	
Description	**Metric**	Baseline	**S1**	**S2**	**S3**	**S4**	**S5**	**S6**
Relevance	ROUGE ↑	0.2701	0.2283	0.2429	**0.3349**	0.3334	0.3080	0.3056
	BLEU ↑	4.1857	2.6787	3.0217	6.0490	**6.1354**	4.2153	4.1381
	METEOR ↑	**0.2791**	0.2459	0.2575	0.2703	0.2676	0.2632	0.2584
	BERTScore ↑	0.8358	0.8239	0.8282	0.8581	**0.8586**	0.8533	0.8534
Readability	FKGL ↓	12.2884	**8.3792**	9.3130	16.6736	16.0718	15.5356	15.5398
	DCRS ↓	7.2444	**6.1730**	6.5255	10.5558	10.3976	10.3787	10.4061
	CLI ↓	11.8690	**8.4256**	9.2667	15.8282	15.3358	14.1439	14.1545
	LENS ↑	65.5266	70.7002	71.8203	74.2810	76.0519	77.2428	**77.5635**
Factuality	AlignScore ↑	0.6061	0.4526	0.4893	**0.6366**	0.6307	0.4483	0.4506
	SummaC ↑	0.5348	**0.6141**	0.6114	0.4456	0.4550	0.4202	0.4186

Table 1: Performance of the 5cNLP team for Subtask 1.1. The baseline was scored from the results of zero-shot prompting on Llama-3.1-8B-Instruct (8-bit quantized). Submissions S1 and S2 correspond to the scores of additional role-prompting experiments performed on the same Llama-3.1 model. Submissions S3 and S4 correspond to the scores of one and few shot prompting respectively on Llama-3.3-70B-Instruct. Similarly, submissions S5 and S6 correspond to the scores of one and few shot prompting on GPT-4.1.

		Llama-3.3-70B-Inst.		GPT-4.1	
Description	**Metric**	**S1**	**S2**	**S3**	**S4**
Relevance	ROUGE ↑	**0.3364**	0.3350	0.3117	0.3089
	BLEU ↑	**5.9982**	5.9029	4.2778	4.1743
	METEOR ↑	**0.2764**	0.2747	0.2733	0.2659
	BERTScore ↑	**0.8576**	**0.8576**	0.8531	0.8533
Readability	FKGL ↓	16.8155	16.2979	15.7437	**15.5391**
	DCRS ↓	10.5226	**10.2896**	10.3104	10.3314
	CLI ↓	15.7708	15.2399	**14.1524**	14.2205
	LENS ↑	73.8590	75.5722	76.9570	**77.4515**
Factuality	AlignScore ↑	**0.6258**	0.6099	0.4431	0.4461
	SummaC ↑	**0.4468**	0.4455	0.4185	0.4154

Table 2: Performance of the 5cNLP team for Subtask 1.2. Submissions S1 and S2 correspond to the scores of RAG one and few shot prompting respectively on Llama-3.3-70B-Instruct. Similarly, submissions S3 and S4 correspond to the scores of RAG one and few shot prompting on GPT-4.1.

		Llama-3.3-70B-Inst.		GPT-4.1		
Description	**Metric**	**S1**	**S2**	**S3**	**S4**	**S5**
Relevance	ROUGE ↑	0.4424	0.5078	0.4679	0.5170	**0.5547**
	BLEU ↑	16.3978	23.4148	19.7649	25.0122	**28.2705**
	METEOR ↑	0.5051	0.5630	0.5169	0.5654	**0.6095**
	BERTScore ↑	0.9196	0.9317	0.9257	0.9332	**0.9371**
Readability	FKGL ↓	12.3058	9.8568	8.8586	8.5402	**8.0463**
	DCRS ↓	10.0489	9.6991	9.2135	**9.1778**	9.2373
	CLI ↓	10.1783	9.1757	8.2113	**8.1571**	8.2250
Factuality	Similarity ↑	0.8309	0.8561	0.8401	0.8591	**0.8717**
	RadGraph ↑	0.2452	0.2759	0.2566	0.2872	**0.3170**
	F1CheXbert ↑	0.7172	0.7348	0.6971	0.7162	**0.7495**

Table 3: Performance of the 5cNLP team for Subtask 2.1 across 5 submissions. **S1:** Structured zero-shot prompt with Llama-3.3-70B-Instruction model. **S2:** Structured few-shot prompt with random examples with Llama-3.3-70B Instruct model. **S3:** Structured zero-shot prompt with GPT-4.1 model. **S4:** Structured few-shot prompt with random examples with GPT-4.1 model. **S5:** Structured few-shot prompt with similarity-based examples with GPT-4.1 model.

LLMs for lay summarization of research articles and radiology reports.

Task 1 was particularly challenging due to the length of the research articles. Models often can only attend to portions of the input, potentially missing critical information—especially in longer documents. Moreover, using RAG with external sources introduces additional complexities. RAG requires a supplementary step: clinical term identification. In our approach, we extracted clinical terms through prompting, which were then used to query UMLS. We believe that explicitly incorporating a dedicated clinical entity recognition or term extraction step could significantly enhance the quality of the generated summaries.

Prompt Structure and Role Specification

Task 1: When compared to the baseline, role specification in Task 1 prompts produced responses with higher readability but lower relevance. Prompts that specified roles such as *"You will act as a teacher"* significantly improved the simplicity of the responses' language; however, the style of writing did not align with the gold standard and resulted in lower relevance scores.

Subtask 2.1: Naive, unstructured prompts, such as *"The following is a radiology report containing medical terms: <radiology-report>. I would like a brief summary of the radiology report that anyone without medical knowledge can understand, i.e., a layman report"*, performed significantly worse than structured prompts incorporating explicit role specification and output guidelines. For instance, prompts beginning with: *"You are an expert medical communicator. Your task is to..."*, consistently produced higher-quality layperson summaries, emphasizing the importance of structured role-focused instructions.

Model Scale and Performance

Task 1 and Subtask 2.1: Across both structured and unstructured prompts, larger parameter models within the same architecture demonstrated superior performance. For example, the Llama-3.3-70B-Instruct model outperformed its smaller counterpart, Llama-3.1-8B. For Subtask 1.1, this also demonstrates the larger models' ability to consider a greater amount of information and instruction. They are not constrained by the token limit as was the case with Llama-3.1-8B. For Subtask 2.1, a similar trend was observed with GPT-4.1o compared to GPT-4.1, underscoring the impact of model scale, as well as context length on translation accuracy.

In-Context Learning

Task 1 and Subtask 2.1: The inclusion of contextual examples within prompts further improved model performance. Few-shot prompting, particularly with dynamically selected examples based on cosine similarity from the training sample embedding space, yielded the best results. This approach ensured that the model received relevant, semantically aligned demonstrations for the given input.

Retrieval Augmented Generation

Subtask 1.2: When analyzing the impact of incorporating external knowledge, we should compare Subtask 1.1 prompts against their Subtask 1.2 counterparts (i.e., Subtask 1.1. S3 against Subtask 1.2 S1, S4 against S2, etc.). Overall, using our methods, we observed no significant performance impact when including external knowledge. This outcome can be attributed to several factors. First, the definitions included in the prompt may have been insufficient or irrelevant. Second, the provided definitions may not have added any new information beyond what the LLMs already contained.

6 Conclusion and Future Work

We proposed a framework for translating medical texts into layperson's language focusing on summarizing biomedical articles and translating radiology reports. Using state-of-the-art LLMs (e.g., Llama-3.3-70B-Instruct and GPT-4.1), our 5cNLP team ranked third in Subtask 1.2 and second in Subtask 2.1. Rankings were based on normalized averages across all evaluation metrics. Our experiments highlighted the importance of structured, role-specific prompting, model scale, and contextual example selection in optimizing LLM performance. Moreover, while LoRA fine-tuning was applied to smaller models, prompt engineering yielded better results.

Future work may include full model training, improved prompt design, and the integration of additional external knowledge sources. For Subtask 2.2, alternative strategies for multimodal fusion could be explored. The proposed framework is also adaptable to other biomedical applications, such as patient question answering, clinical decision support, and summarizing electronic health records for non-expert audiences.

Limitations

Our experiments are limited to English-language radiology reports. Experiments for other languages could reveal more challenges in generating lay summaries. We also had limited time and computational resources; therefore, our conclusions are valid only for a small number of LLMs.

Ethics

The datasets provided by the shared task organizers were carefully prepared to ensure proper use of the data, without information about the patients. We used the datasets solely for research purposes, as expected.

Acknowledgments

Research reported in this publication was supported in part by the Intramural Research Program of the National Institute of Mental Health: ZIC-MH002968 (Francisco Pereira and Juan Antonio Lossio-Ventura) and by the Natural Science and Engineering Research Council of Canada (Diana Inkpen).

References

Olivier Bodenreider. 2004. The unified medical language system (umls): integrating biomedical terminology. *Nucleic acids research*, 32(suppl_1):D267–D270.

Tom B. Brown, Benjamin Mann, Nick Ryder, Melanie Subbiah, Jared Kaplan, Prafulla Dhariwal, Arvind Neelakantan, Pranav Shyam, Girish Sastry, Amanda Askell, Sandhini Agarwal, Ariel Herbert-Voss, Gretchen Krueger, Tom Henighan, Rewon Child, Aditya Ramesh, Daniel M. Ziegler, Jeffrey Wu, Clemens Winter, and 12 others. 2020. Language models are few-shot learners. *Preprint*, arXiv:2005.14165.

Aurelia Bustos, Antonio Pertusa, Jose-Maria Salinas, and Maria de la Iglesia-Vayá. 2020. Padchest: A large chest x-ray image dataset with multi-label annotated reports. *Medical Image Analysis*, 66:1134–1142.

Callum Chan, Sunveer Khunkhun, Diana Inkpen, and Juan Antonio Lossio-Ventura. 2025. Prompt engineering for capturing dynamic mental health self states from social media posts. In *Proceedings of the 10th Workshop on Computational Linguistics and Clinical Psychology (CLPsych 2025)*, pages 256–267, Albuquerque, New Mexico. Association for Computational Linguistics.

Maria de la Iglesia Vayá, Jose Manuel Saborit, Joaquim Angel Montell, Antonio Pertusa, Aurelia Bustos, Miguel Cazorla, Joaquin Galant, Xavier Barber, Domingo Orozco-Beltrán, Francisco García-García, Marisa Caparrós, Germán González, and Jose María Salinas. 2020. Bimcv covid-19+: a large annotated dataset of rx and ct images from covid-19 patients. *Preprint*, arXiv:2006.01174.

Tomas Goldsack, Zheheng Luo, Qianqian Xie, Carolina Scarton, Matthew Shardlow, Sophia Ananiadou, and Chenghua Lin. 2023. Overview of the biolaysumm 2023 shared task on lay summarization of biomedical research articles. In *The 22nd Workshop on Biomedical Natural Language Processing and BioNLP Shared Tasks*, pages 468–477, Toronto, Canada. Association for Computational Linguistics.

Tomas Goldsack, Carolina Scarton, Matthew Shardlow, and Chenghua Lin. 2024. Overview of the BioLaySumm 2024 shared task on the lay summarization of biomedical research articles. In *Proceedings of the 23rd Workshop on Biomedical Natural Language Processing*, pages 122–131, Bangkok, Thailand. Association for Computational Linguistics.

Tomas Goldsack, Zhihao Zhang, Chenghua Lin, and Carolina Scarton. 2022. Making science simple: Corpora for the lay summarisation of scientific literature. In *Proceedings of the 2022 Conference on Empirical Methods in Natural Language Processing*, pages 10589–10604, Abu Dhabi, United Arab Emirates. Association for Computational Linguistics.

A. Johnson, T. Pollard, R. Mark, S. Berkowitz, and S. Horng. 2024. MIMIC-CXR Database (version 2.1.0).

A.E.W. Johnson, T.J. Pollard, and S.J. et al. Berkowitz. 2019. Mimic-cxr, a de-identified publicly available database of chest radiographs with free-text reports. *Sci Data 6, 317.*

Junnan Li, Dongxu Li, Silvio Savarese, and Steven Hoi. 2023. Blip-2: Bootstrapping language-image pre-training with frozen image encoders and large language models. In *International conference on machine learning*, pages 19730–19742. PMLR.

Junnan Li, Dongxu Li, Caiming Xiong, and Steven Hoi. 2022. Blip: Bootstrapping language-image pre-training for unified vision-language understanding and generation. In *International conference on machine learning*, pages 12888–12900. PMLR.

Jiachang Liu, Dinghan Shen, Yizhe Zhang, Bill Dolan, Lawrence Carin, and Weizhu Chen. 2021. What makes good in-context examples for gpt-3? *Preprint*, arXiv:2101.06804.

Zheheng Luo, Qianqian Xie, and Sophia Ananiadou. 2022. Readability controllable biomedical document summarization. In *Findings of the Association for Computational Linguistics: EMNLP 2022*, pages 4667–4680, Abu Dhabi, United Arab Emirates. Association for Computational Linguistics.

Chenghao Xiao, Kun Zhao, Xiao Wang, and Siwei Wu. 2025a. BioLaySumm Shared Task at ACL.

Chenghao Xiao, Kun Zhao, Xiao Wang, Siwei Wu, Sixing Yan, Tomas Goldsack, Sophia Ananiadou, Noura Al Moubayed, Liang Zhan, William Cheung, and Chenghua Lin. 2025b. Overview of the biolaysumm 2025 shared task on lay summarization of biomedical research articles and radiology reports. In *The 24th Workshop on Biomedical Natural Language Processing and BioNLP Shared Tasks*, Vienna, Austria. Association for Computational Linguistics.

Shaolei Zhang, Qingkai Fang, Zhe Yang, and Yang Feng. 2025. Llava-mini: Efficient image and video large multimodal models with one vision token. *arXiv preprint arXiv:2501.03895*.

Kun Zhao, Chenghao Xiao, Chen Tang, Bohao Yang, Kai Ye, Noura Al Moubayed, Liang Zhan, and Chenghua Lin. 2024. X-ray made simple: Radiology report generation and evaluation with layman's terms. *arXiv preprint arXiv:2406.17911*.

Kun Zhao, Chenghao Xiao, Sixing Yan, William K. Cheung, Kai Ye, Noura Al Moubayed, Liang Zhan, and Chenghua Lin. 2025. X-ray made simple: Lay radiology report generation and robust evaluation. *Preprint*, arXiv:2406.17911.

A Appendix

A.1 Task 1 and 2 datasets' splits

Dataset	Training	Validation	Test
PLOS	24,773	1,376	142
eLife	4,346	241	142

Table 4: Training, validation, and test splits for the PLOS and eLife datasets for Task 1.

Dataset	Training	Validation	Test
PadChest	116,847	7,824	7,130
MIMIC-CXR	45,000	5,000	500
BIMCV-COVID19	31,364	2,042	3221
Open-i	2,243	134	186

Table 5: Training, validation, and test splits for the evaluated datasets for Task 2.

A.2 Task 1 Prompt Templates

```
### Task:
Your task is to perform layman summarization of the following biomedical article by succintly summarizing the article in a way that is easy to understand for a general audience and should not contain highly technical terms.
—
### Guidelines for Output:
- The summary should be in layman terms and will not include any technical terms.
- The summary should avoid using acronyms.
- Limit the output summary to 300 words.
- The output should only contain the summary and will not reference the article itself.
- Do not provide sources.
- Do not include any disclaimers.
- Do not include any information that is not relevant to the summarization.
- Do not repeat the guidelines given by the prompt.
—
### Input:
{article}

### Output:
```

Table 6: Zero-Shot Structured Prompt Template for Task 1.1, Baseline.

A.3 Task 2 Prompt Templates

A.4 Task 2.1 Experiments

We conducted additional experiments comparing structured and unstructured prompts using both zero-shot and few-shot approaches with randomly selected examples. Table 16 summarizes these results, which were generated using a subset of the validation data. Due to a technical error, we couldn't compute F1CheXbert scores for experiments E1 and E2.

```
### Task:
Your task is to perform layman summarization of the following biomedical article by succintly summarizing the article in a way that is easy to understand for a general audience and should not contain highly technical terms.
—
### Role:
You will act as a middle school teacher who is explaining the article to a group of grade 7 students who are 12 years old and who require simple language to understand your summarization.
—
### Guidelines for Output:
- The summary should be in layman terms and will not include any technical terms.
- The summary should avoid using acronyms.
- Limit the output summary to 300 words.
- The output should only contain the summary and will not reference the article itself.
- Do not provide sources.
- Do not include any disclaimers.
- Do not include any information that is not relevant to the summarization.
- Do not repeat the guidelines given by the prompt.
—
### Input:
{article}

### Output:
```

Table 7: Zero-Shot Structured Role Prompt Template for Task 1.1, S1.

```
### Task:
Your task is to perform layman summarization of the following biomedical article by succintly summarizing the article in a way that is easy to understand for a general audience and should not contain highly technical terms.
—
### Role:
You will act as a secondary school teacher who is explaining the article to a group of grade 9 students who are 15 years old and who require simple language to understand your summarization.
—
### Guidelines for Output:
- The summary should be in layman terms and will not include any technical terms.
- The summary should avoid using acronyms.
- Limit the output summary to 300 words.
- The output should only contain the summary and will not reference the article itself.
- Do not provide sources.
- Do not include any disclaimers.
- Do not include any information that is not relevant to the summarization.
- Do not repeat the guidelines given by the prompt.
—
### Input:
{article}

### Output:
```

Table 8: Zero-Shot Structured Role Prompt Template for Task 1.1, S2.

```
### Task:
Your task is to perform layman summarization of the following biomedical article by succintly summarizing the article in a way that is easy to understand for a general audience, avoiding technical jargon unless briefly defined.
—
### Instructions:
- Use the example below as a guide, matching its structure and writing style in your summary.
- The summary should be in layman terms.
- Briefly define any technical terms that must be included.
- Do not reference the original article or include disclaimers.
- Exclude any information not relevant to the summary.
- Do not provide sources
- Do not repeat the guidelines given by the prompt
- Avoid repeating information unnecessarily
—
### Guidelines for Output:
- Format: Clear, flowing prose in one paragraph.
- Content: Capture the essential meaning and logic of the article as if the summary itself were a brief version of the full text.
- Audience: General readers without specialized knowledge of the topic.
—
### Example:
##### Article:
{example_article}

##### Summary:
{example_summary}
—
### Now, summarize the following article based on the given criteria and using the same style of the example:

### Article:
{article}

### Summary:
```

Table 9: One-Shot Structured Prompt Template for Task 1.1, S3 and S5.

Task:
Your task is to perform layman summarization of the following biomedical article by succintly summarizing the article in a way that is easy to understand for a general audience, avoiding technical jargon unless briefly defined.
—
Instructions:
- Use the example below as a guide, matching its structure and writing style in your summary.
- The summary should be in layman terms.
- Briefly define any technical terms that must be included.
- Do not reference the original article or include disclaimers.
- Exclude any information not relevant to the summary.
- Do not provide sources
- Do not repeat the guidelines given by the prompt
- Avoid repeating information unnecessarily
—
Guidelines for Output:
- Format: Clear, flowing prose in one paragraph.
- Content: Capture the essential meaning and logic of the article as if the summary itself were a brief version of the full text.
- Audience: General readers without specialized knowledge of the topic.
—
Examples:
-> Example 1
Summary:
{example_summary_1}

-> Example 2
Summary:
{example_summary_2}

-> Example 3
Summary:
{example_summary_3}

-> Example 4
Summary:
{example_summary_4}

-> Example 5
Summary:
{example_summary_5}
—
Now, summarize the following article based on the given criteria and using the same style of the example:

Article:
{article}

Summary:

Table 10: Few-Shot Structured Prompt Template for Task 1.1, S4 and S6.

Task:
Your task is to perform layman summarization of the following biomedical article by succintly summarizing the article in a way that is easy to understand for a general audience, avoiding technical jargon unless briefly defined.
—
Definitions:
Use the following definitions to better understand and summarize the article.
{definitions}
—
Instructions:
- Use the example below as a guide, matching its structure and writing style in your summary.
- The summary should be in layman terms.
- Briefly define any technical terms that must be included.
- Do not reference the original article or include disclaimers.
- Exclude any information not relevant to the summary.
- Do not provide sources
- Do not repeat the guidelines given by the prompt
- Avoid repeating information unnecessarily
—
Guidelines for Output:
- Format: Clear, flowing prose in one paragraph.
- Content: Capture the essential meaning and logic of the article as if the summary itself were a brief version of the full text.
- Audience: General readers without specialized knowledge of the topic.
—
Example:
Article:
{example_article}

Summary:
{example_summary}
—
Now, summarize the following article based on the given criteria and using the same style of the example:

Article:
{article}

Summary:

Table 11: One-Shot Structured Prompt Template for Task 1.2, S1 and S3.

```
### Task:
Your task is to perform layman summarization of the following biomedical article by succinctly summarizing the article in a way that is easy to understand for a general audience, avoiding technical jargon unless briefly defined.

—
### Definitions:
Use the following definitions to better understand and summarize the article.
{definitions}
—
### Instructions:
- Use the example below as a guide, matching its structure and writing style in your summary.
- The summary should be in layman terms.
- Briefly define any technical terms that must be included.
- Do not reference the original article or include disclaimers.
- Exclude any information not relevant to the summary.
- Do not provide sources
- Do not repeat the guidelines given by the prompt
- Avoid repeating information unnecessarily
—
### Guidelines for Output:
- Format: Clear, flowing prose in one paragraph.
- Content: Capture the essential meaning and logic of the article as if the summary itself were a brief version of the full text.
- Audience: General readers without specialized knowledge of the topic.
—
### Examples:
-> Example 1
##### Summary:
{example_summary_1}

-> Example 2
##### Summary:
{example_summary_2}

-> Example 3
##### Summary:
{example_summary_3}

-> Example 4
##### Summary:
{example_summary_4}

-> Example 5
##### Summary:
{example_summary_5}
—
### Now, summarize the following article based on the given criteria and using the same style of the example:

### Article:
{article}

### Summary:
```

Table 12: Few-Shot Structured Prompt Template for Task 1.2, S2 and S4.

Task:
You are an expert medical communicator. Your role is to translate radiology reports, originally written for healthcare professionals, into language that an average person without a medical background can understand. The rewritten report should preserve all essential medical findings and implications suitable for the general public. Note that you must avoid redundancy.

—

Definitions:
- Radiology Report: A medical document that describes findings from imaging studies such as X-rays, CT scans, or MRIs.
- Layman Report: A simplified, non-technical explanation suitable for someone with no formal medical education.

—

Guidelines:
- Generated number of tokens: Try to match the number of tokens of the original Radiology Report, adjusting as needed based on report complexity.
- Avoid Speculation: Do not add interpretations beyond what is stated in the original report.
- Maintain a Reassuring and Neutral Tone: Use clear, calm, and factual language.
- Structure: Present the information in a single, coherent paragraph.
- The single paragraph can be composed of one or several sentences.
- Adhere to Reported Diagnoses: Only summarize what is already reported; do not include diagnoses not explicitly stated. - Avoid redundancy.

—

Guidelines for Output:
- Format: Clear and concise prose.
- Redundancy: Avoid repeating information unnecessarily.
- Length: Should closely match the number of tokens in the original Radiology Report, adjusting as needed based on report complexity.
- Audience: A general reader with no medical background or clinical training.

—

Analyze the Following Radiology Report Based on the Given Criteria:

Radiology Report:
{radiology_report}

Response (Layman Report):

Table 13: Zero-Shot Structured Prompt Template for Task 2.1, S1 and S3

Task:
You are an expert medical communicator. Your role is to translate radiology reports, originally written for healthcare professionals, into plain language that an average person without a medical background can understand. The rewritten report should preserve all essential medical findings and implications suitable for the general public. Note that you must avoid redundancy.

—

Definitions:
- Radiology Report: A medical document that describes findings from imaging studies such as X-rays, CT scans, or MRIs.
- Layman Report: A simplified, non-technical explanation suitable for someone with no formal medical education.

—

Guidelines:
- Generated number of tokens: Try to match the number of tokens of the original Radiology Report, adjusting as needed based on report complexity.
- Avoid Speculation: Do not add interpretations beyond what is stated in the original report.
- Maintain a Reassuring and Neutral Tone: Use clear, calm, and factual language.
- Structure: Present the information in a single, coherent paragraph.
- The single paragraph can be composed of one or several sentences.
- Adhere to Reported Diagnoses: Only summarize what is already reported; do not include diagnoses not explicitly stated.
- Avoid redundancy.

—

Guidelines for Output:
- Format: Clear and concise prose.
- Redundancy: Avoid repeating information unnecessarily.
- Length: Should closely match the number of tokens in the original Radiology Report, adjusting as needed based on report complexity.
- Audience: A general reader with no medical background or clinical training.

—

Examples:

-> Example 1
Radiology Report:
{example_radiology_report_1}
Response (Layman Report):
{example_layman_report_1}

-> Example 2
Radiology Report:
{example_radiology_report_2}
Response (Layman Report):
{example_layman_report_2}

-> Example 3
Radiology Report:
{example_radiology_report_3}
Response (Layman Report):
{example_layman_report_3}

-> Example 4
Radiology Report:
{example_radiology_report_4}
Response (Layman Report):
{example_layman_report_4}

-> Example 5
Radiology Report:
{example_radiology_report_5}
Response (Layman Report):
{example_layman_report_5}

—

Analyze the Following Radiology Report Based on the Given Criteria:

Radiology Report:
{radiology_report}

Response (Layman Report):

Table 14: Few-Shot Structured Prompt Template with random examples for Task 2.1, S2 and S4.

```
### Task:
You are an expert medical communicator. Your task is to translate radiology reports, originally written for healthcare professionals, into plain language that an average person without a medical background can understand. The rewritten report, referred to as Layman Report, should preserve all essential medical findings and implications suitable for the general public.

—
### Definitions:
- Radiology Report: A medical document that describes findings from imaging studies such as X-rays, CT scans, or MRIs.
- Layman Report: A simplified, non-technical explanation suitable for someone with no formal medical education.

—
### Instructions:
- Use the examples below as a guide, matching their structure and writing style in your layman report.
- The rewritten report should be in layman terms.
- Briefly define any technical terms that must be included.
- Try to match the number of tokens of the original Radiology Report, adjusting as needed based on report complexity.
- Maintain a Reassuring and Neutral Tone: Use clear and factual language.
- Structure: Present the information in a single, coherent paragraph.
- The single paragraph can be composed of one or several sentences.
- Adhere to Reported Diagnoses: Only rewrite what is already reported; do not include diagnoses not explicitly stated.

—
### Guidelines for Output:
- Format: Clear, flowing prose in one paragraph.
- Content: Capture the essential meaning and logic of the radiology report.
- Length: Should closely match the number of tokens in the original Radiology Report, adjusting as needed based on report complexity.
- Audience: A general reader with no medical background or clinical training.

—
### Examples:

-> Example 1
##### Radiology Report:
{similar_example_radiology_report_1}
##### Response (Layman Report):
{similar_example_layman_report_1}

-> Example 2
##### Radiology Report:
{similar_example_radiology_report_2}
##### Response (Layman Report):
{similar_example_layman_report_2}

...
...
...

-> Example 10
##### Radiology Report:
{similar_example_radiology_report_10}
##### Response (Layman Report):
{similar_example_layman_report_10}

—
### Now, rewrite the following radiology report based on the given criteria and using the same style of the examples:

### Radiology Report:
{radiology_report}

### Layman Report:
```

Table 15: Few-Shot Structured Prompt Template with Cosine Similarity-Based Examples for Task 2.1, S5.

		Llama-3.3-70B-Instruct				GPT-4.1o	
Description	**Metric**	**E1**	**E2**	**E3**	**E4**	**E5**	**E6**
Relevance	ROUGE ↑	0.3070	0.3593	0.3899	**0.4806**	0.3590	0.4434
	BLEU ↑	8.2984	11.7237	14.8005	**20.3253**	12.8078	17.5307
	METEOR ↑	0.4132	0.4583	0.4645	**0.5352**	0.3949	0.4723
	BERTScore ↑	0.8875	0.8886	0.9141	0.9170	0.9136	**0.9238**
Readability	FKGL ↓	9.3505	10.3292	9.7895	**7.9905**	10.6400	9.5130
	DCRS ↓	**8.9585**	10.3663	9.4053	9.3723	10.3322	10.1358
	CLI ↓	**8.4965**	10.1607	9.1438	9.0414	9.9672	9.8145
Factuality	Similarity ↑	0.7045	0.7320	0.7888	0.7964	0.7697	**0.8112**
	RadGraph ↑	0.1512	0.1722	0.2167	0.1938	0.2109	**0.2252**
	F1CheXbert ↑	-	-	0.7200	0.7100	**0.7700**	0.7450

Table 16: Task 2.1 experiments run on Llama-3.3-70B-Instruct and GPT 4.1o models. **E1:** Unstructured zero-shot prompt. **E2:** Unstructured few-shot prompt with 5 random examples. **E3:** Structured zero-shot prompt. **E4:** Structured few-shot prompt with 5 random examples. **E5:** Structured zero-shot prompt. **E6:** Structured few-shot prompt with 5 random examples.

MIRAGES at BioLaySumm2025: The Impact of Search Terms and Data Curation for Biomedical Lay Summarization

Benjamin Pong **Ju-Hui Chen** **Jonathan Jiang**
Abimael Hernandez Jimenez **Melody Vahadi**
Department of Linguistics, University of Washington, Seattle, WA, USA
{benpong, juhuic, jjiang85, abimaelh, mvahadi}@uw.edu

Abstract

Biomedical articles are often inaccessible to non-experts due to their technical complexity. To improve readability and factuality of lay summaries, we built on an extract-then-summarize framework by experimenting with novel extractive summarization strategies and employing Low Rank Adaptation (LoRA) fine-tuning of Meta-Llama-3-8B-Instruct on data selected by these strategies. We also explored counterfactual data augmentation and post-processing definition insertion to further enhance factual grounding and accessibility. Our best performing system treats the article's title and keywords (i.e. search terms) as a single semantic centroid and ranks sentences by their semantic similarity to this centroid. This constrained selection of data serves as input for fine-tuning, achieving marked improvements in readability and factuality of downstream abstractive summaries while maintaining relevance. Our approach highlights the importance of quality data curation for biomedical lay summarization, resulting in 4th best overall performance and 2nd best Readability performance for the BioLaySumm 2025 Shared Task at BioNLP 2025.

1 Introduction

Biomedical research journals contain the latest findings on public health but highly technical language prevents the general public from understanding their content, which poses a challenge to health literacy (Guo et al., 2021). One solution is creating lay summaries – short, readable versions of scientific texts that use plain language and provide contextual information to bridge knowledge gaps.

This paper presents our submission to the BioLaySumm 2025 shared task 1.1 (Xiao et al., 2025), which focuses on generating lay summaries for biomedical articles. This task builds on previous editions of the shared task introduced in 2023 (Goldsack et al., 2023) and further developed in 2024 (Goldsack et al., 2024), which emphasize the challenges of readability, factuality, and accessibility in biomedical lay summarization. We built on the success of an extract-then-summarize pipeline (You et al., 2024) by developing novel sentence selection strategies that identify the most salient content from each article, prior to summarization, using titles and key words (i.e search terms). Unlike You et al. (2024) who explored the use of keywords for definition retrieval, and (Zhou et al., 2024) who explored title infusion for prompting, we explored the impact of these search terms at the level of extractive summarization. Our system [1] aims to balance relevance, readability, and factuality.

2 Dataset

The datasets used for this task are the PLOS and eLife datasets (Goldsack et al., 2022). The PLOS dataset comprises text from articles from life sciences. The eLife dataset contains articles on life sciences and medicine. The PLOS data set contains 24,773 training instances and 1,376 validation instances, while eLife contains 4,346 training and 241 validation instances.

3 Methods

Our system includes a preliminary retrieval-based extractive summarization process, and model finetuning and inference using Meta-Llama-3-8B-Instruct [2] (AI@Meta, 2024).

3.1 Preliminary Experiment: Preprocessing and Extractive Summarization

We first investigated which extractive summarization strategy would be most useful for finetuning and downstream abstractive summarization. We removed information in parentheses and citations. To

[1] https://github.com/Abimaelh/bio-laysum.git

[2] https://huggingface.co/meta-llama/Meta-Llama-3-8B-Instruct

BioNLP 2025 Shared Tasks, pages 232–239
August 1, 2025

extract salient content, we employed seven extractive strategies using SpaCy's sentence tokenizer (Honnibal et al., 2020), BioBERT's (Lee et al., 2019) embeddings, and cosine similarity for similarity scoring.

Strategy 1 (Control): Selects the first 4096 tokens for abstractive summarization.

Strategy 2: Converts the title to an embedding, ranks sentences by cosine similarity to the title embedding and selects the top 40.

Strategy 3: Enhances Strategy 2 by concatenating keywords into the title to form an embedding before computing similarity and selecting the top 40 sentences.

Strategy 4: Inspired by the utility of singular value decomposition (SVD), for topic modeling and text summarization Steinberger and Jezek (2004), we apply SVD to group sentences by topic and select the top 40 sentences from the topics ranked closest to the gold summary.

Strategy 5: Compute the article's mean embedding and extract the top 40 sentences that are most semantically similar.

Strategy 6: Prepends title and keywords to the article and segment the article into four core sections(abstract, introduction, results, and discussion)[3]. From this condensed content, we rank sentences according to their similarity to the mean embedding of the uncondensed article, and select the top 40 sentences.

Strategy 7: The reverse of 6, where we segment the article to the same four core sections, extract the top 40 sentences and prepend the title and keywords.

The outputs of the following seven extractive strategies were summarized by Meta-Llama-3-8B-Instruct (prompt in Section 3.2) and are evaluated on the eLife validation set using Strategy 1 as a control and comparing their relative performance. The articles were trimmed to 4096 tokens for inference, due to computational constraints. Appendix A shows the evaluation results and analysis. Strategy 2 and 3 showed reasonable potential to influence downstream abstractive summarization.

[3]We simply segmented the article into chunks according to the number of section headings, used these chunks as proxies for sections and removed the chunks corresponding to Materials and Methods since they are less relevant for summarization.

3.2 Baseline: Zero-shot prompt

As our baseline, we prompted Meta-Llama-3-8B-Instruct to generate abstractive lay summaries for articles on Strategy 1 using the following zero-shot prompt template:

```
System: You are a chatbot with expertise in summarizing documents
User: Provide a lay summary of the following text: {article}
```

3.3 Meta-Llama-3-8B-Instruct Finetuning

To evaluate how the best performing extractive strategies influence downstream summarization quality, we finetuned Meta-Llama-3-8B-Instruct on the unprocessed data (Strategy 1), and top-performing Strategies 2 and 3 using Low Rank Adaptation (LoRA) (Hu et al., 2022), and compared these finetuned instances against the baseline.

The data for finetuning was prepared by randomly selecting 650 training instances from both eLife and PLOS, totaling 1300 shuffled samples for finetuning. For evaluation, we used 150 randomly selected validation samples from both datasets, totaling 300 shuffled samples.

We present the set of hyperparameters considered in Appendix B, Table 4, and refer to them as sets 1 to 3 for the rest of this paper. Our experiments are incremental, starting from finetuning on 200 samples across Sets 1 and 2. Based on our results, finetuning on Strategy 1 using Set 1 did not improve over the baseline, but finetuning on Strategy 2 and 3 boosted Readability and Factuality scores. Finetuning on Set 2 did not show improvements.

Following this near-positive results, we performed a sample-size ablation study on 1000 samples and 1300 samples using set 1, to test if sample size further improves model performance. Since our results show that a sample size beyond 1000 does not induce improvements, we conducted further experiments on hyperparameter sets 3 on 1000 training samples.

3.4 Counterfactual Data Augmentation

Prior work (Rajagopal et al., 2022) claim that training on counterfactually augmented data can improve factual consistency of general-domain abstractive summaries by inducing entity-errors, and attempt to extend this hypothesis for lay summarization. To develop the counterfactual data, we used the same 1000 training samples that were

Table 1: Finetuning evaluation scores (Systems 1-16) on 300 randomly sampled data instances from both eLife and PLOS' validation set . Entries under model configuration for systems 1-15 are interpreted as: `llama_{hyperparameter set}_{strategy}_{sample size}`

System	Model Configuration	R-1	R-2	R-L	BERTScore	FKGL	DCRS	CLI	SummaC	AlignScore
Baseline	No-finetuning	0.4316	0.1130	0.4015	0.8500	12.7278	10.5846	13.0331	0.5654	0.6424
1	llama_1_1_200	0.3985	0.1148	0.3701	0.8482	13.3061	11.1538	13.2315	0.5078	0.6125
2	llama_1_2_200	0.4174	0.1069	0.3906	0.8477	12.6564	10.3625	12.4354	0.5692	0.6296
3	llama_1_3_200	0.4221	0.1095	0.3936	0.8488	12.9245	10.5067	12.4350	0.5555	0.6248
4	llama_2_1_200	0.4172	0.1172	0.3846	0.8482	14.0118	11.1758	13.7989	0.5078	0.6346
5	llama_2_2_200	0.4238	0.1116	0.3949	0.8492	13.1023	10.5741	13.0970	0.5509	0.6337
6	llama_2_3_200	0.4252	0.1117	0.3946	0.8496	12.9096	10.6275	13.0385	0.5572	0.6434
7	llama_1_1_1000	0.4130	0.1125	0.3868	0.8399	12.4096	10.6496	12.0954	0.6300	0.7122
8	llama_1_2_1000	0.4057	0.1025	0.3814	0.8448	11.3494	9.9221	11.6261	0.6300	0.6557
9	llama_1_3_1000	0.4045	0.1014	0.3799	0.8441	11.0981	9.7785	11.3806	0.6300	0.6846
10	llama_1_1_1300	0.4153	0.1115	0.3886	0.8464	12.7704	11.0471	12.8860	0.7100	0.7032
11	llama_1_2_1300	0.4156	0.1138	0.3893	0.8453	12.1517	10.5510	12.4498	0.6738	0.6255
12	llama_1_3_1300	0.4112	0.1072	0.3861	0.8424	11.5830	10.1520	11.8491	0.644	0.6094
13	llama_3_1_1000	0.4157	0.1125	0.3892	0.8399	12.3269	10.7378	12.4745	0.6385	0.7514
14	llama_3_2_1000	0.4158	0.1162	0.3880	0.8427	12.8280	10.9584	12.7057	0.6720	0.6223
15	llama_3_3_1000	0.4069	0.1025	0.3814	0.8445	11.2793	9.8721	11.5129	0.6133	0.6066
16	counterfactual	0.4001	0.0989	0.3770	0.8427	11.3365	10.0515	11.6893	0.6469	0.625

used to finetune System 9[4] but selected 250 samples to be modified by employing BERN2 (Sung et al., 2022), a multitask Named Entity Recognition (NER) model to extract biomedical entity mentions from their gold summaries. These entity mentions were masked out with their corresponding categories. We used Meta-Llama-3-8B-Instruct to substitute each category with a random entity belonging to that category, followed by a finetuning experiment on a data mixture of counterfactual data (See Appendix C for training templates and prompt template).

3.5 Postprocessing: Lay definition Insertion using LLM and UMLS

To enhance the readability and relevance of the generated summaries by our best performing model in Table 1 (i.e., System 9) we added a postprocessing strategy by using LLMs and UMLS as external knowledge bases of lay definitions. The goal is to simplify some biomedical terms from the summaries, and provide contextual knowledge through definitions [5].

We used SciSpacy's Biomedical NER model (Neumann et al., 2019) to extract biomedical entity mentions, and their definitions through its connection with the Unified Medical Language System (UMLS) database. For entity mentions that are absent, we employed Meta-Llama-3-8B-Instruct to provide definitions. With this hybrid approach to definition retrieval, we constructed a term-definition dictionary for each generated summary. For each generated summary, we randomly extracted 10 pairs of terms and definitions to be incorporated into a prompt for postprocessing. The prompt templates can be found in Appendix D.

[4]This turned out to be our best performing model. See Results.

[5]Note that this experiment was conducted on test set summaries produced by our highest-achieving model, and was evaluated on the system provided by the organizers.

4 Evaluation

All experiments except postprocessing were done using a subset of the metrics given by the organizers, on 300 randomly chosen validation samples as mentioned. For relevance, ROUGE-1, ROUGE-2, and ROUGE-L (Lin, 2004), and BERTscore (Zhang et al., 2020) were used. For readability, Flesch-Kincaid Grade Level (FKGL) (Kincaid et al., 1975), Dale-Chall Readability Score (DCRS) (Chall and Dale, 1995), and Coleman-Liau index (CLI) (Coleman and Liau, 1975) were used. Lower FKGL, DCRS and CLI scores represent superior readability. Finally, for factuality, SummaC (Laban et al., 2022) and AlignScore (Zha et al., 2023) were used. Postprocessing experiment, as well as our best performing model, were (re)evaluated on the test set using the organizers' evaluation pipeline.

5 Results and Analysis

5.1 Experimental results

We report the results of our experiments in Table 1. The system that we submitted to the leaderboard for BioLaySumm2025 is system 9.

Table 2: Comparison of best system with and without post-processing. These systems were evaluated on the test set using the evaluation pipeline provided by the organizers of Biolaysumm. We submitted our best performing model, (i.e., system 9.)

System	ROUGE	BLEU	METEOR	BERTScore	FKGL	DCRS	CLI	LENS	AlignScore	SummaC
Best (submission)	0.2877	4.6323	0.2305	0.8461	11.7109	8.4596	11.9899	71.2714	0.6811	0.6047
$\text{Best}_{+postprocessing}$	0.2498	3.1827	0.2021	0.8345	12.9000	8.3068	11.7878	61.9381	0.6026	0.5916

5.1.1 Impact of hyperparameters and sample-size ablation study

The results of systems 1 to 3 show that finetuning on filtered articles using hyperparameter set 1 generally outperform the zero-shot baseline in terms of readability, while finetuning on unfiltered ones did not. Hyperparameter set 2 did not improve model performance. Our ablation study on hyperparameter set 1 shows that increasing the sample size to 1000 for finetuning has a larger positive effect on both readability and factuality (Systems 7-9) compared to the baseline, but a sample size beyond that did not (Systems 10-12). However, finetuning does not seem to improve relevance scores across the board.

5.1.2 Impact of extractive summarization strategies

The effect of extractive summarization strategies is compounded on by the effect of sample size. System 2 outperforms system 3 in terms of readability and factuality, suggesting that keywords are inert. However, when the sample size increases to 1000, while they both outperform the baseline, system 9 outperformed system 8 in readability and factuality. This suggests that while the title is capable of extracting pivotal sentences in the article, the impact of keywords scales with data volume.

5.2 Impact of data augmentation

As expected from (Rajagopal et al., 2022), finetuning on counterfactually augmented data showed improvements in SummaC score, but a slight decrease in relevance and readability scores (Compare systems 9 and 16). This experiment verifies the reproducibility of (Rajagopal et al., 2022)'s work on using counterfactual data augmentation improves factuality for summarization with tradeoffs in relevance. In addition, our experiment sparks the promise of extending their methodology to the context of biomedical lay summarization. We leave this exploration to future work.

5.3 Impact of Post-processing using definition insertions

As presented in Table 2, our result for post-processing surprisingly showed marginal improvements in readability scores (DCRS and CLI), and a drop in other evaluation metrics. We speculate that while definition insertions helped with text simplification, the NER model is flawed in that it also extracts non-technical terms like "blood" and "human". Redundant definitions of these terms could have been incorporated into the summary, hence affecting factual consistency, and inducing verbosity.

5.4 Results of Final System Submission

Table 2 shows the results of our best performing model, which we submitted to the leaderboard. Our model was ranked 4th on the leaderboard, and achieved 2nd place in terms of Readability scores.

6 Discussion and Conclusion

Our study highlights the trade-offs in biomedical lay summarization between input selection, model fine-tuning, and postprocessing. Strategically curating input–particularly by leveraging document titles and keywords–can significantly improve the readability of generated summaries. Finetuning Meta-Llama-3-8B-Instruct on such targeted content surpasses using unfiltered inputs.

A comparison of extractive strategies reveals that title-based selection performs better with smaller training sets, while the inclusion of keywords becomes more effective as the models handle more data, suggesting that keywords provide additional semantic information that enhances generalization, particularly in data-rich settings across different topics.

Our ablation study shows that increasing finetuning sample size from 200 to 1000 improves performance across readability scores (FKGL, Dale-Chall), factuality (SummaC and AlignScore), but increasing sample sizes up to 1300 samples plateaus (System 10-12) or slightly reverses gains, possibly due to noise from lower quality training samples. These findings emphasize that high quality extractive pre-processing can have a more positive impact than increasing fine-tuning sample volume alone in domain-specific summarization tasks.

Regarding hyperparameters, Set 1 was consistently effective, especially when used with 1000 samples (e.g., Systems 7-9). Raising LoRA rank to 13 and increasing the effective batch size (Set 3) yielded only marginal improvements (e.g., System 13 vs. System 7), suggesting limited benefit from increasing model capacity under our current setup.

However, we do not see improvements in Relevance scores across the board, possibly due model capacity and hyperparameter issues. Another reason for this is, improved readability may have oversimplified the summaries, resulting in information loss.

Overall, our results demonstrate that thoughtful input design and targeted fine-tuning are critical for effective biomedical lay summarization. Our future work may explore adaptive extractive techniques and multiphase generation pipelines to further enhance summary clarity and trustworthiness.

7 Limitations

Our study has several limitations that inform opportunities for future work. First, we only evaluated decoder-only LLM-based architectures–specifically Meta-Llama-3-8B-Instruct–and did not explore neural encoder-decoder models, such as T5 or BART, which are commonly used for summarization tasks. This architectural constraint explains the limited improvement in BERTscore and Relevance Scores, which often favor outputs more closely aligned with gold summaries at the token or phrase level. Secondly, while our resource-constrained hyperparameter search identified workable configurations, future work should prioritize expanded hyperparameter optimization to fully exploit the model's capacity. Thirdly, our counterfactual data augmentation experiment, requires more complexity and development to investigate the tradeoffs between relevance and factuality. Aforementioned, in our postprocessing step, using NER to extract technical biomedical terms fails to sufficiently exclude non-technical medical terminologies, which may have contributed to redundant additions and edits to the summaries. Furthermore, randomly selecting 10 term-definitions does not circumvent this issue. Future work in this direction should consider more discriminate ways to filter out non-technical terms from biomedical texts, so that actual technical terms can be easily identified for simplification. Finally, while our system did reasonably well for readability, we did not explicitly investigate the effect of readability control (Luo et al., 2022) since the degree of simplicity is subjected to each individual's demands and technical expertise.

References

AI@Meta. 2024. Llama 3 model card.

Jeanne Sternlicht Chall and Edgar Dale. 1995. *Readability Revisited: The New Dale-Chall Readability Formula*. Brookline Books.

Meri Coleman and T. L. Liau. 1975. A computer readability formula designed for machine scoring. *Journal of Applied Psychology*, 60:283–284.

Tomas Goldsack, Zheheng Luo, Qianqian Xie, Carolina Scarton, Matthew Shardlow, Sophia Ananiadou, and Chenghua Lin. 2023. Overview of the biolaysumm 2023 shared task on lay summarization of biomedical research articles. In *The 22nd Workshop on Biomedical Natural Language Processing and BioNLP Shared Tasks*, pages 468–477, Toronto, Canada. Association for Computational Linguistics.

Tomas Goldsack, Carolina Scarton, Matthew Shardlow, and Chenghua Lin. 2024. Overview of the BioLaySumm 2024 shared task on the lay summarization of biomedical research articles. In *Proceedings of the 23rd Workshop on Biomedical Natural Language Processing*, pages 122–131, Bangkok, Thailand. Association for Computational Linguistics.

Tomas Goldsack, Zhihao Zhang, Chenghua Lin, and Carolina Scarton. 2022. Making science simple: Corpora for the lay summarisation of scientific literature. In *Proceedings of the 2022 Conference on Empirical Methods in Natural Language Processing*, pages 10589–10604, Abu Dhabi, United Arab Emirates. Association for Computational Linguistics.

Yue Guo, Wei Qiu, Yizhong Wang, and Trevor Cohen. 2021. Automated lay language summarization of biomedical scientific reviews. *Proceedings of the AAAI Conference on Artificial Intelligence*, 35(1):160–168.

Matthew Honnibal, Ines Montani, Sofie Van Landeghem, and Adriane Boyd. 2020. spaCy: Industrial-strength natural language processing in python.

Edward J Hu, Yelong Shen, Phillip Wallis, Zeyuan Allen-Zhu, Yuanzhi Li, Shean Wang, Lu Wang, and Weizhu Chen. 2022. LoRA: Low-rank adaptation of large language models. In *International Conference on Learning Representations*.

J. Peter Kincaid, Robert P. Fishburne, Richard L. Rogers, and Brad S. Chissom. 1975. Derivation of new readability formulas (automated readability index, fog count, and flesch reading ease formula) for navy enlisted personnel. Technical report, Naval Air Station Memphis: Chief of Naval Technical Training. Research Branch.

Philippe Laban, Tobias Schnabel, Paul N. Bennett, and Marti A. Hearst. 2022. SummaC: Re-visiting NLI-based models for inconsistency detection in summarization. *Transactions of the Association for Computational Linguistics*, 10:163–177.

Jinhyuk Lee, Wonjin Yoon, Sungdong Kim, Donghyeon Kim, Sunkyu Kim, Chan Ho So, and Jaewoo Kang. 2019. Biobert: a pre-trained biomedical language representation model for biomedical text mining. *Bioinformatics*, 36(4):1234–1240.

Chin-Yew Lin. 2004. ROUGE: A Package for Automatic Evaluation of Summaries. In *Text Summarization Branches Out*, pages 74–81, Barcelona, Spain. Association for Computational Linguistics.

Zheheng Luo, Qianqian Xie, and Sophia Ananiadou. 2022. Readability controllable biomedical document summarization. In *Findings of the Association for Computational Linguistics: EMNLP 2022*, pages 4667–4680, Abu Dhabi, United Arab Emirates. Association for Computational Linguistics.

Mark Neumann, Daniel King, Iz Beltagy, and Waleed Ammar. 2019. ScispaCy: Fast and Robust Models for Biomedical Natural Language Processing. In *Proceedings of the 18th BioNLP Workshop and Shared Task*, pages 319–327, Florence, Italy. Association for Computational Linguistics.

Dheeraj Rajagopal, Siamak Shakeri, Cicero Nogueira dos Santos, Eduard Hovy, and Chung-Ching Chang. 2022. Counterfactual data augmentation improves factuality of abstractive summarization. *Preprint*, arXiv:2205.12416.

Josef Steinberger and Karel Jezek. 2004. Using latent semantic analysis in text summarization and summary evaluation. In *Proceedings of the 7th International Conference ISIM*.

Mujeen Sung, Minbyul Jeong, Yonghwa Choi, Donghyeon Kim, Jinhyuk Lee, and Jaewoo Kang. 2022. Bern2: an advanced neural biomedical namedentity recognition and normalization tool.

Chenghao Xiao, Kun Zhao, Xiao Wang, Siwei Wu, Sixing Yan, Sophia Ananiadou, Noura Al Moubayed, Liang Zhan, William Cheung, and Chenghua Lin. 2025. Overview of the biolaysumm 2025 shared task on lay summarization of biomedical research articles and radiology reports. In *The 24nd Workshop on Biomedical Natural Language Processing and BioNLP Shared Tasks*, Vienna, Austria. Association for Computational Linguistics.

Zhiwen You, Shruthan Radhakrishna, Shufan Ming, and Halil Kilicoglu. 2024. UIUC_BioNLP at BioLaySumm: An extract-then-summarize approach augmented with wikipedia knowledge for biomedical lay summarization. In *Proceedings of the 23rd Workshop on Biomedical Language Processing*, Bangkok, Thailand. Association for Computational Linguistics.

Yuheng Zha, Yichi Yang, Ruichen Li, and Zhiting Hu. 2023. AlignScore: Evaluating factual consistency with a unified alignment function. In *Proceedings of the 61st Annual Meeting of the Association for Computational Linguistics (Volume 1: Long Papers)*, pages 11328–11348, Toronto, Canada. Association for Computational Linguistics.

Tianyi Zhang, Varsha Kishore, Felix Wu, Kilian Q. Weinberger, and Yoav Artzi. 2020. Bertscore: Evaluating text generation with bert. In *International Conference on Learning Representations*.

Jieli Zhou, Cheng Ye, Pengcheng Xu, and Hongyi Xin. 2024. Team YXZ at BioLaySumm: Adapting large language models for biomedical lay summarization. In *Proceedings of the 23rd Workshop on Biomedical Natural Language Processing*, pages 818–825, Bangkok, Thailand. Association for Computational Linguistics.

A Evaluation Results for preliminary experiments

Based on our evaluation of the preliminary experiment using the metrics described in Section 4, we observed that Strategy 2 did the best for readability scores, clearly surpassing the baseline, while Strategy 4 (SVD topic modeling) did the best for BERTscore. As for ROUGE-L, ROUGE-1, and SummaC, Strategy 2 and 3 did comparable to the control. The low scores for factuality are to be expected (Zhou et al., 2024) from just LLM prompting techniques without further finetuning. But the scores for Strategies 2 and 3 follow the control. Hence, we chose Strategies 1, 2 and 3 for finetuning. Full evaluation results can be found below in Table 3.

B Hyperparameters

Table 4 shows the hyperparameters that we used for our experiments.

Across all sets, we applied the AdamW optimizer, a LoRA dropout rate of 0.1, a LoRA alpha of 16 and a linear learning rate scheduler.

C Prompt Templates for Counterfactual Data Augumentation

We provide the following prompt templates for counterfactual data augmentation process.

The {text} refers to the gold summary and the * represents the entity mention that has been masked out and replaced with the entity category. The prompt below replaces * with a random entity mention of that category, and its output is an entity-error-induced gold summary:

```
System: You are a chatbot with knowledge in medical terms and their definitions in context.
User: The following text contains words enclosed in *̇These words are categories for biomedical entities. Replace the words with randomly chosen biomedical entities from your wealth of knowledge, and then enumerate a list of the replacements. {text}'
```

The output of the above is used for finetuning, where the model is trained to recognize factual deviance:

Table 3: Preprocessing Methods Performance Metrics Comparison

Preprocessing	R-1	R-2	R-L	BERTScore	FKGL	DCRS	CLI	SummaC	AlignScore
1 (baseline)	0.4031	0.1017	0.3779	0.8412	13.0715	10.9113	13.6171	0.5893	0.4521
2	0.4048	0.0967	0.3788	0.8417	12.8065	10.4800	13.2654	0.3885	0.3988
3	0.4070	0.0978	0.3802	0.8420	13.0214	10.5933	13.5069	0.3767	0.4147
4	0.4089	0.1020	0.3830	0.8423	12.9394	10.7197	13.6397	0.3464	0.3727
5	0.3929	0.0945	0.3675	0.8387	12.8777	10.7964	13.2739	0.3912	0.3625
6	0.3799	0.0894	0.3541	0.8310	13.8133	10.9826	13.1877	0.3523	0.3588
7	0.3851	0.0926	0.3580	0.8366	14.4809	11.0687	13.8176	0.3665	0.3444

Table 4: Hyperparameters

Hyperparameters	Set 1	Set 2	Set 3
Learning Rate	2×10^{-5}	8×10^{-6}	2×10^{-5}
Batch Size	4	8	4
Epochs	3	5	3
Grad. Accumulation	2	1	2
r (Rank)	10	10	13
LoRA dropout	0.1	0.1	0.1
lr scheduler	linear	linear	linear
Optimizer	AdamW	AdamW	AdamW

```
<|begin_of_text|><|start_header_id|>
system<|end_header_id|> You are
a chatbot with expertise in
summarizing documents. <|eot_id|>
<|start_header_id|>user
<|end_header_id|>
Provide a wrong lay
summary of this article:
{preprocessed article} <|eot_id|>
<|start_header_id|>assistant
<|end_header_id|>
Wrong lay Summary:
{entity-error-induced gold summary}
<|eot_id|>
```

Note that the template above is only for the 250 samples that were selected for counterfactual augmentation. For the rest of the 750 samples, we had the <assistant> prompt to indicate "lay summary". A mixture of original data and counterfactual data is used as training data for this finetuning experiment.

D Prompt Templates for Postprocessing Step

The prompt template used to extract definitions of entity mentions from Meta-Llama-3-8B-Instruct is as follows: System: "You are an expert who can provide informative and lay definitions to biomedical terms." User: Provide only the definition of the biomedical term: term'

As mentioned in the main text, term-definition dictionaries were constructed and incorporated into a prompt to generate a postprocessed summary. The prompt template used is:

System: "You are an expert biomedical editor skilled at simplifying complex medical terms for a lay audience. Use the provided dictionary to replace technical terms with their lay definitions while preserving the original meaning."

User: **Biomedical Lay Definitions Dictionary:** {term_dictionary}
*Task:** - Read the following summary: {summary}
- Replace all technical terms in the summary with their lay definitions from the dictionary.
- Do not add or remove key information.
- If a term isn't in the dictionary, retain the original term.
*Return only the paraphrased summary in one line, without any commentary**

SUWMIT at BioLaySumm2025: Instruction-based Summarization with Contrastive Decoding

Priyam Basu, Jose Cols, Daniel Jarvis, Yongsin Park, Daniel Rodabaugh
Department of Linguistics, University of Washington
{pbasu77,jcols,dljarvi,yongsinp,drodaba}@uw.edu

Abstract

In the following paper, we present our team's approach to subtask 1.1 of the BioLaySumm 2025 shared task, which entails the automated generation of lay summaries from biomedical articles. To this end, we experiment with a variety of methods for text preprocessing, extractive summarization, model fine-tuning, and abstractive summarization. Our final results are generated on a fine-tuned Llama 3.1 Instruct (8B) model, notably achieving top scores on two out of four relevance metrics, as well as the highest overall ranking among this year's participating teams on the plain lay summarization subtask.

1 Introduction

Biomedical articles often contain information of interest to audiences beyond the community of medical researchers and practitioners; however, the large volume of content, in combination with domain-specific technical language, often leaves such text unsuited for consumption by non-experts. The automated generation of lay summaries may, therefore, serve as a tool for improving the accessibility of scientific publications to a broader public by offering a non-technical glance to potential readers (Goldsack et al., 2024). Following previous iterations initiated by Goldsack et al. (2023), the BioLaySumm 2025 shared task presents precisely this objective, calling for teams to make use of the `PLOS` and `eLife` datasets (Goldsack et al., 2022; Luo et al., 2022b) to build automated summarization systems with a focus on ease of understanding while maintaining relevance and factuality (Xiao et al., 2025).

Winners of the BioLaySumm 2023 shared task (Turbitt et al., 2023) saw success in generating summaries based on the abstracts of articles and leveraging domain knowledge of GPT-style models, with summaries generated by their system offering better relevance and factuality scores than the fine-tuned BioGPT (Luo et al., 2022a) model they tested against, though at the cost of readability. Winners of the BioLaySumm 2024 (You et al., 2024) subsequently investigated an alternative approach to the fine-tuning of the model, using TextRank (Mihalcea and Tarau, 2004) to extract the most salient content before passing it to a GPT model for summarization, augmented by a BERT-based clustering technique and a keyword-based method to extract definitions from the Wikipedia dataset. Another team, Modi and Karthikeyan (2024), achieved top factuality scores by running preprocessing methods over article abstracts before passing content through an LLM.

Building on the success of these previous teams, we develop and publicly release an open-source,[1] end-to-end pipeline to facilitate rapid experimentation in summarization (Section 3.1). Our best model results from experiments conducted through this pipeline.

2 Data

The shared task organizers have made available two datasets, `PLOS` and `eLife` (Goldsack et al., 2022; Luo et al., 2022b), which include biomedical research articles and their corresponding expert-written lay summaries. Together, these datasets comprise a total of 29,119 training instances and 1,617 validation instances, with approximately 85% of instances sourced from `PLOS`, and the remaining 15% from `eLife`. Additional dataset statistics are provided in Appendix B.

3 Methods

In this section, we provide an overview of the methodology used for our final submission, which is an abstractive summarization model based on Meta's Llama 3.1 Instruct (8B) (Grattafiori et al., 2024). Although this model did not perform the

[1] `https://github.com/whopriyamuw/biolaysumm2025-task`

BioNLP 2025 Shared Tasks, pages 240–248
August 1, 2025 ©2025 Association for Computational Linguistics

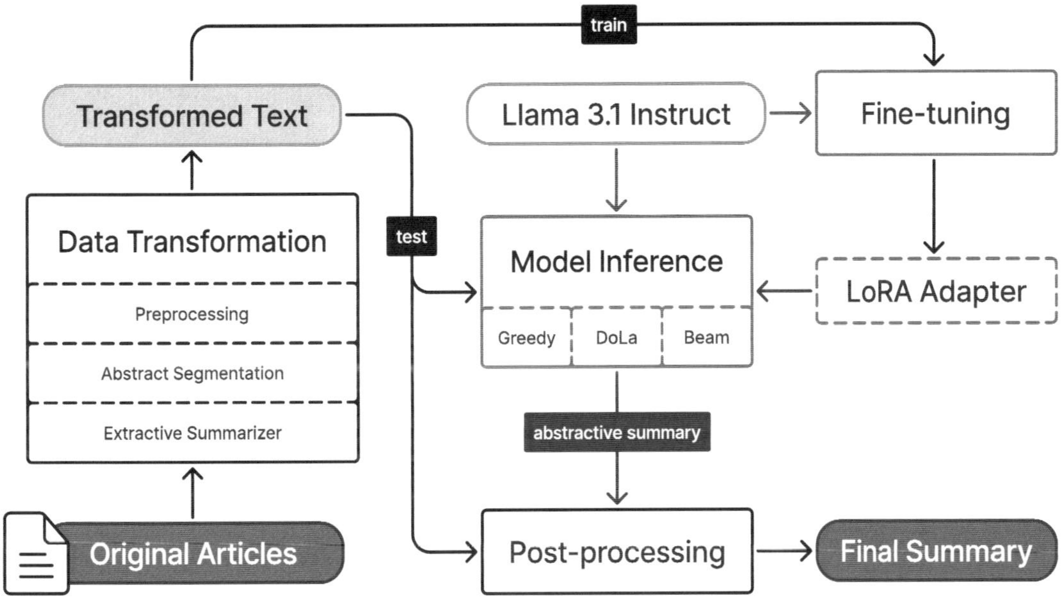

Figure 1: Our proposed pipeline for rapid experimentation comprises four toggleable modules: data transformation, model fine-tuning, model inference, and post-processing. We conducted over 20 experiments using distinct combinations of these modules. Dashed boxes denote optional or composable functionality.

best in all our experiments (Section 4), it offers the most balanced performance across the three groups of evaluation metrics: *relevance*, *readability*, and *factuality* (see Section 3.4).

3.1 Pipeline

Our proposed pipeline, illustrated in Figure 1, is designed to facilitate experimentation through modular and composable functionality, consisting of four components: data transformation, parameter-efficient fine-tuning, model inference, and post-processing. These modules are implemented as Python scripts, on top of the `transformers` (Wolf et al., 2020) and `torchtune` (torchtune maintainers and contributors, 2024) libraries, and can be configured using command-line arguments.

Initially, articles undergo a **data transformation** phase comprising optional preprocessing (Section 4.1), extractive summarization (Section 4.2), and abstract segmentation (Section 4.3). We apply an identical transformation procedure to each of the three splits from the `eLife` and `PLOS` datasets. The resulting transformed texts are then stored as a separate column within a newly derived dataset, alongside the original "article" and "summary" columns. This derived dataset serves as input for all subsequent stages of the pipeline.

The **model inference** module uses the Llama Instruct model, optionally combined with a LoRA adapter (Hu et al., 2021) that was **fine-tuned** on the transformed text to generate abstractive summaries. During inference, multiple decoding strategies are available: greedy decoding, beam search, and DoLa (Chuang et al., 2024).

Finally, the **post-processing** module can be used to refine further the pipeline's output, which can be the abstractive summary or the text resulting from the data transformation stage.

3.2 Fine-tuning

The Llama model was fine-tuned using LoRA (Hu et al., 2021) for 2 epochs, training separate models for the `PLOS` and `eLife` datasets, with varying batch sizes depending on the GPU and input length. When fine-tuning on full articles on an A40 GPU, a batch size of 2 was used for the `PLOS` dataset and 1 for the `eLife` dataset. The model employed `bf16` precision, and activation checkpointing, activation offloading, and `torch.compile` were used to reduce VRAM usage.

LoRA was applied to the query, value, output projection layers within the attention layers, as well as the MLP layers, with a rank of 8, α of 16, and dropout set to 0.0. The model was optimized using fused AdamW (Loshchilov and Hutter, 2019), with a learning rate of 3e-4 and weight decay of 0.01. A cosine learning rate scheduler with 100 warmup steps was used.

The random seed was set to 4 for reproducibility, and prompts from Table 4 were used to instruct the model.

3.3 Abstractive summarization

We add the LoRA adapters trained on full-text articles to the base Llama instruct model to generate the abstractive summaries. The model instructions follow the `system`, `user`, and `assistant` structure defined by the Chat Markup Language. Furthermore, the `system` messages, summarized in Table 4, include specific target grade-level drawing on the instruction-based readability control outlined by Ribeiro et al. (2023).

To decode the output tokens, we apply Decoding by Contrasting Layers (DoLa) (Chuang et al., 2024) on the lower layers, 0, 2, and 20, using a repetition penalty of 1.2. Compared to beam search and greedy decoding, we found DoLa to provide the best balance between *readability* and *factuality*.

Model inference is performed on a single NVIDIA A40 GPU with a batch size of 1, using the `EOS` token for padding, which takes an average runtime of 62 minutes on the `test` split. Furthermore, we limit the maximum number of tokens generated to 384. We selected this value based on the median summary lengths of the training splits and empirical evaluation comparing output lengths of 256 and 512 tokens (see Figure 4). Furthermore, each submission file, `plos.txt` and `elife.txt`, is created using adapter weights tuned to the respective dataset. Except for the `system` message version, all inference parameters remain constant across runs.

3.4 Evaluation

For experimental validation, we train models on the `train` split of the data and evaluate them on the `validation` split using a pipeline made available by the shared task organizers.[2] Summaries are assessed across 11 automated metrics falling into one of three criteria: *relevance*, *readability*, and *factuality*. To compare results, we adopt the ranking approach used in the previous iteration of BioLaySumm (Goldsack et al., 2024). Specifically, we apply min-max normalization to each metric and average the scores within each criterion before calculating an overall average across all criteria. Our model selection is based on achieving the highest average score from this methodology. The metrics are categorized as follows:

Relevance ROUGE (Lin, 2004), BLEU (Papineni et al., 2002), METEOR (Banerjee and Lavie, 2005), and BERTScore (Zhang et al., 2020).

Readability Flesch-Kincaid Grade Level (Kincaid et al., 1975), Dale-Chall Readability Score (Dale and Chall, 1948), CLI (Coleman and Liau, 1975), and LENS (Maddela et al., 2023).

Factuality AlignScore (Zha et al., 2023) and SummaC (Laban et al., 2022).

4 Results and Analysis

In this section, we present our experimental setup and findings obtained through our end-to-end pipeline. Table 1 summarizes the results of these experiments.

4.1 Preprocessing

We replicate the preprocessing approach from Modi and Karthikeyan (2024) to remove content within parentheses, braces, and brackets. Additionally, we apply a number-aware regular expression to collapse additional spacing around punctuation marks and other special characters. In Table 1, we denote experiments that utilized preprocessed inputs with a "`pre`" suffix. Our findings indicate that preprocessing leads to improved *relevance* scores and a better `FKGL` score, especially when combined with fine-tuning. However, these improvements are nullified by lower `LENS` and `SummaC` scores. We hypothesize that removing parentheticals from the input prevents the model from including chunk cues in the output, thereby reducing lexical overlap and potentially lowering entailment scores.

4.2 Extractive summarization

Our extractive summarization method follows from You et al. (2024), using TextRank (Mihalcea and Tarau, 2004) and embedding-based similarity matching. For the latter, we experiment with five pre-trained language embedding models explicitly built for processing biomedical text data, namely: `BioBERT` (Lee et al., 2019), `MedEmbed` (Balachandran, 2024), `PubMedBERT` (Gu et al., 2021), `PubMedBERT-MS-MARCO` (Deka et al., 2022), and `Medical-MiniLM-L6`.[3] Sentence embeddings created using these models are used to measure semantic similarity between them. We also test different embedding models using k-values of 20, 30,

[2] `https://github.com/gowitheflow-1998/BioLaySumm2025`

[3] `https://huggingface.co/Manal0809/medical-term-similarity`

Input	PEFT	k	Relevance				Readability				Factuality	
			ROUGE	BLEU	METEOR	BertS	FKGL	DCRS	CLI	LENS	AlignS	SummaC
Ext	×	10	0.323	4.561	0.242	0.833	10.154	7.917	11.342	71.258	0.532	0.528
	×	20	0.338	5.266	0.256	0.834	10.408	7.965	11.574	68.262	0.532	0.527
	×	30	0.342	5.530	0.261	0.834	10.868	8.040	11.938	66.199	0.526	0.512
	✓	10	0.366	7.253	0.272	0.854	9.334	7.529	10.075	77.782	0.615	0.622
	✓	20	0.373	7.730	0.278	0.856	9.037	**7.523**	9.901	78.760	0.626	0.637
	✓	30	0.373	7.650	0.277	0.856	9.062	7.526	9.954	79.118	0.633	0.640
	✓	40	0.379	8.421	0.285	0.857	9.004	7.534	10.008	78.472	0.643	0.645
Ext_{pre}	×	10	0.328	4.767	0.247	0.834	10.185	10.589	11.316	71.154	0.533	0.529
	×	20	0.337	5.181	0.259	0.834	10.348	10.739	11.501	68.144	0.516	0.517
	×	30	0.341	5.386	0.261	0.835	10.640	10.927	11.770	67.053	0.531	0.513
Abs +Ext	✓	10	0.379	8.279	0.292	0.855	8.924	10.304	9.966	78.128	0.634	0.610
	✓	20	0.380	8.332	0.294	0.856	8.999	10.261	10.033	77.653	0.635	0.614
	✓	30	0.380	8.373	0.293	0.855	8.829	10.226	9.950	76.940	**0.648**	0.614
	✓	40	0.382	8.651	**0.297**	0.855	8.956	10.232	9.934	76.674	0.646	0.608
Abs $+\text{Ext}_{(\text{abs})}$	✓	10	0.356	7.462	0.278	0.848	8.885	10.171	9.728	76.015	0.594	0.604
	✓	20	0.365	7.845	0.282	0.853	8.869	10.326	9.850	77.129	0.637	0.637
	✓	30	0.372	8.109	0.284	0.854	9.020	10.376	9.975	78.025	0.643	0.643
	✓	40	0.372	8.200	0.289	0.852	8.857	10.283	9.847	75.797	0.641	0.614
Abs	✓	–	0.369	7.532	0.277	0.854	8.783	10.278	**9.803**	**79.448**	0.634	0.663
Abs_{pre}	✓	–	0.373	8.126	0.289	0.853	**8.733**	10.250	9.809	77.527	0.637	0.599
Full	✓	–	**0.385**	**8.694**	0.289	**0.859**	9.308	7.674	10.143	78.670	0.643	**0.663**
	×	–	0.344	5.766	0.259	0.840	12.483	8.450	12.896	67.947	0.600	0.483
$\text{Full}_{\text{post}}$	✓	–	0.384	8.523	0.287	**0.859**	9.329	10.455	10.153	79.206	0.644	0.662

Table 1: Performance of our abstractive summarization experiments on the eLife validation split. We use PEFT to denote models fine-tuned with LoRA and k to represent the extractive summary length. Data inputs are: **(Ext)** extractive summary, **(Ext_{pre})** preprocessed extractive summary, **(Abs+Ext)** abstract concatenated with extractive summary, **($\text{Abs+Ext}_{(\text{abs})}$)** abstract concatenated with extractive summary that excluded the abstract during extraction, **(Abs)** abstract only, **(Abs_{pre})** preprocessed abstract, **(Full)** entire article, and **($\text{Full}_{\text{post}}$)** entire article, with post-processing applied to the generated summary.

and 40 for summary length. The results indicate a consistent preference for the BioBERT embedding model, regardless of the number of sentences selected. As shown in Figure 3, the overall evaluation score correlates positively with the summary length.

4.3 Training data

We fine-tuned the base instruct model at different levels of input granularity and transformations.

Extractive summary In these experiments, we use the summaries extracted via BioBERT embeddings as the only input. Our results indicate that performance generally improves with more context, although this leads to longer training times. We found that the model fine-tuned on extracted summaries with $k = 40$ is comparable to our best model while requiring less training time.

Abstract-only In this setting, the model is trained solely on the abstract, which is the first paragraph of the input article and serves as a condensed, high-level overview of the study. Even without additional context, the model demonstrated solid performance in terms of readability and factual accuracy. This combination offered the best balance between summarization quality and computational efficiency (see Appendix C).

Abstract and extractive summary We concatenate abstracts with extractive summaries to enrich the input, aiming to provide the model with additional context to improve the factual accuracy and clarity of the generated summaries. We explore two configurations: in Abs+Ext, the abstract is concatenated with an extractive summary generated from the full article, whereas in Abs+Ext(abs), we first remove the abstract from the article before producing the extractive summary. Our evaluation indi-

Decoding	Runtime	Relevance				Readability				Factuality	
		ROUGE	BLEU	METEOR	BertS	FKGL	DCRS	CLI	LENS	AlignS	SummaC
DoLa	02:35:41	**0.39**	**9.21**	**0.30**	**0.86**	**9.16**	10.39	**10.10**	77.82	**0.67**	**0.65**
Greedy	02:17:50	**0.39**	9.13	0.30	**0.86**	9.23	**10.38**	10.16	78.19	0.66	0.64
Beam search	07:32:55	0.37	6.56	0.29	0.85	11.31	10.39	10.55	**79.61**	0.55	0.49

Table 2: Runtime and evaluation comparison of the three decoding strategies implemented in our pipeline.

cates that repeating key information (as evidenced by comparing `Ext`, `Abs+Ext`, and `Abs+Ext(abs)`) yields improved *relevance* scores; however, we observe a decline in both *readability* and *factuality*. We hypothesize that the concatenation disrupts the logical ordering of information, which is crucial for these criteria.

Full-text The model is trained on the entire article without any data transformation. This setting showed the best performance, possibly due to having more context, and was our model of choice. Our final submission was trained both on the train split and the validation split. The models were trained on `eLife` for 2 epochs and on `PLOS` for 1.4 epochs.

4.4 Decoding strategies

We investigate the effect of three decoding strategies on our evaluation criteria: greedy decoding, beam search, and DoLa (Chuang et al., 2024). As demonstrated in Table 2, beam search performed poorly, showing significantly lower factuality and relevance scores while also requiring additional hours for inference. Summaries generated using DoLa and greedy decoding had comparable performance and runtimes, with the former achieving the best scores in eight out of eleven metrics. Notably, contrastive decoding yielded the highest factuality results.

4.5 Post-processing

In these experiments, we applied the same text processing method detailed in Section 4.1. Additionally, we removed incomplete sentences arising from the decoding limit on the maximum output token length. Specifically, we identified summaries that did not end with a period and discarded all tokens that appeared after the final complete sentence. Surprisingly, this post-processing step resulted in decreased performance across seven of eleven evaluation metrics, including three *readability* scores, despite the intuitive assumption that truncated sentences negatively affect summary quality.

5 Conclusion

In this study, we presented an end-to-end pipeline for generating lay summaries of biomedical articles. Our approach achieved the highest overall rank in subtask 1.1 of BioLaySumm 2025. Our method balances readability and factuality by employing instruction-based readability control and contrastive decoding (Chuang et al., 2024). In particular, we include the Flesch-Kincaid grade-level target in the system message to improve readability, and control over the LoRA weights enabled the application of contrastive decoding for improved factual accuracy.

We posit that investigating more advanced instruction strategies, such as self-reflection and synthesized chain-of-thought (CoT), represents a promising direction for future research. These strategies could incorporate factual claims and lay terminology to improve the model's relevance and factual accuracy. Furthermore, adding a reinforcement learning component, such as Direct Preference Optimization (Rafailov et al., 2023), to our pipeline could help select outputs that better align with the evaluation framework of this task.

Acknowledgments

We thank the organizers of the shared task for their guidance and coordination. We also thank the anonymous reviewers for their valuable feedback and suggestions, which helped improve the quality of this paper.

Parts of this work were completed on Hyak, the University of Washington's high-performance computing cluster. This resource was funded by the Student Technology Fee.

References

Abhinand Balachandran. 2024. Medembed: Medical-focused embedding models.

Satanjeev Banerjee and Alon Lavie. 2005. METEOR: An automatic metric for MT evaluation with improved correlation with human judgments. In *Pro-*

ceedings of the ACL Workshop on Intrinsic and Extrinsic Evaluation Measures for Machine Translation and/or Summarization*, pages 65–72, Ann Arbor, Michigan. Association for Computational Linguistics.

Yung-Sung Chuang, Yujia Xie, Hongyin Luo, Yoon Kim, James R. Glass, and Pengcheng He. 2024. Dola: Decoding by contrasting layers improves factuality in large language models. In *The Twelfth International Conference on Learning Representations*.

Meri Coleman and Ta Lin Liau. 1975. A computer readability formula designed for machine scoring. *Journal of Applied Psychology*, 60(2):283.

Edgar Dale and Jeanne S. Chall. 1948. A formula for predicting readability: Instructions. *Educational Research Bulletin*, 27(2):37–54.

Pritam Deka, Anna Jurek-Loughrey, and P Deepak. 2022. Improved methods to aid unsupervised evidence-based fact checking for online health news. *Journal of Data Intelligence*, 3(4):474–504.

Tomas Goldsack, Zheheng Luo, Qianqian Xie, Carolina Scarton, Matthew Shardlow, Sophia Ananiadou, and Chenghua Lin. 2023. Overview of the biolaysumm 2023 shared task on lay summarization of biomedical research articles. In *The 22nd Workshop on Biomedical Natural Language Processing and BioNLP Shared Tasks*, pages 468–477, Toronto, Canada. Association for Computational Linguistics.

Tomas Goldsack, Carolina Scarton, Matthew Shardlow, and Chenghua Lin. 2024. Overview of the BioLaySumm 2024 shared task on the lay summarization of biomedical research articles. In *Proceedings of the 23rd Workshop on Biomedical Natural Language Processing*, pages 122–131, Bangkok, Thailand. Association for Computational Linguistics.

Tomas Goldsack, Zhihao Zhang, Chenghua Lin, and Carolina Scarton. 2022. Making science simple: Corpora for the lay summarisation of scientific literature. In *Proceedings of the 2022 Conference on Empirical Methods in Natural Language Processing*, pages 10589–10604, Abu Dhabi, United Arab Emirates. Association for Computational Linguistics.

Aaron Grattafiori, Abhimanyu Dubey, Abhinav Jauhri, Abhinav Pandey, Abhishek Kadian, Ahmad Al-Dahle, Aiesha Letman, Akhil Mathur, Alan Schelten, Alex Vaughan, Amy Yang, Angela Fan, Anirudh Goyal, Anthony Hartshorn, Aobo Yang, Archi Mitra, Archie Sravankumar, Artem Korenev, Arthur Hinsvark, and 542 others. 2024. The llama 3 herd of models. *Preprint*, arXiv:2407.21783.

Yu Gu, Robert Tinn, Hao Cheng, Michael Lucas, Naoto Usuyama, Xiaodong Liu, Tristan Naumann, Jianfeng Gao, and Hoifung Poon. 2021. Domain-specific language model pretraining for biomedical natural language processing. *ACM Trans. Comput. Healthcare*, 3(1).

Edward J. Hu, Yelong Shen, Phillip Wallis, Zeyuan Allen-Zhu, Yuanzhi Li, Shean Wang, Lu Wang, and Weizhu Chen. 2021. Lora: Low-rank adaptation of large language models. *Preprint*, arXiv:2106.09685.

J. Peter Kincaid, Jr. Fishburne, Robert P., Richard L. Rogers, and Brad S. Chissom. 1975. Derivation of new readability formulas (automated readability index, fog count, and flesch reading ease formula) for navy enlisted personnel. Research Branch Report 8-75, Institute for Simulation and Training, University of Central Florida.

Philippe Laban, Tobias Schnabel, Paul N. Bennett, and Marti A. Hearst. 2022. SummaC: Re-visiting NLI-based models for inconsistency detection in summarization. *Transactions of the Association for Computational Linguistics*, 10:163–177.

Jinhyuk Lee, Wonjin Yoon, Sungdong Kim, Donghyeon Kim, Sunkyu Kim, Chan Ho So, and Jaewoo Kang. 2019. Biobert: a pre-trained biomedical language representation model for biomedical text mining. *Bioinformatics*, 36(4):1234–1240.

Chin-Yew Lin. 2004. ROUGE: A package for automatic evaluation of summaries. In *Text Summarization Branches Out*, pages 74–81, Barcelona, Spain. Association for Computational Linguistics.

Ilya Loshchilov and Frank Hutter. 2019. Decoupled weight decay regularization. *Preprint*, arXiv:1711.05101.

Renqian Luo, Liai Sun, Yingce Xia, Tao Qin, Sheng Zhang, Hoifung Poon, and Tie-Yan Liu. 2022a. Biogpt: generative pre-trained transformer for biomedical text generation and mining. *Briefings in Bioinformatics*, 23(6):bbac409.

Zheheng Luo, Qianqian Xie, and Sophia Ananiadou. 2022b. Readability controllable biomedical document summarization. In *Findings of the Association for Computational Linguistics: EMNLP 2022*, pages 4667–4680, Abu Dhabi, United Arab Emirates. Association for Computational Linguistics.

Mounica Maddela, Yao Dou, David Heineman, and Wei Xu. 2023. LENS: A learnable evaluation metric for text simplification. In *Proceedings of the 61st Annual Meeting of the Association for Computational Linguistics (Volume 1: Long Papers)*, pages 16383–16408, Toronto, Canada. Association for Computational Linguistics.

Rada Mihalcea and Paul Tarau. 2004. Textrank: Bringing order into text. In *Proceedings of the 2004 Conference on Empirical Methods in Natural Language Processing*, pages 404–411, Barcelona, Spain. Association for Computational Linguistics.

Satyam Modi and T Karthikeyan. 2024. Eulerian at BioLaySumm: Preprocessing over abstract is all you need. In *Proceedings of the 23rd Workshop on Biomedical Natural Language Processing*, pages 826–830, Bangkok, Thailand. Association for Computational Linguistics.

Kishore Papineni, Salim Roukos, Todd Ward, and Wei-Jing Zhu. 2002. Bleu: a method for automatic evaluation of machine translation. In *Proceedings of the 40th Annual Meeting of the Association for Computational Linguistics*, pages 311–318, Philadelphia, Pennsylvania, USA. Association for Computational Linguistics.

Rafael Rafailov, Archit Sharma, Eric Mitchell, Stefano Ermon, Christopher D. Manning, and Chelsea Finn. 2023. Direct preference optimization: Your language model is secretly a reward model. *arXiv preprint arXiv:2305.18290*.

Leonardo F. R. Ribeiro, Mohit Bansal, and Markus Dreyer. 2023. Generating summaries with controllable readability levels. In *Proceedings of the 2023 Conference on Empirical Methods in Natural Language Processing*, pages 11669–11687, Singapore. Association for Computational Linguistics.

torchtune maintainers and contributors. 2024. torchtune: Pytorch's finetuning library.

Oisín Turbitt, Robert Bevan, and Mouhamad Aboshokor. 2023. MDC at BioLaySumm task 1: Evaluating GPT models for biomedical lay summarization. In *The 22nd Workshop on Biomedical Natural Language Processing and BioNLP Shared Tasks*, pages 611–619, Toronto, Canada. Association for Computational Linguistics.

Thomas Wolf, Lysandre Debut, Victor Sanh, Julien Chaumond, Clement Delangue, Anthony Moi, Pierric Cistac, Tim Rault, Rémi Louf, Morgan Funtowicz, Joe Davison, Sam Shleifer, Patrick von Platen, Clara Ma, Yacine Jernite, Julien Plu, Canwen Xu, Teven Le Scao, Sylvain Gugger, and 3 others. 2020. Transformers: State-of-the-art natural language processing. In *Proceedings of the 2020 Conference on Empirical Methods in Natural Language Processing: System Demonstrations*, pages 38–45, Online. Association for Computational Linguistics.

Chenghao Xiao, Kun Zhao, Xiao Wang, Siwei Wu, Sixing Yan, Tomas Goldsack, Sophia Ananiadou, Noura Al Moubayed, Liang Zhan, William CHEUNG, and Chenghua Lin. 2025. Overview of the biolaysumm 2025 shared task on lay summarization of biomedical research articles and radiology reports. In *The 24nd Workshop on Biomedical Natural Language Processing and BioNLP Shared Tasks*, Vienna, Austria. Association for Computational Linguistics.

Zhiwen You, Shruthan Radhakrishna, Shufan Ming, and Halil Kilicoglu. 2024. UIUC_BioNLP at BioLaySumm: An extract-then-summarize approach augmented with Wikipedia knowledge for biomedical lay summarization. In *Proceedings of the 23rd Workshop on Biomedical Natural Language Processing*, pages 132–143, Bangkok, Thailand. Association for Computational Linguistics.

Yuheng Zha, Yichi Yang, Ruichen Li, and Zhiting Hu. 2023. AlignScore: Evaluating factual consistency with a unified alignment function. In *Proceedings of the 61st Annual Meeting of the Association for Computational Linguistics (Volume 1: Long Papers)*, pages 11328–11348, Toronto, Canada. Association for Computational Linguistics.

Tianyi Zhang, Varsha Kishore, Felix Wu, Kilian Q. Weinberger, and Yoav Artzi. 2020. Bertscore: Evaluating text generation with bert. In *International Conference on Learning Representations*.

A Instruction Messages

Table 4 details the system messages used to instruct the model in generating the summaries. We found that including the target domain and grade level contributed to better *readability* scores. The `eLife` summaries were created with version 1, while the `PLOS` summaries were produced with version 2.

B Dataset Statistics

The Public Library of Science (`PLOS`) is a non-profit, open-access publisher launched in 2000 with the goal of providing free access to full-text scientific articles. It currently publishes 14 academic journals in a range of fields such as biology, medicine, and computational biology. `eLife` is likewise a non-profit, peer-reviewed, open-access publisher for articles in the biomedical and life science domains established in 2012. Articles in the two datasets cover various topics and specialties within the biomedical domain. We report length statistics for the `PLOS` and `eLife` datasets in Table 3.

Dataset	# Docs	Doc	Summary	
		# words	# words	# sents
PLOS	27,525	5,366.7	175.6	7.8
eLife	4,828	7,806.1	347.6	15.7

Table 3: Average word and sentence counts for each dataset. Adapted from Goldsack et al. (2022).

C Computational Efficiency

Although using full article texts as model input yielded the highest performance, this approach is significantly more resource-intensive than relying only on extractive summaries or abstracts. This difference is clearly illustrated in Figure 2, which compares average inference runtimes on the `eLife` and `PLOS` datasets. Specifically, inference on full-text inputs required over 30 times the runtime of

#	Message
1	You are a specialist medical communicator responsible for translating biomedical articles into a clear, accurate 1020 sentence summary for non-experts. The summary should be at a FleschKincaid grade level of 1014 and explain any technical terms.
2	You are a specialist medical communicator responsible for translating biomedical articles into a clear, accurate 10 to 20 sentence summary for non-experts. The summary should have a FleschKincaid grade level of 10 to 14, explaining any technical terms in simple language. Ensure factual accuracy by using terminology from the source article, and omit all in-text citations.

Table 4: The two system messages used to generate the abstractive summaries. Generative language models were used to refine the messages.

abstract-only inputs, while providing only a 14.86% improvement in the overall average score.

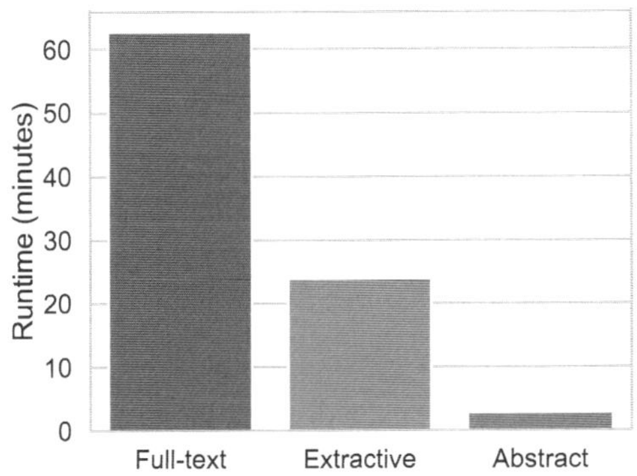

Figure 2: Inference runtime comparison of the summarization model based on different input types: full-text articles, extractive summaries, and abstracts.

D Training Challenges and Workarounds

There is a peculiarity that we would like to mention about our training setup. While University of Washington's high-performance computing cluster Hyak offers powerful hardware, GPU jobs are prone to preemption and can run at most for 8-9 hours before being requeued. However, a full epoch exceeded that limit, sometimes taking over 24 hours. At the time of our experiment, `torchtune` did not support mid-epoch checkpointing, so we had to split the data into smaller sections to ensure each partial epoch could finish within the time limit. The actual split sizes were smaller to accommodate preemption and were dynamically adjusted along with the batch size based on the number and model of the GPU in use. The total number of epochs was set to $\left\lceil \frac{1}{\text{split ratio}} \right\rceil \times$ (number of epochs) to have `torchtune` save the training state between partial epochs. Training processes were killed and restarted after each partial epoch to force `torchtune` to reload the training configuration file with updated data splits. This part is specific to Hyak, and the code will only be included in the `release/class` branch and excluded from the `main` branch and future releases.

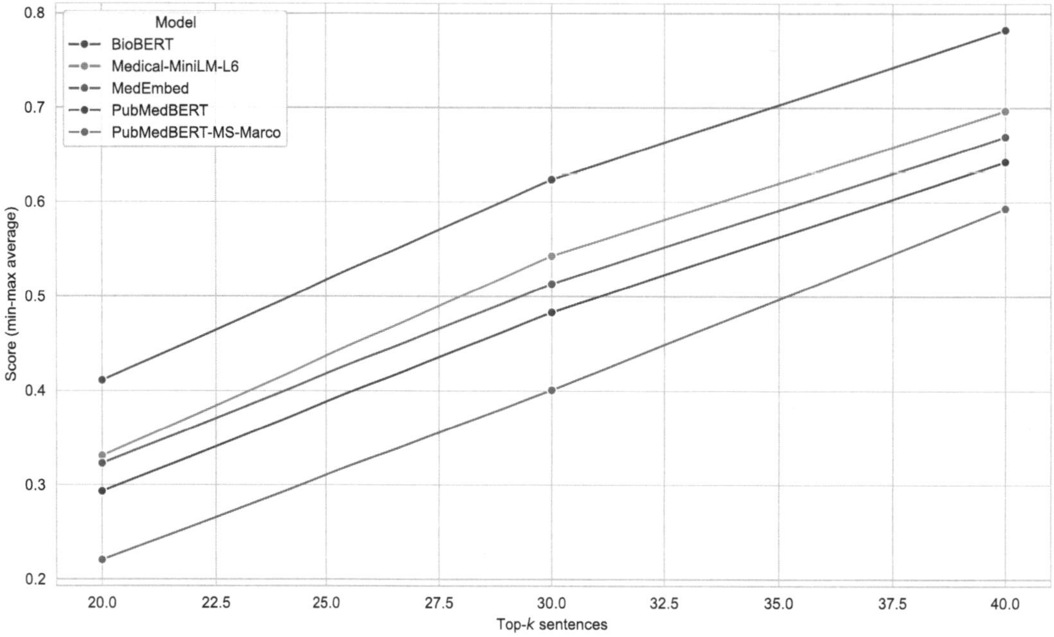

Figure 3: Relative performance of extractive methods on the eLife training data, categorized by embedding model and the top-k sentences extracted using TextRank (Mihalcea and Tarau, 2004).

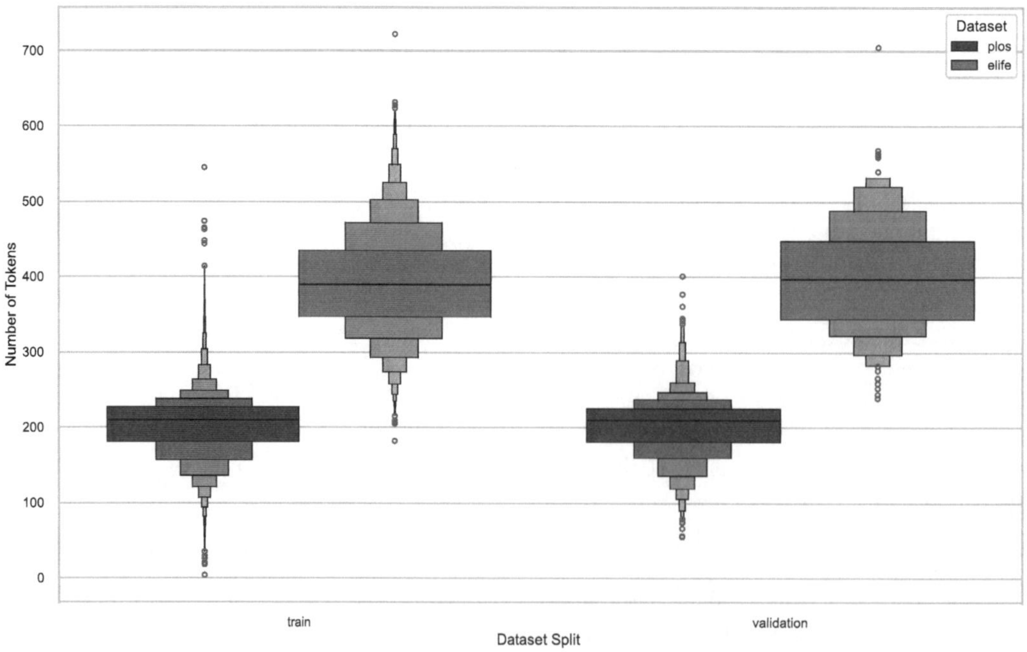

Figure 4: Distribution of token counts across training and validation splits for the PLOS and eLife datasets.

BDA-UC3M @ BioLaySumm: Efficient Lay Summarization with Small-Scale SoTA LLMs

Ilyass Ramzi[1] **and Isabel Segura Bedmar**[2]
[1]Graduate School of Engineering and Basic Sciences, Universidad Carlos III de Madrid
[2]Computer Science and Engineering Department, Universidad Carlos III de Madrid
100510978@alumnos.uc3m.es, isegura@inf.uc3m.es

Abstract

The growing need to make biomedical research accessible to non-expert audiences has motivated the development of effective lay summarization systems. While large language models (LLMs) have set recent benchmarks, their computational demands limit widespread adoption. This paper explores the use of small-scale, state-of-the-art LLMs (4B–7B parameters) for biomedical lay summarization in the BioLaySumm 2025 shared task. Leveraging dynamic 4-bit quantization, extractive preprocessing, prompt engineering, data augmentation, and Direct Preference Optimization, our system, based on Gemma3 4B, Qwen3 4B, and GPT-4.1-mini, ranked second in its category, showing that compact models can deliver high-quality, factually accurate summaries.

1 Introduction

Recent advances in large language models (LLMs) have demonstrated exceptional performance in generating lay summaries of biomedical literature, supporting the critical goal of making complex scientific content accessible to non-expert audiences (Goldsack et al., 2024, 2023). However, most state-of-the-art approaches rely on very large models—often with tens of billions of parameters—posing significant barriers for practical deployment and reproducibility due to hardware and computational requirements.

The BioLaySumm 2025 Shared Task challenges participants to develop automated systems for lay summarization of biomedical research articles, with evaluation based on relevance, readability, and factuality across established benchmark datasets (eLife and PLOS) (Xiao et al., 2025; Goldsack et al., 2022). The official baselines for this task, such as Llama3 8B and Qwen2.5 7B, set a high standard for both scale and performance.

This paper presents the approach developed by BDA-UC3M for BioLaySumm 2025, aiming to demonstrate that small-scale, state-of-the-art LLMs (4B–7B parameters), when carefully optimized and fine-tuned, can achieve competitive—and in some cases, superior—performance to larger baselines. Our system leverages recent advances in LLM efficiency, notably:

- **Parameter-efficient models and training:** Utilizing compact LLMs, including Gemma3 4B (Dynamic 4-bit Instruct) (Team, 2025), Qwen3 4B (Dynamic 4-bit Safetensor, fine-tuned) (Yang et al., 2025), and GPT-4.1-mini (via OpenAI API), all selected for their strong performance-to-size ratio.

- **Accessible compute and deployment:** All model training and inference is performed on consumer-grade GPUs[1], with deployment streamlined using RunPod pods[2].

- **Advanced pipeline building on prior SoTA:** Our approach systematically integrates and improves strategies from the top BioLaySumm 2024 systems (You et al., 2024; Zhao et al., 2024; Kim et al., 2024)—combining robust extract-then-summarize frameworks, advanced prompt engineering, targeted data augmentation, and factuality-aware fine-tuning (DPO).

While retrieval-augmented generation (RAG) has shown promise in biomedical summarization by enriching model inputs with external knowledge sources such as Wikipedia, this work does not incorporate RAG due to timeline constraints. Future iterations may revisit lightweight retrieval solutions for even greater factuality.

[1]https://docs.unsloth.ai/
[2]https://docs.runpod.io/pods/overview

BioNLP 2025 Shared Tasks, pages 249–255
August 1, 2025

2 Methods

2.1 Datasets

We used the official BioLaySumm 2025 task datasets: **eLife** and **PLOS** (Xiao et al., 2025; Goldsack et al., 2022). Table 1 summarizes the dataset splits:

Dataset	Train	Validation	Test
eLife	4346	241	142
PLOS	24773	1376	142

Table 1: Dataset splits for BioLaySumm 2025.

Tokenization (using `cl100k_base`) revealed substantial variation in article lengths, consistent with previous dataset reports (Goldsack et al., 2022):

- **eLife**: Train articles averaged 14,140 tokens (max 46,150), summaries averaged 428 tokens.
- **PLOS**: Train articles averaged 8,925 tokens (max 32,623), summaries averaged 233 tokens.

Test sets do not include reference summaries. eLife summaries are typically longer and more abstracted, while PLOS summaries are shorter and more closely tied to the article content (Luo et al., 2022).

2.2 Preprocessing

TextRank Extraction. To efficiently compress long articles and highlight salient content, we used a custom TextRank implementation (adapted from methods described in (You et al., 2024)):

- Articles were segmented into sentences using spaCy (`en_core_web_sm`), with only sentences over 20 characters retained.
- TF-IDF vectors and cosine similarity were used to construct a similarity matrix.
- Sentences were ranked with PageRank over the similarity graph, and the top N ($N = 50$) were selected.

Chunking + TextRank. For models with smaller context windows (notably Qwen3 4B, 32K tokens (Yang et al., 2025)), we applied chunking:

- Articles were split into chunks of $\sim$40 sentences.
- TextRank was run independently within each chunk, and the top sentences (e.g., 10 per chunk) were extracted.
- If the total number of selected sentences across all chunks exceeded the limit, we applied a global re-ranking step: all previously selected sentences were pooled and TextRank was run again on this subset to select the final top 50, ensuring the most salient content across the full article was retained. This step was used when the combined top sentences from all chunks could not fit in the model's input context.

For GPT-4.1-mini and Gemma3 4B (Team, 2025), chunking and re-ranking were not required due to their larger context capabilities.

Data Augmentation. Training diversity was enhanced by using GPT-4.1-mini to paraphrase and augment summaries, especially where extractive strategies omitted lay-relevant detail. This data augmentation step follows insights from previous top systems (Zhao et al., 2024).

2.3 Model Setup and Training

We focused on small yet state-of-the-art LLMs for efficiency and reproducibility:

- **Gemma3 4B** (Dynamic 4-bit Instruct)
- **Qwen3 4B** (Dynamic 4-bit Safetensor)
- **GPT-4.1-mini** (via OpenAI API)

Fine-tuning and inference for Gemma3 and Qwen3 models leveraged the Unsloth framework, which combines two key techniques for maximal efficiency:

- **Dynamic 4-bit quantization** reduces memory usage by compressing model weights to 4 bits on-the-fly, enabling large LLMs to run on consumer-grade GPUs (Han et al., 2024).
- **LoRA (Low-Rank Adaptation)** introduces lightweight, trainable adapter layers, allowing only a small subset of parameters to be fine-tuned while the core model weights remain frozen (Hu et al., 2021).

Together, these methods allowed efficient training and adaptation of large models on standard hardware (RTX 3090, 24GB VRAM). For comparison, GGUF format is intended only for inference.

2.4 Fine-Tuning and Hyperparameters

- **Gemma3 4B:** Fine-tuned with Unsloth using LoRA adapters (Hu et al., 2021) and default settings: temperature=1.0, top_k=64, top_p=0.95. Training used per-device batch size 2, gradient accumulation 4, max_steps 30, learning rate 2×10^{-4}, weight decay 0.01, AdamW 8-bit optimizer.

- **Qwen3 4B:** Followed Unsloth's effective setup: rank=32, lora_alpha=32, dropout=0, "unsloth" gradient checkpointing. Training used the same batch, learning rate, and optimizer setup as above, with memory optimized for 32K context (Yang et al., 2025).

- **GPT-4.1-mini:** Utilized OpenAI API with recommended temperature and top_p settings. SFT used standard instruction-following templates; context window up to 32K tokens. Prompt design followed OpenAI's best practices[3].

2.5 Prompt Engineering

We systematically developed and tested a suite of prompts, evaluating both zero-shot and few-shot settings as well as dataset-specific refinements. Our approach was influenced by prior competition leaders (You et al., 2024; Zhao et al., 2024; Kim et al., 2024) and included:

- **Baseline Prompts (V1):** Focused on clarity and accessibility for lay readers.

- **Structured/Prescriptive Prompts (V2):** Provided numbered guidelines for better output organization.

- **Competition-Optimized Prompts (V3):** Explicitly referenced BioLaySumm metrics (ROUGE, BLEU, METEOR, BERTScore, LENS, AlignScore, SummaC, FKGL, CLI, DCRS) (Xiao et al., 2025), instructing models to optimize relevance, readability, and factuality.

- **Refined Prompts (V4):** Further emphasized factuality, accuracy, and discouraged speculative language or fabricated author names.

Model-Specific Prompts:

- *Qwen3 4B and Gemma3 4B:* Used instruction-tuned prompts with explicit, structured guidance for one-paragraph, factually accurate lay summaries (Yang et al., 2025; Team, 2025).

- *GPT-4.1-mini:* Incorporated OpenAI's prompt engineering best practices (OpenAI, 2025b), with iterative refinements based on validation.

Prompt selection was finalized for each dataset and model through ablation, guided by the best combination of metric performance and qualitative validation. **For full prompt templates, refer to Appendix A.**

3 Results and Analysis

3.1 Main Results

Table 2 reports the primary evaluation metrics for our three models—GPT-4.1-mini, Gemma3 4B, and Qwen3 4B—on both the eLife and PLOS test sets. Each metric is averaged per dataset, followed by the overall average across datasets.

All three models performed closely, with GPT-4.1-mini slightly outperforming on relevance and semantic similarity, while Qwen3 4B showed a small edge on factuality metrics (AlignScore, SummaC) (You et al., 2024; Kim et al., 2024; Zhao et al., 2024; Team, 2025; Yang et al., 2025).

3.2 Ablation and Component Analysis

We performed ablation studies to analyze the effect of prompt style, DPO training (Kim et al., 2024), and extractive chunking (You et al., 2024).

Prompt Style:

- GPT-4.1-mini achieved best results with a general, clarity-focused prompt.

- Gemma3 4B benefited from refined, constraint-driven prompts.

- Qwen3 4B excelled with explicit, stepwise prompts.

DPO: Direct Preference Optimization (DPO) improved factuality and readability metrics (AlignScore, SummaC, FKGL, DCRS) (Kim et al., 2024), but slightly reduced ROUGE/BLEU due to prioritizing factual alignment over surface-level overlap.

Chunking/Extraction: Chunking was crucial for Qwen3 4B due to its limited context window (Yang et al., 2025), ensuring representation across all article sections.

[3] https://cookbook.openai.com/examples/gpt4-1_prompting_guide

Model	Dataset	ROUGE	BLEU	METEOR	BERTS	FKGL	DCRS	CLI	LENS	Align	SummaC
GPT-4.1-mini	eLife	0.371	8.07	0.298	0.869	9.94	8.28	11.47	70.70	0.619	0.545
	PLOS	0.335	8.08	0.290	0.870	14.71	10.24	14.87	57.49	0.764	0.533
	Avg	**0.353**	**8.08**	**0.294**	0.870	**12.32**	**9.26**	**13.17**	**64.10**	0.691	0.539
Gemma3 4B	eLife	0.370	7.57	0.297	0.869	9.97	8.30	11.60	69.54	0.618	0.555
	PLOS	0.335	7.77	0.284	0.871	14.73	10.39	15.05	56.96	0.767	0.526
	Avg	0.352	7.67	0.290	**0.870**	12.35	9.35	13.32	63.25	0.693	0.541
Qwen3 4B	eLife	0.367	7.16	0.287	0.869	10.32	8.52	11.89	69.41	0.631	0.558
	PLOS	0.334	8.01	0.288	0.871	14.80	10.41	15.10	57.07	0.774	0.530
	Avg	0.351	7.59	0.288	0.870	12.56	9.47	13.49	63.24	**0.702**	**0.544**

Table 2: Performance of our models on eLife and PLOS test sets for BioLaySumm 2025. FKGL, DCRS, and CLI: lower is better (readability). All other metrics: higher is better. For BERTScore, values are rounded to three decimals; Gemma3 4B achieved the highest score at full precision.

3.3 Discussion of Findings

Our experiments confirm that carefully optimized small-scale LLMs (<7B parameters) can approach the performance of much larger models in biomedical lay summarization (Team, 2025; Yang et al., 2025; Xiao et al., 2025). While none of our models surpassed last year's BART/LED-based systems in extractive metrics such as ROUGE and BLEU (You et al., 2024; Goldsack et al., 2024), all achieved high semantic similarity and factuality, with Gemma3 4B posting the highest BERTScore among our submissions.

Ablation studies highlighted that prompt engineering and DPO training have strong, model-specific impacts, introducing a clear trade-off: optimizing for factuality and readability can reduce surface-level overlap with reference summaries, and vice versa (Kim et al., 2024; Zhao et al., 2024). Chunking strategies for models with limited context windows (e.g., Qwen3 4B) proved essential for consistent performance. As the datasets were unchanged year-on-year (Goldsack et al., 2022), our results indicate that further gains with small LLMs may require new architectures or additional external knowledge integration.

4 Conclusion

This work shows that well-optimized, small-scale LLMs can produce high-quality biomedical lay summaries, rivaling larger models in semantic and factual metrics while remaining accessible for training on standard hardware.

Limitations

Despite these strengths, certain limitations remain:

- **Performance Gap to Large Models:** Despite competitive scores, small LLMs still lag behind last year's best large-scale (BART/LED) and generative models on overlap-based metrics (ROUGE, BLEU), which likely benefit from larger pretraining corpora and parameter capacity.
- **Resource and Timeline Constraints:** All training was performed on single consumer GPUs, restricting the scope of hyperparameter search, ablation, and deeper multi-stage fine-tuning that could further boost results.
- **No External Knowledge Integration:** We did not implement retrieval-augmented generation (RAG). As a result, factual consistency may suffer for highly novel, underrepresented topics.

Future Work

Several avenues for further research and improvement are suggested by our findings:

- **Extended Fine-Tuning:** Implementing extended and curriculum-based training, including domain-adaptive pretraining or self-supervised objectives, to bridge the gap with larger models.
- **Hybrid and Ensemble Approaches:** Combining small LLMs with external retrieval modules to maximize both efficiency and factual accuracy.
- **Cross-Domain and Multilingual Expansion:** Testing the generalizability of our methods to other scientific fields and non-English corpora.

Our findings suggest that with further refinement, small and hardware-efficient LLMs can play a key role in making biomedical research broadly accessible, supporting both researchers and the general public.

References

Tomas Goldsack, Zheheng Luo, Qianqian Xie, Carolina Scarton, Matthew Shardlow, Sophia Ananiadou, and Chenghua Lin. 2023. Overview of the biolaysumm 2023 shared task on lay summarization of biomedical research articles. In *The 22nd Workshop on Biomedical Natural Language Processing and BioNLP Shared Tasks*, pages 468–477, Toronto, Canada. Association for Computational Linguistics.

Tomas Goldsack, Carolina Scarton, Matthew Shardlow, and Chenghua Lin. 2024. Overview of the BioLaySumm 2024 shared task on the lay summarization of biomedical research articles. In *Proceedings of the 23rd Workshop on Biomedical Natural Language Processing*, pages 122–131, Bangkok, Thailand. Association for Computational Linguistics.

Tomas Goldsack, Zhihao Zhang, Chenghua Lin, and Carolina Scarton. 2022. Making science simple: Corpora for the lay summarisation of scientific literature. In *Proceedings of the 2022 Conference on Empirical Methods in Natural Language Processing*, pages 10589–10604, Abu Dhabi, United Arab Emirates. Association for Computational Linguistics.

Daniel Han, Michael Han, and Unsloth team. 2024. Unsloth - dynamic 4-bit quantization.

Edward J. Hu, Yelong Shen, Phillip Wallis, Zeyuan Allen-Zhu, Yuanzhi Li, Shean Wang, Lu Wang, and Weizhu Chen. 2021. Lora: Low-rank adaptation of large language models. *arXiv preprint arXiv:2106.09685*.

Hwanmun Kim, Kamal raj Kanakarajan, and Malaikannan Sankarasubbu. 2024. Saama technologies at biolaysumm: Abstract based fine-tuned models with lora. In *Proceedings of the 23rd Workshop on Biomedical Language Processing*, Bangkok, Thailand. Association for Computational Linguistics.

Zheheng Luo, Qianqian Xie, and Sophia Ananiadou. 2022. Readability controllable biomedical document summarization. In *Findings of the Association for Computational Linguistics: EMNLP 2022*, pages 4667–4680, Abu Dhabi, United Arab Emirates. Association for Computational Linguistics.

OpenAI. 2025a. Gpt-4.1 overview.

OpenAI. 2025b. Gpt-4.1 prompting guide. https://cookbook.openai.com/examples/gpt4-1_prompting_guide.

Gemma Team. 2025. Gemma 3.

Chenghao Xiao, Kun Zhao, Xiao Wang, Siwei Wu, Sixing Yan, Sophia Ananiadou, Noura Al Moubayed, Liang Zhan, William Cheung, and Chenghua Lin. 2025. Overview of the biolaysumm 2025 shared task on lay summarization of biomedical research articles and radiology reports. In *The 24nd Workshop on Biomedical Natural Language Processing and BioNLP Shared Tasks*, Vienna, Austria. Association for Computational Linguistics.

An Yang, Anfeng Li, Baosong Yang, Beichen Zhang, Binyuan Hui, Bo Zheng, Bowen Yu, Chang Gao, Chengen Huang, Chenxu Lv, Chujie Zheng, Dayiheng Liu, Fan Zhou, Fei Huang, Feng Hu, Hao Ge, Haoran Wei, Huan Lin, Jialong Tang, and 41 others. 2025. Qwen3 technical report. *arXiv preprint arXiv:2505.09388*.

Zhiwen You, Shruthan Radhakrishna, Shufan Ming, and Halil Kilicoglu. 2024. Uiuc_bionlp at biolaysumm: an extract-then-summarize approach augmented with wikipedia knowledge for biomedical lay summarization. In *Proceedings of the 23rd Workshop on Biomedical Language Processing*, Bangkok, Thailand. Association for Computational Linguistics.

Ruijing Zhao, Siyu Bao, Siqin Zhang, Jinghui Zhang, Weiyin Wang, and Yunian Ru. 2024. Ctyun ai at biolaysumm: Enhancing lay summaries of biomedical articles through large language models and data augmentation. In *Proceedings of the 23rd Workshop on Biomedical Language Processing*, Bangkok, Thailand. Association for Computational Linguistics.

A Prompt Examples

This appendix contains the full prompt templates used in our experiments for PLOS, eLife, and instruction-tuned models.

A.1 PLOS Prompts (GPT-4.1-mini)

Listing 1: PLOS V1 – Baseline Prompt

```
system_prompt = (
    "You are a biomedical science writer tasked with
        rewriting research article summaries for the general
        public. "
    "The original summaries were written by the researchers
        themselves and may include technical language or
        academic phrasing.\n\n"
    "Your goal is to rewrite each summary so it is:\n"
    "1. Clear and easy to understand without specialized
        knowledge\n"
    "2. Focused on the study's background, question,
        findings, and significance\n"
    "3. Free from jargon, unless the term is briefly
        explained\n\n"
    "Do not speculate or exaggerate findings. Aim for
        accuracy, simplicity, and a neutral, informative
        tone."
)
```

Listing 2: PLOS V2 – Structured Style Prompt

```
system_prompt = (
    "You are a professional biomedical writer. Your task is
        to rewrite research article summaries for a public
        audience. "
    "Each summary should:\n\n"
    "1. Start with a plain-language introduction of the
        topic\n"
    "2. Explain the problem or motivation for the research\n"
    "3. Describe the key findings clearly and accurately\n"
    "4. Conclude with a statement about the significance or
        impact\n\n"
    "Use clear language and avoid jargon unless briefly
        explained. "
    "Write in a calm, educational tone that avoids
        exaggeration or speculation."
)
```

Listing 3: PLOS V3 – Competition-Optimized Prompt

```
system_prompt = (
    "You are a language model participating in a biomedical
        summarization competition (BioLaySumm 2025). "
    "You are given compressed scientific article inputs from
        PLOS journals. "
    "Your task is to generate accurate, clear, and concise
        lay summaries that perform well across automated
        evaluation metrics.\n\n"
    "Your summary should be optimized for the following
        metrics:\n"
    "- ROUGE (surface overlap)\n"
    "- BLEU & METEOR (fluency and lexical alignment)\n"
    "- BERTScore (semantic similarity)\n"
    "- LENS, AlignScore, SummaC (faithfulness and factual
        consistency)\n"
    "- FKGL, CLI, DCRS (readability)\n\n"
    "Guidelines for the summary:\n"
    "1. Use the language of the source where appropriate to
        maximize ROUGE and BLEU\n"
    "2. Be faithful to the article and avoid hallucinations
        to improve factual scores (AlignScore, LENS)\n"
    "3. Use simple, fluent language to keep readability
        scores (FKGL, DCRS) low\n"
    "4. Prioritize the article's main research question,
        methods, findings, and relevance\n"
    "5. Avoid speculative language or overstatements\n"
    "6. Stay within ~500 tokens (max 512) for the summary\n"
    "7. Minimize technical terms unless they are clearly
        explained\n\n"
    "You are writing for an educated non-expert audience.
        Your tone should be professional, informative, and
        neutral - avoid promotional language. "
    "The compressed article is provided below."
)
```

Listing 4: PLOS V4 – Refined Prompt (Author Names, Readability Emphasis)

```
system_prompt = (
    "You are a language model assisting in a biomedical
        summarization competition (BioLaySumm 2025). "
    "You are given compressed versions of PLOS journal
        articles and must produce high-quality lay summaries
        for a non-expert audience.\n\n"
    "Key goals:\n"
    "- Maximize ROUGE, BLEU, METEOR (surface-level match and
        lexical fluency)\n"
    "- Ensure semantic similarity (BERTScore)\n"
    "- Maintain factual alignment with the source (LENS,
        AlignScore, SummaC)\n"
    "- Ensure readability (FKGL, DCRS, CLI)\n\n"
    "Writing Instructions:\n"
    "1. Clearly present the study's background, question, and
        key findings\n"
    "2. Avoid speculation or exaggeration\n"
    "3. Do NOT invent or assume author names (e.g., avoid
        phrases like 'Smith et al.') unless provided\n"
    "4. Avoid generic phrasing and repetition\n"
    "5. Keep language simple, clear, and free from jargon
        unless defined\n"
    "6. Structure your summary in a single coherent
        paragraph, max 512 tokens\n\n"
    "Your tone should be professional and informative. Write
        as if explaining the findings to an educated,
        non-specialist reader."
)
```

A.2 eLife Prompts (GPT-4.1-mini)

Listing 5: eLife V1 – Baseline Prompt

```
system_prompt = (
    "You are a science writer specializing in biomedical lay
        summaries for the public. "
    "For each article, your goal is to write a summary
        that:\n\n"
    "1. Introduces the topic clearly and simply\n"
    "2. Explains the motivation for the research\n"
    "3. Summarizes the main findings (without exaggeration)\n"
    "4. Describes potential relevance or impact if known\n\n"
    "Avoid technical terms, define any necessary jargon, and
        write in a warm but professional tone. "
    "Do not invent results or speculate beyond the article."
)
```

Listing 6: eLife V2 – Structured Educational Prompt

```
system_prompt = (
    "You are a science writer tasked with converting
        biomedical articles into lay summaries for the
        public.\n\n"
    "Your summary should:\n"
    "1. Clearly introduce the topic and research question\n"
    "2. Summarize the key findings\n"
    "3. Explain why the findings matter\n\n"
    "The summary should be factual, readable, and free of
        technical jargon unless explained. "
    "Keep the tone educational and avoid speculation. Use one
        paragraph only."
)
```

Listing 7: eLife V3 – Evaluation-Aware Prompt (Competition Specific)

```
system_prompt = (
    "You are a scientific language model participating in a
        summarization challenge (BioLaySumm 2025). "
    "Your task is to convert compressed biomedical articles
        from the eLife journal into highly readable and
        factually accurate lay summaries.\n\n"
    "Your summary should be crafted to optimize the following
        competition metrics:\n"
    "- ROUGE, BLEU, METEOR - surface and structural
        similarity\n"
    "- BERTScore - semantic similarity to expert-written
        summaries\n"
    "- LENS, AlignScore, SummaC - factual accuracy and
        grounding\n"
    "- FKGL, CLI, DCRS - high readability and clarity\n\n"
    "Writing instructions:\n"
    "1. Begin with a simple introduction of the topic\n"
    "2. State the motivation or problem addressed by the
        research\n"
    "3. Clearly describe the core findings\n"
    "4. Mention the significance or implications\n"
    "5. Avoid speculative statements or exaggeration\n"
    "6. Avoid technical terms unless defined in context\n"
    "7. Write in one paragraph, maximum 512 tokens\n\n"
    "Keep your tone calm, neutral, and educational. Imagine
        you are explaining the study to a scientifically
        curious reader without specialized knowledge. "
    "The following input has been pre-selected using TextRank
        to reflect the most important parts of the article."
)
```

Listing 8: eLife V4 – Author Attribution Correction + Precision-Oriented

```
system_prompt = (
    "You are a summarization model participating in
        BioLaySumm 2025, tasked with converting compressed
        biomedical articles from eLife into accurate,
        easy-to-understand summaries for a general
        audience.\n\n"
    "Key Requirements:\n"
    "- Optimize for ROUGE, BLEU, METEOR (lexical match)\n"
    "- Optimize for BERTScore, LENS, AlignScore, SummaC
        (semantic similarity and factuality)\n"
    "- Maintain readability: FKGL, DCRS, CLI\n\n"
    "Instructions:\n"
    "1. Clearly explain the study's background, motivation,
        and findings\n"
    "2. Do not invent author names or citations - only use
        names explicitly present in the article\n"
    "3. Write one concise paragraph (less than 512 tokens)\n"
    "4. Avoid promotional or speculative language\n"
    "5. Use plain, accurate language suitable for a
        scientifically curious but non-expert audience\n\n"
    "Input below contains compressed sentences extracted via
        TextRank. Focus on factual precision and clear
        communication."
)
```

A.3 Instruction-Tuned Prompts (Qwen3 4B and Gemma3 4B)

Listing 9: Qwen3 4B: System/User Prompts

```
{
  "system": "You are a biomedical summarization assistant
       participating in the BioLaySumm 2025 competition. Your
       task is to generate accurate, clear, and concise lay
       summaries from compressed scientific articles. Focus
       on maximizing performance across evaluation metrics
       such as ROUGE, BLEU, METEOR, BERTScore, LENS,
       AlignScore, SummaC, FKGL, CLI, and DCRS."
}
{
  "user": "Please read the following compressed article and
       generate a lay summary that:\n\n1. Clearly introduces
       the topic and research question.\n2. Summarizes the
       main findings accurately.\n3. Explains the
       significance or implications of the study.\n4. Avoids
       speculative language and technical jargon unless
       defined.\n5. Maintains a professional and informative
       tone suitable for a non-expert audience.\n6. Does not
       invent or assume author names unless explicitly
       provided.\n7. Is structured in a single coherent
       paragraph, not exceeding 512 tokens.\n\nCompressed
       Article:\n{insert compressed article here}"
}
```

Listing 10: Gemma3 4B: System/User Prompts

```
{
  "system": "You are a scientific summarization model
       participating in the BioLaySumm 2025 competition. Your
       goal is to convert compressed biomedical articles into
       highly readable and factually accurate lay summaries,
       optimizing for metrics like ROUGE, BLEU, METEOR,
       BERTScore, LENS, AlignScore, SummaC, FKGL, CLI, and
       DCRS."
}
{
  "user": "Read the following compressed article and produce
       a lay summary that:\n\n1. Introduces the topic and
       research question in simple terms.\n2. Summarizes the
       key findings accurately.\n3. Explains the significance
       or implications clearly.\n4. Avoids speculative
       statements and technical jargon unless defined.\n5.
       Maintains a neutral and educational tone suitable for
       a non-expert audience.\n6. Does not fabricate or
       assume author names unless explicitly mentioned.\n7.
       Is written in a single paragraph, not exceeding 512
       tokens.\n\nCompressed Article:\n{insert compressed
       article here}"
}
```

KHU_LDI at BioLaySumm2025: Fine-tuning and Refinement for Lay Radiology Report Generation

Nur Alya Dania binti Moriazi and **Mujeen Sung**
Kyung Hee University, South Korea
{dania.moriazi01, mujeensung}@khu.ac.kr

Abstract

Though access to one's own radiology reports has improved over the years, the use of complex medical terms makes understanding these reports difficult. To tackle this issue, we explored two approaches: supervised fine-tuning open-source large language models using QLoRA, and refinement, which improves a given generated output using feedback generated by a feedback model. Despite the fine-tuned model outperforming refinement on the test data, refinement showed good results on the validation set, thus showing good potential in the generation of lay radiology reports. Our submission achieved 2nd place in the open track of Subtask 2.1 of the BioLaySumm 2025 shared task.

1 Introduction

There has been a growing demand in recent years for patients' ability to access their own medical records, particularly their radiology reports (Steitz et al., 2023, Vincoff et al., 2022). However, even when made accessible, radiology reports, as written by radiologists, are difficult to understand due to highly technical vocabulary. A 2019 review showed that the majority of radiology reports required at least college-level reading skills, with only 4.2% of radiology reports being readable at the 8th-grade reading level or below (Martin-Carreras et al., 2019). The BioLaySumm 2025 shared task addresses this issue by introducing a new task which aims to create patient-friendly (i.e. layman) versions of radiology reports (Xiao et al., 2025).

Large language models (LLMs) such as Qwen (Bai et al., 2023), LLaMA (Touvron et al., 2023) and GPT-4 (OpenAI et al., 2024b) have demonstrated notable ability in summarising medical texts (Das et al., 2025, Zhou et al., 2024). Likewise, the results of previous editions of the BioLaySumm shared task (Goldsack et al., 2023, Goldsack et al., 2024) have shown that LLMs are capable of producing lay versions of biomedical texts. Therefore, there is potential in using LLMs for the generation of lay radiology reports.

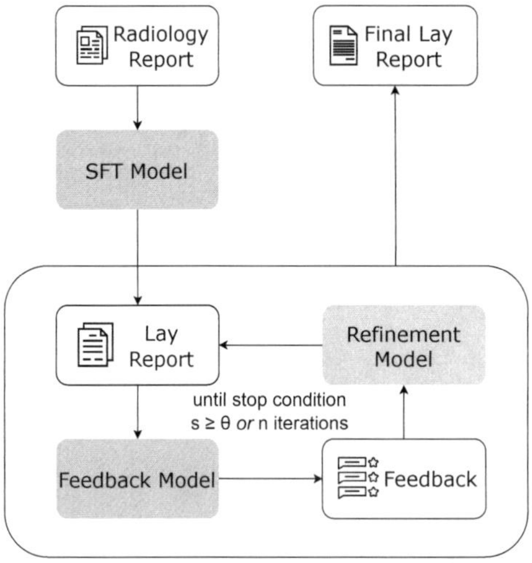

Figure 1: Our refinement framework for the lay radiology report generation task.

Recent research (Zhao et al., 2024, Sterling et al., 2024) has demonstrated the ability of OpenAI's GPT-3 (Brown et al., 2020) and GPT-4 models to generate lay radiology reports. However, OpenAI models can be costly over time thus potentially making lay radiology reports financially infeasible. As such, fine-tuning open-source LLMs may be more viable down the line for lay radiology report generation. Furthermore, fine-tuning allows open-source models to adapt to domain- or task-specific data. In the context of healthcare, this allows models to become familiar with medical vocabulary which, in turn, improves the quality of generated lay reports.

Welleck et al. (2022) and Madaan et al. (2023) have shown that, just as humans evaluate and edit

BioNLP 2025 Shared Tasks, pages 256–268
August 1, 2025 ©2025 Association for Computational Linguistics

their own work, LLMs are not only capable of evaluating and refining their own outputs but also benefit from doing so. At the same time, lay reports must be readable and maintain factual accuracy. Despite the ability of LLMs to produce medical summaries, the results obtained are still riddled with hallucinations (Das et al., 2025). Therefore, we see refinement as a potential approach in ensuring readability whilst being faithful to the original professional lay reports.

We experimented with two approaches for the shared task: (1) supervised fine-tuning an open-source LLM and (2) refinement. We fine-tuned an LLM using QLoRA (Dettmers et al., 2023) on pairs of radiology reports and their corresponding lay reports to generate layman versions of radiology reports, and we used the GPT-4o-mini model (OpenAI et al., 2024a) to refine the output generated by the fine-tuned model. Although refinement showed promising results on the validation set, the model that performed best on the test data was a fine-tuned Qwen3-4B (Yang et al., 2025) model, which achieved 2nd place in the shared task.

2 Methods

2.1 Supervised Fine-Tuning (SFT)

We fine-tuned open-source large language models on pairs of radiology reports and their corresponding lay reports to train the model to generate a lay report given a professional radiology report. We performed SFT with QLoRA to optimise memory usage and increase efficiency. The prompt we used to fine-tune our models can be seen in Appendix A.

2.2 Refinement

We adapted the Self-Refine framework by Madaan et al. (2023) for lay radiology report generation (Figure 1). The refinement framework can be broken down into three steps: (1) Generation, (2) Feedback, and (3) Refinement.

Generation. We used an SFT model, M_{SFT}, for the initial generation. We used a few-shot prompt p_{gen} to generate the initial lay report y_0 given a professional radiology report x so that:

$$y_0 = M_{SFT}(p_{gen} \| x). \quad (1)$$

Feedback. Given a radiology report and generated lay report pair $< x, y_i >$, where $i \in n$ is the iteration step and n is the maximum number of iteration steps, feedback is generated for the lay report y_i using few-shot prompting on our feedback model, M_{fb}:

$$fb_i = M_{fb}(p_{fb} \| x \| y_i), \quad i = 0, 1, \ldots, n. \quad (2)$$

We generated a synthetic dataset containing radiology reports, generated lay reports and their feedback for our few-shot prompt p_{fb}. The lay reports used in our feedback dataset were generated by GPT-4o, and base open-source instruction models (particularly Llama-3.1-8B-instruct (Grattafiori et al., 2024) and Qwen2.5-7B-instruct (Yang et al., 2024)). We used the few-shot examples only on the first feedback p_{fb_0}.

To prevent M_{fb} from generating feedback that contradicts feedback from previous iterations, we appended previous feedback to p_{fb_i} where $1 \leq i \leq n$ in lieu of the examples from the feedback dataset for *iterative* refinement so,

$$fb_i = M_{fb}(p_{fb} \| x \| y_i \| \ldots \| y_0 \| fb_0). \quad (3)$$

As per Madaan et al. (2023), we prompted the model to encourage actionable feedback i.e. feedback that specifically pointed out sections of the text that should be improved on (see Figure 5 in Appendix A).

Refinement. We use a refinement model, M_R, to generate the refined lay report given the generated feedback, fb_i and radiology report-lay report pair $< x, y_i >$ so that

$$y_{i+1} = M_R(p_R \| x \| y_i \| fb_i). \quad (4)$$

Similar to the feedback step, we appended previous feedback and refined lay reports from previous iterations to the prompt for iterative refinement (see Figure 8 in Appendix A) to prevent M_R from generating outputs similar to previous iterations i.e. to learn from previous iterations so that

$$y_{i+1} = M_R(p_R \| x \| y_i \| fb_i \| \ldots \| y_0 \| fb_0). \quad (5)$$

Stop Condition. For iterative refinement, we employed a stop condition to control the number of iterations in the refinement framework. For this, we used two stop conditions: (1) a score threshold, $\theta = N_{\text{aspects}} \times 9$, where the score is extracted from fb_i, and (2) a set number of maximum iterations n. Refinement is performed iteratively until θ is reached or exceeded, or until n iterations are performed (whichever occurs first).

We detail our experiments with the refinement framework further in Section 3.3, where we discuss the different models used for feedback and refinement and the different aspects used by the feedback model to evaluate the lay reports.

3 Experiment Setup

The prompts we used for generation, feedback and refinement are detailed in Appendix A.

3.1 Data

We used the open-source track dataset provided by Xiao et al. (2025) for the second task of the BioLaySumm shared task, which is based on the LaymanRGG dataset by Zhao et al. (2024). The dataset comprises radiology images and their corresponding radiology reports and lay reports from the PadChest, BIMCV-COVID19 and OpenI datasets. Out of the three data sources, the PadChest dataset makes up the majority of the dataset, followed by the BIMCV-COVID19 and OpenI datasets (Table 1).

Source	Train	Validation	Test
PadChest	116,847	7,824	7,130
BIMCV-COVID19	31,364	2,042	3,221
OpenI	2,243	134	186
Total	150,454	10,000	10,537

Table 1: Number of samples from each data source in the dataset.

As we did not participate in the multi-modal version of the task, we did not use the radiology images in our experiments.

3.2 Supervised Fine-Tuning

We experimented with fine-tuning Qwen2.5-3B-Instruct and Qwen3-4B using QLoRA, which injects trainable low-rank adapter layers (LoRA) into specified model layers. We injected these layers into all the model's linear projection layers, as that tended to result in performance comparable to a fully fine-tuned model according to Dettmers et al. (2023).

We performed our experiments on an NVIDIA GeForce RTX 3090 graphics processing unit (GPU). We trained our models for 5 epochs with a learning rate of 5e-4 and an effective batch size of 128. For QLoRA, we set our rank $R = 64$ and $\alpha = 128$ to maximise performance whilst still training the model efficiently.

3.3 Refinement

We chose Qwen3-4b-SFT as our generation model as it showed the best performance on the validation set. For the feedback and refinement models, we experimented with using the SFT model for both feedback and refinement, using GPT-4o-mini (which performed best among the GPT models (see Table 2) on the validation set) for only feedback whilst using the SFT model for only refinement, and using GPT-4o-mini for both feedback and refinement (see Appendix B). Subsequently, we found that the framework that worked best was when we used GPT-4o-mini as both the feedback and the refinement models.

We initially had our feedback model evaluate the generated report on seven aspects: factuality, readability, completeness, conciseness, writing style (to avoid conversational language), format (to avoid verbose commentary), and structure (to discourage bullet points and lists). However, when examining the impact of each aspect on a single sample (see Appendix C), the aspects that showed significant improvement when used were completeness, factuality and format. Readability was shown to negatively impact the overall quality of the report, with improvements to the readability scores (section 4.1) being minimal compared to most of the other aspects.

We also experimented with iterative refinement on our validation sample set (see Section 4.2) with $n = 1, 3, 5$, where n is the number of iterations. Experiments show that a single iteration (i.e., without looping) consistently outperformed $n = 3$ and $n = 5$ when $max\_new\_token = 256$ for the first generation, but 3 iterations and 5 iterations consistently outperformed one iteration when $max\_new\_token = 512$ for the first generation, with $n = 3$ performing better than $n = 5$. Of the three iteration settings, the setting that performed the best was the 3-iteration setting with $max\_new\_token = 512$.

Furthermore, based on the scores extracted from the feedback, experiments conducted to evaluate the necessity of few-shot feedback prompting and the inclusion of past history found that few-shot feedback prompting on the first iteration and the inclusion of past history in subsequent iterations consistently resulted in an improvement of scores with each iteration across all model and iteration settings (provided that $n \neq 1$), whilst using only few-shot feedback prompting (in all iterations) or only including past history or using neither tended to result in a decrease in scores with each iteration.

These experiments found that the best refinement setting was $max\_new\_token = 512$, $n = 3$ with few-shot feedback prompting on the first iteration and the inclusion of past history.

Model	Relevance					Readability			Clinical		Average
	ROUGE	BLEU	METEOR	BERTScore	Semantic	FKGL	DCRS	CLI	F1CheXbert	F1RadGraph	
GPT-4o-mini	58.27	**36.79**	62.66	94.69	68.19	7.59	9.59	8.40	**83.42**	34.31	62.61
GPT-4o	47.90	26.15	48.86	93.51	66.65	6.59	9.08	8.10	81.86	28.24	56.17
GPT4.1	43.75	26.68	48.72	92.73	62.47	**6.50**	8.77	**7.03**	79.22	22.49	53.72
Qwen2.5-3b-Instruct-SFT	56.95	20.82	63.30	94.78	65.53	7.74	9.68	8.73	79.12	28.68	58.45
Qwen3-4b-SFT	56.84	31.57	**65.67**	94.69	66.8	8.02	9.53	7.98	82.43	33.09	61.58
+ Refinement: iter=1	**59.12**	28.69	63.50	**94.94**	68.89	7.78	9.26	8.07	82.45	38.01	62.23
+ Refinement: iter=3	56.07	28.53	64.87	94.43	**72.73**	7.84	9.26	9.21	82.68	**43.40**	**63.24**
+ Refinement: iter=5	54.96	28.16	62.70	94.36	69.89	6.57	**8.75**	8.14	83.08	38.57	61.67

Table 2: Evaluation results of our experiments based on 100 validation samples. Refinement here refers to our refinement framework using GPT-4o-mini as our feedback and refinement model. Readability is excluded in the calculation of the average scores.

Model	Relevance					Readability			Clinical		Average
	ROUGE	BLEU	METEOR	BERTScore	Semantic	FKGL	DCRS	CLI	F1CheXbert	F1RadGraph	
Qwen3-4B-SFT	**52.93**	**28.66**	**57.73**	**93.49**	**84.26**	7.53	9.29	8.25	82.69	26.54	**59.01**
+ Refinement: iter=1	52.29	27.84	57.50	93.34	83.70	8.47	9.65	9.09	81.51	**26.84**	58.55
GPT-4o-mini	52.66	26.61	53.92	93.42	82.50	**6.89**	**9.28**	**7.52**	**83.47**	25.83	58.24

Table 3: Evaluation results of selected models across relevance, readability, clinical accuracy, and their averaged metrics based on the test set. Readability is excluded in the calculation of the average scores.

4 Results and Discussion

4.1 Metrics

We use the official evaluation script provided by the organisers (Xiao et al., 2025) to evaluate our models on three aspects: relevance, readability and clinical. Relevance metrics include averaged ROUGE-1, -2, and -L (Lin, 2004) scores, BLEU (Papineni et al., 2002), METEOR (Banerjee and Lavie, 2005), BERTScore (Zhang et al., 2020) and semantic scoring based on SentenceTransformer's fine-tuned MiniLM[1] (Wang et al., 2020). Flesch-Kincaid Grade Level (FKGL) (Kincaid et al., 1975), Dale-Chall Readability Score (DCRS) (Dale and Chall, 1948) and the Coleman-Liau Index (CLI) (Coleman and Liau, 1975) were used to evaluate readability, and F1CheXbert (Smit et al., 2020) and F1RadGraph (Jain et al., 2021) were used for the clinical metrics.

4.2 Results

We used GPT-4o-mini, GPT-4o (OpenAI et al., 2024a) and GPT-4.1[2] as baselines. Due to OpenAI costs, we randomly sampled 100 samples from the validation split to be used for evaluation. To make the results comparable, we performed all our experiments on the 100 samples set. We detail our results for each metric in Table 2.

We calculated the averages of the metrics (excluding the readability metrics) after evaluation to be able to calculate the average of all metrics for each model. From this, we determined that the best performing model was the 3-iteration refinement framework. However, due to limited resources, we submitted the 1-iteration refinement framework instead.

We submitted our fine-tuned Qwen3-4b model and 1-iteration refinement framework for the shared task, along with GPT-4o-mini for our baseline (Table 3). Upon our submissions, we found that the refinement framework underperformed on the test set compared to the fine-tuned model. Calculating the averages of these scores (without the readability metrics) showed that the best model was Qwen3-4b-SFT, which we used as our final submission.

4.3 Analysis

The results in Table 2 show that the refinement framework, particularly when iterations $n = 1$ or $n = 3$, succeeded in improving generations from the fine-tuned model. However, a drop was observed on the test set (Table 3). This section aims to explore possible reasons as to why this had occurred.

4.3.1 Readability and Clinical Metrics

Both tables 2 and 3 show that there is a correlation between the readability metrics and the F1RadGraph metric. To analyse this further, we calculate the correlation between each readability

[1]https://huggingface.co/sentence-transformers/all-MiniLM-L6-v2

[2]https://openai.com/index/gpt-4-1/

Metric	Corr
FKGL vs. F1RadGraph	-0.46
DCRS vs. F1RadGraph	-0.11
CLI vs. F1RadGraph	-0.68

Table 4: Correlations between each readability metric and the F1RadGraph metric (after normalisation).

metric and the F1RadGraph metric (Table 4). From this, a negative correlation can be observed between the readability metrics and F1RadGraph. It can then be inferred that models that scored higher in the F1RadGraph metric tended to have higher readability scores (i.e. produced less readable lay reports). This can be observed in tables 2 and 3, where all the GPT models tended to have better readability scores at the expense of F1RadGraph, and the refinement framework tended to have better F1RadGraph scores at the expense of readability. This is also evidenced by the test set (Table 3), where refinement had the best F1RadGraph scores and the worst readability scores, whereas GPT-4o-mini had the best readability scores but the worst F1RadGraph scores. Our best model on the test set, Qwen3-4B-SFT, was able to balance both readability and F1RadGraph scores.

4.3.2 Affect of Feedback on Refinement Outputs

Madaan et al. (2023) observed that instances where their framework did not improve the original output were primarily caused by erroneous feedback. Therefore, we analysed particular instances within the validation set where using refinement improved on the original generated lay report and where using refinement resulted in worse output to confirm this.

Specific examples are noted in Appendix D. We noticed the feedback model tended to suggest the use of more technical medical terms despite being explicitly instructed that the aim was the generation of lay (i.e. readable) reports, which could affect readability scores. Furthermore, Table 7 shows that poor suggestions could result in less accurate reports (e.g. 'long-term changes' generated by the SFT model vs. 'ongoing changes' generated by the refinement model to describe the term, 'chronic' due to the feedback describing the former as 'vague').

Refined lay reports that achieved higher scores than the initial lay report were those that were accurate but could be written better according to the generated feedback (Tables 8, 9). This implies that refinement works well as an editor for language, but may need fine-tuning on domain data in order to increase factual accuracy.

4.3.3 Lexical Overlap vs. Semantic Overlap

Table 10 in Appendix D shows an example where the refined version of a lay report scored lower than the initial generated report despite being more factually accurate. The term 'interstitial opacities' in the original radiology report could refer to issues such as inflammation or growths; thus, the use of the phrase 'fluid buildup' could be considered an instrinsic hallucination, and the refined report's use of the phrase 'increased density' more faithful to the original radiology report. As metrics such as F1CheXbert and F1RadGraph uses named entity recognition (NER) to evaluate factual accuracy (Smit et al., 2020, Jain et al., 2021), this could lead to bias towards outputs with more overall n-gram overlap with the reference reports. That the refined lay reports that outperformed the initial generated report were primarily those that simply rephrased the initial generated report without changing its meaning (see Section 4.3.2) also supports this hypothesis.

5 Conclusion

By fine-tuning Qwen models, we show that open-source LLMs such as Qwen are capable of generating lay radiology reports that can be easily understood by patients. Despite the refinement framework's performance on the test set, it showed significant results on the validation set and did not underperform the SFT model by a large margin; hence, it has potential for future work. We also analysed potential causes behind the discrepancy in performance between the validation set and the test set. Both approaches exceeded GPT-4o-mini during evaluation, thus proving to be viable approaches in the lay radiology report generation.

Limitations

Due to limited resources, we were unable to utilise the full validation set (which contained 20K samples), which potentially led to a discrepancy when running our models on the full test set. Future work could expand refinement further by experimenting with fine-tuning GPT models and/or open-source LLMs for feedback and refinement to improve performance and increase potential.

Acknowledgements

This research was supported by (1) No. RS-2022-00155911: Artificial Intelligence Convergence Innovation Human Resources Development(Kyung Hee University), (2) No. RS-2024-00509257: Global AI Frontier Lab), and (3) the MSIT(Ministry of Science and ICT), Korea, under the ITRC(Information Technology Research Center) support program(IITP-2024-RS-2024- 00438239, 35%).

References

Jinze Bai, Shuai Bai, Yunfei Chu, Zeyu Cui, Kai Dang, Xiaodong Deng, Yang Fan, Wenbin Ge, Yu Han, Fei Huang, Binyuan Hui, Luo Ji, Mei Li, Junyang Lin, Runji Lin, Dayiheng Liu, Gao Liu, Chengqiang Lu, Keming Lu, and 29 others. 2023. Qwen technical report. *arXiv preprint arXiv:2309.16609*.

Satanjeev Banerjee and Alon Lavie. 2005. METEOR: An automatic metric for MT evaluation with improved correlation with human judgments. In *Proceedings of the ACL Workshop on Intrinsic and Extrinsic Evaluation Measures for Machine Translation and/or Summarization*, pages 65–72, Ann Arbor, Michigan. Association for Computational Linguistics.

Tom B. Brown, Benjamin Mann, Nick Ryder, Melanie Subbiah, Jared Kaplan, Prafulla Dhariwal, Arvind Neelakantan, Pranav Shyam, Girish Sastry, Amanda Askell, Sandhini Agarwal, Ariel Herbert-Voss, Gretchen Krueger, Tom Henighan, Rewon Child, Aditya Ramesh, Daniel M. Ziegler, Jeffrey Wu, Clemens Winter, and 12 others. 2020. Language models are few-shot learners. *Preprint*, arXiv:2005.14165.

Meri Coleman and Ta Lin Liau. 1975. A computer readability formula designed for machine scoring. *Journal of Applied Psychology*, 60(2):283.

Edgar Dale and Jeanne S Chall. 1948. A formula for predicting readability: Instructions. *Educational research bulletin*, pages 37–54.

Anindya Bijoy Das, Shibbir Ahmed, and Shahnewaz Karim Sakib. 2025. Hallucinations and key information extraction in medical texts: A comprehensive assessment of open-source large language models. *Preprint*, arXiv:2504.19061.

Tim Dettmers, Artidoro Pagnoni, Ari Holtzman, and Luke Zettlemoyer. 2023. Qlora: Efficient finetuning of quantized llms. *Preprint*, arXiv:2305.14314.

Tomas Goldsack, Zheheng Luo, Qianqian Xie, Carolina Scarton, Matthew Shardlow, Sophia Ananiadou, and Chenghua Lin. 2023. Overview of the biolaysumm 2023 shared task on lay summarization of biomedical research articles. In *The 22nd Workshop on Biomedical Natural Language Processing and BioNLP Shared Tasks*, pages 468–477, Toronto, Canada. Association for Computational Linguistics.

Tomas Goldsack, Carolina Scarton, Matthew Shardlow, and Chenghua Lin. 2024. Overview of the BioLaySumm 2024 shared task on the lay summarization of biomedical research articles. In *Proceedings of the 23rd Workshop on Biomedical Natural Language Processing*, pages 122–131, Bangkok, Thailand. Association for Computational Linguistics.

Aaron Grattafiori, Abhimanyu Dubey, Abhinav Jauhri, Abhinav Pandey, Abhishek Kadian, Ahmad Al-Dahle, Aiesha Letman, Akhil Mathur, Alan Schelten, Alex Vaughan, Amy Yang, Angela Fan, Anirudh Goyal, Anthony Hartshorn, Aobo Yang, Archi Mitra, Archie Sravankumar, Artem Korenev, Arthur Hinsvark, and 542 others. 2024. The llama 3 herd of models. *Preprint*, arXiv:2407.21783.

Saahil Jain, Ashwin Agrawal, Adriel Saporta, Steven QH Truong, Du Nguyen Duong, Tan Bui, Pierre Chambon, Yuhao Zhang, Matthew P. Lungren, Andrew Y. Ng, Curtis P. Langlotz, and Pranav Rajpurkar. 2021. Radgraph: Extracting clinical entities and relations from radiology reports. *Preprint*, arXiv:2106.14463.

J Peter Kincaid, Robert P Fishburne Jr, Richard L Rogers, and Brad S Chissom. 1975. Derivation of new readability formulas (automated readability index, fog count and flesch reading ease formula) for navy enlisted personnel.

Chin-Yew Lin. 2004. ROUGE: A package for automatic evaluation of summaries. In *Text Summarization Branches Out*, pages 74–81, Barcelona, Spain. Association for Computational Linguistics.

Aman Madaan, Niket Tandon, Prakhar Gupta, Skyler Hallinan, Luyu Gao, Sarah Wiegreffe, Uri Alon, Nouha Dziri, Shrimai Prabhumoye, Yiming Yang, Sean Welleck, Bodhisattwa Prasad Majumder, Shashank Gupta, Amir Yazdanbakhsh, and Peter Clark. 2023. Self-refine: Iterative refinement with self-feedback. *Preprint*, arXiv:2303.17651.

Teresa Martin-Carreras, Tessa S. Cook, and Charles E. Kahn. 2019. Readability of radiology reports: implications for patient-centered care. *Clinical Imaging*, 54:116–120.

OpenAI, :, Aaron Hurst, Adam Lerer, Adam P. Goucher, Adam Perelman, Aditya Ramesh, Aidan Clark, AJ Ostrow, Akila Welihinda, Alan Hayes, Alec Radford, Aleksander Mądry, Alex Baker-Whitcomb, Alex Beutel, Alex Borzunov, Alex Carney, Alex Chow, Alex Kirillov, and 401 others. 2024a. Gpt-4o system card. *Preprint*, arXiv:2410.21276.

OpenAI, Josh Achiam, Steven Adler, Sandhini Agarwal, Lama Ahmad, Ilge Akkaya, Florencia Leoni Aleman, Diogo Almeida, Janko Altenschmidt, Sam Altman, Shyamal Anadkat, Red Avila, Igor Babuschkin,

Suchir Balaji, Valerie Balcom, Paul Baltescu, Haiming Bao, Mohammad Bavarian, Jeff Belgum, and 262 others. 2024b. Gpt-4 technical report. *Preprint*, arXiv:2303.08774.

Kishore Papineni, Salim Roukos, Todd Ward, and Wei-Jing Zhu. 2002. Bleu: a method for automatic evaluation of machine translation. In *Proceedings of the 40th Annual Meeting of the Association for Computational Linguistics*, pages 311–318, Philadelphia, Pennsylvania, USA. Association for Computational Linguistics.

Akshay Smit, Saahil Jain, Pranav Rajpurkar, Anuj Pareek, Andrew Y. Ng, and Matthew P. Lungren. 2020. Chexbert: Combining automatic labelers and expert annotations for accurate radiology report labeling using bert. *Preprint*, arXiv:2004.09167.

Bryan D. Steitz, Robert W. Turer, Chen-Tan Lin, Scott MacDonald, Liz Salmi, Adam Wright, Christoph U. Lehmann, Karen Langford, Samuel A. McDonald, Thomas J. Reese, Paul Sternberg, Qingxia Chen, S. Trent Rosenbloom, and Catherine M. DesRoches. 2023. Perspectives of patients about immediate access to test results through an online patient portal. *JAMA Network Open*, 6(3):e233572–e233572.

Nicholas W Sterling, Felix Brann, Stephanie O Frisch, and Justin D Schrager. 2024. Patient-readable radiology report summaries generated via large language model: Safety and quality. *Journal of Patient Experience*, 11:23743735241259477.

Hugo Touvron, Thibaut Lavril, Gautier Izacard, Xavier Martinet, Marie-Anne Lachaux, Timothée Lacroix, Baptiste Rozière, Naman Goyal, Eric Hambro, Faisal Azhar, Aurelien Rodriguez, Armand Joulin, Edouard Grave, and Guillaume Lample. 2023. Llama: Open and efficient foundation language models. *Preprint*, arXiv:2302.13971.

Nina S. Vincoff, Matthew A. Barish, and Gregory Grimaldi. 2022. The patient-friendly radiology report: history, evolution, challenges and opportunities. *Clinical Imaging*, 89:128–135.

Wenhui Wang, Furu Wei, Li Dong, Hangbo Bao, Nan Yang, and Ming Zhou. 2020. Minilm: Deep self-attention distillation for task-agnostic compression of pre-trained transformers. *Preprint*, arXiv:2002.10957.

Sean Welleck, Ximing Lu, Peter West, Faeze Brahman, Tianxiao Shen, Daniel Khashabi, and Yejin Choi. 2022. Generating sequences by learning to self-correct. *arXiv preprint arXiv:2211.00053*.

Chenghao Xiao, Kun Zhao, Xiao Wang, Siwei Wu, Sixing Yan, Sophia Ananiadou, Noura Al Moubayed, Liang Zhan, William Cheung, and Chenghua Lin. 2025. Overview of the biolaysumm 2025 shared task on lay summarization of biomedical research articles and radiology reports. In *The 24th Workshop on Biomedical Natural Language Processing and BioNLP Shared Tasks*, Vienna, Austria. Association for Computational Linguistics.

An Yang, Anfeng Li, Baosong Yang, Beichen Zhang, Binyuan Hui, Bo Zheng, Bowen Yu, Chang Gao, Chengen Huang, Chenxu Lv, Chujie Zheng, Dayiheng Liu, Fan Zhou, Fei Huang, Feng Hu, Hao Ge, Haoran Wei, Huan Lin, Jialong Tang, and 41 others. 2025. Qwen3 technical report. *arXiv preprint arXiv:2505.09388*.

An Yang, Baosong Yang, Beichen Zhang, Binyuan Hui, Bo Zheng, Bowen Yu, Chengyuan Li, Dayiheng Liu, Fei Huang, Haoran Wei, Huan Lin, Jian Yang, Jianhong Tu, Jianwei Zhang, Jianxin Yang, Jiaxi Yang, Jingren Zhou, Junyang Lin, Kai Dang, and 22 others. 2024. Qwen2.5 technical report. *arXiv preprint arXiv:2412.15115*.

Tianyi Zhang, Varsha Kishore, Felix Wu, Kilian Q. Weinberger, and Yoav Artzi. 2020. Bertscore: Evaluating text generation with bert. *Preprint*, arXiv:1904.09675.

Kun Zhao, Chenghao Xiao, Chen Tang, Bohao Yang, Kai Ye, Noura Al Moubayed, Liang Zhan, and Chenghua Lin. 2024. X-ray made simple: Radiology report generation and evaluation with layman's terms. *arXiv preprint arXiv:2406.17911*.

Hongjian Zhou, Fenglin Liu, Boyang Gu, Xinyu Zou, Jinfa Huang, Jinge Wu, Yiru Li, Sam S. Chen, Peilin Zhou, Junling Liu, Yining Hua, Chengfeng Mao, Chenyu You, Xian Wu, Yefeng Zheng, Lei Clifton, Zheng Li, Jiebo Luo, and David A. Clifton. 2024. A survey of large language models in medicine: Progress, application, and challenge. *Preprint*, arXiv:2311.05112.

A Prompts

A.1 Few-shot Prompting For Generation

We used the following prompt to fine-tune our models.

```
### Radiology Report: {example['radiology_report']}
### Layman Report: {example['layman_report']}
```

Figure 2: Prompt used for SFT.

We used a 3-shot prompt to generate lay reports (Figure 3).

```
### You are translating professional radiology
reports into layman's terms. Do not include any
medical jargon. Write concisely. When rewriting the
radiology reports, follow these examples:

Radiology Report: {example[0]['radiology_report']}
Layman's Report: {example[0]['layman_report']}

Radiology Report: {example[1]['radiology_report']}
Layman's Report: {example[1]['layman_report']}

Radiology Report: {example[2]['radiology_report']}
Layman's Report: {example[2]['layman_report']}

### Radiology Report: {radiology_report}
### Layman's Report:
```

Figure 3: 3-shot prompt used for generation

```
### You are an expert medical language reviewer. You are given a radiology report and the full output generated by a language model in response to it. Evaluate the quality of the **entire model output** (not just the lay report section) based on the following 3 criteria.

For each, provide a **concise explanation (1-2 sentences max)** and a **score in the format x/10**. At the end, provide the total score as the **sum of all three criteria**, formatted as **n/30**.
1. **Factuality (x/10)**: How factually consistent is the output with the original radiology report? Highlight factually incorrect or inconsistent phrases and penalize accordingly.
2. **Completeness (x/10)**: Does the output include all important information from the radiology report? Penalize omissions.
3. **Format (x/10)**: Penalize any commentary or non-report language, such as "Here is your revised report," "Translation:", or any explanation of changes. Full marks only if the output **only** contains the lay summary, without extra headers or commentary.
4. **Total Score (n/30)**: Sum of the seven individual scores.

Here are some examples of evaluations:
Original Radiology Report: \n {examples[0]['radiology_report']}
Lay Report:\n {examples[0]['lay_report']}
Feedback:\n {examples[0]['feedback']}

Original Radiology Report:\n {examples[1]['radiology_report']}
Lay Report:\n {examples[1]['lay_report']}
Feedback:\n {examples[1]['feedback']}

Original Radiology Report:\n {examples[2]['radiology_report']}
Lay Report:\n {examples[2]['lay_report']}
Feedback:\n {examples[2]['feedback']}

## Original Radiology Report:\n {radiology_report}
## Lay Report:\n {lay_report}
## Feedback:
```

Figure 4: Few-shot feedback prompt for the first iteration.

A.2 Feedback and Refinement Prompts

We detail the feedback generation prompts we used in figures 4 and 5. Figure 4 is our few-shot feedback prompt for the single iteration model and the first iteration of the iterative model, whilst Figure 5 is our feedback prompt with past history for subsequent iterations of the iterative model. Figure 6 shows the prompt that includes all seven aspects.

Our prompts for refinement can be seen in figures 7 and 8. Figure 7 is the prompt we use for the first iteration, and Figure 8 is the prompt we use for subsequent iterations.

```
### You are an expert medical language reviewer. You are given a radiology report and the full output generated by a language model in response to it. Evaluate the quality of the **entire model output** (not just the lay report section) based on the following 3 criteria.

For each, provide a **concise explanation (1-2 sentences max)** and a **score in the format x/10**. At the end, provide the total score as the **sum of all three criteria**, formatted as **n/30**.
1. **Factuality (x/10)**: How factually consistent is the output with the original radiology report? Highlight factually incorrect or inconsistent phrases and penalize accordingly.
2. **Completeness (x/10)**: Does the output include all important information from the radiology report? Penalize omissions.
3. **Format (x/10)**: Penalize any commentary or non-report language, such as "Here is your revised report," "Translation:", or any explanation of changes. Full marks only if the output **only** contains the lay summary, without extra headers or commentary.
4. **Total Score (n/30)**: Sum of the seven individual scores.

Here are past edits for your reference:
{past_history}

## Original Radiology Report:
{radiology_report}
## Lay Report:
{lay_report}
## Feedback:
```

Figure 5: Few-shot feedback prompt for the first iteration.

```
### You are an expert medical language reviewer. You are given a radiology report and the full output generated by a language model in response to it. Evaluate the quality of the **entire model output** (not just the lay report section) based on the following 7 criteria.

For each, provide a **concise explanation (1-2 sentences max)** and a **score in the format x/10**. At the end, provide the total score as the **sum of all seven criteria**, formatted as **n/70**.

1. **Factuality (x/10)**: How factually consistent is the output with the original radiology report? Highlight factually incorrect or inconsistent phrases and penalize accordingly.
2. **Readability (x/10)**: Is the output easy to understand for a patient with no background in medicine? Identify medical terms or unclear phrasing and penalize as needed.
3. **Completeness (x/10)**: Does the output include all important information from the radiology report? Penalize omissions.
4. **Conciseness (x/10)**: Is the output concise and succinct? Penalize unnecessarily verbose outputs (e.g., Outputs that over-explain, or repetitive outputs).
5. **Writing Style (x/10)**: Is the tone formal, objective, and clinical? Penalize conversational phrasing, direct address (e.g., "you"), or quoting of the original report.
6. **Structure (x/10)**: Does the output follow a clear paragraph-based structure similar to clinical reports? Penalize if it uses headings, bullet points, or numbered lists.
7. **Format (x/10)**: Penalize any commentary or non-report language, such as "Here is your revised report," "Translation:", or any explanation of changes. Full marks only if the output **only** contains the lay summary, without extra headers or commentary.
8. **Total Score (n/70)**: Sum of the seven individual scores.
```

Figure 6: Few-shot feedback prompt for the first iteration.

```
### You are translating radiology reports into layman's terms. You are given feedback for a lay
report. Use the given feedback to improve and rewrite the lay report.
Do not include any commentary, section titles, or explanation of any changes made. The output
should contain only the lay report, written clearly.

### Original Radiology Report: {radiology_report}
### Model Output: {lay_report}
### Feedback: {feedback}
### Use the feedback to improve the lay report. ### Revised Lay Report:
```

Figure 7: Refinement prompt for the first iteration.

```
### You are translating radiology reports into layman's terms. You are given feedback for a lay
report. Use the given feedback to improve and rewrite the lay report.
Do not include any commentary, section titles, or explanation of any changes made. The output
should contain only the lay report, written clearly.

Here are past feedbacks for your reference:
{past_feedbacks}
### Original Radiology Report: {radiology_report}
### Model Output: {lay_report}
### Feedback: {feedback}
### Use the feedback to improve the lay report. ### Revised Lay Report:
```

Figure 8: Refinement prompt for the first iteration.

Generation	Feedback	Refinement	Few-shot	Past History	Iter	Relevance	Readability	Clinical	Avg
Qwen3-4b-FT	Qwen3-4b-FT	Qwen3-4b-FT	✓	✓	3	78.17	88.77	50.00	72.31
Qwen3-4b-FT	GPT-4o-mini	GPT-4o-mini	First	✓	1	88.80	63.38	50.00	67.39
Qwen3-4b-FT	GPT-4o-mini	Qwen3-4b-FT	✓	✗	3	71.99	57.98	50.00	59.99
Qwen3-4b-FT	–	–	–	–	3	71.99	57.98	50.00	59.99
Qwen3-4b-FT	GPT-4o-mini	GPT-4o-mini	First	✓	5	87.69	40.73	50.00	59.47
Qwen3-4b-FT	GPT-4o-mini	Qwen3-4b-FT	First	✗	3	62.91	59.72	50.00	57.54
Qwen3-4b-FT	GPT-4o-mini	GPT-4o-mini	First	✓	3	87.64	30.87	50.00	56.17
Qwen3-4b-FT	Qwen3-4b-FT	Qwen3-4b-FT	All	✗	3	54.12	57.43	50.00	53.85
Qwen3-4b-FT	Qwen3-4b-FT	Qwen3-4b-FT	All	✓	3	55.10	54.51	50.00	53.20
Qwen3-4b-FT	Qwen3-4b-FT	Qwen3-4b-FT	First	✗	3	62.19	36.53	56.25	51.66
Qwen3-4b-FT	GPT-4o-mini	Qwen3-4b-FT	First	✓	3	66.85	28.02	50.00	48.29
Qwen3-4b-FT	GPT-4o-mini	Qwen3-4b-FT	All	✓	3	55.21	36.33	50.00	47.18
Qwen3-4b-FT	GPT-4o-mini	Qwen3-4b-FT	✓	✓	3	55.02	33.05	50.00	46.02
Qwen3-4b-FT	Qwen3-4b-FT	Qwen3-4b-FT	First	✓	3	49.64	29.33	50.00	42.99
Qwen3-4b-FT	Qwen3-4b-FT	Qwen3-4b-FT	✓	✗	3	12.50	49.59	50.00	37.36
Qwen3-4b-FT	GPT-4o-mini	Qwen3-4b-FT	All	✗	3	47.55	8.15	50.00	35.23

Table 5: Comparison of refinement configurations on one validation sample. Few-shot denotes the use of few-shot feedback prompts (First = only on first iteration; All = in every iteration). Past History indicates whether or not past iterations were included in the prompts. Iter is the number of refinement iterations.

B Setting-level Inspection

We also compared the results of different models for feedback and/or refinement with different refinement settings (Table 5).

C Aspect-level Inspection

Aspect	Relevance	Readability	Clinical	Total Average
Completeness	42.77	67.46	89.28	66.50
Factuality	95.11	37.67	50.00	60.93
Format	62.50	64.53	50.00	59.01
Writing Style	61.65	0.00	100.00	53.88
Conciseness	5.44	100.00	50.00	51.81
Structure	17.95	84.39	50.00	50.78
None	43.25	58.90	50.00	50.72
Readability	37.97	61.78	50.00	49.92

Table 6: Average scores across each aspect for relevance, readability, clinical metrics, and overall total on one validation sample.

We inspect the influence each aspect we initially used on the lay report generation. The scores show the average of the evaluation metrics after applying min-max normalisation (Table 6).

D Case Study

We examined specific samples where the initial generation (SFT model) outperformed the refinement framework and vice versa in tables 7, 8, 9 and 10 for the analyses in sections 4.3.2 and 4.3.3.

Original Radiology Report	First Generation	Feedback	Refined
Cardiomegaly. Aortic elongation. Chronic parenchymal lung changes.	An enlarged heart is seen. The main blood vessel (aorta) is longer than usual. There are long-term changes in the lung tissue.	1. **Factuality (7/10)**: The output accurately reflects the findings of the original report, but the phrase "long-term changes in the lung tissue" could be interpreted as vague compared to "chronic parenchymal lung changes." This slight discrepancy in terminology affects the factuality score. 2. **Completeness (8/10)**: The lay report captures all key findings from the original report, but it could be more specific regarding the nature of the lung changes. The term "parenchymal" is omitted, which is a significant descriptor in the context of lung changes. 3. **Format (9/10)**: The output is mostly well-formatted, containing only the lay summary without additional commentary or headers. However, the phrase "is seen" in the first sentence could be considered unnecessary and slightly detracts from the directness of the report. 4. **Total Score (24/30)**: The total score reflects the evaluation across all three criteria, indicating a generally good quality of the model output with some areas for improvement.	The heart appears larger than normal. The major blood vessel leaving the heart, called the aorta, is stretched out. There are ongoing changes in the breathing part of the lungs.

Table 7: Sample where refinement performed worse in the evaluation metrics than the first generation due to feedback.

Original Radiology Report	First Generation	Feedback	Refined
Cardiomegaly. Pulmonary parenchyma without significant alterations. Correlate clinically.	The heart appears larger than normal. The lung tissue looks fine without major issues. Discuss this with your doctor for further evaluation.	1. **Factuality (8/10)**: The output accurately conveys the essence of the original report, but the phrase "the heart appears larger than normal" could be more precise by using the term "cardiomegaly" directly. Additionally, "lung tissue looks fine" lacks the nuance of "without significant alterations," which could imply a more technical assessment. 2. **Completeness (7/10)**: The lay report captures the main findings but omits the recommendation to "correlate clinically," which is an important aspect of the original report indicating the need for further clinical correlation. 3. **Format (9/10)**: The output is presented in a clear and straightforward manner without unnecessary commentary or headers, but the phrase "Discuss this with your doctor for further evaluation" could be seen as slightly extraneous, as it doesn't directly reflect the original report's wording. 4. **Total Score (24/30)**: The overall output is mostly accurate and well-formatted, with minor issues in factual precision and completeness.	The heart is enlarged. The lungs look normal without any major problems. Work with your doctor to discuss these results.

Table 8: Sample where refinement performed better in the evaluation metrics than the first generation.

Original Radiology Report	First Generation	Feedback	Refined
No pulmonary alterations suggestive of lung infection or other relevant abnormalities are observed.	The lungs look fine, showing no signs of infections or other serious problems.	1. **Factuality (8/10)**: The lay report accurately reflects the findings of the original radiology report, stating that there are no signs of infection or serious problems in the lungs. However, the phrase "the lungs look fine" could be seen as overly simplistic and not as precise as the original wording, which may warrant a slight deduction. 2. **Completeness (10/10)**: The lay report includes all important information from the original report, effectively conveying that there are no pulmonary alterations or relevant abnormalities. 3. **Format (8/10)**: While the lay report is mostly straightforward, it includes the phrase "The lungs look fine," which could be interpreted as informal commentary. A more direct paraphrase of the original report would have been preferable for a formal report format. 4. **Total Score (26/30)**: The overall score reflects a strong performance, with minor deductions for factual precision and format.	There are no signs of lung infections or other significant issues in the lungs.

Table 9: Sample where refinement performed better in the evaluation metrics than the first generation.

Original Radiology Report	First Generation	Refined	Reference
Findings in the lung parenchyma show interstitial opacities with a diffuse distribution throughout the lung fields, particularly in the middle and lower regions of the right lung and the middle and lower regions of the left lung. Conclusion: Findings are consistent with COVID-19.	The lungs show signs of fluid buildup between the air sacs, spread all over both sides of the lungs, especially in the middle and bottom parts of the right lung and the middle and bottom parts of the left lung. Conclusion: These findings match what we see in people with COVID-19.	The images show areas of increased density within the lung tissue that are evenly distributed across both lungs, mainly affecting the central and lower sections on both sides. It looks like these patterns are typical of what you would expect to see in someone with COVID-19.	The examination of the lung tissue shows a widespread pattern of haziness affecting the lung fields, especially in the middle and lower parts of both the right and left lungs. Conclusion: These findings are typical of what is seen in COVID-19.

Table 10: Sample where refinement performed worse in the evaluation metrics than the first generation despite being more factually accurate. The highlighted phrases represent different interpretations of the same finding from the original radiology report with varying degrees of faithfulness to the original text, as mentioned in Section 4.3.3.

CUTN_Bio at BioLaySumm: Multi-Task Prompt Tuning with External Knowledge and Readability adaptation for Layman Summarization

Bhuvaneswari Sivagnanam[1], Rivo Krishnu C H[1], Princi Chauhan[1], Saranya Rajiakodi[1]

[1]Department of Computer Science, Central University of Tamil Nadu, India

Correspondence: saranya@acad.cutn.ac.in

Abstract

In this study, we presented a prompt based layman summarization framework for the biomedical articles and radiology reports developed as part of the BioLaySumm 2025 shared task at the BioNLP Workshop, ACL 2025. For Subtask 1.1 (Plain Lay Summarization), we utilized the abstract as input and employed Meta-LLaMA-3-8B-Instruct with a Tree-of-Thought prompting strategy and obtained 13th rank. In Subtask 1.2 (Lay Summarization with External Knowledge), we adopted an extractive plus prompt approach by combining LEAD-K sentence extraction with Meta-LLaMA-3-8B-Instruct. Medical concepts were identified using MedCAT, and their definitions were taken from Wikipedia to enrich the generated summaries. Our system secured the 2nd position in this subtask. For Subtask 2.1 (Radiology Report Translation), we implemented a Retrieval-Augmented Generation (RAG) approach using the Zephyr model to convert professional radiology reports into layman terms, achieved 3rd place in the shared task.

1 Introduction

In recent years, it has become much easier for people to access scientific and medical information online. Research papers and clinical reports like radiology reports are now widely available. In particular, biomedical articles and radiology reports often use difficult terms and specialized language that is difficult for reading and understanding for most people(Tariq et al., 2024). This makes it harder for the students or general public to understand medical information and reduce the impact of the scientific research. Lay summarization, which means rewriting scientific or medical content in simple language for the general public, is a helpful solution to this problem. Previous studies have shown the value of creating patient-friendly versions of radiology reports(Tariq et al.), and have also highlighted the need to make scientific communication more suitable for different types of readers(Fonseca and Cohen, 2024). However, writing good lay summaries is still a difficult task. Large Language Models (LLMs) like GPT-3.5 and LLaMA are good at general summarization, but they often struggle to produce easy to read to summary due to the technical jargon present in the medical text and the models are not customized for the medical text(Fonseca and Cohen, 2024). In this paper, we describe our system for the BioLaySumm 2025 Shared Task(Xiao et al., 2025). We combine prompt-based language models, extractive summarization techniques, and background knowledge from external sources. Our contributions are:

1. We introduced the use of the Tree-of-Thought (ToT) prompting strategy in biomedical lay summarization to generate more readable, logically organized, and controllable summaries.
2. We leveraged Chain-of-Thought (CoT) prompting and role based prompting for lay summarization of biomedical articles improving the clarity and factual consistency without requiring large scale supervised data.
3. We developed a retrieval-augmented summarization pipeline for radiology reports by storing medical concepts and definitions from the dataset and Wikipedia in ChromaDB, enabling definition retrieval to improve clarity for lay readers.

2 Related Work

Recent studies in biomedical text summarization have focused on making complex medical artciles to be easier to understand for general audience, especially because of the fast growing number of scientific articles. Researchers have explored both extractive methods and abstractive methods to create summaries that are easier for non-experts

BioNLP 2025 Shared Tasks, pages 269–274
August 1, 2025

to read. Traditional extractive techniques such as Lead-K, TextRank, and TF-IDF help pick the most important sentences from a text. Tools like SciSpacy (Neumann et al., 2019) and MedCAT (Kraljevic et al., 2021) are useful for identifying medical terms that might need simpler explanations in the summaries. New large language models (LLMs) such as Meta's LLaMA-3-Instruct (Touvron et al., 2023) and OpenAI's GPT-3.5/4 (Achiam et al., 2023) have shown strong abilities to write summaries using prompts, even without much extra training. Models that are fine-tuned to follow instructions produce better results and clearer outputs that matches with the result user wants. Tree-of-thought prompting (Yao et al., 2023) and chain-of-thought reasoning (Wei et al., 2022) have demonstrated improvements in factuality and coherence for complex text generation tasks, including medical content. In the context of radiology report translation, recent shared tasks (e.g., BioLaySumm, MEDIQA) and benchmark datasets such as MIMIC-CXR (Johnson et al., 2019) and PadChest(Bustos et al., 2020) have helped the development of models that generate layman friendly summary of professional reports. Prior studies (You et al., 2024) have used Retrieval-Augmented Generation (RAG) to supplement missing background knowledge and improve factual accuracy. To evaluate how good the summaries are, common tools include ROUGE, BLEU, METEOR, and BERTScore for relevance, as well as readability scores like FKGL and DCRS. To check if the summary facts match the source text, tools like AlignScore and SummaC are often used.

3 Methodology

For this biomedical articles summarization task, we used the PLOS and eLife dataset provided by the organizers as described in (Goldsack et al., 2022; Luo et al., 2022; Goldsack et al., 2023, 2024). The PLOS dataset is the larger of the two, comprising 24,773 training and 1,376 validation instances, while the eLife dataset contains 4,346 training and 241 validation instances. For the radiology report summarization task, we used the close track setup which includes Open-i, PadChest, BIMCV-COVID19, along with MIMIC-CXR (Zhao et al., 2025; Xiao et al., 2025).

3.1 Plain Lay Summarization

We used the Meta-LLaMA-3-8B-Instruct model to simplify medical texts into layperson-friendly summaries along with Tree of Thought algorithm(1). To ensure high-quality simplification, we designed a prompt that instructed the model to (i) shorten and simplify long sentences, (ii) replace complex medical terms with everyday language and (iii) keep the original meaning accurate. After generating the summaries, we cleaned them using regular expressions to remove extra characters, model tags, and formatting issues. This gave us a neat version of the simplified text. For each input, we asked the model to generate two versions (n = 2). Although creating more versions (like 3–5) can give better results, we chosen two to save time. We then picked the best one using a custom scoring formula called wrb(Readbility-Bertscore based): wrb = 0.55 × br_score + 0.45 × avg_read. Here, br_score is (1 – BERTScore) × 100, and avg_read is the average of two readability scores: Flesch-Kincaid Grade Level and Dale-Chall Readability Score. We used BERTScore to check how close the simplified summary was to the original abstract. After choosing the best summary in the first round, we fed it back into the model to generate more refined outputs. This was done for up to two rounds (m = 2). If the first summary already scored well (above a quality threshold of 12), we skipped the second round to save time.

3.2 Lay Summarisation with External Knowledge

We presented a layman summary system[Figure 1] designed for biomedical articles, using two different extractive–abstractive hybrid models. The first model used Meta's LLaMA-3-8B-Instruct, combining section-wise leading sentence extraction and medical term definitions from MedCAT. The second model used GPT-3.5-turbo with TF-IDF-based sentence selection and medical terms identified using SciSpacy. Both models used specially written prompts and the GPT model used a step-by-step reasoning prompt to improve accuracy and structure. In the first setup, each article from the BioLaySumm dataset is splitted into its usual sections (like Introduction, Methods, Results, Discussion) using newlines as markers. From each section, the first 10 sentences are picked (Lead-10 method), since these usually carry the main ideas. These sentences are joined with the abstract to make a

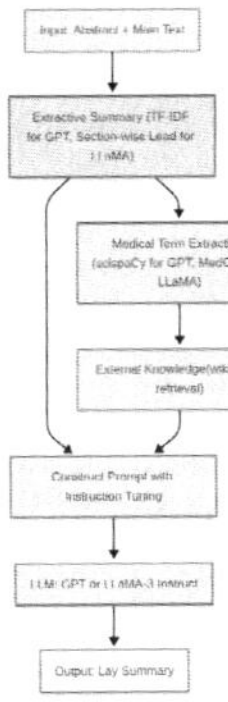

Figure 1: This flowchart illustrates the subtask 1.2 pipeline for biomedical lay summarization using external knowledge. Extracted summaries and abstract are combined with medical term definitions from sources like Wikipedia, and a LLaMA-3-Instruct model generates simplified summaries for the general public.

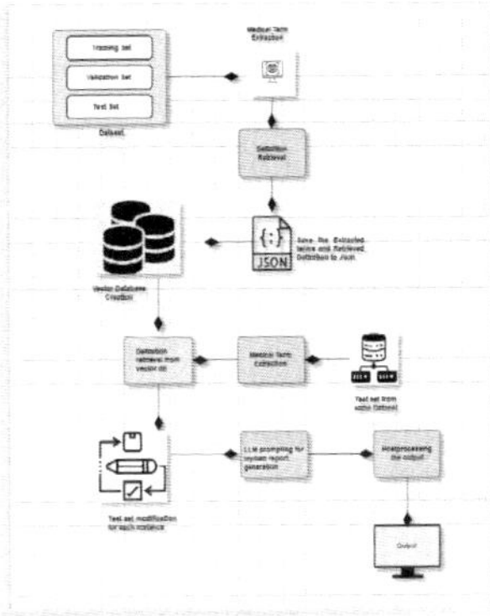

Figure 2: Pipeline for the radiology report translation

shorter input for the summarization model. To get the cleaned text, we preprocessed the text data by removing the contents in brackets like citations. To make the summary easier to understand for non-experts, we extract medical terms from the abstract and Lead-10 sebtences using MedCAT. This tool links terms to standard medical databases. We removed common English words using a dictionary and fetched simple definitions for up to 10 terms from Wikipedia. These definitions are added to the prompt to give extra background. The final prompt includes: The abstract, sentences picked from each section, up to 10 definitions of medical terms. The second setup uses GPT-3.5-turbo through OpenAI's API. Instead of section wise extractive summary, this setup used TF-IDF to pick the top 40 most important sentences from the full text. We then used SciSpacy model to find medical terms, based on known databases like UMLS and MeSH. Definitions are again retrieved from Wikipedia and added to the prompt. The GPT model uses a step-by-step reasoning prompt (Chain-of-Thought) to improve its output. This prompt asks the model to think through the abstract and sentences, explain hard terms, and then write a clear summary in simple words. This helps reduce errors and improves clarity.

3.3 Radiology Report Translation

To generate layman-friendly summaries from complex radiology reports, we developed a structured pipeline [Figure 2] integrating biomedical entity recognition, semantic definition retrieval, large language model prompting and post-processing. We began with the BioLaySumm/LaymanRRG-closed track dataset and used the full dataset to construct a comprehensive dictionary of medical definitions. Each report was processed using the SciSpacy model to extract medical terms and their definition was retrieved from Wikipedia using their API. We then encoded each definition using the BAAI/bge-base-en sentence transformer and indexed them in a ChromaDB vector database, facilitating fast and semantically meaningful retrieval. For inference, we focused on the test split of the dataset. Each test report underwent entity extraction, and the corresponding terms were matched with the most semantically relevant definitions from our vector database. These definitions, combined with the original report, were formatted into a context-rich prompt and passed to the zephyr-7b-beta model using Hugging Face's text-generation pipeline. The model, deployed in half-precision (FP16) on GPU, produced concise and coherent lay summaries. To ensure clarity and fluency, we applied a post-processing step to the generated summaries. This included removing redundant phrases, fixing formatting inconsistencies, and refining grammar and sentence flow. Each report's full pipeline output including the original report, extracted terms, definitions used, generated summary, and the final cleaned summary was saved in a CSV file. This allowed for incremental saving and supported recovery in case of interruptions, ensuring robustness and reproducibility of the process.

4 Result and Discussion

The performance of our submission were presented in the Table 1

Task	Model used	ROUGE	BLEU	METEOR	BERTScore	FKGL	DCRS	CLI	LENS(Task1)	AlignScore(Task1)	SummaC(Task1)
SubTask 1.2	GPT 3.5	0.2961816563	4.081113931	0.2281900631	0.8549233545	13.36619718	10.25193662	14.74401408	80.00152005	0.6890650382	0.5070385472
SubTask 1.2	LLaMA-3-8B-Instruct	0.2894793205	4.127686909	0.2682454109	0.8340255209	9.892605634	7.869929577	11.38640845	75.58860065	0.5547797396	0.737774146
SubTask 1.1	LLaMA-3-8B-Instruct with TOT	0.2681919676	3.24775969	0.2263706053	0.8484312852	10.52429577	8.835915493	11.43105634	84.14457219	0.5888629015	0.5489283793
SubTask 1.1	Preprocessed Abstract	0.3281246986	7.120012357	0.2833030102	0.8612545194	16.90774648	11.35848592	17.51320423	40.15254495	0.9937086519	0.9464885324
Task	**Model used**	**ROUGE**	**BLEU**	**METEOR**	**BERTScore**	**Similarity(Task2)**	**FKGL**	**DCRS**	**CLI**	**F1chexbert(Task2)**	**Radgraph(Task2)**
SubTask 2.1_close	zephyr-7b-beta	0.4038681644	14.89689754	0.427866722	0.9128268815	0.7975429296	7.358711512	8.527076782	7.360385788	0.7041964506	0.2162386679

Table 1: Evaluation results demonstrates the performance of our submission across Subtasks 1.1 and 1.2 of biomedical lay summarization, and Subtask 2.1 of radiology report translation.

4.1 Lay Summarisation with External Knowledge

We tested our system on 142 biomedical articles from the BioLaySumm2025-PLOS test set. Both models (LLaMA-3 with Lead-10 + MedCAT and GPT-3.5 with TF-IDF + SciSpacy + Chain-of-Thought) were run on the same dataset for comparison. We evaluated the results using relevance, readability and factuality metrics. GPT-3.5 performed better than LLaMA-3 in readability, factual correctness, and overall ROUGE-L scores, especially when it explained complex terms using step-by-step reasoning. In general, GPT-3.5 created summaries that explained key findings and terms more clearly for general readers. LLaMA-3 sometimes skipped important context. The combination of TF-IDF extraction with reasoning-based prompts worked especially well when extra background knowledge was needed for understanding.

4.2 Radiology Report Translation

In the radiology report translation task, we developed an approach using a retrieval-augmented generation (RAG) strategy with the Zephyr model to produce summaries that are not only accurate but also easier to understand for non-experts. The summaries generated through this method gave high relevance. The scores of readability metrics confirmed that the simplified texts were written at an accessible level, making them more understandable to the general public. From a clinical perspective, our approach maintained a good balance between simplifying the language and retaining important medical content.

5 Conclusion

In this work, we explored prompt-based summarization techniques to convert complex biomedical articles and radiology reports into simple, easy-to-understand summaries for general readers. We used a combination of extractive methods (such as Lead-10 and TF-IDF) and large language models like Meta-LLaMA-3-Instruct and GPT-3.5. For Subtask 1.1, we focused on section-wise summarization, while in Subtask 1.2, we added medical definitions retrieved using MedCAT and Wikipedia to enrich the knowledge gaps. In Subtask 2.1, we applied a Retrieval-Augmented Generation (RAG) approach with the Zephyr model to generate layman summaries from professional radiology reports. Our approach produced strong results across the shared task subtasks, showing the effectiveness of combining external knowledge, extractive summarization, and instruction-tuned language models. Even though our system produced good results, it still has has some limitations. First, the quality of extracted summaries using Lead or TF-IDF depends on the structure of the original article. If the article is not organized properly, important information might be missed. Second, retrieving accurate definitions from Wikipedia or other public sources may introduce inconsistencies, not having clear explanations or no explanations. Each and every retrieval of wikipedia definitions takes too much time and that limited the medical definition to only 10 terms. Finally, while chain-of-thought prompting improved factuality in GPT-based generation, it occasionally produced longer or slightly off-topic outputs when trying with llama that requires further refinement. In future work, we plan to improve summarization by fine-tuning the models instead of prompt tuning on more diverse medical datasets. We will also be focusing on controllable summarization and exploring the ways to get definitions from other medical resources that can increase the terms count and layman definitions. We want to incorporate user feedback mechanisms to assess how helpful the generated summaries are for real patients and the general public.

6 Declaration of AI usage

We used generative AI tools like chatGPT for paraphrasing, grammar checking while writing this article. After using this tool, the author(s) reviewed and edited the content as needed and take(s) full responsibility for the content of the published article.

References

Josh Achiam, Steven Adler, Sandhini Agarwal, Lama Ahmad, Ilge Akkaya, Florencia Leoni Aleman, Diogo Almeida, Janko Altenschmidt, Sam Altman, Shyamal Anadkat, and 1 others. 2023. Gpt-4 technical report. *arXiv preprint arXiv:2303.08774*.

Aurelia Bustos, Antonio Pertusa, Jose-Maria Salinas, and Maria De La Iglesia-Vaya. 2020. Padchest: A large chest x-ray image dataset with multi-label annotated reports. *Medical image analysis*, 66:101797.

Marcio Fonseca and Shay B Cohen. 2024. Can large language model summarizers adapt to diverse scientific communication goals? *arXiv preprint arXiv:2401.10415*.

Tomas Goldsack, Zheheng Luo, Qianqian Xie, Carolina Scarton, Matthew Shardlow, Sophia Ananiadou, and Chenghua Lin. 2023. Overview of the biolaysumm 2023 shared task on lay summarization of biomedical research articles. In *The 22nd Workshop on Biomedical Natural Language Processing and BioNLP Shared Tasks*, pages 468–477.

Tomas Goldsack, Carolina Scarton, Matthew Shardlow, and Chenghua Lin. 2024. Overview of the BioLaySumm 2024 shared task on the lay summarization of biomedical research articles. In *Proceedings of the 23rd Workshop on Biomedical Natural Language Processing*.

Tomas Goldsack, Zhihao Zhang, Chenghua Lin, and Carolina Scarton. 2022. Making science simple: Corpora for the lay summarisation of scientific literature. In *Proceedings of the 2022 Conference on Empirical Methods in Natural Language Processing*, pages 10589–10604, Abu Dhabi, United Arab Emirates. Association for Computational Linguistics.

Alistair EW Johnson, Tom J Pollard, Nathaniel R Greenbaum, Matthew P Lungren, Chih-ying Deng, Yifan Peng, Zhiyong Lu, Roger G Mark, Seth J Berkowitz, and Steven Horng. 2019. Mimic-cxr-jpg, a large publicly available database of labeled chest radiographs. *arXiv preprint arXiv:1901.07042*.

Zeljko Kraljevic, Thomas Searle, Anthony Shek, Lukasz Roguski, Kawsar Noor, Daniel Bean, Aurelie Mascio, Leilei Zhu, Amos A Folarin, Angus Roberts, and 1 others. 2021. Multi-domain clinical natural language processing with medcat: the medical concept annotation toolkit. *Artificial intelligence in medicine*, 117:102083.

Zheheng Luo, Qianqian Xie, and Sophia Ananiadou. 2022. Readability controllable biomedical document summarization. In *Findings of the Association for Computational Linguistics: EMNLP 2022*, pages 4667–4680, Abu Dhabi, United Arab Emirates. Association for Computational Linguistics.

Mark Neumann, Daniel King, Iz Beltagy, and Waleed Ammar. 2019. Scispacy: fast and robust models for biomedical natural language processing. *arXiv preprint arXiv:1902.07669*.

Amara Tariq, Sam Fathizadeh, Gokul Ramaswamy, Shubham Trivedi, Aisha Urooj, Nelly Tan, Matthew T Stib, Bhavik N Patel, and Imon Banerjee. 2024. Patient centric summarization of radiology findings using large language models. *medRxiv*, pages 2024–02.

Amara Tariq, Shubham Trivedi, Aisha Urooj, Gokul Ramasamy, Sam Fathizadeh, Matthew Stib, Nelly Tan, Bhavik Patel, and Imon Banerjee. Patient-centric summarization of radiology findings using two-step training of large language models. *ACM Transactions on Computing for Healthcare*.

Hugo Touvron, Thibaut Lavril, Gautier Izacard, Xavier Martinet, Marie-Anne Lachaux, Timothée Lacroix, Baptiste Rozière, Naman Goyal, Eric Hambro, Faisal Azhar, and 1 others. 2023. Llama: Open and efficient foundation language models. *arXiv preprint arXiv:2302.13971*.

Jason Wei, Xuezhi Wang, Dale Schuurmans, Maarten Bosma, Fei Xia, Ed Chi, Quoc V Le, Denny Zhou, and 1 others. 2022. Chain-of-thought prompting elicits reasoning in large language models. *Advances in neural information processing systems*, 35:24824–24837.

Chenghao Xiao, Kun Zhao, Xiao Wang, Siwei Wu, Sixing Yan, Sophia Ananiadou, Noura Al Moubayed, Liang Zhan, William Cheung, and Chenghua Lin. 2025. Overview of the biolaysumm 2025 shared task on lay summarization of biomedical research articles and radiology reports. In *The 24nd Workshop on Biomedical Natural Language Processing and BioNLP Shared Tasks*, Vienna, Austria. Association for Computational Linguistics.

Shunyu Yao, Dian Yu, Jeffrey Zhao, Izhak Shafran, Tom Griffiths, Yuan Cao, and Karthik Narasimhan. 2023. Tree of thoughts: Deliberate problem solving with large language models. *Advances in neural information processing systems*, 36:11809–11822.

Zhiwen You, Shruthan Radhakrishna, Shufan Ming, and Halil Kilicoglu. 2024. UIUC_BioNLP at BioLaySumm: An extract-then-summarize approach augmented with Wikipedia knowledge for biomedical lay summarization. In *Proceedings of the 23rd Workshop on Biomedical Natural Language Processing*, pages 132–143, Bangkok, Thailand. Association for Computational Linguistics.

Kun Zhao, Chenghao Xiao, Sixing Yan, William K. Cheung, Kai Ye, Noura Al Moubayed, Liang Zhan, and Chenghua Lin. 2025. X-ray made simple: Lay radiology report generation and robust evaluation. *Preprint*, arXiv:2406.17911.

A Appendix A: Algorithm

Algorithm 1 Tree of Though Based Iterative Text Simplification with Readability and Content Preservation

```
Input: Abstract
max_rounds ← 2
branches_per_round ← 2
current_text ← complex_text
for round_num = 0 to max_rounds - 1 do
    candidates ← [ ]
    for i = 0 to branches_per_round - 1 do
        generated ← SIMPLIFY_TEXT_WITH_LLAMA(current_text)
        simplified ← CLEAN_SIMPLIFIED_OUTPUT(generated)
        avg_read ← READABILITY_SCORES(simplified)
        input_fc ← BERTSCORE(simplified, complex_text)    ▷ or AlignScore / SummaC
        br_score ← (1 − input_fc) × 100
        wrb ← 0.55 × br_score + 0.45 × avg_read
        candidates.append([simplified, wrb, readability, input_fc])
    end for
    best_candidate ← candidate with lowest wrb
    if best_candidate.wrb ≤ target_grade then
        return best_candidate.simplified
    else
        current_text ← best_candidate.simplified
    end if
end for
return current_text
```

Team XSZ at BioLaySumm2025: Section-Wise Summarization, Retrieval-Augmented LLM, and Reinforcement Learning Fine-Tuning for Lay Summaries

Pengcheng Xu
Shanghai Jiao Tong University
xu_pengcheng@sjtu.edu.cn

Sicheng Shen
University of Michigan
demodemo@umich.edu

Jieli Zhou
Shanghai Jiao Tong University
zhoujieli@sjtu.edu.cn

Hongyi Xin
Shanghai Jiao Tong University
hongyi.xin@sjtu.edu.cn

Correspondence: hongyi.xin@sjtu.edu.cn

Abstract

We present a multi-stage pipeline for BioLaySumm 2025 Subtask 1.1 that improves readability, relevance, and factuality. First, we select the top-5 relevant sections and generate summaries with BioBART. Next, we retrieve a K-shot demonstration using BGE embeddings to prompt Llama 3 8B and fine-tune it with LoRA. We then merge section summaries via a second BioBART pass. Finally, we apply reinforcement learning (PPO and GRPO) with a composite reward combining factuality (AlignScore, SummaC), relevance (ROUGE-L, BERTScore), and readability (LENS, FKGL, DCRS, CLI). On PLOS and eLife validation sets, our pipeline reduces DCRS from 9.23 to 8.56 and CLI from 12.98 to 12.65, and boosts AlignScore from 0.722 to 0.862, demonstrating balanced gains in lay-summary quality.

1 Introduction

Biomedical articles are rife with technical jargon and complex discourse that hinder comprehension by non-specialist readers (Goldsack et al., 2023). Lay summaries—concise paraphrases in accessible language—play a critical role in democratizing scientific knowledge for patients, policy-makers, and the general public. The BioLaySumm shared task (ACL 2023–2025) has steadily advanced methodologies for abstractive biomedical summarization, evolving from pure encoder–decoder models to modern large language model (LLM)–based systems with controllable generation capabilities (**?**).

Recent years have seen three major trends in lay summarization: (1) *Section-wise summarization*, which breaks long articles into manageable chunks (Zhang and Roberts, 2021), (2) *Few-shot prompting* of LLMs to leverage in-context learning without full fine-tuning (Dong et al., 2022), and (3) *Reinforcement learning (RL)* to directly optimize non-differentiable metrics such as readability indices and factuality scores (Kryscinski et al., 2020; Henderson et al., 2022). Parallel advances in parameter-efficient adaptations—LoRA (Hu et al., 2021) and adapters (Pfeiffer et al., 2020)—have made LLM fine-tuning practical under compute constraints.

In this work, we integrate these strands into a cohesive pipeline: structured section selection, BioBART summarization, Llama 3 8B prompting with K-shot retrieval, LoRA adaptation, summary merging, and final RL-based fine-tuning. Our contributions are:

- A detailed, modular architecture that combines supervised and RL stages to address readability, relevance, and factuality.
- A retrieval-augmented K-shot prompting strategy using BGE embeddings for demonstration selection.
- An RL fine-tuning regimen employing both PPO and the lightweight GRPO algorithm with a multi-component reward aligned to shared task criteria.
- Empirical validation on PLOS and eLife showing significant improvements in readability indices (e.g., DCRS ↓0.67), CLI ↓0.33, and factuality (AlignScore ↑0.14).

2 Related Work

2.1 Biomedical Lay Summarization

Biomedical lay summarization focuses on translating complex scientific content into language that

BioNLP 2025 Shared Tasks, pages 275–280
August 1, 2025 ©2025 Association for Computational Linguistics

is understandable to non-expert audiences. Early approaches to this task leveraged encoder–decoder architectures such as BART and BioBART, fine-tuned on biomedical literature (Beltagy et al., 2020; GanjinZero, 2023). These models demonstrated promising results on short texts but struggled with full-length documents. To address this, section-level summarization strategies were introduced, which broke down scientific articles into segments and generated summaries for each part (Cohan et al., 2020). Recent developments have led to benchmark efforts such as BioLaySumm, which provide standardized evaluation settings to advance the generation of accessible biomedical summaries.

2.2 Prompting and Few-Shot LLMs

In-context learning with large language models such as GPT-3 and LLaMA variants has shown that providing carefully selected task demonstrations within the input prompt can enable strong performance on new tasks without the need for additional fine-tuning (Brown et al., 2020). Retrieval-augmented generation (RAG) enhances language model outputs by incorporating relevant external knowledge retrieved from a large corpus (Lewis et al., 2020). RAG systems improve factual accuracy and adaptability, addressing limitations in static model parameters.

2.3 Parameter-Efficient Fine-Tuning

Parameter-efficient fine-tuning methods such as LoRA (Hu et al., 2021) and AdapterFusion (Pfeiffer et al., 2020) have emerged as effective strategies for adapting large pretrained models to new tasks while reducing the number of trainable parameters. These approaches introduce small, trainable modules to integrate into the model's architecture. Recent work has demonstrated the effectiveness of these techniques in domain-specialized summarization, particularly in biomedical settings (Pakull et al., 2024).

2.4 Reinforcement Learning

Reinforcement learning (RL) has been widely adopted in text generation tasks to optimize ROUGE scores (Rennie et al., 2017), factual consistency (Kryscinski et al., 2020), and controllable text attributes like simplicity and politeness (Liu et al., 2022). Among various RL algorithms, Proximal Policy Optimization (PPO) has gained popularity for its stability during fine-tuning (Schulman et al., 2017). More recently, GRPO has been introduced as a memory-efficient alternative that eliminates the need for a separate critic network by grouping and scoring sampled outputs together, halving memory usage while maintaining competitive performance (Stooke and Abbeel, 2021).

2.5 BioLaySumm2024

In the previous iteration of BioLaySumm, Goldsack et al. provided an overview of the 2023 competition (Goldsack et al., 2023), and in 2024 they extended this with an in-depth summary of that year's results and tasks (Goldsack et al., 2024). Top teams found that while direct prompting of LLMs improves readability, it may reduce factual accuracy and relevance. To address this, several adaptation techniques were incorporated—including title infusion, K-shot prompting, LLM rewriting, and instruction fine-tuning—that effectively balance these quality aspects and secured first place in readability at the 2024 BioLaySumm competition.

2.6 BioLaySumm2025

Xiao et al. present an overview of the 2025 shared task, which now also includes radiology-report summarization in addition to standard biomedical articles (Xiao et al., 2025). They highlight how the community moved toward more retrieval-augmented pipelines and multi-objective optimization for readability and factuality.

3 Problem Formulation

Given article $x = (x_1, \ldots, x_n)$ and reference lay summary $y = (y_1, \ldots, y_m)$, we learn f_θ to maximize the conditional log-likelihood:

$$\theta^* = \arg\max_\theta \sum_{t=1}^{m} \log p_\theta(y_t \mid y_{<t}, x).$$

4 Method

4.1 Section Selection

To systematically assess relevance, we parse each article into J distinct structural sections denoted as s_j, where $j \in \{1, \ldots, J\}$ (e.g., *Abstract*, *Introduction*, *Methods*, *Results*, *Discussion*). Each section s_j is encoded into a high-dimensional vector representation using a pre-trained sentence-transformer model. We then compute the cosine similarity between each section's embedding and a predefined domain-specific query that captures the target domain, quantifying how relevant each section's content is. After computing similarity scores for all J

sections, we rank them in descending order. Finally, we select the top $J' = 5$ sections with the highest similarity scores as the most domain-relevant content for downstream processing.

4.2 Section-Wise Summarization

Each selected section s_j of the biomedical document is summarized independently. Specifically, the summarization for each section is performed by applying the BioBART model, denoted as

$$z_j = \text{BioBART}(s_j; \phi)\,,$$

where ϕ represents the set of model parameters of BioBART-v2-base that have been fine-tuned on the training fold of the dataset. By summarizing sections individually, this approach mitigates the challenges posed by input length limitations of transformer-based models and allows the model to focus on the unique content and semantic structure of each section.

4.3 K-Shot Demonstration Retrieval

For a given test article x^*, we first compute its embedding using the BGE M3 encoder, denoted as $e_{\text{BGE}}(x^*)$. To leverage relevant contextual information, we retrieve the single most similar training instance (x_i, y_i) by finding the training example whose embedding has the highest cosine similarity:

$$i^* = \arg\max_i \cos\bigl(e_{\text{BGE}}(x^*), e_{\text{BGE}}(x_i)\bigr).$$

The retrieved pair $\mathcal{D}_1 = (x_{i^*}, y_{i^*})$ is then served as the input prompt of the large language model (LLM) to provide an example demonstration for in-context learning.

4.4 LoRA Fine-Tuning

We inject adapters into the LLaMA 3 8B model to enable parameter-efficient fine-tuning. Specifically, for each weight matrix $W \in R^{d\times k}$ within the model, we learn a low-rank update defined as

$$W' = W + AB,$$

where $A \in R^{d\times r}$ and $B \in R^{r\times k}$ are trainable matrices with a small rank $r = 8$. This low-rank decomposition significantly reduces the number of parameters that must be updated during training. We train the adapter parameters for 3 epochs using a learning rate of 5×10^{-5} and a batch size of 16. Early stopping based on performance on the validation fold is employed to prevent overfitting and to select the best-performing model checkpoint.

4.5 Summary Merging

After independently summarizing each selected section to obtain the set of partial summaries $\{z_j\}_{j=1}^{J'}$, we concatenate them into a single combined representation Z. This concatenated input serves as the basis for a second pass through the BioBART model, expressed as

$$\hat{y} = \text{BioBART}(Z; \phi'),$$

where ϕ' denotes the parameters of BioBART fine-tuned specifically for this second-stage summarization task. By leveraging this two-step process, the approach addresses the challenges posed by lengthy biomedical texts while improving the consistency and readability of the final output.

4.6 Reinforcement Learning Fine-Tuning

After completing the supervised training stages, we further refine the model using reinforcement learning (RL) to directly optimize multiple quality metrics. For each input, we generate $m = 4$ candidate summaries and compute a composite reward R that balances several evaluation metrics:

$$\begin{aligned} R = &\underbrace{\text{AlignScore} + \text{SummaC}}_{\text{factual}} + \\ &\underbrace{\text{ROUGE-L} + \text{BERTScore}}_{\text{relevance}} + \\ &\underbrace{\text{LENS} - \alpha(\text{FKGL} + \text{DCRS} + \text{CLI})}_{\text{readability}}. \end{aligned}$$

Each individual metric score is normalized to the range $[0, 1]$ via min-max scaling based on the train-validation distributions, ensuring balanced contributions across diverse metrics. We perform RL fine-tuning using two algorithms: Proximal Policy Optimization (PPO) with clipping parameter $\epsilon = 0.2$ and KL-penalty coefficient $\beta = 0.1$, and Grouped Reward Policy Optimization (GRPO) with a group size of 4. Both methods are run for one epoch over the training set with a learning rate of 1×10^{-6}.

4.7 Pseudo-Code

Here is our RL implementation:

5 System Architecture

The figure 1 shows the flow chart of our method.

Algorithm 1: Two-stage training: LoRA fine-tuning followed by RL optimization

Input: Training corpus $\mathcal{D}_{\text{train}}$
Output: Fine-tuned parameters $\boldsymbol{\theta}$

for each *article* $x \in \mathcal{D}_{train}$
  $S \leftarrow$ `select_sections`(x) `// top 5 relevant sections`
  $z \leftarrow [\text{BioBART}(s) \mid s \in S]$ `// latent embeddings`
  prompt $\leftarrow$ `retrieve_demo`$(x) \,\|\,$ `concat`(z)
  $\boldsymbol{\theta} \leftarrow$ `LoRA_finetune`$(\boldsymbol{\theta}, \textit{prompt}, y_{\text{ref}})$

for each *article* $x \in \mathcal{D}_{train}$
  $\{\hat{y}^{(i)}\}_{i=1}^{m} \leftarrow$ `generate`$(x, m, \boldsymbol{\theta})$
  $\mathcal{R} \leftarrow \{\, R(\hat{y}^{(i)})\}_{i=1}^{m}$ `// compute rewards`
  $\boldsymbol{\theta} \leftarrow$ `update_rl`$(\boldsymbol{\theta}, \mathcal{R}, \text{PPO/GRPO})$

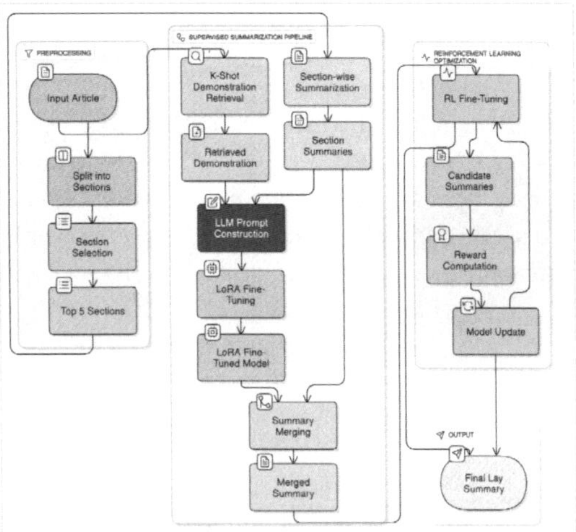

Figure 1: Overview of our method

6 Experiments

6.1 Setup

Datasets We evaluate on PLOS (24,773 train/1,376 val) and eLife (4,346 train/241 val) as per Goldsack et al. (2022).

Metrics We report relevance (ROUGE-1/2/L, BERTScore), readability (FKGL, DCRS, CLI, LENS), and factuality (AlignScore, SummaC) using the shared task evaluation scripts (Goldsack et al., 2024).

Baselines We compare against Few-shot Llama3-8B and BioBART-only, and Baseline-qwen2.5-7B-sft, plus our supervised pipeline without RL ("Ours (no RL)").

	FKGL↓	DCRS↓	CLI↓	LENS↑
Baseline Llama 3 8B	**12.21**	9.23	12.98	**72.86**
Baseline-qwen2.5-7B-sft	12.71	9.65	13.70	60.22
Ours (method1: Llama3 ft)	12.59	**8.56**	**12.65**	63.22

Table 1: Readability on test set (↓ better except LENS↑).

	AlignScore↑	SummaC↑
Baseline Llama 3 8B	0.722	**0.644**
Baseline-qwen2.5-7B-sft	0.754	**0.644**
Ours (method2: section_sum + BioBART)	**0.862**	0.528

Table 2: Factuality on test set (↑ better).

	FKGL↓	DCRS↓	CLI↓	LENS↑
Baseline Llama 3 8B	12.21	9.23	12.98	72.86
Ours (no RL)	12.59	8.56	12.65	63.22
Ours + RL	**11.78**	**8.32**	**12.40**	**74.71**

Table 3: Readability on validation set (↓ better except LENS↑).

	AlignScore↑	SummaC↑
Baseline Llama 3 8B	0.722	**0.644**
Ours (no RL)	0.862	0.528
Ours + RL	**0.891**	0.613

Table 4: Factuality on validation set (↑ better).

6.2 End-to-End Performance

Table 1 shows readability improvements: our fine-tuned Llama3 without RL (method1: Llama3 fine-tune) reduces DCRS from 9.23 to 8.56 and CLI from 12.98 to 12.65. This method obtains high readability, ranking top 3 among all teams this year. Table 2 reports factuality of our method2: section-wise summarization + BioBART: our system boosts AlignScore from 0.722 to 0.862 and maintains high AlignScore. This method reaches top 5 in factuality among all teams this year.

6.3 Ablation Study: Impact of RL Fine-Tuning

Table 3 and Table 4 quantify gains from RL: it further reduces FKGL by 0.81 points and increases LENS by 11.49, while factuality AlignScore improves from 0.862 to 0.891 and SummaC from 0.528 to 0.613.

7 Conclusion

We present a comprehensive pipeline that systematically improves biomedical lay summaries through section-wise summarization, retrieval-augmented prompting, LoRA fine-tuning, and RL fine-tuning. Experimental results and ablations confirm balanced gains in readability, relevance, and factuality over both baselines and leading LLMs.

References

Iz Beltagy, Matthew E Peters, and Arman Cohan. 2020. Longformer: The long-document transformer. *arXiv*.

Tom B Brown, Benjamin Mann, Nick Ryder, Melanie Subbiah, Jared Kaplan, Prafulla Dhariwal, and 1 others. 2020. Language models are few-shot learners. *Advances in Neural Information Processing Systems*, 33:1877–1901.

Arman Cohan, Iz Beltagy, and Kyle Lo. 2020. Long document summarization by coarse-to-fine gated self-attention. In *Proceedings of the 58th Annual Meeting of the ACL*, pages 2318–2330. ACL.

Qingxiu Dong, Lei Li, Damai Dai, Ce Zheng, Zhiyong Wu, Baobao Chang, Xu Sun, Jingjing Xu, and Zhifang Sui. 2022. A survey on in-context learning. In *arXiv*.

GanjinZero. 2023. Biobartv2: Domain-adaptive pre-training of a biomedical generative language model. *GitHub repo*.

Tomas Goldsack, Zheheng Luo, Qianqian Xie, Carolina Scarton, Matthew Shardlow, Sophia Ananiadou, and Chenghua Lin. 2023. Overview of the biolaysumm 2023 shared task on lay summarization of biomedical research articles. In *The 22nd Workshop on Biomedical Natural Language Processing and BioNLP Shared Tasks*, pages 468–477.

Tomas Goldsack, Carolina Scarton, Matthew Shardlow, and Chenghua Lin. 2024. Overview of the biolaysumm 2024 shared task on lay summarization of biomedical research articles. In *23rd Workshop on Biomedical NLP and BioNLP Shared Tasks*, Bangkok, Thailand. ACL.

Tomas Goldsack, Zhihao Zhang, Chenghua Lin, and Carolina Scarton. 2022. Making science simple: Corpora for the lay summarisation of scientific literature. In *EMNLP*, Abu Dhabi, UAE. ACL.

Peter Henderson, Ananya Ramaswamy, Xue Chen, and 1 others. 2022. A survey of reinforcement learning methods for controllable text generation. *arXiv preprint arXiv:2203.00001*.

Edward J Hu, Yelong Shen, Phillip Wallis, Zeyuan Allen-Zhu, Yuanzhi Li, Shean Wang, Lu Wang, and Weizhu Chen. 2021. Lora: Low-rank adaptation of large language models. *arXiv*.

Wojciech Kryscinski, Bryan McCann, Caiming Xiong, and Richard Socher. 2020. Evaluating the factual consistency of abstractive text summarization. In *Proceedings of the 58th Annual Meeting of the ACL*, pages 9336–9349. ACL.

Patrick Lewis, Ethan Perez, Aleksandra Piktus, Vladimir Karpukhin, Naman Goyal, and 1 others. 2020. Retrieval-augmented generation for knowledge-intensive nlp tasks. In *Advances in Neural Information Processing Systems*, volume 33, pages 9459–9474.

Yang Liu, Ani Nenkova, and Kathleen McKeown. 2022. Incorporating controllability in neural abstractive summarization. In *Proceedings of the 60th Annual Meeting of the ACL*, pages 876–890. ACL.

Tabea M. G. Pakull, Hendrik Damm, Ahmad Idrissi-Yaghir, and et al. 2024. Wispermed at biolaysumm: Adapting autoregressive llms for lay summarization. *arXiv*.

Jonas Pfeiffer, Julia Pfeiffer, Timo Schick, Daniel Dahlmeier, and Sebastian Ruder. 2020. Adapterfusion: Non-destructive task composition for transfer learning. In *Proceedings of the 5th Workshop on Representation Learning for NLP*, pages 91–107. ACL.

Steven J Rennie, Paul Marcheret, Y-Lan Mroueh, Jerret Ross, and Vinay Goel. 2017. Self-critical sequence training for image captioning. *Proceedings of the IEEE Conference on Computer Vision and Pattern Recognition*, pages 7008–7024.

John Schulman, Filip Wolski, Prafulla Dhariwal, Alec Radford, and Oleg Klimov. 2017. Proximal policy optimization algorithms. In *Proceedings of the 35th International Conference on Machine Learning*, pages 1920–1928. PMLR.

Adam Stooke and Pieter Abbeel. 2021. Group proximal policy optimization. In *Advances in Neural Information Processing Systems*, volume 34, pages 14825–14835.

Chenghao Xiao, Kun Zhao, Xiao Wang, Siwei Wu, Sixing Yan, Sophia Ananiadou, Noura Al Moubayed, Liang Zhan, William Cheung, and Chenghua Lin. 2025. Overview of the biolaysumm 2025 shared task on lay summarization of biomedical research articles and radiology reports. In *24th Workshop on Biomedical NLP and BioNLP Shared Tasks*, Vienna, Austria. ACL.

Eva Zhang and Adam Roberts. 2021. Section-level summarization of scientific documents with discourse-aware encoding. In *Proceedings of the 2021 Conference on Empirical Methods in NLP*, pages 1234–1245. ACL.

A Implementation Details

All training stages are implemented using the HuggingFace Transformers framework and executed on a cluster of 8 NVIDIA A100 GPUs. During supervised fine-tuning, we use a batch size of 16 to maximize GPU utilization, while for reinforcement learning stages, the batch size is reduced to 8 to accommodate the additional computational overhead incurred by sampling multiple outputs per input. We plan to release all code and configuration files publicly upon acceptance to facilitate reproducibility and further research.

VeReaFine: Iterative Verification Reasoning Refinement RAG for Hallucination-Resistant on Open-Ended Clinical QA

Pakawat Phasook[1,†] **Rapepong Pitijaroonpong**[1,†]
Jiramet Kinchagawat[2] Amrest Chinkamol[2] Kiartnarin Udomlapsakul[2]
Tossaporn Saengja[2] Jitkapat Sawatphol[2,*] Piyalitt Ittichaiwong[2,*]
[1]KMUTT [2]PreceptorAI Tech

Abstract

Large language models (LLMs) can generate medical responses, but they often "hallucinate" unsupported or incorrect clinical assertions, risking patient safety and trust. We introduce **VeReaFine**, a "Verifier-RAG" pipeline, an iterative fact-checking – retrieval process: (1) Given a medical query, we fetch the top-k passages from a large biomedical corpus (e.g., PubMed, StatPearls) using a two-stage dense retriever and reranker, (2) employ a small LLM verifier to extract a concise "ground-truth" context from the retrieved data, (3) dynamically issue up to three targeted retrieval queries whenever evidence is lacking, (4) draft an answer with a 7-B generator grounded solely in groundtruth context, and (5) re-verify and refine the Generator LLM response to purge any remaining hallucinations. By iteratively fetching only the missing facts, VeReaFine ensures that every generated response is grounded, yielding performance uplifts with minimal extra cost. On the BioNLP 2025 ClinIQLink "LLM Lie-Detector" challenge, our 7-B generator augmented with VeReaFine rivals or surpasses a 32-B medical model on open-ended reasoning, reduces multi-hop inverse step-identification errors by 26%. These results demonstrate that moderate-size LLMs and our proposed pipeline can improve the result in open-ended Question Answering in clinical QA.

1 Introduction

Open-ended question answering in medical domain demands two aspects in answers: coherence and factuality. Large language models (LLMs) are usually coherent and able to produce human-like responses, but common issues found in their responses are *hallucinations*. Hallucinated responses can look convincing while misrepresenting clinical facts, which compromise patient safety and clinical decision-making (Maynez et al., 2020). Existing studies have proposed several strategies to help reduce the hallucination issue. Retrieval-Augmented Generation (RAG) mitigates some of these risks by providing relevant documents to the model, yet it cannot ensure that the LLM correctly incorporates all retrieved facts or refrains from utilizing incorrect contextual information (Lewis et al., 2020b). Chain-of-Thought (CoT) prompting results in intermediate reasoning text and improves multi-step problem-solving (Zhang et al., 2022), but remains vulnerable when its internal knowledge is incomplete or outdated (Madaan et al., 2023). Likewise, self-verification approaches - where the model critiques its own outputs help post-hoc error detection but lack systematic integration of external evidence, limiting their efficacy in specialized domains such as medicine (Dhuliawala et al., 2023; Manakul et al., 2023).

One key driver of hallucinations in medical LLMs is simply a shortage of domain knowledge: if the model's internal parameters don't "know" enough about specific drugs, anatomy, or clinical guidelines, it will confidently fabricate plausible-sounding—but wrong—information (e.g., see M1-32B's analysis in (UCSC-VLAA, 2024; Huang et al., 2025; UCSC-VLAA, 2024)). A naive RAG approach attempts to compensate by overloading the generator with large bundles of retrieved text, but this often backfires: too much loosely related information can confuse the LLM, leading it to latch onto irrelevant or outdated facts. Prior work has tried three main remedies—pure RAG grounding, chain-of-thought prompting, and self-verification loops—but none simultaneously guarantees that (a) the generator truly receives "just enough" high-precision medical context, and (b) each claim is checked against external evidence before being

[†]Equal Contributions
[*]Corresponding Authors

BioNLP 2025 Shared Tasks, pages 281–288

emitted. To address these challenges, we introduce **VeReaFine**, a "Verifier-RAG" pipeline that alternates between retrieval, verification, and collection medical groundtruth in up to three attempts (Figure 1). At each iteration, VeReaFine performs:

1. **Query-Driven Retrieval.** Embed the input question and retrieve top k biomedical passages from a curated corpus (PubMed abstracts (U.S. National Library of Medicine, 2023, 2024), StatPearls (MedRAG Team, 2024; StatPearls Publishing, 2024) etc.) using BM-Retriever-410M(Hugging Face, 2024b; Xu et al., 2024), then rerank them with a BM-Retriever-2B(Hugging Face, 2024a).
2. **Relevance Verification.** Use an 8B medical reasoning verifier (MedReason-8B(Hugging Face, 2024c; Wu et al., 2025)) to assess direct relevancy of each retrieved passage to the question. Passages deemed germane are marked as the "ground-truth" context; irrelevant ones are discarded.
3. **Adaptive Context Expansion.** If the current ground-truth set is insufficient to answer the query, the verifier formulates a focused "feedback query" identifying exactly what evidence is missing. This feedback drives another retrieval round. We repeat this at maximum of three iterations.
4. **Answer Generation.** Condition a 7B generator (Qwen2.5-7B-Instruct) on the final ground-truth context to draft an answer free of unsupported facts (Qwen Team, 2025, 2024; Yang et al., 2024).
5. **Answer Re-Verification & Refinement.** The verifier re-checks the generated draft against the ground-truth context. If any residual hallucinations are flagged, the generator is prompted to refine and/or excise those hallucinated claims. This final pass ensures every assertion is evidence-backed.

By fusing targeted retrieval with in-loop verification and refinement, VeReaFine guarantees that each claim in the answer is sanctioned by the curated biomedical evidence.

We evaluate VeReaFine on the BioNLP 2025 ClinIQLink "LLM Lie-Detector" shared task(BioNLP Shared Task Organizers, 2025), focusing on open-ended formats–short answer, short-inverse, multi-hop, and multi-hop-inverse–where hallucinations are most prevalent. Our experiments show that, despite using a moderate-size 7B generator, VeReaFine achieves recall gains of +60–100% at the 75th percentile (P75) over the same model without verification, and recovers over 90% of the step-identification fidelity of a 32B baseline (Sub2: M1-32B (UCSC-VLAA, 2024; Huang et al., 2025; UCSC-VLAA, 2024)) in multi-hop inverse questions. These results highlight that carefully orchestrated retrieval and verification can allow smaller models to match or surpass much larger ones in clinical factuality.

1.1 Our Contributions

VeReaFine advances open-ended medical QA by embedding an explicit verifier into every stage of the RAG cycle. Specifically, we contribute:

1. **Tri-Loop Verifier-RAG Architecture.** We introduce a tightly integrated three-stage feedback loop (Figure 1) whereby:
    - *Retrieval:* A bi-encoder (BM-Retriever-410M) retrieves top-k passages, which are then precisely ranked by a cross-encoder (BM-Retriever-2B(Hugging Face, 2024a) (Karpukhin et al., 2020).
    - *Verification:* An 8B medical reasoning model (MedReason-8B(Hugging Face, 2024c)) examines each passage for relevance and sufficiency, discarding irrelevant snippets and—when evidence is lacking—issuing focused "feedback queries" to retrieve missing context.
    - *Generation:* A 7B LLM (Qwen2.5-7B-Instruct) produces the final answer conditioned only on the fully vetted "ground-truth" context (Yang et al., 2024).

    By interleaving verification with both retrieval and generation, every claim in the output is explicitly sanctioned by external evidence.
2. **Iterative Verification Refinement.** We cast VeReaFine's operation as an Expectation–Maximization analogue:
    - *Verification step:* The verifier extracts constraints by flagging unsupported assertions in the draft answer.
    - *Refinement step:* The generator revises the answer to satisfy those constraints, thereby increasing evidence alignment with ground truth.

    We show that, assuming a verifier with non-negative correction fidelity, each iteration cannot reduce the system's overall factuality score.
3. **Improve performance Open-Ended QA with Modest Models.** On the BioNLP 2025 ClinIQLink "LLM Lie-Detector" shared task,

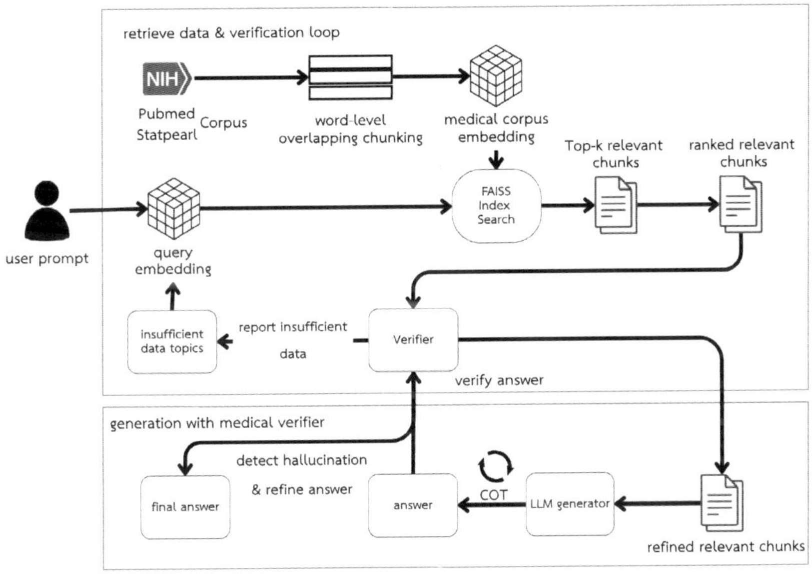

Figure 1: The VeReaFine tri-loop pipeline.

VeReaFine's 7B-parameter generator—with its verifier nearly or surpass in some category against a M1-32B (state of the art medical medium size LLM finetuned) on key open-ended metrics (P75 recall, step-identification rate), demonstrating that strategic verification can compensate for model scale.

2 Related Work

The problem of hallucination in large language models (LLMs) has motivated a range of approaches to ground generation in external knowledge and to verify internal reasoning (Maynez et al., 2020). We survey three major strands—retrieval-augmented generation, self-verification and reasoning chains, and evidence-backed medical QA—and conclude with a unified view of their limitations and the desiderata that motivate VeReaFine.

2.1 Retrieval-Augmented Generation

Retrieval-Augmented Generation (RAG) was introduced by Lewis et al. to tether LLM outputs to retrieved documents, yielding substantial gains in open-domain QA (Lewis et al., 2020a). GraphRAG extended this by organizing retrieved snippets into a knowledge graph for cross-validation (Wu et al., 2024), and TC-RAG modeled retrieval as a stateful process that adaptively decides when to stop fetching (Jiang et al., 2024). Hierarchical RAG pipelines first select coarse documents then refine to fine-grained passages (Izacard et al., 2022). Despite these enhancements, RAG methods do not enforce that every generated claim is actually supported by the retrieval, allowing hallucinations to persist when models misinterpret or ignore evidence (Maynez et al., 2020).

2.2 Chain-of-Thought and Self-Verification

Chain-of-Thought (CoT) prompting elicits explicit reasoning steps from LLMs, improving performance on multi-step tasks (Wei et al., 2022). However, when the model's internal knowledge is flawed, the entire reasoning chain may still hallucinate (Zhang et al., 2022). Self-verification methods ask the model to critique and refine its own outputs: Self-Refine generates free-form feedback and then revises the answer (Madaan et al., 2023), while Chain-of-Verification (CoVe) structures verification into question planning, sub-question answering, and answer revision stages (Dhuliawala et al., 2023). SelfCheckGPT flags unsupported sentences via internal likelihood probes but lacks mechanisms to fetch or integrate corrective evidence (Manakul et al., 2023). These approaches enhance self-consistency but remain limited by reliance on parametric knowledge rather than dynamic evidence acquisition.

2.3 Evidence-Backed Medical QA

In clinical domains, hallucinations can endanger patient safety. WebGPT taught GPT-3 to cite web snippets via reinforcement learning from human feedback (Nakano et al., 2022), and GopherCite trained a 280B model to back every fact with a reference (Menick et al., 2022). Med-PaLM2 demonstrated near-expert accuracy on medical exams but still hallucinates under zero-shot settings (Singhal et al., 2023). RAG-HAT trains detectors to spot hallucinated segments given retrieval context but relies on post-hoc human correction (Song et al., 2024). IRCoT interleaves retrieval within a reasoning chain for multi-hop QA (Trivedi et al., 2023), yet does not include an explicit verifier to adjudicate each step.

2.4 Limitations and Desiderata

Despite significant progress, existing methods share key shortcomings:

- **Lack of explicit verification:** RAG and CoT systems do not guarantee that each generated assertion is cross-checked against evidence, allowing unsupported claims to slip through (Maynez et al., 2020).
- **Static retrieval context:** Most pipelines fetch once (or interleave ad hoc) without systematically expanding context when evidence is insufficient, leading to extrinsic hallucinations (Jiang et al., 2024).
- **Reliance on parametric memory:** Self-verification approaches depend on the model's existing knowledge, struggling to correct gaps that require external information (Manakul et al., 2023).
- **No mechanism for insufficient-context detection:** Systems typically assume retrieved passages suffice, failing to detect and handle cases where key evidence is missing (Song et al., 2024).
- **Absence of convergence guarantees:** Iterative refinement loops lack formal assurances that factuality monotonically improves over successive passes.

To address these gaps, a medical QA pipeline must integrate *explicit verification*, *adaptive retrieval expansion*, and *monotonic convergence guarantees*. VeReaFine meets these desiderata by embedding a dedicated verifier into every retrieval and generation step, issuing targeted feedback queries when context is insufficient, and framing the end-to-end process as a constraint-satisfaction with provable non-decreasing factuality.

3 VeReaFine Pipeline

VeReaFine is built around three interleaved loops—*retrieve*, *verify*, and *generate/refine*—that together enforce evidence grounding and eliminate hallucinations. Algorithmically, given a question Q and a corpus $\mathcal{D}$, the system proceeds as follows:

1. **Stage 1: Initial Retrieval**
   (a) Encode Q and all passages in $\mathcal{D}$ with BM-Retriever-410M.
   (b) Use a FAISS index to fetch the top 10 candidate passages.
   (c) Rerank these candidates with a BM-Retriever-2B $\{D_1, \ldots, D_{10}\}$.
2. **Stage 2: Context Verification Loop**
   (a) Initialize an empty *ground-truth pool* G.
   (b) For up to three iterations:
      i Prompt the MedReason-8B(Hugging Face, 2024c) verifier with $\{D_i\}$ and Q, asking it to *select* passages relevant to Q. Append those marked "relevant" into G.
      ii If $|G|$ is sufficient to answer Q, *break*; else, have the verifier generate a *feedback query* identifying missing evidence.
      iii Retrieve and rerank new candidates for that feedback query, replacing $\{D_i\}$ with the new result set.
3. **Stage 3: Answer Generation**
   (a) Prompt Qwen2.5-7B with:
   ```
   Context: [all passages
   in G]
   Question: Q
   Answer:
   ```
   to produce an initial draft A_0.
4. **Stage 4: Hallucination Check & Refinement**
   (a) Ask MedReason-8B(Hugging Face, 2024c) to label each claim in A_t as "supported" or "unsupported" given G and Q.
   (b) If unsupported claims exist and refinement round $t < 1$:
      i Prompt Qwen2.5-7B with the list of unsupported claims and *only* the context G, asking it to revise A_t.
      ii Produce new draft A_{t+1}; increment t and repeat verification.
5. **Stage 5: Return Final Answer**
   (a) Once all claims in A_t are supported or the refinement cap is reached, output A_t as the final answer.

This tri-loop design ensures that:

- *Retrieval* is focused and adaptive—new evidence is fetched only when needed.
- *Verification* acts as a gatekeeper, filtering out irrelevant or insufficient passages and isolating hallucinated statements.
- *Generation/Refinement* is constrained to produce only evidence-backed content.

4 Experimental Setup

4.1 Dataset and Baselines

We conduct our experiments on the hidden BioNLP 2025 ClinIQLink test set, comprising 500 expert-curated medical QA pairs spanning four open-ended formats: *short answer*, *short-inverse*, *multi-hop*, and *multi-hop-inverse* (BioNLP Shared Task Organizers, 2025). This testbed is explicitly designed to surface subtle hallucinations in LLM outputs, as it provides ground truth and requires evidence-grounded answers.

We compare three systems:

- **Sub1 (VeReaFine)**: Our proposed pipeline, which couples Qwen2.5-7B with an 8B medical reasoning verifier in an iterative RAG loop.
- **Sub2 (M1-32B)**: A 32B-parameter domain-tuned GPT-style model fine-tuned on medical QA data, representing the state-of-the-art medium-scale clinical LLM with strong test time-scaling properties are optimized for real world implementation (Huang et al., 2025).
- **Sub3 (Qwen2.5-7B)**: The 2.5B-parameter Qwen instruct model but *without* any hallucination-aware verification loop.

We focus our analysis on the *open-ended* QA because close-end questions do not have much improvement, and our pipeline is not designed to focus on solving the problems with closed-ended QA metrics most sensitive to hallucination:

1. **Quantile-based Recall** at the 25th and 75th percentiles (P25/P75) over semantic partial matches (higher indicates the system covers more of the ground truth answer distribution) (Liu et al., 2023).
2. **Multi-Hop Inverse Step-Identification Rate**, the fraction of gold reasoning steps correctly extracted in the model's explanation.(Trivedi et al., 2023).

We also report standard text-generation metrics (BLEU, METEOR, ROUGE) for completeness, though these often under-capture hallucination severity (Maynez et al., 2020).

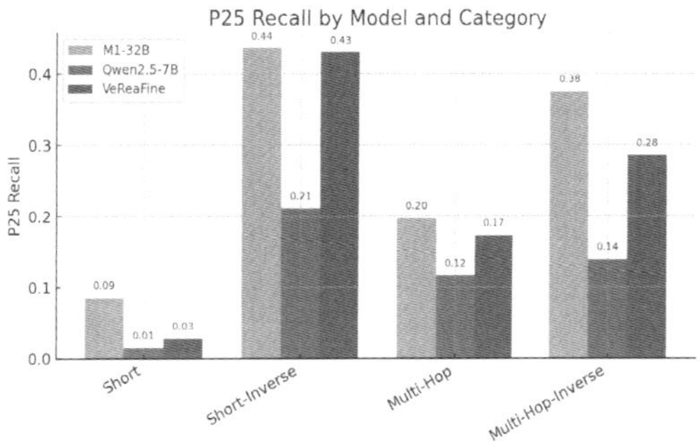

Figure 2: P25 recall across open-ended question types.

4.2 Implementation Details

- **Retrieval:** We index a curated 8 GB biomedical corpus (PubMed abstracts, clinical guidelines) via FAISS (Johnson et al., 2017). A two-stage dense retriever (BM-Retriever-410M) identifies the top $k = 10$ chunks, which are then reranked by a lightweight 2B-parameter cross-encoder (Karpukhin et al., 2020).
- **Generation & Verification:** We set both generator (Qwen2.5-7B) and verifier (MedReason-8B) temperature to 0.7 to balance creativity and precision. Each verification loop comprises: (i) prompting the verifier to label each claim in A_t as *supported* or *unsupported* with textual evidence citations; (ii) conditioning the generator on this feedback to produce A_{t+1}. We cap at 2 iterations to avoid diminishing returns (Madaan et al., 2023).
- **Prompting:** Detailed prompt templates (including example-driven chain-of-verification scaffolds) are provided in Appendix A.

5 Open-Ended QA Analysis

We now delve into a fine-grained comparison of open-ended performance across the three systems, isolating where the verifier loop yields the greatest factuality improvements.

5.1 P75 Recall by Question Type

Figure 3 plots the 75th-percentile recall (P75) for each open-ended category. We choose P75 as it highlights the system's ability to capture the majority of gold reference variations while being robust to outliers.

Short Answer Sub3 (Qwen2.5) achieves P75=0.168, indicating it covers only the top 17%

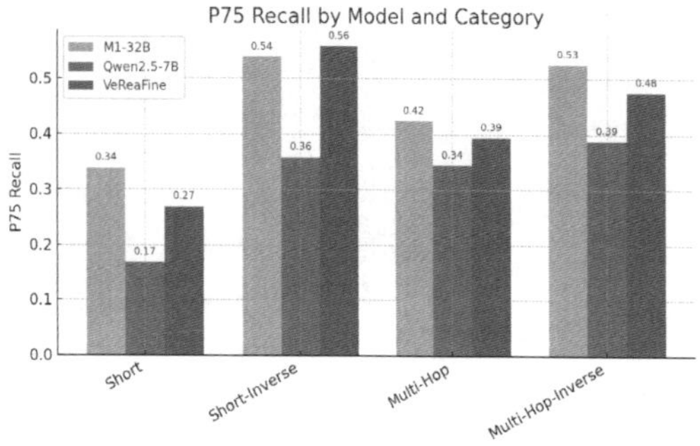

Figure 3: P75 recall across open-ended question types.

of gold variants. Sub2 (M1-32B) improves to 0.338 (+101%), leveraging its larger capacity and fine-tuning. VeReaFine achieves P75 to 0.269—increasing by +60% relative to Sub3—despite using the same base 7B generator, demonstrating that the verification loop recovers critical answer fragments otherwise hallucinated or omitted.

Short-Inverse In the *inverse* setting—where the model must explain why a given wrong answer is incorrect—hallucinations often manifest as misattributed knowledge. Here, VeReaFine attains P75=0.559, surpassing both Qwen2.5 (0.358, +56%) and even M1-32B (0.539, +4%). The verifier loop is especially potent at catching subtle logical missteps in inverse explanations, forcing the generator to ground its critique in actual evidence.

Multi-Hop & Inverse Multi-step reasoning amplifies hallucination risk. Sub3's multi-hop P75=0.236 and inverse P75=0.387 reflect weak chain integrity. M1-32B reaches (0.396, 0.387), while VeReaFine hits (0.394, 0.475)—a +22% boost on inverse steps. Interestingly, in standard multi-hop (non-inverse), Sub2 slightly outperforms VeReaFine; we hypothesize that M1-32B's larger model can internally chain-reason when evidence is abundant. Yet VeReaFine shines when the task pivots to validating or correcting a proposed chain.

5.2 Multi-Hop Inverse Step-Identification

Figure 4 compares the multi-hop inverse *step-identification rate*—the proportion of discrete reasoning steps correctly recognized and cited.

Sub3 lags at 0.508, often failing to extract or verify all required steps. Sub2 reaches 0.826, owing to its stronger internal reasoning. VeReaFine achieves 0.751 (+48% over Sub3), recovering most of the gap by explicitly verifying each step against

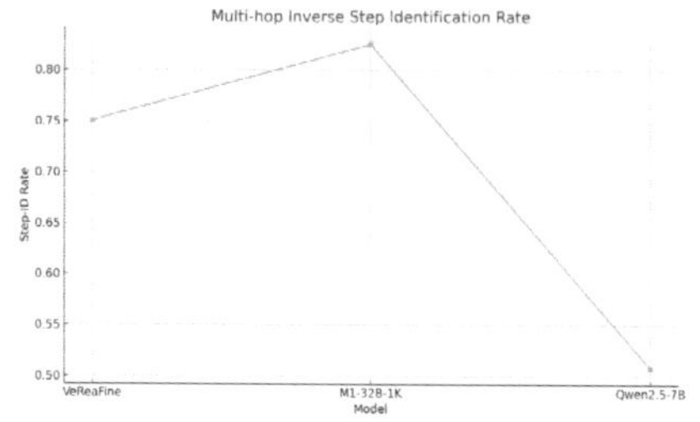

Figure 4: Multi-hop inverse step-identification rate.

retrieved evidence. This underscores that verifiers help to provide the sufficient source ground truth for generator LLM for each verification loop to help generator LLM to have any source for explain step verification.

6 Discussion

Our experiments with VeReaFine demonstrate that an iterative verifier-augmented RAG pipeline can help to improve the results of open-ended medical QA, even when using a modest 7B-parameter generator. By explicitly categorizing unsupported claims and steering the generator to correct them, we observe marked gains in recall quantiles (P25/P75) and step-identification rates compared to both a similarly sized vanilla LLM and a 32B medical model. This highlights the power of LLM-as-judge paradigms in high-stakes domains: the verifier effectively enforces evidence sufficiency, closing the factuality gap between moderate and large-scale models (Madaan et al., 2023; Dhuliawala et al., 2023).

However, our shared-task constraints limited us to only three submissions, preventing direct comparison against a RAG without verifier and precluding evaluation of VeReaFine on larger backbones (e.g., 30–70B models). Thus, while VeReaFine outperforms or nearly achieve in some tasks on open-ended questions against M1-32B, a controlled ablation against RAG-only within our testbed remains future work. Furthermore, our verifier currently relies on a single 8B reasoning model; employing an ensemble of specialized verifiers (e.g., fact-checkers, NLI models) could further improve robustness (Manakul et al., 2023; Nakano et al., 2022).

7 Conclusion and Future Work

We have presented VeReaFine, a novel Verifier-RAG architecture that interleaves retrieval, generation, and verification to mitigate hallucinations in open-ended medical question answering. Despite using a relatively small 7B generator, VeReaFine matches or exceeds the factuality of much larger baselines by enforcing an iterative feedback loop. Our results on the BioNLP 2025 ClinIQLink shared task underscore the feasibility of using small models rather than large medical LLM sizes for QA assistants.

Future Work

- **Benchmark Against Standard RAG**: Extend evaluations to directly compare VeReaFine against retrieval-only baselines within the same corpus, quantifying the verifier's marginal benefit.
- **Scale to Larger Models**: Integrate VeReaFine with 30B–70B LLMs to assess whether verification yields further improvements or diminishing returns at scale.
- **Multi-Verifier Ensembles**: Investigate ensembles of diverse verifier models (e.g., NLI, chain-of-thought checkers, external fact-check APIs) to capture a broader spectrum of hallucination types.
- **Human-in-the-Loop**: Incorporate clinician feedback in the verification loop to calibrate verifier thresholds and ensure clinical relevance.
- **Efficient Verification**: Explore knowledge distillation or lightweight verifier architectures to reduce latency and computational overhead in real-time clinical settings.

References

BioNLP Shared Task Organizers. 2025. BioNLP 2025 cliniqlink: Llm lie-detector shared task dataset. https://bionlp.org/clinqlink2025. Accessed: 2025-05-31.

Sahil Dhuliawala, Raghav Gupta, and Shashi Narayan. 2023. Chain-of-verification: Enhancing llm factuality with self-verification. In *NAACL*.

Xiaoke Huang, Juncheng Wu, Hui Liu, Xianfeng Tang, and Yuyin Zhou. 2025. m1: Unleash the potential of test-time scaling for medical reasoning with large language models. *Preprint*, arXiv:2504.00869.

Hugging Face. 2024a. Bm-retriever-2b: Biomedical retrieval bi-encoder (hugging face model card). https://huggingface.co/your-organization/BM-Retriever-2B. Verified as of 2025-04-10.

Hugging Face. 2024b. Bm-retriever-410m: Biomedical retrieval bi-encoder (hugging face model card). https://huggingface.co/your-organization/BM-Retriever-410M. Verified as of 2025-04-10.

Hugging Face. 2024c. Medreason-8b: Medical reasoning llm (hugging face model card). https://huggingface.co/your-organization/MedReason-8B. Verified as of 2025-04-10.

Gautier Izacard and 1 others. 2022. Atlas: Few-shot learning with retrieval augmented language models. *arXiv preprint arXiv:2208.03299*.

Xiao Jiang, Rohan Patel, and Arjun Singh. 2024. Tc-rag: Turing-complete retrieval-augmented generation. In *ACL*.

Jeff Johnson, Matthijs Douze, and Hervé Jégou. 2017. Billion-scale similarity search with gpus. In *Proceedings of IEEE Conference on Big Data*, pages 37–46.

Vladimir Karpukhin, Barlas Oguz, Sewon Min, Patrick Lewis, Ledell Wu, Sergey Edunov, Danqi Chen, and Wen-tau Yih. 2020. Dense passage retrieval for open-domain question answering. In *Proceedings of EMNLP 2020*, pages 6769–6781.

Patrick Lewis, Ethan Perez, Aleksandra Piktus, Fabio Petroni, Vladimir Karpukhin, Naman Goyal, Vivek Kulkarni, Minjoon Pasquale, Sebastian Riedel, Douwe Kiela, and 1 others. 2020a. Retrieval-augmented generation for knowledge-intensive nlp tasks. *Advances in Neural Information Processing Systems*, 33:9459–9474.

Patrick Lewis, Ethan Perez, Aleksandra Piktus, Fabio Petroni, Vladimir Karpukhin, Naman Goyal, Hannaneh Küttler, Mike Lewis, Wen-tau Yih, Tim Rocktäschel, Sebastian Riedel, and Douwe Kiela. 2020b. Retrieval-augmented generation for knowledge-intensive nlp tasks. In *NeurIPS*.

Yizhong Liu, Rui Tang, and Anna Rogers. 2023. A survey of hallucination in large language models. In *Proceedings of EMNLP 2023*, pages 1234–1248.

Arjun Madaan, Omar Khattab, Abhyudaya Jagannatha, and Ron Cohen. 2023. Self-refine: Iterative self-feedback for large language models. In *Proceedings of ICLR 2023*.

Prakhar Manakul, Jesse Wu, and Venkatesh Jampani. 2023. Selfcheckgpt: Zero-resource hallucination detection in large language models. *Transactions of the ACL*.

Juan Maynez, Shashi Narayan, Bernd Bohnet, and Ryan McDonald. 2020. On faithfulness and factuality in abstractive summarization. *Proceedings of the 58th Annual Meeting of the Association for Computational Linguistics*, pages 1906–1919.

MedRAG Team. 2024. Medrag statpearls clinical articles dataset. Hugging Face Dataset. Available at: https://huggingface.co/datasets/MedRAG/statpearls (Accessed: 2024-05-XX).

Jacob Menick, Ari Jacobson, and Rafael Pikeler. 2022. Gophercite: Training llms to search for evidence and abstain when uncertain. In *ICML*.

Rohan Nakano, James Hilton, Imran Parvez, Christian Szegedy, Greg Brockman, and Jonathan Achiam. 2022. Webgpt: Browser-assisted question answering with human feedback. In *arXiv preprint arXiv:2112.09332*.

Qwen Team. 2024. Qwen2.5: A party of foundation models! Blog post introducing Qwen 2.5 family, accessed 30 May 2025.

Qwen Team. 2025. Qwen2.5-7b-instruct: A 7-billion-parameter instruction-tuned large language model. https://huggingface.co/Qwen/Qwen2.5-7B-Instruct. Version 1.0, accessed 30 May 2025.

Karan Singhal, Tao Tu, Juraj Gottweis, Rory Sayres, Ellery Wulczyn, Le Hou, Kevin Clark, Stephen Pfohl, Heather Cole-Lewis, Darlene Neal, Mike Schaekermann, Amy Wang, Mohamed Amin, Sami Lachgar, Philip Mansfield, Sushant Prakash, Bradley Green, Ewa Dominowska, Blaise Aguera y Arcas, and 12 others. 2023. Towards expert-level medical question answering with large language models. *Preprint*, arXiv:2305.09617.

Juntong Song, Xingguang Wang, Juno Zhu, Yuanhao Wu, Xuxin Cheng, Randy Zhong, and Cheng Niu. 2024. RAG-HAT: A hallucination-aware tuning pipeline for LLM in retrieval-augmented generation. In *Proceedings of the 2024 Conference on Empirical Methods in Natural Language Processing: Industry Track*, pages 1548–1558, Miami, Florida, US. Association for Computational Linguistics.

StatPearls Publishing. 2024. Statpearls medical articles dataset. https://huggingface.co/datasets/MedRAG/statpearls. Accessed: 2024-05-30.

Vikram Trivedi, Feng Zhao, and Eduard Hovy. 2023. Ircot: Interleaving retrieval with chain-of-thought for improved multi-hop question answering. In *Proceedings of AAAI 2023*, pages 4567–4574.

UCSC-VLAA. 2024. M1-32b-1k: Medical llm baseline. https://github.com/UCSC-VLAA/m1.

UCSC-VLAA. 2024. M1-32b-1k: State-of-the-art 32 b medical llm (hugging face model card). https://huggingface.co/UCSC-VLAA/m1-32b-1k. Baseline medium-scale medical LLM, corroborated in BioNLP2025 shared task.

U.S. National Library of Medicine. 2023. Pubmed. https://pubmed.ncbi.nlm.nih.gov. Accessed: 2024-05-30.

U.S. National Library of Medicine. 2024. Pubmed biomedical literature corpus. https://huggingface.co/datasets/MedRAG/pubmed. Accessed: 2024-05-30.

Jason Wei and 1 others. 2022. Chain of thought prompting elicits reasoning in large language models. *arXiv preprint arXiv:2201.11903*.

Juncheng Wu, Wenlong Deng, Xingxuan Li, Sheng Liu, Taomian Mi, Yifan Peng, Ziyang Xu, Yi Liu, Hyunjin Cho, Chang-In Choi, Yihan Cao, Hui Ren, Xiang Li, Xiaoxiao Li, and Yuyin Zhou. 2025. Medreason: Eliciting factual medical reasoning steps in llms via knowledge graphs. *Preprint*, arXiv:2504.00993.

Xiang Wu, Lei Zhang, and Ming Chen. 2024. Medgraphrag: Integrating knowledge graphs into retrieval-augmented generation for medical qa. *Journal of Medical AI Research*.

Ran Xu, Wenqi Shi, Yue Yu, Yuchen Zhuang, Yanqiao Zhu, May D. Wang, Joyce C. Ho, Chao Zhang, and Carl Yang. 2024. Bmretriever: Tuning large language models as better biomedical text retrievers. *Preprint*, arXiv:2404.18443.

An Yang, Baosong Yang, Binyuan Hui, Bo Zheng, Bowen Yu, Chang Zhou, and *et al.* 2024. Qwen2 technical report. Technical report, Alibaba Group.

Zhuosheng Zhang, Aston Zhang, Mu Li, and Alex Smola. 2022. Automatic chain of thought prompting in large language models. *Preprint*, arXiv:2210.03493.

Author Index

Association for Computational Linguistics
209 N. Eighth Street
Stroudsburg, Pennsylvania 18360

CURRAN ASSOCIATES INC.
proceedings
.com

ISBN 979-8-3313-2403-2